Fungal Antigens

Isolation, Purification, and Detection

Fungal Antigens

Isolation, Purification, and Detection

Edited by

Edouard Drouhet

Institut Pasteur
Paris, France

Garry T. Cole

University of Texas
Austin, Texas

Louis de Repentigny

Sainte Justine Hospital
Montreal, Canada

Jean-Paul Latgé

Institut Pasteur
Paris, France

and

Bertrand Dupont

Institut Pasteur
Paris, France

Plenum Press • New York and London

Library of Congress Cataloging in Publication Data

International Symposium on Fungal Antigens (1st: 1986: Paris, France)
    Fungal antigens: isolation, purification, and detection: [proceedings of the First International Symposium on Fungal Antigens, held in November 1986, in Paris, France] / edited by Edouard Drouhet . . . [et al.].
        p.   cm.
    Bibliography: p.
    Includes index.

ISBN-13: 978-1-4612-8075-0   e-ISBN-13: 978-1-4613-0773-0
DOI: 10.1007/978-1-4613-0773-0

    1. Fungal antigens—Congresses. 2. Fungi, Pathogenic—Congresses. I. Drouhet, Edouard, 1919–   . II. Title.
QR 186.6.F85167   1986   88-36998
589.2′04292—dc19   CIP

Proceedings of the First International
Symposium on Fungal Antigens, held in November 1986,
in Paris, France

© 1988 Plenum Press, New York

Softcover reprint of the hardcover 1st edition 1988

A Division of Plenum Publishing Corporation
233 Spring Street, New York, N.Y. 10013

PREFACE

Three years ago when Professor Garry Cole visited our
Mycology Unit at the Pasteur Institute we discussed the
possibility of organizing a small International Symposium on
"Isolation, Purification and Detection of Fungal Antigens"
limited to 8 American/Canadian scientists and to 8 French
participants. The location chosen was the Pasteur Institute
because of the historical and current importance of the
Institute as a Center for Research in Immunology and Medical
Mycology. The interest demonstrated by all medical mycolo-
gists we contacted led us to expand the small original
meeting to an international symposium in which all aspects of
antigens of pathogenic and allergenic fungi and actinomycetes
related to man, animals, and even plants would be discussed.
Our wish was also to hold this Symposium in the same week as
the Anniversary meeting of the French Society of Medical
Mycology which was founded at the Pasteur Institute 30 years
ago with my colleagues Gabriel Segretain and Francois Mariat.
The aim of the "First International Symposium on Fungal
Antigens" was to exchange information on the isolation, purification and detection of fungal antigens in view of further
standardization and applications of immunological methods in
several fields of Mycology. The term antigen commonly refers
to preparations containing a multiplicity of antigenic determinants. Emphasis is now on purification and characterization of specific, biologically active antigens which can be
used in serodiagnosis and immunotherapy of mycoses. A variety
of techniques are now used for the isolation of fungal
antigens from culture filtrates, whole or disrupted cells,
cell walls and cytoplasm. The successful application of
serological tests for diagnosis of fungal diseases largely
depends on the quality of antigen preparations. Characterization of different cell types of pathogenic fungi on the
basis of their antigenic determinants provides a mechanism for
exploring differences in host response during progressive
stages of infection.

For these reasons, it was necessary at this Symposium to
first address current methodologies including chemical techniques used to study the structure of fungal antigens, cytochemical and immunochemical techniques used to study antigens
of the fungal cell at the electron-microscopic level, enzyme
immunological techniques used to detect antibodies and antigens and hydridoma technology used to produce monoclonal antibodies. All these methods were discussed in detail by experienced researchers. The most important and recent results

obtained from the application of these methodologies to the fungi and actinomycetes responsible for human and animal mycotic infections or allergic manifestations, including _Candida_ _albicans_, _Cryptococcus_ _neoformans_, _Aspergillus_ _fumigatus_, _Coccidioides_ _immitis_ and other filamentous fungi, dermatophytes, dimorphic forms, and thermophilic actinomycetes were presented. The major contributions (lectures) and short contributed papers in this book represent the communications presented at this meeting.

We hope that this "First International Symposium of Fungal Antigens" will be followed by additional symposia devoted to other aspects of fungal antigens which have been only mentioned in this book, and to new developments in the exciting and promising field of Medical Mycology.

The Editors

ACKNOWLEDGEMENTS

The organizers are grateful to the Pasteur Institute which sponsored this Symposium, as well as the Research Foundations and Laboratories which provided assistance for the participation of numerous lecturers and helped in the organization of this meeting. We are particularly grateful to the Janssen Research Foundation, Squibb, Bayer, Pfizer Companies, and Diagnostics Pasteur.

Thanks are also due to chairmen, speakers, and all participants who contributed so much to the discussions and success of this symposium.

Particular thanks are due to the following members of the Unit of Mycology of the Pasteur Institute, whose efforts contributed so much to the success of this meeting: Claude de Bievre, Eveline Gueho, Sophie Paris, N'Guyen van Huong, Patrick Boiron, Jean-Michel Delga, Oumaima Granet, Anne Beauvais, Luce Improvisi, Alain Pietfroid, Olivier Ronin and especially the secretarial staff, Ginette Monlleo, Marinette Cormier, and Irene Herve-Gruyer.

Professor R. Dedonder, Director of Pasteur Institute, welcoming participants to the Symposium on Fungal Antigens in presence of Professor E. Drouhet, Chairman of the Organizing Committee. Behind Dr. Dedonder is a painting of Elie Metchnikoff. Photograph taken in the library of the Pasteur Institute.

Symposium Participants

CONTENTS

INTRODUCTION

Overview on Fungal Antigens 3
    E. Drouhet

METHODOLOGY

Chemical Methods to Study the Structure of
       Fungal Polysaccharide Antigens. 41
    B. Fournet

Cytochemical Techniques for the Study of
       Fungal Polysaccharides. 57
    J. Schrevel

Immunochemistry of Fungal Antigens at the
       Electron Microscopy Level 71
    J. Müller and R. Kappe

Enzyme-Linked Immunosorbent Assay and Related
       Sensitive Assays for Antigen and
       Antibody Detection. 81
    J. Guesdon, Z. Pires de Camargo,
       E. Drouhet, and S. Avrameas

Techniques Available to Prepare Monoclonal Antibodies
       Directed Toward Fungal Polysaccharide Antigens . . 99
    E. Drouhet and J.P. Latgé

Fungal Exoantigens. 111
    L. Kaufman and P. Standard

INVERTEBRATE PATHOGENS

Fungal Elicitors of Invertebrate Cell Defense System. . . 121
    D. Boucias and J.P. Latgé

CANDIDA

Immunochemistry of <u>Candida albicans</u> Antigens. 141
    S.L. Salhi and J.M. Bastide

Cytological, Chemical and Serological Immunodetection
    of a Candida albicans Antigen. 149
  D. Poulain, G. Strecker, J.F. Dubremetz,
  B. Fortier, R. Rousseau, and J. Van Cutsem

Candida Antigens and Their Role in Adherence Mechanisms. . 157
  R. Robert, J.M. Senet, G. Tronchin, and A. Bouali

Antigen and Metabolite Detection in Invasive Candidiasis. . 169
  L. de Repentigny

Mannan Antigen of Candida albicans and Cellular
    Immune Responses in vitro and in vivo 185
  A. Durandy, A. Fisher, E. Drouhet, and C. Griscelli

CONTRIBUTED PAPERS

Antigenicity of Candida albicans Cell Wall after
    Enzymatic Digestion. 197
  I. Berdicevsky and J. Muller

Differential Expression of Secretory Proteinase by
    Candida albicans and C. parapsilosis Invading
    Human Mucosa 198
  M. Borg and R. Ruchel

Production of Anti-Candida Monoclonal Antibodies:
    Immunopurification of C. albicans Cell
    Wall Antigens. 200
  T. Chardès, J.C. Lebecq, B. Pau, and J. Bastide

Candida Antigen and Arabinitol Levels in the Sera of
    Patients with Proven or Probable
    Invasive Candidosis. 202
  A.G. Deacon and M.D. Richardson

Mannan Detection by ELISA-Inhibition in Desseminated
    Candidiasis: Experimental Study in Rabbits 203
  J. Garcia de Lomas, C. Morales, M.A. Grau,
  J.M. Delga, E. Drouhet, B. Dupont, and L. Improvisi

Evaluation of the Candidate Super Latex Agglutination
    Test for the Diagnosis of Vaginal Candidosis . . . 205
  V. Hopwood, D.W. Warnock, J.D. Milne, T. Crowley,
  C.T. Horrocks, and P.K. Taylor

An Attempt to Obtain Candida Skin Antigens for the
    Diagnosis of Systemic Candidosis 206
  K. Iwata, Y. Yamamoto, T. Yamashita, H. Uehara,
  T. Kamoshida, and T. Yanai

Candida Mannan Antigen Detection versus Quantitative
    Culture in Urine, Sputum, Stool, and
    Vaginal Swabs. 209
  R. Kappe, D. Kubitza, I. Scheidecker, and J. Muller

Systematic Use of a Latex Test to Detect Soluble Candida
    Antigens in Serodiagnosis of Systemic Candidiasis. 211
  Y. Le Fichoux, H. Chan, and P. Marty

Evaluation of Immune Complexes and Detection of a
        M_r48,000 Protein in Candidiasis. 212
    B.T. Maida and H.R. Buckley

Monoclonal Antibodies: Anti-Candida albicans Mannan and
        Anti-Aspergillus fumigatus Galactomannan 213
    C. Munoz, J.C. Mazie, J.M. Delga, B. Dupont,
        J.P. Latge, and E. Drouhet

Yeast Killer Toxin Monoclonal Antibodies. 216
    L. Polonelli, M. Castagnola, and G. Morace

A Quantitative Immunofluorescence Test: An Evaluation
        of Candida Antigen in the Diagnosis of
        Invasive Candidiasis 217
    A. Sanchez-Sousa, J.M. Aguiar, C. Torres,
        H. Cercenado, A.M. Malo, and F. Baquero

Diagnosis of Systemic Candidosis by Latex
        Particle Agglutination 219
    A.M. Sharp, E.G.V. Evans, and J.A. Carney

Antigens for the Detection of Candida guilliermondii
        var. guilliermondii Infection in Ruminants 220
    P. Sutka and K. Sutka

ALLERGENIC MOULDS

Allergens of Aspergillus fumigatus. 223
    J.L. Longbottom

Allergens of Alternaria and Cladosporium 237
    J.P. Latge and S. Paris

CONTRIBUTED PAPERS

Distinct Cell-Wall and Cytoplasmic Proteinic Antigenic
        Fractions from Candida albicans Inducing IgE
        Antibodies in Hypersensitive Patients. 259
    F. Grange, B. de la Parra, P.I. Gumowski,
        and J.P. Girard

Serial Specific IgG Measurements to Common Moulds 261
    B. Guerin, A.M. Chaissac and B. Gouyon

The Antigenic and Allergenic Properties of
        Botrytis cinerea 262
    H.F. Kauffman, S. van der Heide, F. Beaumont,
        and K. de Vries

CRYPTOCOCCUS

Correlates of Serotype in Cryptococcus neoformans 265
    J.E. Bennett and K.J. Kwon-Chung

Characterization of Cross-Reactivity between cryptococcal
        Polysaccharides by Use of Enzyme-Linked
        Immunosorbent Assays and Monoclonal Antibodies . . 273
    T.R. Kozel and T.F. Eckert

Cytoplasmic Antigens of <u>Cryptococcus</u> <u>neoformans</u> 283
    G. Reyes, J.P. Dandeu, E.Drouhet, and J.P. Latgé

CONTRIBUTED PAPERS

Characterization, Specificity and Protective Effect
    of a Murine Monoclonal Antibody Reactive with
    <u>Cryptococcus</u> <u>neoformans</u> (CN) Capsular
    Polysaccharide (CNPS). 301
    F. Dromer, J. Salamero, J. Charreire, A. Contrepois,
    C. Carbon, and P. Yeni

Kinetic Study of Humoral and Cellular Immune Responses
    to <u>Cryptococcus</u> <u>neoformans</u> Measured by ELISA
    and Immunocyte Migration Inhibition. 303
    S. Gauthier-Rahman, S. Wahab, and E. Drouhet

Serotypes of Clinical and Saprophytic Isolates of
    <u>Cryptococcus</u> <u>neoformans</u>. 306
    E.G.V. Evans, D. Hall, D. Swinne, and C. de Vroey

About Soluble Cryptococcal Antigen (SCA): Interest in
    Screening of Patients with AIDS. 307
    P. Roux, J.L. Touboul, C. Mayaud, D. Basset,
    J.L. Poirot, M. Denis, and F. Lancastre

DERMATOPHYTES

Antigens of Dermatophytes and Their Characterization
    Using Monoclonal Antibodies. 311
    L. Polonelli and B. Morace

Antigens Involved in the Pathology of <u>Fonsecaea</u> <u>pedrosoi</u>:
    Immunochemical Study 319
    O. Ibrahim-Granet and C. de Bievre

CONTRIBUTED PAPERS

Immunological Aspects of a Keratinase Isolated from
    <u>Trichophyton</u> <u>mentagrophytes</u>. 334
    J. Abbink and M. Plempel

Diversity of Antigenic Extracts from the Dermatophyte
    <u>Trichophyton</u> <u>rubrum</u>. 336
    P. de Haan and J. Wilker

Monoclonal Antibodies Against <u>Trichophyton</u> <u>rubrum</u> and
    their Cross-Activity with Related and Non-Related
    Antigens 338
    P. de Haan, E.M.H. van der Raay-Helmer, and J. Wikler

Enzyme-Linked Immunosorbent Assay (ELISA) for Detection
    of Antibodies Against <u>Pityrosporum</u> <u>orbiculare</u> . . 339
    J. Faergemann and S. Johansson

Standardization of Dermatophyte Antigens (Trichophytins)
    for Skin Test Purposes 340
    T. Kaaman

Studies on the Antigens of Pityrosporum (Malassezia)
     Species. 342
     G. Midgley

ASPERGILLUS

Antigenic Differences Between Exoantigens, Mycelial and
     Conidial Antigens in Aspergillus 345
     E.J. Bardana, S. Craig, K. Strangfeld, and
     A. Montanaro

Characterization of Purified Wall Antigens Obtained
     from Aspergillus Species 355
     V.M. Hearn, E.V. Wilson, and D.W.R. Mackenzie

Antigen Detection in Invasive Aspergillosis 367
     B. Dupont

CONTRIBUTED PAPERS

Aspergillosis in Immunocompromised Patients - Part I:
     The Problem of Diagnosis 376
     R.A. Barnes and T.R. Rogers

Aspergillosis in Immunocompromised Patients - Part II:
     Evaluation of an Enzyme-Linked Immunosorbent
     Assay (ELISA) for the Detection of Aspergillus
     Antigen. 377
     K.A. Haynes, R.A. Barnes, E.V. Wilson, V.M. Hearn,
     D.W.R. Mackenzie, and T.R. Rogers

Immunochemical Studies of Candida and Aspergillus
     Antigens: Taxonomic and Diagnostic Significance. . 379
     R.M.F. Guinet, S.M. Bruneau, and H. de Montclos

Interest in Serological Techniques for the Detection
     and Identification of Phoma exigua, the
     Agent of Potato Gangrene 381
     L. Hingand, M. Aguelon, S. Le Coz, C. Kerlan,
     B. Jouan, and J. Dunez

Detection of Aspergillus Glycoprotein in Sera of
     Patients with Invasive Aspergillosis by an
     Enzyme-Linked Immunosorbent Assay. 382
     P. Le Pape and J. Deunff

An Enzyme Immunoassay for the Detection of Tenuazonic
     Acid, a Toxin from the Rice Fungal Pathogen
     Pyricularia oryzae 383
     M.H. Lebrun, P. Ponceto, F. Gaudemer, M. Boutar,
     and A. Gaudemer

Value and Limits of ELISA in Aspergillosis: Serological
     Diagnosis in Clinical Forms and Control of a
     High Risk Population 385
     C. Pinel, R. Grillot, and B. Lebeau

A Rapid Double Antibody Biotin-Streptavidin ELISA for
    Aspergillus fumigatus Glycoprotein Antigen 386
  M.D. Richardson, L.A. McTaggart, and G.S. Shankland

Peroxidase Anti-Peroxidase Staining for the Specific
    Identification of Fungal Antigens in Tissue
    Sections 387
  G.S. Shankland and M.D. Richardson

An Assessment of the Assay for Serum Antibodies to
    Aspergillus fumigatus Catalase in the
    Diagnostic Laboratory. 388
  H.C. Schönheyder

Detection of Aspergillus fumigatus Antigenic Components
    Recognizing IgG, IgM, IgA or IgE Specific
    Antibodies by ELIFA. 390
  H.Thoannes, J. Poirriez, and J.M. Pinon

Antigens and Allergens of the Genus Chrysosporium
    (Emmonsia) Agent of Adiaspiromycosis 391
  A. Tomsikova

COCCIDIOIDES AND HISTOPLASMA

Wall-Associated Antigens of Coccidioides immitis. 395
  G.T. Cole, S.H. Sun, J. Dominguez, L. Yuan,
  M. Franco, and T.N. Kirkland

Immunochemical Analysis of Histoplasmin Proteins
    and Polysaccharide 417
  E. Reiss and S.L. Bragg

Immune Responses to Subcellular Antigens of
    Histoplasma capsulatum 431
  R.P. Tewari, R.B. Kohler, and L.J. Wheat

CONTRIBUTED PAPER

Recognition of Coccidiodes immitis Antigens with
    Monoclonal Antibodies. 445
  S.J. Kraeger, D.J. P. Gennevois, R.A. Cox,
  and A.E. Karu

ACTINOMYCETES

Antigens of Thermophilic Actinomycetes. 448
  P. Boiron

INDEX 463

Introduction

OVERVIEW OF FUNGAL ANTIGENS

Edouard Drouhet

Institut Pasteur, Unité de Mycologie, Paris

Fungal antigens (Ags) are of great interest not only for the study of their fundamental properties, but also because of their use as reagents for the diagnosis of fungal diseases. The successful application of immunological tests for diagnosing mycotic infections is very important because the incidence and mortality of the mycoses, particularly those caused by opportunistic fungi has shown a marked increase in recent years, especially in patients whose immune defense mechanisms have been compromised by antibiotics, immunosuppressive therapy, heroin addiction, or severe underlying diseases (cancer, AIDS, hematologic disorders). In fact, invasive fungal infections have been reported in recent years in 25 % of patients who are chronically and intensively immunosuppressed by reason of underlying diseases and drug therapy. The incidence of aspergillosis has increased from 1.91 to 4.8 per million persons (+ 158 %) and the incidence of cryptococcosis from 1.3 per million to 2.3 per million (+ 78 %) (Bullock and Deepe, 1983). The incidence rate was substantially higher in 1986 due to opportunistic infections such as cryptococcosis, candidosis and histoplasmosis occuring in AIDS patients, and to the new septicemic syndrome of candidosis observed in heroin addicts (Dupont and Drouhet, 1985). This emphasizes the utility of standardized antigens as diagnostic tools in the clinical and laboratory evaluation of patients with suspected fungal infections. Reviewing the scientific literature on the immunology of the mycoses for the last twelve years (Table 1), we are impressed by the increasing number of papers dealing with antigens, allergens, serological tests and, immunity to the opportunistic fungi Candida, Aspergillus, Cryptococcus and the systemic dimorphic fungi Histoplasma and Coccidioides. Recently, exhaustive reviews were published on fungal antigens (Huppert, 1983 ; Longbottom and Austwick, 1986 ; Poulain et al., 1986), on applications of immunological methods in mycology (Longbottom, 1986), on serodiagnosis of mycoses (Mackenzie 1983), on cell-mediated immunity (Cox, 1983) and on molecular immunology of mycotic and actinomycotic infections (Reiss, 1986).

Pathogenic fungi and fungal antigens

The difficulties encountered in immunologic studies of

TABLE 1 - PUBLISHED PAPERS ON IMMUNOLOGY OF MYCOSES (1972-1985)*

	ASPERGILLUS	CANDIDA	CRYPTOCOCCUS	BLASTOMYCES	COCCIDIOIDES	HISTOPLASMA
IMMUNIZATION (general, clinical, Ig classes, immuno deficiences)	144	164	30	17	25	48
ALLERGY.SKIN TESTS	185	96	9	11	36	77
IMMUNIZATION	3	30	29	5	20	12
IMMUNOTHERAPY	2	62	3	4	16	5
CELL MEDIATED IMMUNITY	8	87	101	14	27	20
SEROLOGY	203	264	33	15	33	95
ANTIGENS	87	85	37	43	43	50
TOTAL	532	788	252	109	190	317

* 1972-1981 : statistics Mackenzie ISHAM 1981
1982-1985: Rev.Med.Vet.Mycol.(C.A.B.International, Kew, Surrey.U.K.)

human, animal and plant pathogenic fungi are due to the great complexity of their ultrastructural and physicochemical organization and their continuous change during in vitro or in vivo development.

The definition of fungi by Albrech van Haller, 200 years ago, as "a mutable and treacherous tribe" is still valid (Longbottom, 1986). Fungi represent one of the five kingdoms of organisms classified based, according to Whittaker (1969) on three levels of organization : the prokaryotes (kingdom Monera including bacteria), the unicellular eukaryotes (kingdom Protista) and multicellular and multinucleate eukaryotes (Plantae, Fungi, Animalia). The actinomycetes which are true bacteria are included in this Symposium because like the fungi, they can exist in mycelial forms and because the diseases they cause are similar to those caused by true fungi. The dimensions of fungal cells vary from 1 to 300 μm and stand in contrast to the size of other microorganisms such as actinomycetes (0.5-1.4 μm), bacteria ($<$ 1 μm) rickettsia (0.3-0.6 μm), bacteriophages (0.02-0.100 μm) or viruses (0.015-0.300 μm) ; this may explain in part the observation that a simple yeast cell (2-4 μm) of Candida albicans has up to 78 antigenic components revealed by crossed immunoelectrophoresis (Axelsen, 1973). A great number of antigens are also detected in other fungi, for example 65 antigens for Aspergillus fumigatus (Guinet, 1986), 21 for Cryptococcus neoformans (Reyes et al., 1985), and 26 for Coccidioides immitis (Huppert et al., 1979).

Complexity of fungal structure and polymorphism

Among the 50,000 to 500,000 species of fungi present in nature (soil, air, water) more than 180 are pathogenic or potentially pathogenic in particular conditions (opportunistic fungi), and are responsible for systemic, subcutaneous, osteo-articular or superficial mycoses (Table 2).

They exist in several morphological forms : yeasts (budding cells), mycelium (filaments or hyphae with or without septa), and spores of different forms and dimensions (Fig. 1). Sometimes, two forms are observed as for the dimorphic systemic fungi : a parasitic form developed at 37°C as yeasts (Histoplasma capsulatum, H.duboisii, Blastomyces dermatitidis, Paracoccidioides brasiliensis) or spherules (C.immitis), and saprophytic form at 25°C as mycelium and conidia. From this

TABLE 2 - PRINCIPAL PATHOGENIC FUNGI

A - DEEP, SYSTEMIC OR VISCERAL MYCOSES

a) <u>Opportunistic</u> - cosmopolitan

　　　　. yeasts : Candida, Cryptococcus,
　　　　　　　　　Torulopsis, Trichosporon

　　　　. filamentous fungi : Aspergillus
　　　　　　　　　　　　　　Mucor, Rhizopus, Absidia,
　　　　　　　　　　　　　　Fusarium,
　　　　　　　　　　　　　　Penicillium, etc...

endogenous : <u>C.albicans</u>
exogenous saprophytes : soil, air, water

b) <u>Highly</u> pathogenic dimorphic fungi - limited geographically
　　　　　　　　　　　　　　　　　　　　tropical & subtropical

　　　　　histoplasmosis
　　　　　blastomycosis
　　　　　coccidioidomycosis

exogenous : soil

c) <u>Fungi</u> agents of subcutaneous or osteoarticular mycoses

　　　　　sporotrichosis, mycetoma,
　　　　　chromoblastomycoses (+ cerebral abscesses)
　　　　　pheohyphomycoses　　　　　"　　　　　"

exogenous : soil, vegetation, traumatism

B - SUPERFICIAL MYCOSES

<u>Fungi</u> : Dermatophytes - tinea capitis, pedis...
　　　yeasts : Candida, Malassezia, Torulopsis

exo-endogenous

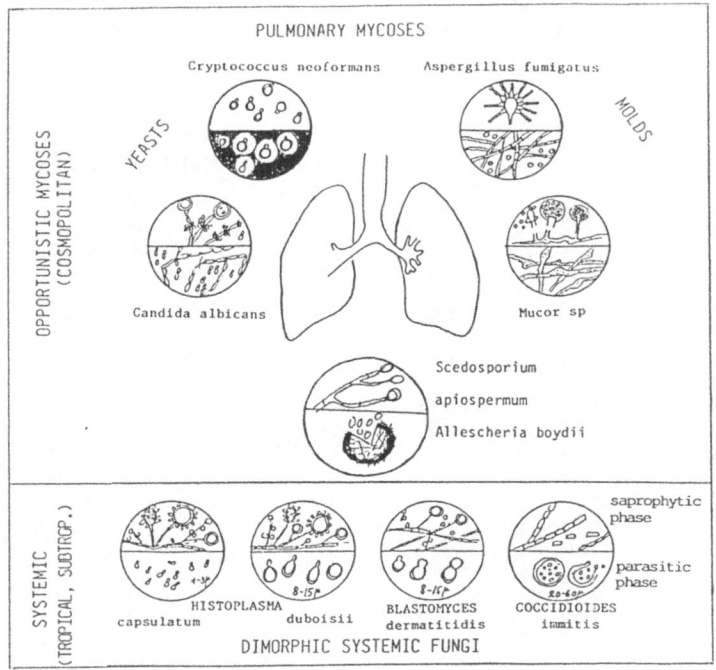

Fig. 1 - Principal fungi of pulmonary mycoses

arises the question : which are the best antigens for detecting the immune response ? The conidia introduced in the respiratory tract in the case of the pulmonary infections, or the yeasts and the spherules which are formed in the host tissues ? Because all these fungi are more easily cultured at 25°C in semi-synthetic (asparagin) or complex media (Sabouraud), the antigens used for immunodiagnosis were usually prepared from culture filtrate of sporulated mycelial forms (histoplasmin, coccidioidin, etc). However, more recent studies seem to demonstrate that the antigens from the parasitic phase such as spherulin from C. immitis may give the best response (Cox et al., 1981 ; Huppert, 1983 ; Levine et al., 1977 ; Wheat et al., 1986).

Another type of dimorphism is represented by Candida albicans. The yeast cells are easily cultured on classical media (Sabouraud's medium) at 25° or 37°C and represent the colonizing saprophytic form on the mucosae of the digestive tract. The mycelial form germinating tube, pseudomycelium or true mycelium is considered as the "aggressive", pathogenic form, invading the deep tissues ; it is obtained with difficulty (most cultures are of both Y and M form) at 37°C in serum or aminoacids medium (Marichal et al., 1986). Although most antigenic preparations of C.albicans used for diagnosis are from yeast cells, some investigators recommended the antigens from mycelium or germ tubes (Syverson et al., 1975; Munoz et al., 1980). However, there is not yet agreement as to which is the most relevant antigen for detection of the immune response (Ponton et al., 1986).

The size of the airborne conidia causing mycotic infections or allergic phenomena of the respiratory tract (Fig. 2) plays an important role in the pathology and allergoimmunology of the pulmonary diseases (Austwick, 1966). Only the conidia of

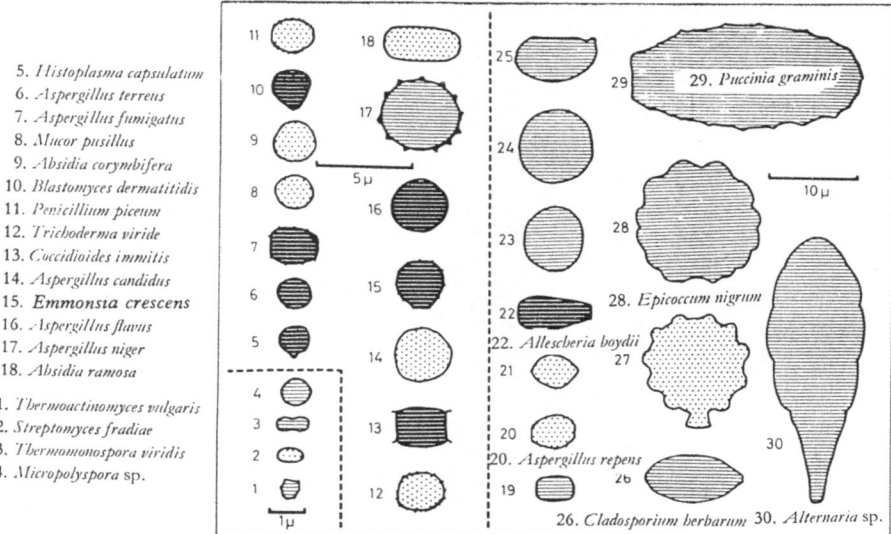

5. *Histoplasma capsulatum*
6. *Aspergillus terreus*
7. *Aspergillus fumigatus*
8. *Mucor pusillus*
9. *Absidia corymbifera*
10. *Blastomyces dermatitidis*
11. *Penicillium piceum*
12. *Trichoderma viride*
13. *Coccidioides immitis*
14. *Aspergillus candidus*
15. **Emmonsia crescens**
16. *Aspergillus flavus*
17. *Aspergillus niger*
18. *Absidia ramosa*

1. *Thermoactinomyces vulgaris*
2. *Streptomyces fradiae*
3. *Thermomonospora viridis*
4. *Micropolyspora* sp.

29. *Puccinia graminis*
28. *Epicoccum nigrum*
22. *Allescheria boydii*
20. *Aspergillus repens*
26. *Cladosporium herbarum* 30. *Alternaria* sp.

Fig. 2 - Spores of fungi and actinomycetes causing allergies and mycoses. Directly pathogenic species-dark shading ; allergenic species-light shading ; species from animal lungs-dotted (from Austwick, 1966).

less than 5 μm diameter can penetrate the alveoli (H.capsulatum, C.immitis, B.dermatitidis, Aspergillus fumigatus), whereas large conidia from Puccinia graminis, Alternaria sp, Epicoccum sp, Cladosporium herbarum penetrate only the upper respiratory tract and cause allergic phenomena. The conidia of A.fumigatus (3 μm) are agents of both infections and allergies. The smallest spores (less than 1 μm) belong to the thermophilic actinomycetes, responsible for hypersensitivity pneumonitis such as farmer's lung (Micropolyspora faeni) or bagassosis (Thermoactinomyces sacchari).

The complexity of the structure of the fungal cell (Fig. 3) emphasizes the difficulties inherent to the study of numerous antigens localized in the cell wall and cytoplasm (Cole, 1986). The rigid cell wall is composed of polysaccharides and proteins largely responsible for the immune response. The wall is sometimes surrounded (C.neoformans) by a large polysaccharide capsule which is related to virulence (Drouhet et al., 1950), producing inhibition of cellular migration (Drouhet and Segretain, 1951) and immunologic tolerance or paralysis. The cytoplasmic ribosomes obtained from mechanically disrupted cells are highly immunogenic and protective in experimental candidosis (Segal et al., 1980 ; Levy et al., 1981) and histoplasmosis (Tewari, 1975).

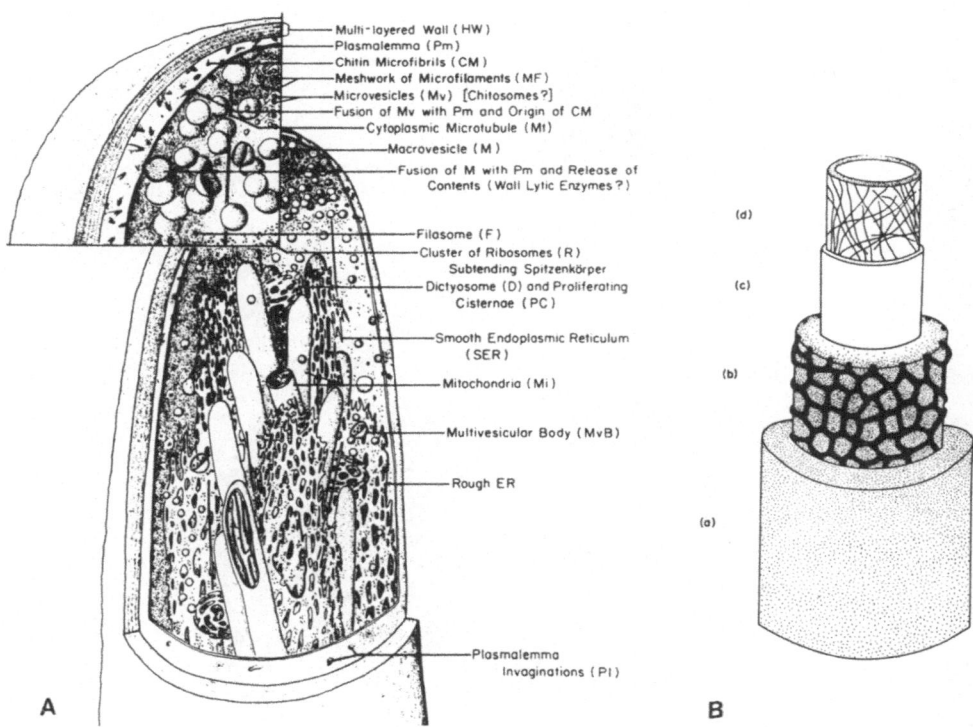

Fig. 3 - Structures of hyphal cells. A. Cytological structure of hyphae (from Cole, 1986). B. Complex ultrastructural of a fungal filament : a) outer mixed polysaccharides (glucans), b) the reticulum : glucans + proteins, c) proteins, d) proteins with embedded chitin microfibrils (from Burnett, 1976). In most pathogenic fungi, glucans from this fungal model (Neurospora) are substituted by mannans which are the immunogenic polysaccharides of the cell wall.

7

The fungal cell wall plays an important part in host-parasite relationships (San Blas, 1982). For example, superficial layers of the cell wall composed essentially of mannan, are responsible for the adherence of Candida yeasts to the epithelial cells of the host, a necessary preliminary stage for saprophytic or parasitic colonization. The role of Candida antigens in adherence mechanisms was recently, investigated (Rotrosen and Calderone, 1986 ; Robert and Senet, 1986).

Antigenic preparations

Despite the large number of fungal products that have been used for immunological investigations, there are no recognized standards or standardized methods for the production of fungal antigens except for a WHO candidate histoplasmin antigen preparation (Pine et al, 1981). The antigenic preparations are obtained from whole or disrupted cells or from culture media with or without autolysis (Fig. 4). Both crude extracts and purified preparations have been used.

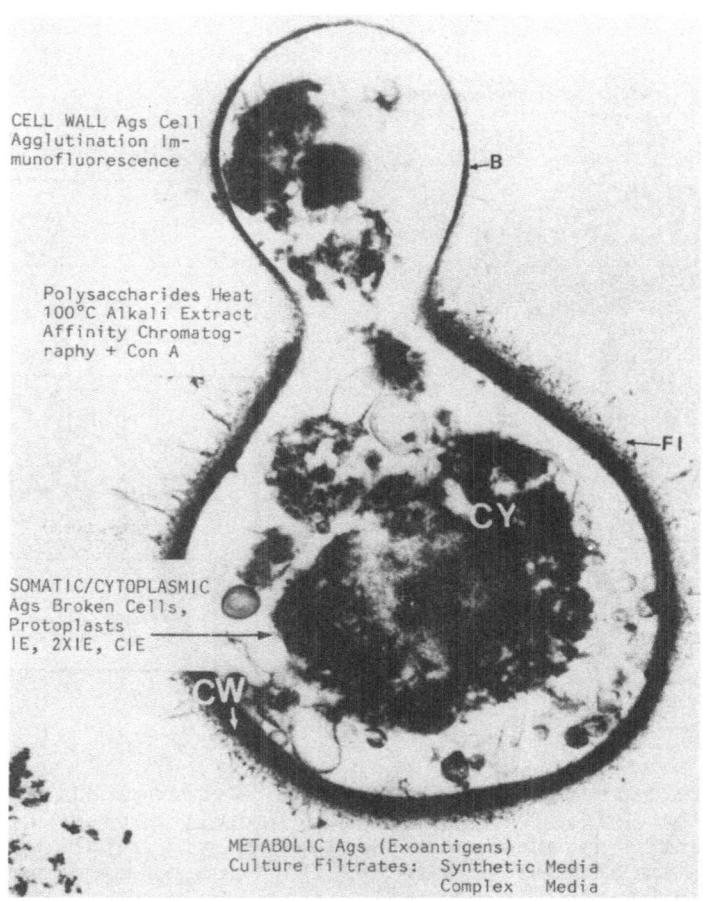

Fig. 4 -   Various antigenic preparations from the fungal cell.
CW : cell wall ; FI : capsular polysaccharide of
fimbriae ; B : bud ; CY cytoplasm

1. _Particulate antigens_. The particulate antigens represented by yeasts, conidia, and mycelium are important for some
serological tests or for the production of antisera. Agglutination tests (Tsuchiya et al., 1974) require homogenous
suspensions of conidia or yeast cells (Longbottom, 1986).
Immunofluorescence requires fixed material on slides or on
histological sections (Drouhet et al, 1972 ; Gordon et al;
1977 ; Kaplan, 1979 ; Othman and Drouhet, 1980 ; Drouhet et
al., 1983). Inert particles, e.g. latex particles of 0.3-0.8 µ
(Bloomfield et al., 1963 ; Gentry et al., 1983, Kaufman et
al., 1981) red blood cells (Meunier-Carpentier and Armstrong,
1986) may also be coated with soluble extracted antigens or
with antibodies and used for serodiagnosis.

2. _Soluble antigens_. The soluble antigens are largely
used for immunological tests particularly for immunoprecipitation, and immunoelectrophoresis,enzymo- and radioimmunoassays.
Three kinds of soluble antigens are currently used :

a) the extracellular antigens from culture filtrate, also
called "metabolic" antigens. Methods of preparation are standardized and the reference antigens are known under the names
of coccidioidin, histoplasmin, blastomycin, candidin (older
name of oidiomycin), sporotrichin and are used for the study
of immediate or delayed hypersensitivity. These preparations
are also used for research and most are commercially available.

b) intracellular, cytoplasmic (cytosol) or "somatic"
antigens representing the supernatant from disrupted whole
cells. These antigens are used largely for immunoelectroprecipitation studies. The term "cytoplasmic" antigens is reserved
by some authors (Poulain et al., 1986) for the intracellular
antigens prepared from protoplasts obtained by treatment with
lytic enzymes, exemplified by the work of Hearn and Mackenzie
(1979) for A.fumigatus and by Reyes et al. (1986) for Cr.neo-
formans.

The extraction of intracellular, somatic antigens neces-
sitates the disruption of the cell wall. Various methods are
available : grinding (dry material) with a pestle and mortar,
homogenizing in different types of blenders, ultrasonication.
From our own experience, the Braun homogenizer (MSK) using
glass beads gives the most efficient disruption even for
Cr.neoformans cells which are the most difficult to break.
Extraction must be carried out promptly and extracts must be
freeze dried to avoid enzymatic degradation of their immunological properties. Few commercially available antigens meet
these standardized criteria.

c) soluble cell wall antigens have been obtained by
chemical extraction (dilute alkali, phenol, acetone, ethylene
glycol, alcohol precipitation, formamide, chloral hydrate,
Triton X-100, etc..).

It is important to distinguish "crude" and "purified"
antigens. Electrophoresis and chromatography techniques have
been widely used to separate and purify antigens (Huppert,
1983).

Exoantigens

Exoantigens are soluble antigens, extracted from the
mycelia of fungi to be identified, which precipitate with
homologous antiserum to form a band of identity with that
formed by a reference antigen and antiserum. This method
developed by Kaufman et al., (1983), allows the identification
of dimorphic fungi such as P.brasiliensis, B.dermatitidis,

even H.capsulatum or C.immitis when mycelial morphology does not permit rapid identification and requires conversion of the mycelial to the distinctive yeast form. Numerous other fungi (Sporothrix schenckii, Penicillium marneffei, Phialophora jeanselmei, etc..) were identified by the exoantigen test.

Standardized cultural conditions for obtaining fungal antigens

A detailed review of this problem was recently carried out by Longbottom and Austwick (1986). The choice for simple or complex, solid or liquid media, for aerated or non-aerated conditions, surface or shaker cultures depend on the material required and the type of cells to be produced. The synthetic and semi-synthetic chemically defined media such as YMA (Difco), YNB + asparagin + glucose, Smith's asparagin medium, synthetic medium for Y phase of H.capsulatum (Pine and Drouhet, 1962) are appropriate especially when the antigens are used as allergens or vaccines, to avoid contamination of the immunoreactive components. However, complex media containing peptone (Sabouraud's medium), casein hydrolysate, beef extract, etc.. are often used, due to their improved yields. A double dialysis technique has been used for antigens of Micropolyspora faeni and Thermoactinomyces sacchari (Boiron, 1986). Addition of a yeast extract to a synthetic medium enhances the IgE binding capacities of fungal extracts (A.fumigatus, Alternaria alternata, Cladosporium herbarum) as measured by RAST (Heide et al., 1985). Shaken cultures in fermentors present the advantages of reproducibility, homogeneity, excellent mycelial production in absence of sporulation, and growth control. For somatic antigens, rapid growth is necessary : 48-72 h for yeasts such as C.albicans and Cr.neoformans and 3-4 days for A.fumigatus. The temperature selected is usually that which will produce optimal growth of the fungus : 25°-30°C for the majority of fungi, 37°C for the yeast phase of dimorphic fungi (H.capsulatum), 50°C for thermophilic actinomycetes (M.faeni, T.sacchari). Heat shock and heat stroke phenomena occur in C.albicans (Dabrowa and Howard, 1984) and Fonsecaea pedrosoi (Ibrahim-Granet and de Bièvre, 1986) with changes in the protein profiles due to modification of temperature conditions, with possible changes in antigens and in the immune response of the host organism.

The size of the inoculum is important for obtaining an early or late growth phase and the content of glucose of the medium must be controlled to avoid autolysis at the end of growth.

Production of experimental antisera

The rabbit is the best animal for the production of antibodies against fungi, particularly New Zealand rabbits for Cr.neoformans. We have obtained excellent reference sera against C.albicans and A.fumigatus antigens utilizing the goat. Mice can also be used for the production of antibodies against fungi. Live or killed (by heat, merthiolate, formol), intact or disrupted fungi or soluble antigens, with or without complete or incomplete Freund's adjuvant, are administered by various routes (Fig. 5).

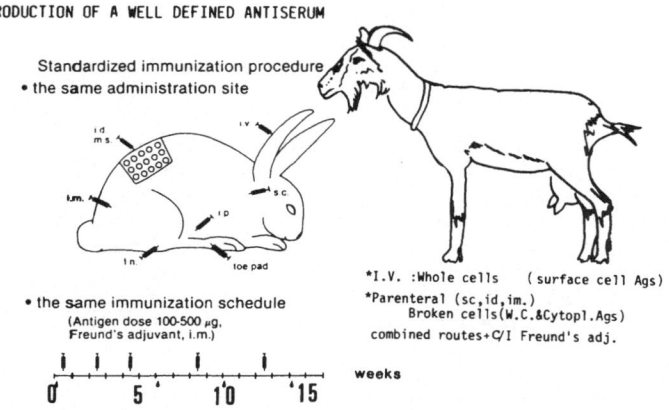

PRODUCTION OF A WELL DEFINED ANTISERUM

Standardized immunization procedure
• the same administration site

i.d.
m.s.
i.v.
i.m.
s.c.
i.p.
i.n.
toe pad

*I.V. :Whole cells   (surface cell Ags)
*Parenteral (sc,id,im.)
    Broken cells(W.C.&Cytopl.Ags)
combined routes+C/I Freund's adj.

• the same immunization schedule
(Antigen dose 100-500 µg,
Freund's adjuvant, i.m.)

0   5   10   15   weeks

Fig. 5 - Procedures for obtaining hyperimmune antisera in experimental animals. Repeated i.v. injections with whole cells
(w.c.) produce antibodies against surface antigens. Parenteral
injections by multiple routes, i.e. subcutaneous (s.c.), intradermal (i.d.), intramuscular (i.m.), intraperitoneal (i.p.)
with complete or incomplete Freund's adjuvant raise especially
precipitating antibodies.

By the intravenous route, whole cells produce antibodies a-
gainst the cellular surface antigens detected by agglutination
and immunofluorescence tests. By parenteral routes (subcutaneous, intradermal, intramuscular, intraperitoneal), broken
cells (cell walls, cytoplasmic antigens) in association or not
with complete or incomplete Freund's adjuvant produce precipitating antibodies easily detected by immunoelectrophoretic
techniques. Antisera are also obtained by active infection
with appropriate fungi. Antibodies are obtained within 2-4
weeks, but a maximum number of precipitins requires a longer
period of immunization. In order to obtain the 78 antigenic
fractions of C.albicans somatic antigen by crossed immunoelectrophoresis in intermediate gel, Axelsen (1974) immunized a
rabbit by multiple routes more than 2 years. Short courses of
immunization tend to give more specific antisera with weak
cross-reactivity, whereas long intensive courses give numerous
bands with intensive cross-reactivity. When the antiserum is
intended for "fingerprinting" of fungi using crossed immunoelectrophoresis, an intensive immunization is necessary. Monospecific antisera for specific diagnostic tests have been
prepared using a relevant immunoprecipiting arc in IE tests
for the C_2 antigen of A.fumigatus (Yarzabal et al., 1973), the
E_2 antigen of P.brasiliensis (Yarzabal et al., 1977), the H
and M antigens of H.capsulatum (Pine, 1963) and B.dermatitidis
A and D antigens (Green et al., 1979). Sera from immunosuppressed (cortisone and cyclophosphamide) rabbits infected with
A.fumigatus conidia were used as a source of circulating
antigens, in order to produce an antiserum to A.fumigatus
galactomannan (Lehmann and Reiss, 1978).

11

Immunologic response to fungal antigens

Fungal antigens produce responses similar to bacterial and viral antigens but some features are proper to fungi. It has been known from the time of Metchnikoff (1892) that similar phenomena occur when a fungus invades an invertebrate or a vertebrate host. The polymorphonuclear (PMN) leukocyte plays an important role in the inflammatory response to fungi and in resistance to fungal disease. One of the mechanisms by which the PMN acts is by adherence and engulfment by pseudopods of the fungal cell which is followed by myeloperoxidase dependent fungicidal activity. Candida and Aspergillus produce catalase and are more resistant to destruction especially when the PMN leukocytes are deficient in myeloperoxidase (chronic granulomatous disease).

Humoral immunity dependent on B lymphocytes and cell-mediated immunity (CMI) dependent on T cells play an important role in the immune response to fungal antigens. The role of antibody in host resistance to fungal diseases in not yet fully understood. CMI contributes significantly to host defence in most fungal diseases. Humoral antibodies are readily demonstrable during the course of disease but their protective effect has not been demonstrated (Cox, 1983). The B cells either by themselves or evolving into plasma cells produce immunoglobulins of five classes : IgA, IgD, IgE, IgG and IgM. IgM and IgG are the principal circulating immunoglobulins, the former as a response to polysaccharide, the latter to protein antigens. IgM appears in the first days of fungal infections, followed immediately by IgG which gradually increase and persist while the IgM titer is diminishing. Secretory IgA is observed in mucous infections (candidosis).

Specific antifungal antibodies of the immunoglobulin E class are detectable in some mycoses, independently of allergic manifestations, particularly in candidosis (Mathur et al., 1977), coccidioidomycosis and histoplasmosis (Cox and Arnold, 1979, 1980) and their presence was related to immunosuppressor cells. IgE antibodies have been observed in chronic diseases in association with depressed CMI (Hay and Shennan, 1984).

Methods of detection of antibodies

Immunodiffusion in gel and immunoelectroprecipitation. They are presently among the most utilized methods for studies on antigens and antibodies in the mycoses. These methods were developed at the Pasteur Institute, first with Oudin's (1946) gel precipitation technique using simple and double diffusion (DD) in tubes, followed in 1948 by Ouchterlony's practical DD plate technique performed in Petri dishes. An important gain was obtained at Pasteur Institute by Grabar and Williams (1952) with immunoelectrophoresis (IE), combining gel electrophoresis with immunodiffusion (ID). In 1952, Bussard also at Pasteur Institute described the electrosyneresis, a very simple, highly sensitive and rapid method based on the migration of antigens towards the anode, and of the antibodies towards the cathode in gel (agar, actually agarose) at pH 8.2. This technique was later named immunoosmophoresis or immunoelectroosmophoresis (Gordon et al., 1971) and is currently better known under the name of counterimmunoelectrophoresis (CIE). It became one of the most employed tests for routine serology of the mycoses (Fig. 6). The two-dimensional immunoelectrophoresis (2DIE) technique described by Ressler (1960)

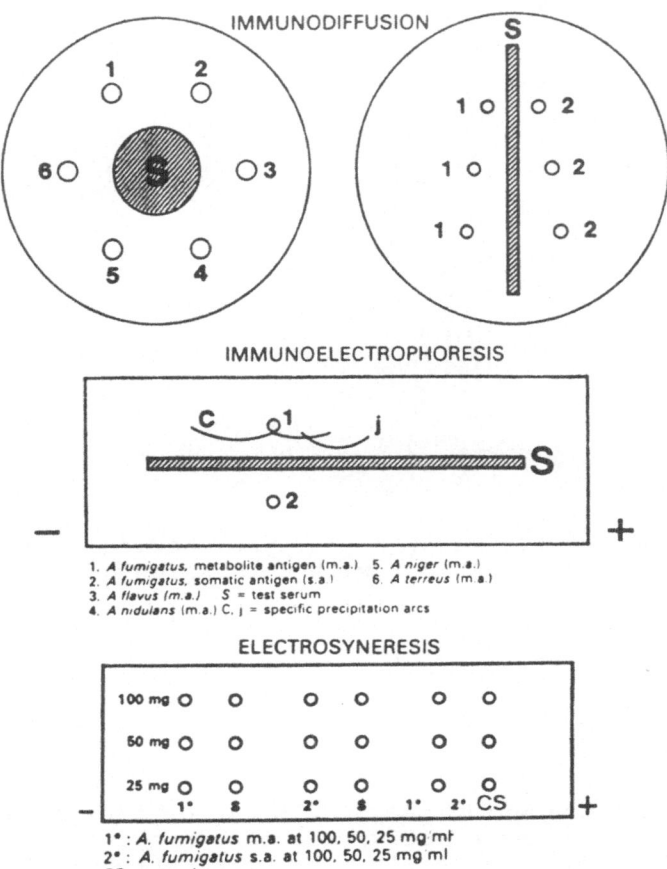

IMMUNODIFFUSION

IMMUNOELECTROPHORESIS

1. *A. fumigatus,* metabolite antigen (m.a.)   5. *A. niger* (m.a.)
2. *A. fumigatus,* somatic antigen (s.a.)   6. *A. terreus* (m.a.)
3. *A. flavus* (m.a.)   S = test serum
4. *A. nidulans* (m.a.) C, j = specific precipitation arcs

ELECTROSYNERESIS

1° : *A. fumigatus* m.a. at 100, 50, 25 mg·ml
2° : *A. fumigatus* s.a. at 100, 50, 25 mg·ml
CS : control serum

Fig. 6 Schematic representation of immunodiffusion in gel and immunoelectroprecipitation techniques applied for routine immunodiagnosis of <u>Aspergillus</u> infections.

and developed by Laurell (1965) under the name of crossed immunoelectrophoresis opened the way to the quantitative immu- noelectrophoretic (QIE) methods. This method provided a most significant contribution to the identification and characteri- zation of the complex antigens of the fungi and the antigenic interrelations that exist between antigenic preparations strains, and genera (Axelsen, 1983, Longbottom, 1986).Axelsen first applied 2DIE (Fig. 7) to the antigenic mosaic of <u>C.albi- cans</u> (Axelsen, Kirkpatrick, 1973, Axelsen et al., 1974, 1975). The numerous variants (crossed IE with intermediate gel, cros- sed line IE, crossed affinity IE, fused rocket IE, tandem crossed IE, crossed radio IE), are widely utilized, well standardized and were the subject of recent reviews (Axelsen, 1983 ; Longbottom, 1986 ; Ouchterlony and Nilsson, 1986) (Fig. 8).

<u>Enzymatic activities associated with fungal antigens</u>

The specific enzyme staining techniques applied to immu- noprecipitation tests of pathogenic fungi originated from the work of Uriel (1963) at Pasteur Institute. The French school of Biguet (Biguet et al., 1967 a, b ; Tran Van Ky et al., 1969

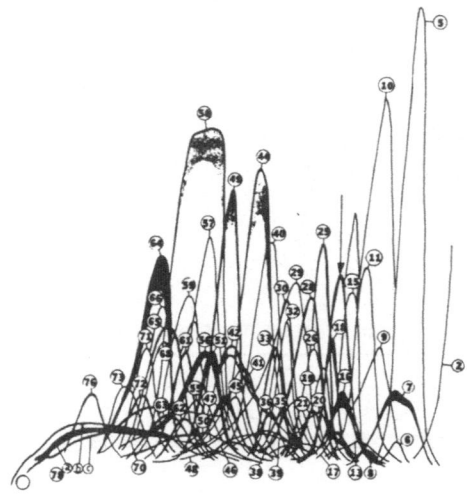

Fig. 7 - 2DIE of somatic Ag of C.albicans showing 78 antigenic fractions (from Axelsen, 1973 ; with permission of Scand.J.Immunol.)

a, b) showed that the principal precipitating antigen-antibody complexes have enzymatic activities of diagnostic interest such as chymotrypsin and catalase activities for A.fumigatus (Fig. 9 A), chymotrypsin and malic dehydrogenase for A.flavus, malic dehydrogenase activity for Candida spp., and catalase and glucuronidase activities for H.capsulatum. Subsequently, the antigenic preparations of Candida (Drouhet, 1978)and Aspergillus (Drouhet et al., 1972) used for serological diagnosis in our laboratory and for commercial distribution (Diagnostics Pasteur, 92430 Marnes la Coquette, France) were rigorously standardized and are lyophilized extracts from live fungal cells with controlled enzymatic activities and immunoprecipitating properties.

Standardization of each batch of antigen is controlled by the number of precipitating lines determined on 1 % agarose (barbital buffer pH 8.2 and sodium tris-glycine barbital buffer pH 8.8) by immunodiffusion, immunoelectrophoresis, CIE and crossed IE (Fig. 9 A), and also by determining the enzymatic activities. Chymotrypsin is evaluated by the Mancini technique and revealed by staining with diazoblue B in presence of specific substrate. Catalase is determined by spectrophotometric measurement and is revealed in a few seconds with H_2O_2 producing bubbles (Fig. 9B).

The mean titers of enzymatic activities of the somatic antigen of A.fumigatus are 68.10 mU/ml for chymotrypsin and 1186.60 mU/ml for catalase. The mean titers for metabolic

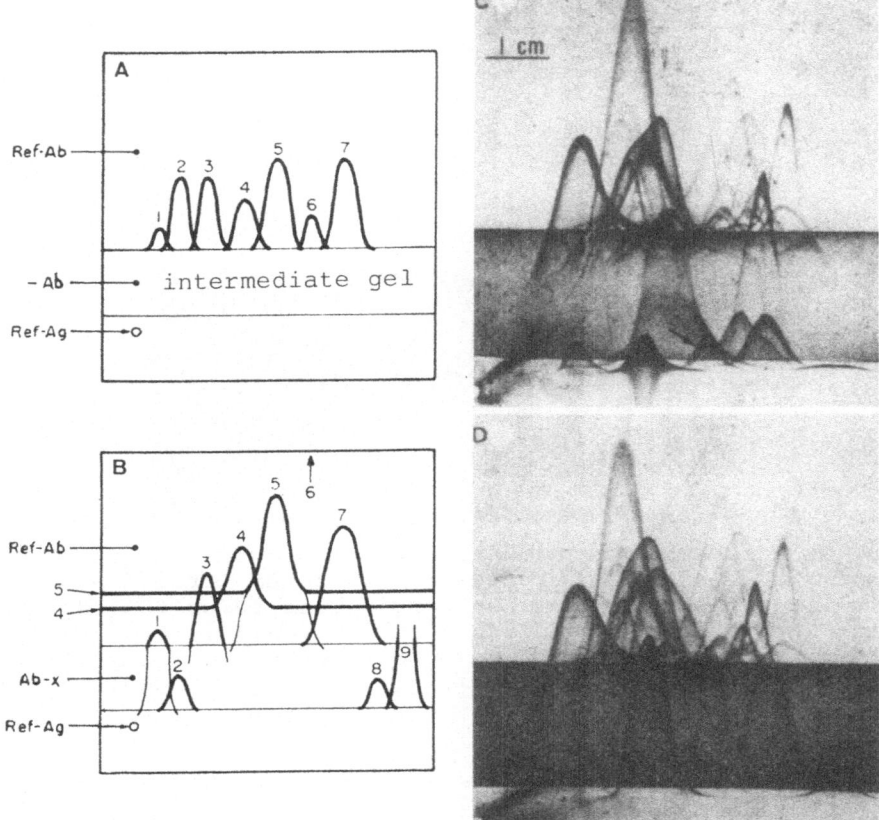

Fig. 8 - Quantitative 2DIE with intermediate gel used for
estimation of titers of antibodies of an unknown serum to be
studied (Ab-x) tested against the reference antigen (Ref-Ag)
comparatively to a known reference antiserum. First dimension
electrophoresis of the Ref-Ag placed in the low gel (anode to
the right). Second dimension electrophoresis (with anode at
the top) for the reference antiserum (Ref-Ab) and Ab-x. A :
Control plate, no Ab in the intermediate gel. B : Test plate
Ab-x in the intermediate gel. C.D. : 2DIE with intermediate
gel using C.albicans extract as antigen, rabbit anti-C.albi-
cans as reference and sera from candidosis patients. C : Serum
(D-1.a) from a patient in whom candidaemia had been demonstrated ; the blood was drawn at the end of Amphotericin B
treatment. Note the patient's strong precipitin response,
indicated by the many precipitates in the intermediate gel.
The arrow points to a precipitate produced by a reaction
between antigen no. 33 and the patient's antibodies. D : Same
patient as A, but the sample (D-1,b) was drawn one month
later. Compare B with A and note the increase in areas deli-
mited by most precipitates of the intermediate gel, signifying
the titre fall of most precipitins. (from Axelsen et al.,
1975, 1983 with permission of Scand.J.Immunol.).

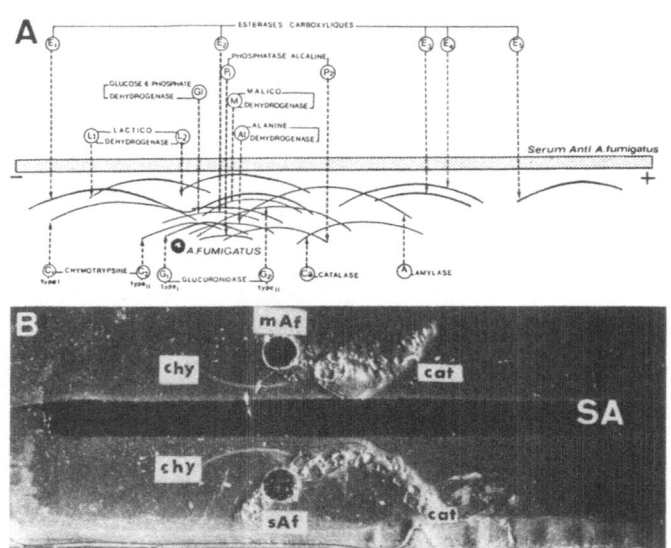

Fig. 9 A - Enzymatic characterization of precipitating Ag-Ab complexes in an anti-A.fumigatus rabbit serum (from Tran Van Ky et al., 1969). B - Human aspergillosis serum (SA) with a precipitating line with catalase activity (CAT) ; the second precipitating line (CHY) has chymotrypsin activity (from Drouhet, 1982). MAF : metabolic A.fumigatus antigen. SAF : somatic A.fumigatus Ag.

antigens of A.fumigatus, A.flavus, A.terreus, A.niger, A.nidulans, are respectively 96, 173, 38, 111 and 10 mU/ml for chymotrypsin and 1311, 133, 708, 190, and 698 for catalase activity. Chymotrypsin and catalase lines appear early in Aspergillus infections revealed by IE and CIE in agarose or cellulose acetate. The latter support is ideally suited to a rapid, sensitive, and easily reproductible method for routine laboratory diagnosis of the deep mycoses. The M band of histoplasmin antigen has catalase activity and the H band a glucuronidase activity, permitting a rapid characterization of these specific precipitating lines.

Enzyme immunoassays and related sensitive assays

The first enzyme immunoassay (EIA) techniques were described independently by Avrameas and Guilbert (1971) at the Pasteur Institute and by Engvall and Perlmann (1971) who gave the name ELISA to the enzyme immunosorbent assay. Subsequently this test was rapidly and extensively applied to for the immunology of pathogenic fungi and their respective mycoses. Enzyme immunoassays employ antibodies or antigens linked to enzymes in such a way that the immunological and catalytic activity of each moiety are preserved. These tests provide an objective endpoint, can be automated, and are simple to perform. In addition, they have a sensitivity and specificity comparable to radioimmunoassays (RIA), and are capable of detecting ng amounts of antigen or antibody, without health risks and with a long shelf life of reagents. Various adaptations of EIA have been reported and compared with other immunological tests for diagnosis of aspergillosis (de Repentigny and Reiss, 1985 ; Richardson and Warnock, 1983) candidosis (Araj et al., 1982 ; Fujita et al., 1986 ; Kostiale et al., 1981 ; Meckstroth et al., 1981 ; de Repentigny et al., 1985, Richardson and Warnock, 1983), cryptococcosis (Scott et al., 1980), extrinsic allergic alveolitis in the form of bagassosis (Boiron et al., 1987) as well as for allergen specific IgE antibodies in allergic aspergillosis (Longbottom, 1986). Enzyme immunoassay has been especially useful for the detection of circulating fungal antigens. In our laboratory, Othman et al. (1978, 1979), utilized a simplified solid phase technique based on the magnetic support Magnogel (Guesdon and Avrameas, 1977), which obviate the need for the time-consuming multiple centrifugations for washings. Magnogel is composed of magnetic polyacrylamide-agarose beads and was used to covalently bind antibodies of C.neoformans (Othman et al., 1979) and C.albicans (Othman et al., 1978). With the same technique, called MELISA, Camargo et al. (1984a) detected IgG to P.brasiliensis and compared it with CIE. Sensitivity was increased by activation with Con A of the polyacrylamide agarose beads coated with somatic Ag instead of metabolic Ag. The erythroimmunoassay described by Guesdon et al. (1983) utilizing red blood cells and chimera antibodies showed a higher sensitivity than classical techniques (complement fixation, immunoprecipitation, in gel) for detecting P.brasiliensis antibodies with patients' sera diluted up to 1/102400. Both MELISA and erythroimmunoassay detect the same kind of antibodies, different from those detected in CIE tests. MELISA using purified metabolic Ag is a promising alternative to CF in the diagnosis of paracoccidioidomycosis, and allows the testing of sera with anticomplementary activity (Camargo et al., 1986).

Immunoblotting (IB) techniques

The recent enzyme linked immunoelectrotransfer blot or Western blot technique (Towbin et al., 1979 ; Burnett, 1981) is based on the electrophoretic separation of the crude antigen mixture in agarose polyacrylamide or SDS-polyacrylamide gel followed in a second step by the transfer blotting of the separated fractions to a nitrocellulose sheet in which they are immobilized. In our laboratory (Delga et al., 1985 ; Ibrahim-Granet et al., 1986) we have studied by polyacrylamide gel electrophoresis (PAGE) and IB the somatic antigens of serotypes A and B of C.albicans with whole and specific antisera. Immunoblotting showed a major antigen of 68,000 daltons which is probably a surface glycoprotein and a protein antigen common to serotypes A and B (Delga et al., 1985). A program in Basic Microsoft allowing the determination of MW and linear graphic representation of an electrophoretic protein pattern was applied to the study of the two serotypes of C.albicans (Ibrahim-Granet et al., 1986). Fourteen bands specific to serotype A and 8 bands specific to serotype B were identified (Fig. 10).

Both serotypes contain an intense band of Mr 100,000, only weakly visible following Coomassie Blue staining. This may indicate that the fraction consists mainly of polysaccharide with only little protein. The intensity of the IB reaction strongly suggests that this fraction is highly immunogenic.

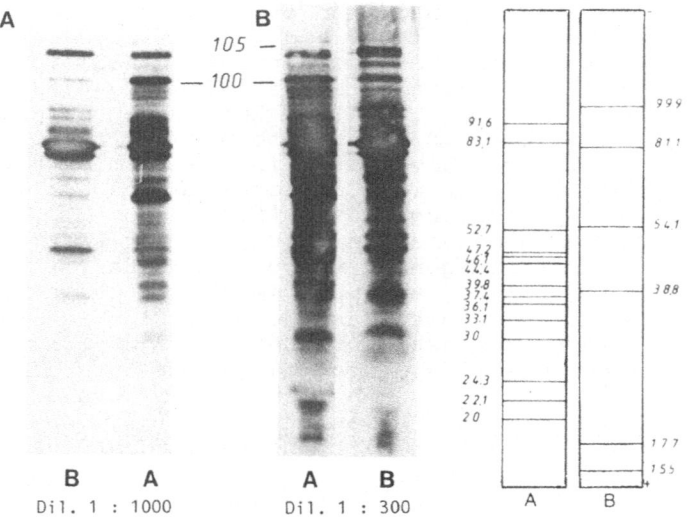

Fig. 10 - Immunoblotting of C.albicans serotypes A and B with (A) the whole anti-C.albicans serum obtained by immunizing a goat with C.albicans serotype A (serum dilution 1 : 1000) and (B) the specific antiserum (adsorbed with serotype B cells) diluted at 1:300. Graph : schematic representation of specific fractions and their corresponding Mr. A : serotype A ; B : serotype B (from Ibrahim-Granet et al., 1986).

Immunoblot analysis was used to characterize the antibody response to C.albicans Y and M antigens in human sera (Manning-Zweerink et al., 1986) ; for characterization of monoclonal antibodies against C.albicans antigens (Strockbine et al., 1984) and M and carbohydrate antigens of histoplasmin (Reiss et al., 1986).

Immunofluorescence (IF)

IF tests, largely used in microbiology since 1956 as a result of Coon's labelling of proteins with fluorescent substances, have been successfully applied to medical mycology by Kaplan and Kaufman (1961), Drouhet et al. (1972), Gordon (1977) and Kaplan (1979). Principles and applications have been the subject of several recent reviews (Kaufman and Reiss, 1986 ; Longbottom, 1986 ; Mackenzie, 1983). The direct fluorescent antibody test is used for rapid visualization and identification of fungi in cultures and clinical specimens and gives excellent results (Kaufman and Kaplan, 1961). The indirect IF method is used to measure class and subclass specific antibodies in serum and body fluids. The application of indirect IF to the diagnosis of aspergillosis, first used by Drouhet et al. (1972, followed by Warnock (1974), is a good complement to precipitin tests. Young germlings of spores are now used in IF tests(Gordon et al., 1977 ; Drouhet et al., 1983) (Fig. 11).

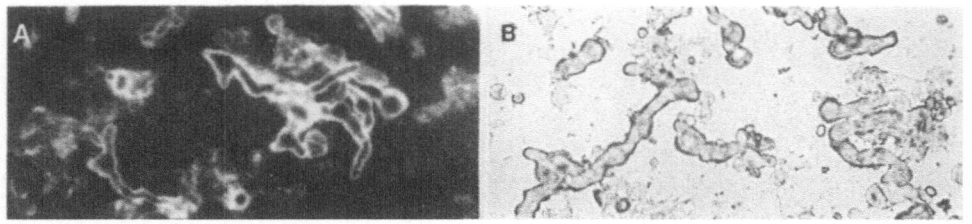

Fig. 11 A - Immunofluorescence on smears of germlings of A.fumigatus. Serum of a patient with bronchopulmonary aspergilloma. Titer 1/160.x100.B - Immunoperoxidase. Same smear (from Drouhet et al., 1983).

Indirect IF with rabbit anti-Trichophyton mentagrophytes var.asteroides, T.interdigitale and T.rubrum sera was utilized for identification of dermatophytes (Drouhet et al., 1976). The principal species from the 3 genera of dermatophytes gave group responses and adsorption of monospecific sera with heterologous species of dermatophytes eliminated the specific IF. Cross-reactions were observed between dermatophytes and various filamentous fungi (Aspergillus, Penicillium, Scopulariopsis, but not with yeasts (Candida, Cryptococcus, Torulopsis). Mucor sp. and yellow pigmented Aspergilli produce an autofluorescence which complicates the interpretation of the results. Serum titers obtained by IF were different for H.capsulatum and H.duboisii, and confirmed the immunological identity of H.duboisii of human and simian origin (Georges et al., 1980).

Immunoperoxidase (IPO)

The utilization of an anti-IgG serum labelled to peroxidase instead of fluorescein isothiocyanate proved to be of the same value as the immunofluorescence method with the advantage of necessitating only an ordinary microscope. Simple techniques were described in our laboratory utilizing smears of Cr.neoformans, C.albicans, A.fumigatus and using direct or indirect techniques (Othman and Drouhet, 1980 ; Othman and Drouhet, 1979 ; Drouhet et al., 1983) (Fig. 11).

Monoclonal antibodies (mAbs) against fungal antigens

Kohler and Milstein first reported in 1975 on the immortalization of mice antibodies produced by clones derived from a single B-lymphocyte. Until 1981, only one report was published on the use of monoclonal antibodies (mAb) in mycology (Hall and Blackstock, 1981) but more recently a great number of mAbs have been obtained against pathogenic fungi and actinomycetes (Table 3). Most mAbs are directed towards yeasts (Candida, Cryptococcus), systemic dimorphic fungi (Histoplasma, Blastomyces, Coccidioides), dermatophytes (Microsporum), molds (Aspergillus) or actinomycetes (Nocardia, Actinomyces). The majority of these mAb have been obtained after immunization of mice with whole fungal structures and only a few with

TABLE 3 – MONOCLONAL ANTIBODIES AGAINST PATHOGENIC FUNGI

GROUP OF FUNGI	ORGANISM	TARGET ANTIGEN	CLASS OF Ig	NUMBER OF HYBRIDOMAS	CHARACTERIZATION BY	REFERENCES
MOLDS	Aspergillus fumigatus	Galactomannan	IgM	1 (1)	ID, IELISA, IF, EM	I.Pasteur group (1986) Munoz et al. (1986)
	Candida albicans	mannan	IgM	5 (1)	ID, IELISA, IF, IF, Aggl., EM	I.Pasteur group (1986) Munoz et al. (1986)
	"	mannan	IgM	17 (2)	IELISA, IF, Aggl	Miyakawa et al. (1986)
	"	cell wall Y	IgM	21	IELISA	Chardès et al. (1986)
	"	cell wall Y	IgM	3	ID, Aggl., IF	Brawner, Cutler (1984)
	"	peptidoglucomannan				Kerkering et al. (1983)
YEASTS	"	cytoplasmic proteins	IgG$_1$	3	StAIP, ITB,	Strockbine et al. (1984)
	"	cell wall Y	IgM$_1$	1	RIA, Aggl, StAIP	Barbfer et al. (1985)
	"	cell wall Y	IgM	3	Aggl, IF, ID, IP, EM	Hopwood et al. (1986)
	"	glucoprotein Ag	IgG	9	ID, ITB, Aggl	Polonelli, Morace (1986)
	Candida tropicalis	mannan	IgM IgG	41 (4) (1)	IELISA	Reiss, de Repentigny (1984)
	Cryptococcus neoformans	pseudo-mycelium	IgM(?)	9	RIA	Hall, Blackstock (1981)
		glucuronoxylomannan	IgG IgM	21 (7) (14)	IELISA, ID, Aggl	Dromer et al. (1987)
SYSTEMIC	Blastomyces dermatitidis	A protein Y cytoplasmic protein	IgM IgM	14	IELISA IP, AC	Green, Harrel (1981) Young, Larsh (1982)
DIMORPHIC	Coccidioides immitis	endospores/spherules	IgM	7(2)	IELISA, La, IF	Karu et al. (1985)
FUNGI	Histoplasma capsulatum	M-protein polysaccharide	IgM	156 (5)	ITB, IP	Reiss, Knowles (1986)
		Y cells	lymphokine activating macrophages to suppress intracellular growth			Wu-Hsieh, Zlotnick, Howard (1984)
DERMATO-PHYTES	Microsporum canis	metabolic Ag	IgG	30 (3mAb)	ID	Polonelli, Morace (1985)

ID : immunodiffusion in gel ; Aggl : agglutination ; IP : immunoprecipitation ; StAIP : Staphylococcus aureus immunoprecipitation
ITB : immunoelectrotransfer blotting ; IELISA : indirect ELISA ; EM : electron microscopy ; AC : affinity chromatography
Y : yeast, La : latex agglutination ; IF : immunofluorescence, RIA : radioimmunoassay

purified, well characterized polysaccharides (mannan, galacto-mannan) or cytoplasmic proteins. Only a few mAbs are IgG, and the majority belong to the IgM class. Indeed, it is well known that numerous polysaccharide antigens are thymus dependent (Durandy et al., 1983, 1986) and that IgM antibodies represent the immune response to these antigens. Our desire to produce mAbs against fungal polysaccharides was motivated by the fact that circulating antigens in both invasive aspergillosis and candidosis are high molecular weight polysaccharides. Our Mycology Unit group in collaboration with the Hybridolab group (J.C. Mazié) have obtained two mAbs of the IgM class against mannan of C.albicans and galactomannan of A.fumigatus, by immunization of high antibody producing Biozzi mice by the i.p. route (Munoz et al., 1986).

The mAbs have considerable potential in medical mycology. Because of their high specificity, they minimize the risk of cross reactions in fungal or antigen detection. They can replace polyclonal Abs in immunodiagnosis (ELISA, histological detection of Ags). They are useful for the study of a protective immune response, passive transfer of immunity, development of vaccines ; antifungal drugs can be conjugated to mAbs to achieve high selective toxicity. The main advantages of monoclonal antibodies over polyclonal antibodies are reflected in the ability to obtain unlimited quantities of the same antibody with defined or selected specificity and affinity.

Nature of fungal antigens

Of the large numbers of antigens contained in the pathogenic fungi, few are considered major antigens as revealed by immune response in experimental animals or humans. A short tabulation is presented in Table 4 and most will be discussed in greater detail in this Symposium. Polysaccharides containing mannan are immunogenic and represent a constant feature of the pathogenic fungi. Several recent reviews summarized these aspects of immunogenic antigens (Huppert, 1983 ; Reiss et al., 1985 ; Reiss, 1986).

Circulating antigens

Since the demonstration of antigenemia, produced in animals experimentally infected with Streptococcus pneumoniae (Dochez & Avery, 1917), there have been numerous reports on the detection of antigens in the blood, urine and CSF of patients and animals infected with other bacteria, fungi, viruses and parasites. In the case of the fungi, with the exception of the capsular polysaccharide of Cr.neoformans detected since 1950 (Neil et al., 1950, 1951), advances have required sensitive immunoassay procedures that became widely applied more recently. In immunocompromised patients with invasive candidosis or aspergillosis, the detection of circulating antigens can provide rapid diagnosis and prompt treatment in the first days of infection when antibodies are not yet produced or are at a low level (Drutz, 1986).

Cryptococcus neoformans. The latex agglutination test (LAT) using particles of 0.8 mm coated with a rabbit anti-Cr.neoformans IgG (Bloomfield et al., 1962) is extremely sensitive and reliable when standard conditions are respected. The cryptococcal polysaccharide is liberated easily from the capsule into the body fluids of 99 % of cryptococcosis patients particularly in the CSF, serum and urine and persists in direct proportion to disease severity. In AIDS, we have

TABLE 4 – POLYSACCHARIDES AND PROTEINS AS MAJOR IMMUNOGENIC ANTIGENS

GROUP OF FUNGI	FUNGI	POLYSACCHARIDE Ag	REF	PROTEIN Ag	REF
MOLDS	Aspergillus fumigatus	galactomannan	1	"chymotrypsic" wall-associated	3
			2	glycoproteins	4
YEASTS	Candida albicans	α mannan	5	acidic carboxyl proteinase	6
	Cryptococcus neoformans	galactoxylomannan glucuronoxylomannan	7 8		
DERMATO-PHYTES	Trichophyton mentagrophytes	galactomannan peptides	9	keratinases I, II, III	10
ZYGO-MYCETES	Rhizopus oryzae	peptidofucomannan	11	alkaline proteinase	12
	Blastomyces dermatitidis	cell wall extracts	13	A protein factor cell wall extracts	14 15
DIMORPHIC SYSTEMIC FUNGI	Histoplasma capsulatum	galactomannan	13	"h" and "m" factors	14
	Paracoccidioides brasiliensis	galactomannan	13	E 2 factor	15
	Coccidioides immitis	3-0-methyl-mannose polymer	16	coccidioidin factor 2 & 11	17
	Sporothrix schenckii	peptido-L-rhamno-D-mannan	18		

Ref. 1 : Bennett (1985) ; 2 : Reiss (1979) ; 3 : Biguet (1964) ; 4 : Hearn (1979) ; 5 : Hasenclever (1961) ; 6 : Rüchel (1981) ; 7&8 : Drouhet (1950) ; Bennett (1965) ; Bhattacharjee (1984) ; 9 : Barker (1967) ; 10 : Collins (1973) ; 11 : Miuzaki (1980) ; 12 : Reinhardt (1985) ; 13 : Azuma (1974) ; 14 : Pine (1985) ; 15 : Yarzabal (1977) ; 16 : Wheat (1977) ; 17 : Huppert(1979) ; 18 : Travassos (1980).

detected in CSF the highest concentrations ever reported until now, up to 2.5×10^5 ng/ml. Its replacement by free antibody is a sign of favorable response to therapy. Occasionally, false-positive results related to the presence of rheumatoid factor (eliminated by dithiothreitol) are observed as well as false-negatives. Antigen detection by ELISA is also a very sensitive method (Scott et al., 1974) sometimes more sensitive than LAT.

Candida albicans. Free C.albicans antigens were observed for the first time concomitantly with precipitating antibodies in chronic mucocutaneous candidosis (CMCC) by Axelsen et al. (1973) by means of crossed IE in intermediate gel containing Con A. The main antigen of the C.albicans cell wall, was detected by Weiner and Yount (1976) in the sera of patients with invasive candidosis hemagglutination inhibition with a minimum sensitivity of 300 ng/ml. Fischer et al. (1978) were the first to prove that the plasma inhibitor (Paterson et al., 1971) of in vitro lymphocyte proliferation by Candida anti-gens is a circulating mannan. Numerous studies based on very sensitive radio and enzymoimmunoassays and on LAT showed de-tection limits between 0.1 and 100 ng/ml but although specifi-city is close to 100 % in most studies, the sensitivity for candidosis is between 47 and 100 %. Antibodies to Candida antigen interfere with the detection of antigen and pretreat-ment of the sera is necessary to dissociate immune complexes (heat, pronase). A new heat-labile antigen, probably a glyco-protein, other than mannan was shown by Gentry et al. (1985) to be detectable by latex agglutination in invasive candidosis but is also present in colonized patients.

Aspergillus fumigatus. A circulating polysaccharide, probably galactomannan (GM), is known to circulate in invasive aspergillosis (Lehmann and Reiss, 1978 ; Reiss and Lehmann, 1979 ; Shaffer et al., 1979a, b ; Weiner and Coats-Stephen, 1979 ; Dupont et al., 1987). The GM isolated from the homogenate of broken mycelium was detected in serum (Weiner, 1980 ; Schaffer et al., 1979), bronchoalveolar lavage fluid and in urine (Dupont et al., 1987) by RIA and ELISA. The GM in serum is complexed, probably to immunoglobulin and dissociation of immune complexes increases the sensitivity of Ag detection. False-negative occur often as well as false positive reactions, and a reliable diagnostic test is urgently needed.

Histoplasma capsulatum. A polysaccharide antigen was detected by RIA by Wheat et al. (1986) in serum, urine and CSF specimens of most of the patients with disseminated histoplasmosis. The urinary Ag appears to be a polysaccharide of low molecular weight, resistant to heat (100°C), to proteinases, and nucleases, but destroyed by periodate, and binding to Con A. The antiserum utilized for the test was raised against the Y phase of H.capsulatum.

Coccidioides immitis. A circulating antigen was detected by Galgiani et al. (1984) in sera of patients with coccidioidal pneumonia using an inhibition ELISA in solid phase with sensitivity of 4 ug/ml. This Ag seems to be associated with IgM activity, but not IgG and IgA. The antiserum utilized in ELISA was raised against the parasitic phase (spherulin) of C.immitis.

Antigens used for immunological diagnosis of mycoses

Immunological tests in current use are summarized in Table 5.

TABLE 5 - CURRENT IMMUNOLOGICAL TESTS FOR DIAGNOSIS OF THE MYCOSES

CURRENT IMMUNOLOGICAL TESTS IN THE DIAGNOSIS OF MYCOSES

MYCOSES	Circulating Ag			IgG, IgM Abs					IgE	SKIN TEST		FUNGUS IDENTIFICATION (culture exoAg)
	La (1)	ELISA RIA	LyT	CF	IF	ID IE,CIE	ELISA	WC/La aggl	RAST	30m	48h	
Aspergillosis												
ABPA (2)					±				+			
aspergilloma					+	+	+					
invasive		+			±	±	±					
Ext.aller.alveolitis (3)												
candidosis												
invasive	+	+			+	+	+	+	±	-	+	+
chronic MCC	+		-		+	+	+	+	±	+	-	
Cryptococcosis	+	+	-		+		+	+				
Histoplasmosis		+			+	+	+	+		-	+	+
Coccidioidomycosis		+			+	+	+	+		-	+	+
Blastomycosis					+	+	+	+				+
Paracoccidioidomycosis					+	+	+	+				+
Sporotrichosis					+		+			-	+	
Dermatophytosis									±	-	+	
Fungal mycetoma						+	+					
Actinomycetoma						+	+					

(1) La : Latex aggl. ; RIA : Radioimmunoassay ; LyT : Lymphocyte transformation ; CF : Complement fixation ; ID :Immunodiffusion ; IE : Immunoelectrophoresis ; CIE : Counterimmunoelectrophoresis ; WC : Whole cells agglutination ; (2) ABPA : allergic bronchopulmonary aspergillosis ; (3) Extrinsic allergic alveolitis e.g. Farmer's lung disease, bagassosis ; MCC : Mucocutaneous candidosis

Fungal antigens as immunosuppressors

The antigens produced by the fungal cell play an important role in the immune response. Although small amounts stimulate the immune response, on the contrary an excess of Ag may block cellular immunity. T-lymphocyte stimulation requires Ag presentation on the surface of the macrophage. Specifically sensitized T-cells respond to Ag by proliferation and release soluble lymphokines which, in turn, function to a) localize and activate macrophages : migration inhibitory factor (MIF), b) act directly on lymphocytes (blastogenic factor) or c) mediate killing or inactivation of target cells. Delayed-type hypersensitivity (DTH) and chronic granulomas are expressions of T-cell mediated immunity. Fungal antigens may be in excess in acute fungal infections, in the invasive forms of aspergillosis, candidosis, cryptococcosis and the dimorphic systemic mycoses, or in chronic infections caused by the dermatophytes, antigens which are known to circulate include mannan in candidosis, glucuronoxylomannan in cryptococcosis, and galactomannan in aspergillosis.

C.albicans. Mannan exerts an immunosuppressive effect in most patients with chronic mucocutaneous candidosis (Durandy et al., 1986, 1987 ; Drouhet, Dupont, 1980, 1983, 1985). Lymphocytes from patients with active disease were unable to produce anti-mannan Abs because of a lack of T-cell Ag specific proliferation. Furthermore, patients' T-lymphocytes exerted a suppressive effect on the anti-mannan antibody production. In vivo, the Ag specific T-cell mediated suppression correlates well with activity of the disease since it disappeared in treated patients with an appropriate effective therapy with ketoconazole (Drouhet and Dupont, 1979, 1983). Serum mannan inhibition of lymphocyte transformation to C.albicans Ag (candidin) disappeared after treatment, and the DTH response to candidin, negative before treatment, became positive (Table 6). The increased number of eosinophils observed in the skin window test (Dikeacou and Drouhet, 1981) after specific stimulation with C.albicans and non-specific (anti-IgE) stimulation also became normal. The role of circulating mannan is important for the perpetuation of disease in CMCC and for the unresponsiveness of cell-mediated immunity. Suppression of mannan production restored the depressed cellular immunity.

TABLE 6 - CELLULAR RESPONSIVENESS TO CANDIDA ANTIGENS AND MITOGENS OF 6 PATIENTS WITH CMCC BEFORE AND AFTER KETOCONAZOLE (Ktz) TREATMENT (from Drouhet et al., 1983, 1985).

Patient	Circulating inhibitory factor (mannan)		Before Ktz intradermal reaction to candidin		After Ktz intradermal reaction to candidin		Transformation lymphocyte test			
							C. albicans		PHA Con A	
	Before Ktz	After Ktz	30'	48 h	30'	48 h	Before Ktz	After Ktz	Before Ktz	After Ktz
1 G. H.	+	−	+	−	+	+	−	+	+	+
2 S. G.	−	−	−	−	+	+	−	−	+	+
3 A. V.	+	−	+	−	+	+	−	+	+	+
4 B. C.	+	−	+ +	−	+	+	−	+	+	+
5 G. F.	+	−	+	−	+	+	−	+	+	+
6 L. F.	−	−	+	−	+	+	−	+	+	+

Cr.neoformans. The cryptococcal capsular polysaccharide (glucuronoxylomannan), recognized as a virulence factor (Drouhet et al., 1950), as an inhibitor of leukocyte migration (Drouhet, Segretain, 1951) and of phagocytosis (Bulmer, 1968, Kozel, 1977) induces an immunological paralysis or tolerance. When the optimally immunogenic dose is exceeded, splenic antibody producing cells (Murphy and Cozad, 1972) and circulating antibodies are suppressed (Kozel et al., 1977). Decreased lymphocyte transformation in response to Cr.neoformans is observed in cryptococcosis in man (Diamond and Bennett, 1973). The capsular polysaccharide is a T-independent elictor of IgM (Reiss et al., 1984), but the antibody level is regulated by T-suppressor cells (Murphy and Moorhead, 1982). Inhibition of migration of spleen cells of mice injected with Cr.neoformans showed a possible effect of MIF on development of Cr.neoformans (Wahab et al., 1986).

Dermatophytes. The immunosuppressive effect of dermatophyte antigens by induction of T-suppressor activity was shown in experimental murine dermatophytosis due to Trichophyton quinckeanum (Calderone and Hay, 1984). Patients with chronic trichophytosis (T.rubrum, T.concentricum) who fail to show DTH to trichophytin, produce specific IgE, immediate hypersensitivity to Ag, and have a poor n vitro cell-mediated response (Hay, 1982 ; Hay and Shennan, 1984 ; Kaaman, 1985 ; Ahmed, 1982). Serum blocking factors of cell-mediated immunity have been demonstrated in chronic dermatophytic infection (Walters et al., 1974).

Other fungal Ags. Serum inhibitory factors suppressing in vitro lymphocyte proliferation have been observed in other mycoses such as histoplasmosis, coccidioidomycosis, and paracoccidioidomycosis (Cox, 1983). Increased non-T suppressor cells have been documented by Stobo (1976) in blastomycosis, CMCC, cryptococcosis and histoplasmosis.

Immunomodulation and fungal antigens: Vaccines

Attempts to obtain protective immunity against experimental mycoses such as murine candidosis have utilized whole cells, cell walls or cytoplasmic antigen (Drouhet and Dupont, 1976 ; Dupont and Drouhet, 1977) with or without non-specific immunostimulation (P_{40} fraction of Corynebacterium granulosum). Prolongation of survival was observed. Fungal ribosomes have been shown to lend protection against experimental histoplasmosis (Tewari et al., 1978) and candidosis (Levy et al., 1981; Segal et al., 1980). In histoplasmosis, optimal protection was achieved with a dose of 100 mg of ribosomal protein. In an experimental candidosis model in mice immunocompromised with cyclophosphamide, ribosomal vaccination gave 60-70 % survival while 100 % mortality was observed in non-vaccinated animals. Study of cell-mediated responses showed that ribosomes primarily stimulate the T-lymphocytes. Recent data (Arvidsson, Abst.Intern.Congress Immunol., Toronto, 1986) showed that polysaccharides from the cell surface complexed with ribosomal RNA are responsible for the protection. In experimental murine coccidioidomycosis, Levine et al. (1977), obtained significant protection with a formalin-killed spherule vaccine. Trials of spherule vaccine in humans are under way (Williams et al., 1984).

Extensive studies (Petrovich et al., 1980) on the vaccination of cattle against Trichophyton verrucosum infections in the USSR and East European countries led to the development of powerful protective preparations (TF 130, LTF 130) but more information is needed on the immunogenic properties of such vaccines which might be extended to human ringworm infections.

Future trends in fungal antigens

The principal trends for further investigations on fungal antigens are : 1) <u>Standardization</u> of internationally recognized reference reagents and procedures leading to serological tests of good specificity, low cross reactions, and high reproductibility. In the past, numerous proposals for such standardization have been outlined (Drouhet, 1965 ; Seeliger, 1965 ; Huppert et al., 1973 ; Axelsen et al., 1975 ; Mackenzie et al., 1980) but only an international collaborative study on the immunodiffusion test for histoplasmosis, sponsored jointly by WHO and CDC, succeeded (Pine et al., 1981). The statements of the ISHAM Standing Committee on serology (Mackenzie et al., 1980) were followed by growing support to this view. Our aims are that this "First International Symposium on Fungal Antigens" may contribute to reinforce the exchange of current knowledge in this area, thus the favoring the desired standardization of fungal antigens. 2) <u>Exoantigens</u>. The use of soluble exoantigens to obtain presumptive identification of unknown cultures (Kaufman et al., 1983) with specific antisera must be extended to numerous pathogenic or opportunistic fungi other than the systemic dimorphic fungi, in spite of the difficulties in obtaining good reference sera for such identifications. 3) <u>Circulating antigen detection</u>. In acute invasive mycotic infections, particularly in immunocompromised patients, there is a great need for reliable, sensitive, and easily reproductible tests for detection of circulating antigens. These methods may also prove useful in chronic mycotic infections. Fundamental knowledge of the nature of circulating antigens is scarce and more investigations are necessary. 4) <u>Monoclonal antibodies</u>. A new generation of mAbs will provide useful tools for antigenic characterization and immunodiagnosis leading to further applications in immunotherapy and chemotherapy. 5) <u>Immunomodulation. Immunoprotection</u>. We need to discover what antigens are able to mediate protective cellular and humoral responses, leading to fungal vaccines which have not yet gone beyond the experimental stage.

<u>Acknowledgements</u>: We thank the authors and the respective journals where they papers were published for the permission to reproduce the figures 2, 3, 6, 7, 8, 9, 10, table 6. This study was supported in part by a grant INSERM-CNAMTS 1984-87 CSS N°3.

References

Ahmed, A.R., 1982, Immunology of human dermatophyte infections, <u>Arch. Dermatol.</u>, 118:521.

Araj, G.F., Hopfer, R.L., Chesnut, S., Fainstein, V., and Bodey, G.P., 1982, Diagnostic value of the enzyme-linked immunosorbent assay for detection of <u>Candida albicans</u> cytoplasmic antigen in sera of cancer patients, <u>J.Clin. Microbiol.</u>, 16:46.

Austwick, P.K.C., 1966, The role of spores in the allergies and mycoses of man and animals, 321, <u>in</u>: Proceedings of Eighteen Symposium of the Colston Research Society University of Bristol. Butterworths, Sci.Publ.London.

Avrameas, S., and Guilbert, B., 1971, Dosage enzymo-immunologique de protéines à l'aide d'immunoabsorbents et d'antigènes marqués aux enzymes, C.R.hebd.Acad.Sci., Paris, 273:2705.

Azuma, I., Kanetsuna, F., Tanaka, Y., Yamamura, Y., and Carbonell, L., 1974, Chemical and immunological properties of galactomannans obtained from Histoplasma duboisii, Histoplasma capsulatum, Paracoccidioides brasiliensis and Blastomyces dermatitidis, Mycopath.Mycol.Appl., 54:111.

Axelsen N.H., 1973, Quantitative immunoelectrophoretic methods ais tools for a polyvalent approach to standardization in the immunochemistry of Candida albicans, Infect.Immun, 7:949.

Axelsen N.H., and Kirkpatrick, C.H., 1973, Simultaneous characterization of free Candida antigens and Candida precipitins in a patient's serum by means of crossed immunoelectrophoresis with intermediate gel, J.Immunol.Meth., 2:245.

Axelsen, N.H., Kirkpatrick, C.H., and Buckley, R.H., 1974, Precipitins to Candida albicans in chronic mucocutaneous candidiasis studied by crossed immunoelectrophoretic findings with intermediate gel: correlation with clinical and immunological findings, Clin.Exp.Immunol., 17:385.

Axelsen, N.H., Buckley, H.R., Drouhet, E., Budtz-Jorgensen, E., Hattel, T., and Lehm Andersen, P., 1975, Crossed immunoelectrophoretic analysis of precipitins to Candida albicans in deep Candida infection. Possibilities for standardization in diagnostic serology, Scand.J.Immunol., 2(suppl.1):57.

Axelsen, N.H., 1983, Handbook of immunoprecipitation in gel techniques, Scand.J.Immunol., 17, (suppl.10), Blackwell Scientific Publications, Oxford.

Barbier, E., Sarthou P., and Le Guern C., 1985, Obtention d'un anticorps monoclonal anti-Candida albicans, Bull.Soc. Fr.Mycol.Méd., 14:13.

Barker, S.A., Basarab, O., and Cruiskshank, C.N.D., 1967, Galactomannan peptides of Trichophyton mentagrophytes. Carbohyd.Res., 3 :325.

Bennett, J.E., and Bhattacharjee, A.K., 1984, Capsular polysaccharides of Cryptococcus neoformans, Rev.Infect.Dis., 5:619.

Bennett, J.E., Bhattacharjee, and A.K., Glaudemans, C.P.J., 1985, Galactofuranosyl groups are immunodominant in Aspergillus fumigatus galactomannan. Mol.immunol., 22, 251-254.

Bennett, J.E., Hasenclever, and H.F.,1965, Cryptococcus neoformans polysaccharide studies of serologic properties and role in infection, J.Immunol., 94:916.

Biguet, J., Tran Van, Ky, P., Andrieu, S., and Fruit, J., 1964, Analyse immunoélectrophorétique d'extraits cellulaires et de milieux du culture d'Aspergillus fumigatus par des immunosérums expérimentaux et des sérums de malades atteints d'aspergillome bronchopulmonaire, Ann.Inst. Pasteur, Paris, 107:72.

Biguet, J., Tran Van Ky, P., Andrieu, S., and Vaucelle, T., 1967, Premières caractérisations d'activités enzymatiques sur les immunoélectrophorégrammes des extraits antigéniques de Histoplasma capsulatum. Consequences diagnostiques pratiques, Ann.Soc.Belge.Méd.Trop., 47:425.

Biguet, J., Tran Van Ky, P., Fruit, J., and Andrieu, S., 1967, Identification d'une activité chymotrypsique au niveau de fractions remarquables de l'extrait antigénique d'Aspergillus fumigatus, Répercussions sur le diagnostic immunologique de l'aspergillose, Revue Immunol., 31:317.

Bloomfield, N., Gordon, M.A., and Elmendorf, D.F., 1963, Detection of Cryptococcus neoformans antigens in body fluids by latex particle agglutination, Proc.Soc.Exp. Biol.Méd, 114:64.

Boiron, P., Drouhet, E., and Dupont, B., 1987, Enzyme-linked immunosorbent assay (ELISA) for immunoglobulin in bagasse workers sera :comparison with counter-immunoelectrophoresis Clinical Allergy, 17:

Brawner, D.L., and Cutler, J.E., 1984, Variability in expression of a cell surface determinant on Candida albicans as evidenced by an agglutinating monoclonal antibody, Infect.Immun., 43:966.

Bullock, W.E., and Deepe, G.S., 1983, Medical Mycology in crisis, J.Lab.Clin.Med., 102:685.

Bulmer, G.S., and Sans, M.D., 1968, Cryptococcus neoformans, III Inhibition of phagocytosis, J.Bacteriol., 95:5.

Burnett, J.H., 1976, Fundamentals of Mycology. Edward Arnold Publ.London.

Burnett, W.N., 1981, "Western blotting" electrophoretic transfer of proteins from sodium dodecyl sulphate polyacrylamide gels to a modified nitrocellulose and radiographic detection with antibody and radio-iodinated protein A, Analyt.Biochem., 212:195.

Bussard, A., 1959, Description d'une technique combinant simultanément l'électrophorèse et la précipitation immunologique dans un gel : l'électrosynérèse, Biochem.Biophys.Acta, 34:258.

Calderon, R.A., and Hay, R.J., 1984, Cell mediated immunity in experimental murine dermatophytosis, Immunol., 53:465.

Camargo, Z.P., Guesdon, J.L., Drouhet, E., and Improvisi, L., 1984a, Magnetic enzyme-linked immunosorbent assay for determination of specific IgG in paracoccidioidomycosis, Sabouraudia, J.Med.Vet., 22:291.

Camargo, Z.P., Guesdon, J.L., Drouhet, E., and Improvisi, L., 1984b, Enzyme-linked immunosorbent assay (ELISA) in the paracoccidioidomycosis. Comparison with counterimmunoelectrophoresis and erythroimmunoassay, Mycopathologia, 88:31.

Camargo, Z.P., Unterkischer, C., and Drouhet, E., 1986, Comparison between magnetic enzyme-linked immunosorbent assay (MELISA) and complement fixation test (CF) in the diagnosis of paracoccidioidomycosis, J.Med.Vet.Mycol., 24:77.

Chaparas, S.D., Morgan, P.A., Holobaugh, P., and Kim, S.J., 1986, Inhibition of cellular immunity by products of Aspergillus fumigatus, J.Med.Vet.Mycol., 24:67.

Chardès, T., Piechaczyk, M., Cavaillès, V., Sahli, S.L., Pau, B., and Bastide, J.M., 1986, Production and partial characterization of anti-Candida monoclonal antibodies, Ann.Inst.Pasteur Immunol., 137C:117.

Cole, G., 1986, Cell differentiation in conidial fungi, Microbiol.Rev., 50:95.

Collins, J.P., Grappel, S.F., and Blank, F., 1973, Role of keratinases in dermatophytosis. Fluorescent antibody studies with keratinases of Trichophyton mentagrophytes, Dermatologica, 146:95.

Cox, R., and Arnold, D.R., 1979, Immunoglobulin E in coccidioidomycosis, J.Immunol., 123:194.

Cox, R., and Arnold, D.R., 1980, Immunoglobulin E in histoplasmosis, Infect.Immunity, 29:290.

Cox, R.A., 1983, Cell-mediated immunity pp 61-98, in: Fungi pathogenic for human and animals. part B, Pathogenicity and detection D.H. Howard (ed): 1, Marcel Dekker Inc.Publisher New York, Basel.

Cox, R.A., 1985, Immunodiagnostic procedures for fungal diseases, in : Clinical immunology Newsletter, 6:17, Elsevier Sci.Publ.C.Inc., Amsterdam.

Dabrowa, N., and Howard, D.H., 1984, Heat shock and heat stroke proteins observed during germination of the blastoconidia of Candida albicans, Infect.Immunity, 44:537.

De Repentigny, L., and Reiss, E., 1984, Current trends on immunodiagnosis of candidiasis and aspergillosis, Rev.Infect.Dis., 6:301.

De Repentigny, L., Marr, L.D., Keller, J.W., Carter, A.W., Kuykendall, R.J., Kaufman, L., and Reiss, E., 1985, Comparison of enzyme immunoassay and gas-liquid chromatography for the rapid diagnosis of invasive candidiasis in cancer patients, J.Clin.Microbiol., 21:972.

Delga, J.M., Dupont, B., and Drouhet, E., 1984, Dosage du mannane de Candida albicans par une technique immuno-enzymatique, Bull.Soc.Fr.Mycol.Méd., 13:395.

Delga, J.M., Ibrahim-Granet, O., Velez-Arango, H., and Drouhet, E., 1985, Serotypes A et B de Candida albicans. Etude comparative par électrophorèse et immunoempreinte, Bull.Soc.Fr.Mycol.Méd., 14:119.

Dikeacou, T, and Drouhet, E., 1981, Cellular inflammatory response to fungal antigen studied with the skin window technique. Correlation of the exudate with the evolution of the mycotic infection, Agents and Actions, 11:6.

Diamond, R.D. and Bennett, J.E., 1973, Disseminated cryptococcosis in man : decreased lymphocyte transformation in response to Cryptococcus neoformans. J.Infect.Dis., 127:694.

Dochez, A.R. and Avery, O.T., 1917, The elaboration of specific soluble substance by Pneumococcus during growth. J.Exper.Med., 26:477.

Dromer, F., Salamero, J., Contrepois, A., Carbon, C., Yeni, P., 1987, Production, characterization, and antibody specificity of a mouse monoclonal antibody reactive with Cryptococcus neoformans capsular polysaccharide. Infect.Immun., 55:742.

Drouhet, E., Segretain, G., and Aubert, J.P., 1950, Polyoside capsulaire d'un champignon pathogène, Torulopsis neoformans, relation avec la virulence, Ann.Inst.Pasteur, 79:891.

Drouhet, E., and Segretain, G., 1951, Inhibition de la migration leucocytaire in vitro par un polyoside capsulaire du Torulopsis neoformans, Ann.Int.Pasteur, 79A:891.

Drouhet, E., 1965, Etude des allergènes (antigènes) fongiques destinés au diagnostic des infections mycosiques, Progr.Immunobiol.Stand. S.Karger Publisher, New York, Amsterdam, pp 170.

Drouhet, E., Camey, L., and Segretain, G., 1972, Valeur de l'immunoprécipitation et de l'immunofluorescence indirecte dans les aspergilloses bronchopulmonaires, Ann.Inst.Pasteur, 123:379

Drouhet, E., Tabet-Derraz, O., Sanchez-Sousa, A., and Viviani, M.A., 1973, Application de l'électrosynérèse et de l'immunoélectrophorèse bidimensionnelle au diagnostic des aspergilloses et à la standardisation des antigènes aspergillaires, Bull.Soc.Fr.Mycol.Méd., 2:7.

Drouhet, E., 1976, Antigènes de Candida, application à l'étude de l'immunité humorale et du diagnostic immunologique, Rev.Fr.Allergol., 5:241.

Drouhet, E., and Dupont, B., 1976, Immunisation par Candida albicans et immunostimulation non spécifique par Corynebacterium granulosum et BCG dans la candidose expérimentale, C.R.Acad.Sci.Paris,, série D, 283:1385.

Drouhet, E., Papachristou-Moraiti, A., and Minas, A., 1976, Essais d'identification des dermatophytes par l'immunofluorescence. Bull.Soc.Fr.Mycol.Med., 5:133.

Drouhet, E., 1978, Systemic candidosis and its immunology, Mykosen, suppl.1:175, Grosse Verlag,.

Drouhet, E, and Dupont, B., 1980, Chronic mucocutaneous candidosis and other superficial and systemic mycoses successfully treated with ketoconazole, Rev.Infect.Dis., 2:606.

Drouhet, E., 1981, Ecology of the serotypes of Candida albicans and phenotypes of resistance to 5-fluorocytosine, p 197,in: Sexuality and Pathogenicity of Fungi, Vanbreuseghem R. and de Vroey C. (Eds), Masson Paris.

Drouhet, E., and Dupont, B., 1983, Laboratory and clinical assessment of ketoconazole in deep seated mycoses, Am.J.Med., 78, Suppl.1B:30.

Drouhet, E., Mogahed, A., and Younes, K.B., 1983, Etude comparative des techniques d'immunoperoxydase et d'immunofluorescence sur frottis et coupes histologiques, Bull. Soc.Fr.Mycol.Méd., 12:121.

Drouhet, E., Dupont, B., and Dikeacou, T., 1985, Antifungal agents and immunity. Zbl.Bakt, Suppl.13,1, in: Chemotherapy and immunity, G.Pulverer, Jeliaszewicz J. (eds) Gustave Fischer Verlag, Stuttgart, New York.

Drutz, D.J., 1986, Antigen detection in fungal infections, New England J.Med., 314:115.

Dupont, B., Bennett, J.E., and Huter, M.A., 1987, Galactomannan antigenemia and antigenuria in aspergillosis : studies in patients and experimentally infected rabbits. J.Infect.Dis., 155:1.

Dupont, B., and Drouhet, E., 1985, Cutaneous, ocular and osteoarticular candidiasis, in heroin addicts : new clinical and therapeutic aspects in 38 patients, J.Infect.Dis., 152:577.

Durandy, A., Fischer, A., and Griscelli, C., 1983, Specific in vitro antimannan-rich antigen of Candida albicans antibody production by sensitized human blood lymphocytes, J.Clin.Invest, 71:1602.

Durandy, A., Fischer, A., Drouhet, E., and Griscelli, C., 1986, Mannan antigen of Candida albicans and cellular immune responses in vitro and in vivo, First International Symposium on fungal antigens, 17-20 november, Paris.

El-Zaatari, F.A., Reiss, E., Yakrus, M.A., Bragg, S.L., and Kaufman,L., 1986, Monoclonal antibodies against isoelectrically focused Nocardia asteroides proteins characterized by enzyme linked immunoelectrotransfer blot method, Diagnostic Immunology, 1986, 4:97.

Engvall, E., and Perlmann, P., 1971, Enzyme-linked immunosorbent assay (ELISA), Quantitative assay of IgG, Immunochemistry, 8:871.

Fischer, A., Ballet, J.J., and Griscelli, C., 1978, Specific inhibition of in vitro Candida induced lymphocyte proliferation by polysaccharide antigens present in the serum of patients with chronic mucocutaneous candidiasis, J.Clin.Invest., 62:1005.

Fujita, S., Matsubara, F., and Matsuda, T., 1986, Enzyme-
    linked immunosorbent assay measurement of fluctuation
    antibody titer and antigenemia in cancer patients with
    and without candidiasis, J.Clin.Microbiol., 23:569.
Galgiani, J.N., Dugger, K.O., Ito, J.I., and Wieden, M.A.,
    1984, Antigenemia in primary coccidioidomycosis, Amer.
    J.Trop. Med.Hyg., 33:645.
Gentry, L.O., Wilkinson, J.D., Lea, A.S., and Price, M.F.,
    1983, Latex agglutination test for detection of Candida
    antigen in patients with disseminated disease, Eur.J.Clin.
    Microbiol., 2:122.
Georges, A.J., Ravisse, P., and Drouhet, E., 1986, Identifica-
    tion spécifique d'Histoplasma duboisii par immunofluores-
    cence sur coupes histologiques, Bull.Soc.Fr.Mycol.Méd.,
    9:121.
Gordon, M.A., Almy, R.E., Greene, C.H., and Fenton, J.W.,
    1971, Diagnostic mycoserology by immunoelectro-osmopho-
    resis, Amer.J.Clin.Pathol., 56:471.
Gordon, M.A., Lapa, E.W., and Kane, J., 1977, Modified indi-
    rect fluorescent-antibody test for aspergillosis, J.Clin.
    Microbiol., 6:161.
Grabar, P., and Williams, C.A., 1953, Méthode permettant l'é-
    tude conjuguée des propriétés électrophorétiques et immu-
    nochimiques d'un mélange de protéines. Application au
    sérum sanguin, Biochim.Biophys.Acta, 10:193.
Green, J.H., Harrell, W.K., Johnson, J.E., and Benson, R.,
    1980, Isolation of an antigen from Blastmyces derma-
    titidis that is specific for diagnosis of blastomycosis,
    Curr.Microbiol., 4:293.
Green, J.H., Harrell, W.K., and Aloisio, C. 1981, The prepara-
    tion of monoclonal antibodies for use as control reagents
    in the serological diagnosis of blastomycosis, Abst.Ann.
    Mtg.Am.Soc.Microbiol.Atlanta Abstr., 167:337.
Guesdon, J.L., and Avrameas, S., 1977, Magnetic solid phase
    enzyme-immunoassay. Immunochemistry, 14:443.
Guesdon, J.L., and Avrameas, S., 1980, Sensitive titration of
    antibodies and antigens using erythroimmunoassay, Ann.Im-
    munol.(Institut Pasteur), 131:389
Guinet, R.M.F., Bruneau, S.M., and de Montclos H., 1986,
    Immunochemical studies of Candida and Aspergillus antigens:
    taxonomic and diagnosis significances. First International
    Symposium on Fungal Antigens, 17-20 nov., Institut Pasteur
    Paris. (this volume).
Hall, N.K., and Blackstock, R., 1981, Production of specific
    antibody to Cryptococcus neoformans by hybridomas in vi-
    tro, Sabouraudia, 19:157.
Harding, S.A., Brody, J.P., and Normansell, D.E., 1980, Anti-
    genemia detected by enzyme-linked immunosorbent assay in
    rabbits with systemic candidiasis, J.Lab.Clin.Méd.,
    95:959.
Hay, R.J., and Shennan, G. 1984, Antibody responses in tinea
    imbricata : the role of immunoglobulin E, Trans.Royal
    Soc.Trop.Med.Hyg., 78:653
Hearn, V.M., and Mackenzie, D.W.R., 1979, The preparation and
    chemical composition of fractions from Aspergillus fumi-
    gatus wall and protoplasts possessing antigenic activity,
    J.Gen.Microbiol., 112:35.
Heide, S. van der, Kauffman, H.F., and Vries, K. de, 1985,
    Cultivation of fungi in synthetic and semi-synthetic
    liquid medium I. Growth characteristics of the fungi and
    biochemical properties of the isolated antigenic mate-
    rial. Allergy, 40:586.

Hopwood, V., Evans, E.G.V., and Carney, J.A., 1985, A comparison of methods for the detection of Candida antigens. Evaluation of a new latex reagent, J.Microbiol.Methods, 80:199.

Hopwood, V., Poulain, D., Fortier, B., Evans, G., and Vernes, A., 1986, A monoclonal antibody to a cell wall component of Candida albicans, Infect.Immun., 54:222.

Huppert, M., Adler, J.P., Rice, E.H., Sun, and S.H., 1979, Common antigens among systemic disease fungi analyzed by two dimensional immunoelectrophoresis, Infect.Immun., 23:479.

Huppert, M., 1983, Antigens used for measuring immunological reactivity, pp121, in: Fungi pathogenic for human and animals, Part B Pathogenecity and Detection, Howard D.H. (ed) Marcel Dekker Publisher, New York, Basel.

Ibrahim-Granet, O., and Bièvre C., de, 1986, Etude des protéines heat shock et heat-stroke chez Fonsecaea pedrosoi: agent de chromomycose, Bull.Soc.Fr.Mycol.Méd., 15:213.

Ibrahim-Granet, O., Delga, J.M., Granet, C., and Drouhet, E., 1986, A program in BASIC for the determination of molecular weights and linear graphic representation of an electrophoretic protein pattern : Application to serotypes A and B of Candida albicans studied by polyacrylamide gel electrophoresis and immunoblotting, Electrophoresis, 7:316.

Jones, J.M., 1980, Kinetics of antibody responses to cell wall mannan and a major cytoplasmic antigen of Candida albicans in rabbits and humans, J.Lab.Clin.Med., 96:845.

Kaaman, T., 1985, Dermatophyte antigens and cell-mediated immunity in dermatophytosis, pp 117-134. in: Current topics in Medical Mycology, vol.1 M.R. McGinnins (ed) Spinger-Verlag Publisher, New York.

Kaplan, W., 1961, The application of fluorescent antibody techniques to medical mycology, A review, Sabouraudia, 1:137.

Kaplan, W., 1979, The fluorescent antibody technique in the diagnosis of mycotic diseases. In : Proceedings International Symposium on Mycoses, Scientific Publication PAHO, N° 205, p 86.

Karu, A.E., Gennevois, D.J.P., Koffman, J.W., Kraeger, S.J., and Levine, H.B., 1985, Research and diagnostic applications of monoclonal antibodies to Coccidioides immitis. Abst.Ninth.Intern.Congr.Intern.Soc.Human.Animal.Mycol. (ISHMA) Atlanta (USA) Abst. N°19-3.

Kauffman, H.F., and De Vries, K., 1980, Antibodies against Aspergillus fumigatus. (I) Standardization of the antigenic composition, Int.Archs.Allergy.Appl.Immunol., 62:252.

Kauffman, H.F., Heide S. van der, Beaumont F., Monchy, J.G.R.de, and Vries K. de, 1984, The allergenic and antigenic properties of spore extracts of Aspergillus fumigatus : A comparative study of spore extracts with mycelium and culture filtrate extracts, J.Allergy.Clin. Immunol., 73:567.

Kaufman, L., Standard, P.G., and Padhye, A.A., 1983, Exoantigen tests for the identification of fungal cultures, Mycopathologia, 82:3.

Kaufman, L., and Reiss, E., 1986, Serodiagnosis of fungal diseases, pp 446. In : Manual of clinical immunology, 3rd ed, Rose, N.R., Friedman H. (eds), American Society for microbiology, Washington D.C.

Kerkering, T.M., Espinel-Ingroff, A., and Shadomy, S., 1979,
    Detection of Candida antigenemia by counterimmunoelectro-
    phoresis in patients with invasive candidiasis, J.Infect.
    Dis., 140:659.
Kozel T.R., Gulley, W.F., and Cazin, J.Jr, 1977, Immune response
    to Cryptococcus neoformans soluble polysaccharide :
    immunological unresponsiveness. Infect.Immun., 18:701.
Kostiala, A.A.I., and Kostiala, I., 1981, Enzyme-linked immu-
    nosorbent assay (ELISA) for IgM, IgG and IgA class anti-
    bodies against Candida albicans antigens : development
    and comparison with other methods, Sabouraudia, 19:123.
Laurrel, C.B., 1965, Antigen-antibody crossed electrophoresis,
    Ann.Biochem., 10:358.
Lehmann, P.F., and Reiss, E, 1978, Invasive aspergillosis:
    antiserum for circulating antigen produced after immuni-
    zation with serum from infected rabbits, Infect.Immun.,
    20:570.
Lehmann, P.F., and Reiss, E., 1980, Detection of Candida albi-
    cans mannan by immunodiffusion, counterimmunoelectropho-
    resis and enzyme-linked immunoassay, Mycopathologia,
    70:83.
Levine, H.B., Scalaronte, G.M., and Chaparas, S.D., 1977,
    Preparation of fungal antigens and vaccines: studies on
    Coccidioides immitis and Histoplasma capsulatum, pp 106-
    125, in: Host-Parasite relationships in systemic mycoses.
    Part I Beemer A.M. et al. (eds),Contribution to Microbio-
    logy and immunology, Vol.4, Proc 21 st OHOLO Biol.Conf.
    Ma'alot, , Israel, S. Karger Publisher, Basel
Levy, R., Segal, F., and Eylan, E., 1981, Protective immunity
    against murine candidiasis elicited by Candida albicans
    ribosomal fractions, Infect.immun., 31:874.
Lew, M.A., Siberg, R., Donahue, D.M., and Maiorca, F., 1982,
    Enhanced detection with an enzyme-linked immune sorbent
    assay of Candida mannan in antibody containing serum
    after heat extraction, J.Infect.Dis., 145:45.
Longbottom, J.L., 1974, Physico-chemical properties and anti-
    genicity of Aspergillus fumigatus depending on different
    culture conditions. in: Aspergillosis and Farmer's Lung
    Disease in Man and Animals. de Haller R., Suter, F.,
    (eds), p.41 Hans Huber Publishers, Bern, Stuttgart, Vienna.
Longbottom, J.L., 1984, Antigens allergens of Aspergillus fu-
    migatus: Characterization by quantitative immunoelectro-
    phoresis methods (QIE). J.Allergy Clin.Immunol., 74:113.
Longbottom, J.L., 1986, Applications of immunological methods
    in mycology, in: Handbook of experimental immunology,
    4.Applications of immunological methods in biomedical
    sciences, D.M. Weir (ed) Blackwell Sci.Publ., Oxford, P.
    121.1.
Longbottom, J.L., and Austwick, P.K.C., 1986, Fungal antigens
    pp 7.1-7.11, in: Handbook of experimental immunology
    1.Immunochemistry. D.M. Weir (ed), Blackwell Sci.Publ.Oxford,
Mackenzie, D.W.R., Kaufman, L., Drouhet, E., and Segal, E.,
    1980, Round table : Standardization of mycoserologic
    reagents and procedures, pp 141, in: Human and animal
    mycology, Proc 7th Congress ISHAM, Kuttin E.S. and Baum
    G.L. (eds) Excerpta Medica, Amsterdam.
Mackenzie, D.W.R., 1982, Current trends in our knowledge of
    the immunology of fungal infections. X Proc. 8th Congress
    ISHAM, M. Baxter (ed) Massey, University New Zealand,
    187-198.

Mackenzie, D.W.R., 1983, Serodiagnosis, pp 121, in: Fungi
    Pathogenic for humans and animals. Part B. Pathogenicity
    and Detection: I Howard D.H. (ed), Marcel Dekker, Publis-
    her, New York, Basel.
Manning-Zweerink, M., Maloney, C.S., Mitchel, T.B., and Weston,
    H., 1986, Immunoblott analyses of Candida albicans
    associated antigens and antibodies in human sera, J.Cli-
    nical Microbiol., 23:46.
Marichal, P., Soreus, J., Van Cutsem, J., and Vanden Bossche,H.,
    1986, Culture media for the study of the effects of azole
    derivatives on germ tube formation and hyphal growth of
    Candida albicans, Mykosen, 29:76.
Mathews, C.R., Burnie, J.P., and Tabqchali, S., 1984, Immuno-
    blot analysis of the serological response in systemic
    candidosis, Lancet, ii:1415.
Mathur, S., Goust, I.M., Horger E.O., and Fundenberg, H.H.,
    1977, Immunoglobulin E anti-Candida albicans antibodies
    and candidiasis, Infect.Immun., 18:257.
Metchnikoff, E., 1892, Leçons sur la Pathologie comparée de
    l'inflammation, Masson, (ed), Paris.
Meunier-Carpentier, F., and Armstrong, D., 1981, Candida anti-
    genemia, as detected by passive hemagglutination inhibi-
    tion, in patients with disseminated candidiasis or Can-
    dida colonization, J.Clin.Microb., 13:10.
Miazaki, T., Yadoma, T., Yadoma, H., Hayashi, O., Suzuki, I.,
    and Ohshima, Y., 1980, Immunochemical examination of the
    polysaccharides of mucorales, pp 96-111, In : Fungal
    polysaccharides, Sandford P.A., Matsuda K. (eds),
    Washington, D.C. American Chemical Society.
Miyakawa, Y., Kagaya, K., Fukazawa, Y., and Sol, G., 1986,
    Production and characterization of agglutinating monoclo-
    nal antibodies against predominant antigenic factors for
    Candida albicans, J.Clin.Microbiol., 23:881.
Munoz, M., Estes, G., Kirpatrick, M., Di Salvo, A., and Virella,
    G., 1980, Purification of cytoplasmic antigens from the
    mycelial phase of Candida albicans: possible advantages
    of its use in Candida serology, Mycopathologia, 72:47.
Munoz, C. Mazié J.C., Delga, J.M., Dupont, B., Latgé, J.P.,
    and Drouhet E., 1986. Monoclonal antibodies anti-Candida
    albicans mannan and anti Aspergillus fumigatus galacto-
    mannan. In : Fungal antigens. First International Sympo-
    sium. Paris. Plenum Press. New York.
Murphy, J.W., and Cozad, G.C., 1972, Immunological unresponsi-
    veness induced by cryptococcol capsular polysaccharide
    assayed by the hemolytic plaque technic. Infect.Immun.,
    5:896.
Murphy, J.W., and Moorhead, J.W., 1982, Regulation of cell me-
    diated immunity in cryptococcosis. Induction of specific
    afferent T suppressor cells by cryptococcal antigen.
    J.Immnol., 128:276.
Neill, J.M., and Kapros, C.E. 1950. Serological tests on
    soluble antigens from mice infected with Cryptococ-
    cus neoformans and Sporotrichum schenkii. Proc.Soc.Exper.
    Biol.Med., 73:557
Neill, J.M., Sugg, J.Y., and Mac Cauley, 1951, Serologicelly
    reactive material in spinal fluid, blood and urine from a
    human case of cryptococcosis (torulosis). Proc.Soc.Exper.
    Biol.Med. 77:775
Odds, F.C., Ryan, M.D., and Sneath, P.H.A., 1983, Standardiza-
    tion of antigens from Aspergillus fumigatus, J.Biol.
    Stand., 11:157.

Othman, T., Guesdon, J.L., and Drouhet, E., 1978, Dosage enzymoimmunologique des anticorps humains anti-Candida albicans en utilisant des perles de polyacrylamide agarose magnétiques. Bull.Soc.Fr.Mycol.Méd., 7:249.

Othman, T., Guesdon, J.L., and Drouhet, E., 1979, Dosage enzymoimmunologique des anticorps anti-Cryptococcus neoformans, Bull.Soc.Fr.Mycol.Méd., 8:35.

Othman, T., and Drouhet, E., 1980, La technique enzymoimmunologique utilisant la peroxydase sur frottis de Candida albicans appliquée à la recherche des anticorps dans les candidoses. Comparaison avec la technique d'immunofluorescence, Bull.Soc.Mycol.Méd., 9:117.

Othman, T., and Drouhet, E., 1980, Recherche des anticorps anti-Cryptococcus neoformans par la technique immunoenzymatique sur frottis utilisant la péroxydase de raifort, Bull.Soc.Fr.Mycol.Méd., 9:111.

Ouchterlony, O., and Nilsson, L.A., 1985, Immunodiffusion and immunoelectrophoresis, pp 32.1-32.49, in: Handbook of experimental immunology, 1.Immunochemistry, Blackwell, D.M. Weir (ed),Scientific Publications, Oxford.

Oudin, J., 1946, Méthode d'analyse immunochimique par précipitation spécifique en milieu gélifié.C.R.Hebd.Séance Acad. Sci., 222:115.

Paterson, P.Y., Semo, R., Blumeuschein, G., and Swelstad, J., 1971, Mucocutaneous candidiasis anergy and a plasma inhibitor of cellular immunity : neversal after amphotericin B. Clin.Experim.Immunol., 1971, 9, 55.

Petrovich, S.V., Golovina, N.P., Ivanova, L.G., and Polyakov, I.D., 1980, Vaktsina LTF-130, izgotovnennaya v SSHA po sovetskoi litsenzii. Veterinariya, Moscou, USSR, 9:35.

Pine, L., Green, J.H., Gross, H., Malcolm, G.B., and Harrell, W.K., 1985, Purified histoplasmin reagents for the identification of h and m antigens and homologous antibodies by immunodiffusion or complement fixation tests, Mycopathologia, 91:39.

Pine, L., Gross, H., Malcolm, G.B., Green, J.H., Barbaree, J.M., Harrell, W.K., Suggs, M.T., Blumer, S.O., Kaufman, L., Ajello, L., Smith, S.J., and May, J.C., 1981, Evaluation of candidate international reference reagents and a microimmunodiffusion test for the identification of precipitins to the H and M antigens of histoplasmin, J.Biol. Stand., 9:513.

Polonelli, L., and Morace, G., 1985, Serological analysis of dermatophyte isolates with monoclonal antibodies produced against Microsporum canis, J.Clin.Microbiol 21:138.

Polonelli, L., and Morace, C., 1986, Specific and common antigenic determinants of Candida albicans isolated detected by monoclonal antibody. J.Clin.Microbiol., 23:366.

Ponton, J., Regulez, P., and Cisterna, R., 1986, Immune responses to yeast and mycelial forms of Candida albicans in intraperitoneally infected mice, Mycopathologia, 94:11.

Poulain, D., Hopwood, V., and Vernes, A., 1986, Antigenic variability of Candida albicans, CRC Critical Reviews in Microbiology, 12:223.

Reinhardt, D.J., Hon, P.J., Abdelab, A.T., 1985, Purification and characterization of an alkaline protease from Rhizopus oryzae. Appl.Env.Microbiol. (in press).

Reiss, E., Stone, S.H., and Hasenclever, H.F., 1974, Serological and cellular immune activity of peptidoglucomannan fractions of Candida albicans cell walls, Infect.Immun., 9:881.

Reiss, E., and Lehmann, P.F., 1979, Galactomannan antigenemia in invasive aspergillosis, Infect.Immun., 25:357.

Reiss, E., Patterson, D.G., Yert, L.W., Holler, J.S., and Ibrahim, B.K., 1981, Structural analysis of mannans from Candida albicans serotypes A and B and from Torulopsis glabrata by methylation gas chromatography mass spectrometry and exo- - mannanase, Biomed.Mass.Spectr., 8:252.

Reiss, E.L., Stockman, R.J., Kuykendall, R.J., and Smith, S.J., 1982, Dissociation of mannan serum complexes and detection of Candida albicans mannan by enzyme immunoassay variations, Clin.Chem., 28:306.

Reiss, E., Huppert, M., and Cherniak, R., 1985, Characterization of protein and mannan polysaccharide antigens of yeasts, molds and actinomycetes, p. 172-207, in: Current topics in Medical Mycology, Vol.1, M.R. McGinnis (ed), Springer Verlag, New York.

Reiss, E., 1986, Molecular immunology of mycotic and actinomycotic infections, Elsevier Science, Publishing Co.Inc, New York.

Reiss, E., Knowles, J.B., Bragg, S.L., and Kaufman, L., 1986, Monoclonal antibodies against the M-protein and carbohydrate antigens of histoplasmin characterized by the enzyme-linked immunoelectrotransfer blot method, Infect.Immun., 53:540.

Reyes, G., Dandeu, J.P., and Drouhet, E. 1985. Study of Cryptococcus neoformans antigens by chromatographic and imunological methods. Abstracts IXth Congress of International Society for Human and Animal Mycology, R4-4, Atlanta.

Richardson, M.D., White, L.O., and Warren, R.C., 1979, Detection of circulating antigen of Aspergillus fumigatus in sera of mice and rabbits by enzyme linked immunosorbent assay, Mycopathologia, 67:83.

Richardson, M.D., and Warnock, D.W., 1983, Enzyme-linked immunosorbent assay and its application to the serological diagnosis of fungal infection, Sabouraudia, 21:1.

Rotrosen, D., Edward J.E. Jr., Gibson T.R., Moore J.C. Cohen A.M., and Green I., 1985. Adherence of Candida to cultured vascular endothelial cells : mechanisms of attachment and endothelial cell penetration. J.Infect.Dis.152:1264.

Ruchel, R., 1981, Properties of a purified proteinase from the yeast Candida albicans, Biochim.Biophys.Acta, 659:99.

Sabetta, J.R., Miniler, P., and Andriole, V.T., 1985, The diagnosis of invasive aspergillosis by an enzyme linked immunosorbent assay for circulating antigen, J.Infect.Dis., 152:946.

San Blas, G., 1982, The cell wall of fungal human pathogens : its possible role in host parasite relationships, Mycopathologia, 79:159.

Scott, E.N., Felton, F.G., and Muchmore, H.G., 1980, Development of an enzyme immunoassay for cryptococcal antibody, Mycopathologia, 70:55.

Scott, E.N., Muchmore, H.G., and Felton, F.G., 1980, Comparison of enzyme immunoassay and latex agglutination methods for detection of Cryptococcus neoformans antigen, Amer.J. Clin.Pathol., 73:790.

Seeliger, H., 1965, Standardization and assay of skin test antigens for mycotic diseases, Progr.Immunobiol.Stand., 2:154, S. Karger Publisher, New York.

Seeliger, H.P.R., 1958, Mykologische serodiagnostik, Barth.Leipzig.

Segal, E., Berg, R.A., Pizzo, P.A., and Bennett, J.E., 1979, Detection of Candida antigen in sera of patients with candidiasis by an enzyme-linked immunosorbent assay-inhibition technique, J.Clin.Microbiol., 10:116.

Segal, E., Levy, R., and Eylan, E., 1980, Experimental immunization with Candida albicans ribosomal fractions: protection and immunological aspects, pp 114-117, in: Proceedings of the seventh Congress of the International Society of Human and animals Mycology, E.S. Kuttin and G.L. Baum (eds), Excerpta Medica, Amsterdam.

Shaffer, P.J., Kobayashi, G.S., and Medoff, G., 1979, Demonstration of antigenemia in patients with invasive aspergillosis by solid phase (protein A rich Staphylococcus aureus) radioimmunoassay, Amer.J.Med., 67:627...P12

Sherwin, W.K., Ross, T.H., Rosenthal, C.M., and Petrozzi, J.W., 1979, An immunosuppressive serum factor in wide spread cutaneous dermatophytosis, Arch.Dermatol., 1979, 115:600.

Stevens, P., Huang, S., Young, L.S., and Berdischewsky, M., 1980, Detection of Candida antigenemia in human invasive candidiasis by a new solid phase radioimmunoassay, Infection 8:334.

Stobo, J.D., Paul, S., Van Scoy, R.E., and Hermans, P.E., 1976, Suppressor thymus derived lymphocytes on fungal infection, J.Clin.Invest., 57:319.

Strockbine, N.A., Largen, M.T., and Buckley, H.R., 1984, Production and characterization of three monoclonal antibodies to Candida albicans proteins, Infect.Immunity, 43:1012.

Strockbine, N.A., Largen, M.T., Zweibel, S.M., and Buckley, H.R., 1984, Identification and molecular weight characterization of antigens from Candida albicans that are recognized by human sera, Infect.Immun., 43:715.

Syverston, R., Buckley, H.R., and Campbell, I, 1975, Cytoplasmic antigen unique to the mycelial or yeast phase of Candida albicans, Infect.Immunity, 12:1184.

Tewari, R.P., 1975, Immunization against histoplasmosis, pp. 441-452, 4th International convocation on immunology, Buffalo, New York, Karger Publisher, Basel.

Tornquist, M. Bendixen, P.H., and Pehrson, B., 1985. Vaccination against ringworm of calves in specialized beef production, Acta.Vet.Scand., 26:29.

Towbin, H., Staehelin, T., and Gordon, J., 1979, Electrophoretic transfer of proteins from polyacrylamide gels to nitrocellulose sheets, Procedure and some applications, Proc.Nat.Acad.Sci., 76:4350.

Tran Van Ky, P., Vaucelle, T., and Biguet, J., 1969, Etude comparée de la structure antigénique par analyse immuno-électrophorétique et par les réactions de caractérisation des activités enzymatiques des extraits antigéniques des champignons pathogènes du genre Aspergillus (A.fumigatus, A.flavus, A.terreus et A.nidulans), Revue Immunol., 34:357.

Travassos, L.A., and Lloyd K.O., 1980, Sporothrix schenckii and related species of Ceratocystis. Microbiol.Rev., 44:683

Tsuchiya, T., Fukazawa, Y., Tuguchi, M., Nakase, T., and Shinoda, T., 1974, Serologic aspects of yeasts classification, Mycopathologia, 53:77.

Uriel, J., 1963, Characterization of enzymes in specific immunoprecipitates, Ann.N.Y.Acad.Sci., 103:956.

Wahab, S., Rahman-Gauthier, S. Siddiqui, M.U., Reyes, G., and Drouhet, E. 1986. Immunité cellulaire aux antigènes de Cryptococcus neoformans mesurée par l'inhibition de la

migration des cellules de rate de souris infectées par
Cryptococcus neoformans. Effet possible du M.I.F. sur le
développement de Cryptococcus neoformans, Bull.Soc.
Fr.Mycol.Méd., 15:55.

Walters, B.A.J., Chick, J.E.D., and Halliday, W.J., 1964,
Cell-mediated immunity and serum blocking factors in
patients with chronic dermatophytic infections.
Inter.Arch.Allergy, 46:849.

Warnock, D.W., and Hann, E.M., 1981, Further evaluation of
indirect immunofluorescence methods for detection of
antibodies against Aspergillus fumigatus. Sabouraudia,
19:49.

Warren, R.C., Bartlett, A., Bidwell, D.E., Richardson, M.D.,
Voller, A., and White, L.O., 1977, Diagnosis of invasive
candidosis by enzyme immunoassays of serum antigen,
Br.Med.J., 1:1183.

Weiner, M.H., and Yount, W.J., 1976, Mannan antigenemia in the
diagnosis of invasive Candida infections. J.Clin.Invest.,
58:1045.

Weiner, M.H., and Coats-Stephen, M., 1979, Immunodiagnosis of
systemic candidiasis: mannan antigenemia detected by
radioimmunoassay in experimental and human infections,
J.Infect.Dis., 140:989.

Weiner, M.H., 1980, Antigenemia detected by radioimmunoassay
in systemic aspergillosis, Ann.Intern.Med., 92:793.

Weiner, M.H., 1983, Antigenemia detected in human coccidioidomycosis, J.Clin.Microbiol., 18:136.

Wheat, L.J., Kohler, R.B., and Tewari, R.P., 1986, Diagnosis
of disseminated histoplasmosis by detection of Histoplasma capsulatum in serum and urine specimens, N.Engl.
J.Med., 314:83.

Whiteker, R.H., 1969, New concepts of kingdoms of organism,
Science, 163:150.

Williams, P.L., Sable, D.L., Sorgen, S.P., Pappagiani, D., and
Levine, K.B., Brodine, S.K., Brower, B.W., Grumet, F.C.,
Stevens, D.A., 1984, Immunologic responsiveness and safety
associated with the Coccidioides immitis spherule vaccine in
volunteers of white, black and filipino ancestry. Amer.J.
Epidemiol., 119:591.

Wu-Hsich, B., Zlotnik, A., and Howard, D.H., 1984, T-cell hybridoma produced lymphokine that activates macrophages to suppress intracellular growth of Histoplasma capsulatum, In-
fect.Immunity, 42:380.

Yarzabal, L.A., Bout, D., Naguira, F., Fruit, J., and Andrieu,
S., 1977, Identification and purification of the specific
antigen of Paracoccidioides brasiliensis, Sabouraudia,
15:79.

Young, K.D., and Larsh, H.W., 1982, Production and characterization of a hybridoma derived antibody to Blastomyces dermatitidis, J.Clin.Microbiol., 15:204.

Methodology

CHEMICAL METHODS TO STUDY THE STRUCTURE OF FUNGAL POLYSACCHARIDE ANTIGENS

Bernard Fournet

Laboratoire de Chimie Biologique de l'Université
des Sciences et Techniques de Lille Flandres
Artois et Unité Associée au CNRS no 217
59655 Villeneuve D'Ascq Cedex, France

INTRODUCTION

Fungi are remarkable for the variety of polysaccharide structures they produce, such as α-D-mannans, β-D mannans, phosphomannans, galactans, phosphogalactans, chitin starch-like polymers, glycogen, pullulan, mycodextran, glucans having $\alpha(1\rightarrow3)$ linkages, $\beta-(1\rightarrow3)$ D- and $\beta-(1\rightarrow6)$ D- glucans. Various polysaccharides containing N-acetylgalactosamine, xylose, arabinose, fucose, glucuronic acid and rhamnose have also been described (1-2). Knowledge of the primary structure of polysaccharides gives the basis for their classification and antienic properties as well as for studies of their three-dimensional structure and biosynthetic pathways.

Elucidations of the primary structure of polysaccharide antigens requires knowledge of the nature and ring sizes (pyranose or furanose) for each sugar residue, the linkage type for each glycosidic linkage, the anomeric configuration (α or β) of glycosidic linkages and the sequence of sugar residues. Several chemical and spectroscopic methods must be used to obtain the primary structure of polysaccharides.

CHEMICAL COMPOSITION OF POLYSACCHARIDES

The first step in determination of the structure of a purified polysaccharide is to identify the monosaccharide components. General composition of polysacchardies in terms of neutral sugars, hexuronic acids, and hexosamines can be performed by colorimetric assays (3-6). Determination of neutral monosaccharides and hexuronic acids is carried out by treatment with acid without prior hydrolysis of the polysaccharides but colorimetric determination of hexosamines requires the prior liberation of the amino sugars. Hydrolysis is accompanied by decomposition leading to chromogens which condense with specific reagents to give colored products.

Identification and molar determination of monosaccharides
present in polysaccharides require the complete liberation of
each monosaccharide and the separation of sugars by chromatographic procedures. In fact, all the glysosidic bonds must
be split without any destruction of the liberated sugars, but
in some cases, it will be necessary to use different conditions for the hydrolysis of different sugars present in the
same polysaccharide; 4N trifluoroacetic acid, 2H at 100°, or,
1M sulfuric acid, 4H at 100°C (7) for aldose containing polysaccharides such as glucans and galactomannans. Uronic acid
and hexosamine-containing polysaccharides are considerably
more difficult to hydrolyse and complete liberation is rarely
possible. Treatment with 72% sulfuric acid, 5 minutes at room
temperature brought to normality at 100°C for 6H releases
disaccharides called aldobiouronic acid (8). Furanose-and
ketose-containing polysaccharides are hydrolysed under comparatively mild conditions: fructans, 0.1M oxalic acid,
80°C, 1H (9); sialic acid polymers, 0.05M sulfuric acid, 80°C,
1H (10). Liberated monosaccharides can be analysed directly
by high performance liquid chromatography using ion exchange
and partition chromatography (11,12), or after transformation
into volatile derivates (trimethylsilyl ethers and acetate or
trifluoroacetate esters), by gas-liquid chromatography (GLC)
(13). These procedures do not permit distinction between D
and L enantiomorphs. This determination can be achieved by
their melting points and optical relations on milligram
quantities; in smaller amounts, enantiomers can be identified
by capillary gas-liquid chromatography of glycosides of chiral
alcohols (14).

    Many non-glucosidic residues can be linked to polysac-
charides. The chemical analysis of these substituents is
dependent on their complete removal and on the specificity of
the analytical procedure. Pyruvic acid can be liberated by
mild acid hydrolysis while the O-acyl groups are labile in
basic solutions. The O-sulfate and phosphate groups are also
analysed after liberation by acid hydrolysis (15). Liberated
substitutents (pyruvate, acetate, succinate) are now identified and estimated by HPLC on polystyrene columns (16).
Finally, organic substituents can be analysed by NMR spectroscopy using intact polysaccharide.

LINKAGE AND SEQUENCE

    When the monosaccharides present in the polysaccharide
are known, the next problem is to determine the type of
linkage for each glucosidic linkage: 1→2, 1→3, 1→4, or 1→6.

Methylation

    Methylation analysis is one of the most powerful proce-
dures for structural analysis of polysaccharides and glycoconjugates. It inovles the permethylation of all free hydroxyl groups of sugars, liberation of methylated monosaccharides by hydrolysis or methanolysis of the full methylated
polysaccharide, and qualitative and quantitative analysis of
the methylated derivatives. Many methylation methods have
been reported since the first procedure described in 1903 by
Puride and Irvine but most of the proposed methods led to an

incomplete methylation of the saccharides. In this connection, the Hakomori procedure (18) using sodium hydride and methylsulfoxide represents a real progress by producing a more powerful nucleophilic reagent than those previously used. Because a sodium hydride required to make the base contains impurities, cleaner reagents were proposed to perform the methylation of small amounts of polysaccharides. In this connection, we developed a procedure based on the assocation of dimenthyl-sulfoxide and n-butyl lithium which gives less background in GLC and, thus, can be used to methylate micro-quantities of polysaccharides (10 to 50 ug of total sugars) (18).

The association of gas-liquid chromatography and mass-spectrometry presents easier identification of methylated monosaccharides obtained by hydrolysis or methanolysis of permethyl-polysaccharides. After methanolysis of permethyl-polysaccharides and acetylation of partially methylated methyl-glycosides, all the methylated sugars, as their acetylated methylglycosidic forms, can be identified by GLC-MS on capillary columns (19). Thus, from the partial structure of extracellular B - D glucans from Botrytis cinerea, methanolysis of the permethyl-polysaccharide yields the products (20) described in Fig. 1.

Partial Depolymerization

Polysaccharides can be depolymerized by various procedures to break the weakest linkages. The liberated oligosaccharides can be separated by chromatography (gel filtration, partition chromatography adsorption chromatography on charcoal, ion exchange chromatography). Analysis of oligosaccharides are also carried out by HPLC (11,12).

Partial acid hydrolysis. The (1→6) linkages are more stable to mild hydrolysis with mineral acid than are the (1→4) linkages. Glycosiduronic acids are much more resistant to hydrolysis at low pH than the corresponding neutral glycosides. For example, hydrolysis of glycuronic acid-containing glucuronoxylomannan of Cryptococcus neoformans (22) leads to the isolation of acidic disaccharides (aldobiouronic acid).

Acetolysis. Partial acetolysis involves treatment of a polysaccharide, or preferably its acetylated derivative, with acetic anhydride containing about 5% sulfuric acid. In contrast to partial hydrolysis, the (1→6) linkages are preferentially split during acetolysis. Accordingly, acetolysis has been extensively utilized to characterize the profile of side chains attached to a (1→6) linked core in mannans from yeasts and fungi (23). Fig. 3 and Table 1 give mannose-containing oligosaccharides obtained by partial acetolysis of mannans with (1→6) linked main chains.

Periodate oxidation. Periodate reacts with polysaccharides to cleave the linkage between carbons carrying vicinal hydroxyl groups. For each linakge cleaved, one mole of periodate is consumed. Primary hydroxyl groups are oxidized to methanol and secondary hydroxyl groups give rise to methanoic acid. The amounts of periodate consumed and the products that are formed will depend on the position of the carbon atom involved in the linkages with the neighboring residues. Smith degradation (26) is a sequence of reactions based on periodate oxidation of polysaccharides followed by reduction and controlled acid hydrolysis. Fig. 4 gives the results of the Smith degradation of the branched β-D glucan

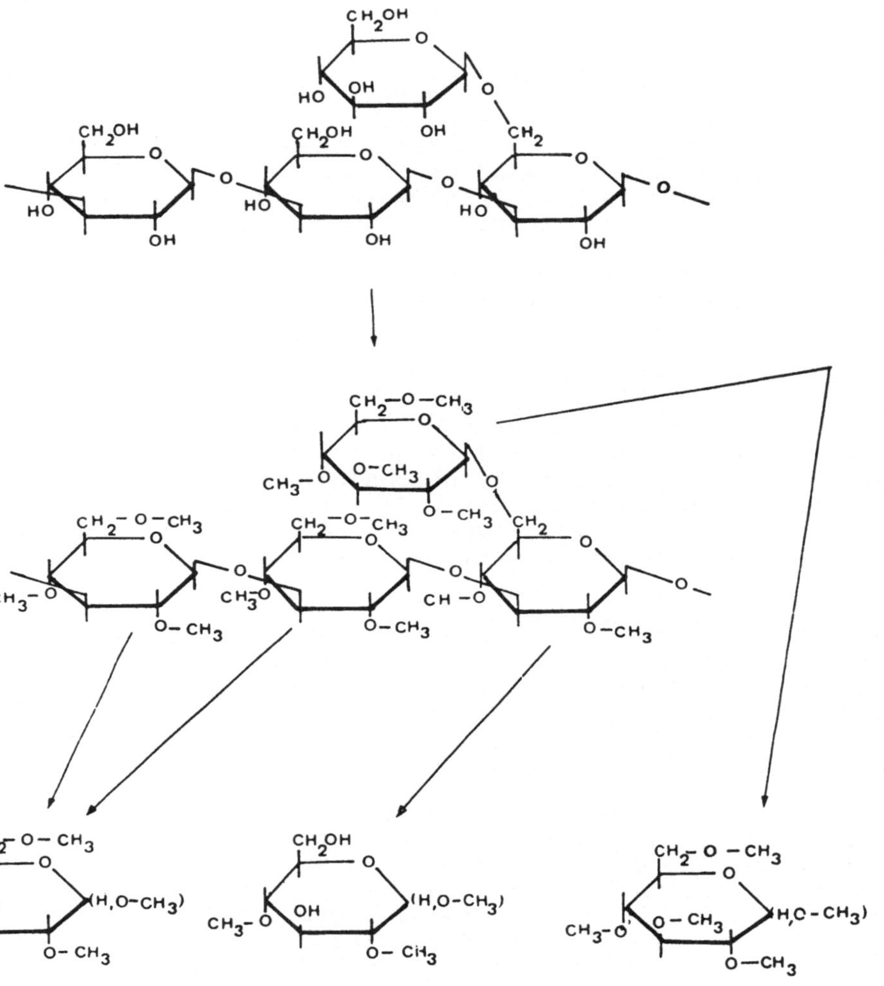

Fig. 1.   Methylation and methanolysis of permethyl β-D glucan from <u>Botrytis</u> <u>cinerea</u> to give permethyl glucose and partially methylated glucose.

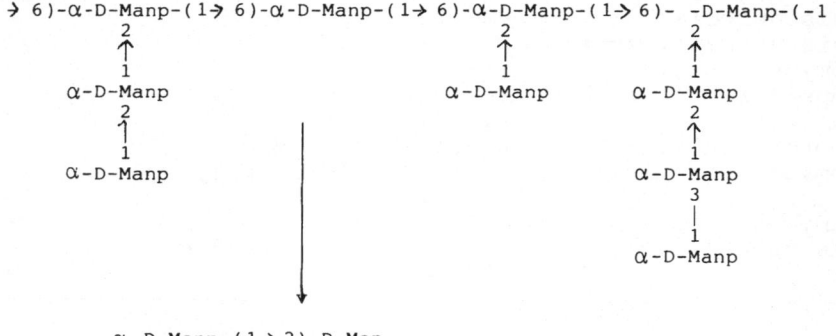

```
→ 6)-α-D-Manp-(1→ 6)-α-D-Manp-(1→ 6)-α-D-Manp-(1→ 6)-  -D-Manp-(-1
              2                              2              2
              ↑                              ↑              ↑
              1                              1              1
         α-D-Manp                      α-D-Manp         α-D-Manp
              2                                             2
              ↑                                             ↑
              1                                             1
         α-D-Manp                                       α-D-Manp
                                                            3
                                                            |
                                                            1
                                                       α-D-Manp
```

```
         α-D-Manp-(1→ 2)-D-Man
         α-D-Manp-(1→ 2)-α-D-Manp-(1→ 2)-D-Man
         α-D-Manp-(1→ 3)-α-D-Manp-(1→ 2)-α-D-Manp-(1→ 2)-D-Man
```

Fig. 2. Preferential formation of disaccharides (aldobiouronic acid) upon acid hydrolysis of glucuronoxylomannan of Cryptococcus neoformans (GXM) Type A.

```
3)-α-D-Manp-(1→3)-α-D-Manp-(1→3)α-D-Manp-(1→3)α-D-Manp-(1→3)-α-D-Manp-(1
            2                         2                        2
            ↑                         ↑                        ↑
            1                         1                        1
       β-D-Xylp                  β-D-GlcUpA                β-D-Xylp

                          β-D-GlcUpA-(1→2)-D-Manp
```

Fig. 3. Selective cleavage by acetolysis of (1 → 6) bonds in yeast mannans with liberation of oligosaccharides.

45

Table 1.   Mannose-containing Oligosaccharides Obtained by
            Partial Acetolysis of Mannopyranans with $(1\rightarrow6)$-
            Linked Main Chains

Parent yeast	Oligosaccharide products	Ref.
Torulopsis bombicola	$[M\text{-}(1\rightarrow2)]_{1-7}\text{-}M$	24
Torulopsis magnoliae	$[M\text{-}(1\rightarrow2)]_{1-9}\text{-}M$	24
Torulopsis apicola	$[M\text{-}(1\rightarrow2)]_{1-5}\text{-}M$	24
Torulopsis gropengiesseri	$[M\text{-}(1\rightarrow2)]_{1-2}\text{-}M$	24
Saccharomyces rouxii	$[M\text{-}(1\rightarrow2)]_{1-2}\text{-}M$	24
Endomycopsis fibuliger	$M\text{-}(1\rightarrow2)\text{-}M, \overline{M}^{2}\text{-}(1\rightarrow3)\text{-}M\text{-}$ $(1\rightarrow2)\text{-}M$	24
Trichosporon aculeatum	$[M\text{-}(1\rightarrow2)]_{1-7}\text{-}M$	24
Saccharomyces cerevisiae	$M\text{-}(1\rightarrow3)\text{-}[\overline{M}\text{-}(1\rightarrow2)]_2\text{-}M,$ $M\text{-}(1\rightarrow2)\text{-}M$	24
Saccharomyces fragilis	$M\text{-}(1\rightarrow2)\text{-}M$	25
Hansenula subpelliculosa	$M\text{-}(1\rightarrow3)\text{-}[M\text{-}(1\text{-}2)]_3\text{-}M,$ $M\text{-}(1\rightarrow3)M, [M\text{-}(1\rightarrow2)]_{1-3}\text{-}M$	24

from Botrytis cinerea (20) with removal of side chains and
formation of an otherwise unmodified 3-linked β-D glucan.

   Enzyme-catalysed hydrolysis. Glycosides are excellent
tools for the partial depolymerization of polysaccharides, for
the determination of the anomeric linkage of each conjugated
monosaccharide and for determining the primary structure of
polysaccharides by sequential degradation. Two types of
enzymes are used: exoglycosidases which hydrolyse glycosidic
bonds of monosaccharides in a terminal non-reducing position
and lead to a stepwise degradation of polysaccharides, and
endoglycosidases which hydrolyse glycosidic bonds in internal
positions. Examples of exoglycosidases are phosphorylase and
B-amylase which act on amylose, amylopectin and glycogen with
the liberation of α-D glucopyranosylphosphate and maltose,
respectively. Exo-$(1\rightarrow3)$ B-D glucanase, from a basidiomycete
(QM 806) prepared acording to the procedure of Peterson and
Kirkwood (27), digests the extracellular β-D glucan from
Botrytis cinerea resulting in complete degradation of the
polysaccharide to gentiobiose and D-glucose in the molar ratio
3:2 (Fig. 5) (20). Cellulase from Streptomyces sp. (QM B814)
is an example of a endoglycosidase which cleaves the glycosidic bond between two unbranched, 4-linked β-D glycopyranose
residues of B-D glucans from oats and barley (28).

   Partial depolymerization of permethylated polysaccharides. Partial hydrolysis of permethylated polysaccharides gives valuable information on primary structure of
polysaccharides. In this way, Valent et al. (20) developed a
procedure based on partial hydrolysis of permethyl-polysaccharide (85% formic acid, 39 min, 80°C, followed by reduction
with $NaBD_4$ [0.2M] and ethylation of the free hydroxyl-
groups). The partially methylated ethylated products can be
analysed by GLC-MS or fractionated by HPLC on reversed-phase
columns.

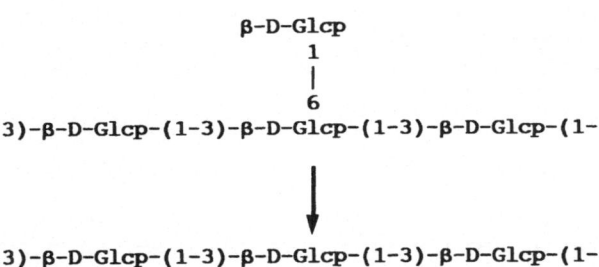

$$\beta\text{-D-Glc}p$$
$$1$$
$$|$$
$$6$$
$$3)\text{-}\beta\text{-D-Glc}p\text{-}(1\text{-}3)\text{-}\beta\text{-D-Glc}p\text{-}(1\text{-}3)\text{-}\beta\text{-D-Glc}p\text{-}(1\text{-}$$

$$\downarrow$$

$$3)\text{-}\beta\text{-D-Glc}p\text{-}(1\text{-}3)\text{-}\beta\text{-D-Glc}p\text{-}(1\text{-}3)\text{-}\beta\text{-D-Glc}p\text{-}(1\text{-}$$

Fig. 4. Smith degradation (periodate oxidation, reduction and acid hydrolysis) of the branched β-D glucan from <u>Botrytis</u> <u>cinerea</u> yields the linear 3-linked main chain.

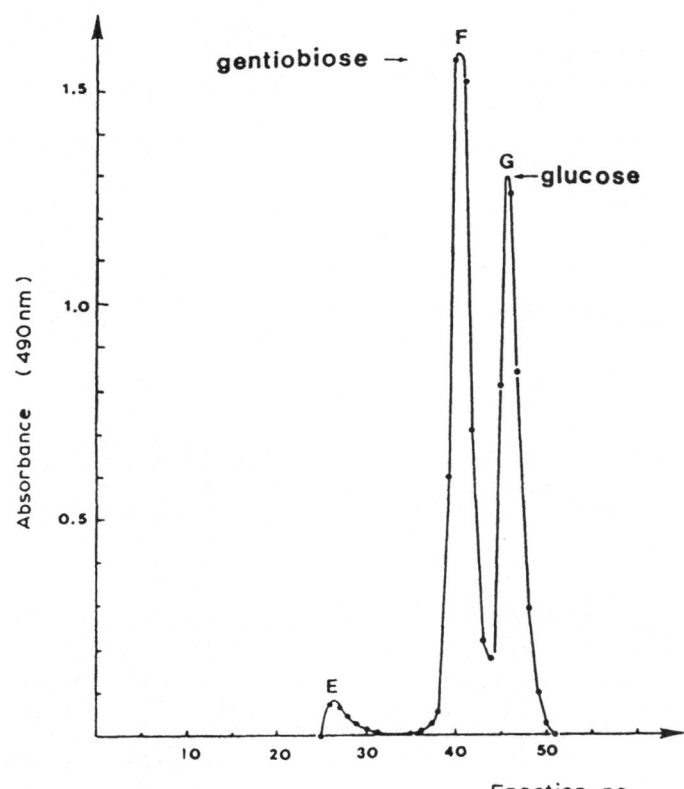

Fig. 5. Separation of the product obtained after treatment of the glucan of <u>Botrytis</u> <u>cinerea</u> with exo-(1→3)-β-D-glucanase on a Sephadex G-15 column (1.6 x 120cm).

Anomeric Nature of Sugar Residues

Angyal and James (30) have shown that fully acetylated aldopyranosides with equatorially attached aglycons in the most stable arrangement (generally the β-anomers, Fig. 6, compound I) are oxidized by chromium trioxide in acetic acid to 5-aldulosonates (Fig. 6, compound II).

The anomers with axially attached aglycons (generally the α-anomers, Fig. 6, compound III) are oxidized only slowly under these same conditions. This reagent has been used to determine the anomeric nature of sugar residues in oligo- and polysaccharides (31). The samples are fuly acetylated, oligosaccharides having first been reduced to their alditols, and the products are then treated with chromium trioxide in acetic acid in the presence of an internal standard. A comparison of sugar analysis, performed before and after oxidation, reveals which sugar residues have been oxidized (32).

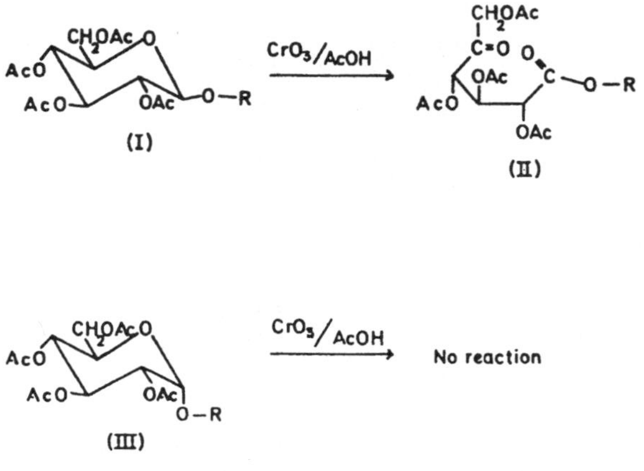

Fig. 6. Oxidation with chromium trioxide in acetic acid of fully acetylated aldopyranosides.

SPECTROSCOPIC METHOD

Nuclear Magnetic Resonance Spectroscopy

Nuclear magnetic resonance spectroscopy has a number of characteristics that make this technique especially advantageous for studies of polysaccharides (33-36). This technique studies the molecular absorption in the domain of the radio-frequency (megahertz). The energy absorbed relates to the nucleus in presence of an external magnetic field.

In common with other physical methods, NMR spectroscopy is nondestructive. It is possible, therefore, to examine a polymer without modifying or degrading it and then recover the material in its intact form.

Until very recently, high resolution NMR studies were restricted to solutions, so that it was not possible to deal directly with polysaccharides of low solubility other than by

wideline NMR techniques. These limitations have been substantially overcome (37) by new methods and instrumentation
that give high-resolution ^{13}C spectra with solid samples.
Similarly, the pulsed Fourier transform (FT) technique (38) by
increasing the ratio of signal/noise permits enhancement in
sensitivity. These techniques allow examination of the
anomeric nature of sugar residues by assignment of chemical
shifts and evaluation of coupling constants (39-41).

   ^{1}H-NMR spectra. To simplify the analysis, the hydroxyl
groups are changed to deuterioxyl groups by exchange treatment
with deuterium oxide. Nevertheless, a strong peak due to
residual water (HOD signal) as well as substantial side bands
of the peak are often obtained, especially if the concentration of polysaccharide is less than 1mM. This peak is close
to the H-1 signal of β-glucopyranosol residues. This problem
can be solved by raising the temperature to approximately
70oC.

   Table II gives the chemical shifts ($\int$) and the coupling
constants $^3J_{H,H}$ between the anomeric and the carbon- 2
protons.

Table II.   1/3H chemical shifts ($\int$) and coupling
              constants $J_{H,H}$ for nuclei of
              polysaccharides.

^{1}H	(ppm)	$^3J_{H,H}$ (H_3)
CH$_3$-C (deoxyhexoses)	1.2 - 1.3	
CH$_3$-C (Pyruvyle)	1.5 - 1.6	
CH$_3$-CON (Acetyle)	1.8 - 2.1	
CH$_3$COO (Acethyle)	2.0 - 2.2	
H-2 to H-6	3.5 - 4.5	
H-5	4.5 - 4.6	
H-2 and H-3 of furanic isomers H-5 of uronic acids	4.2 - 4.5	
H-1 axial (β)	4.5 - 4.8	7
H-1 equatorial (α)	5.3 - 5.8	4
OH (Hydroxyl)	5.0 - 5.4	
HCO$_2$	5.9	

   The anomerica proton signals are well separated from
signals produced by most of the other protons. This fact
greatly simplifies the determination of the number of different kinds of residues in a polysaccharide and allows estimates of their relative proportions. It also makes it easier
to obtain from these signals the chemical shift and coupling
parameters necessary for assigning anomeric configurations by
NMR spectroscopy.

^{13}C-NMR spectra. The use of the FT mode is obligatory for ^{13}C because of the low natural abundance (1.1%) and intrinsically low sensitivity of the ^{13} nucleus. The well-defined ^{13}C spectra obtained under conditions of proton decoupling have provided a quantity of information on chemical and physiocochemical properties of polysaccharides (42). For example, a hexopyranan that is homogeneous in structure or has a two-unit, or three-unit repeating sequence can give 6, 12, or 18 resonances, respectively. Values of $J_{C,H}$ can indicate glycosidic configuration. ^{13}C NMR spectroscopy is also useful in monitoring the purity of polysaccharide preparations.

Table III gives the ^{13}C chemical shifts for nuclei of polysaccharides. We can see that the anomeric carbon C-1 (α and β) give chemical shifts which are clearly defined. The methyl groups appear under highfield conditions (C-6 of 6-deoxy hexoses at 16 ppm). Hexoses of the furan form give charactersitic signals (C-2, C-3, C-4 between 83 and 87 ppm).

Table III

^{13}C	δ (ppm)
CH_3C (CH_3 of 6 deoxy-hexoses)	16-18
CH_3C (Pyruvyle)	22-25
CH_3COH $\quad CH_3-CO_2$	20-23
CH_2C	38
CH_3O	55-61
CH(NH)	58-61
C_6 (CH_2OH)	60-65
C-2 to C5	65-75
C-2, C-3, C-4 furane form	83-87
C-1 (ax, free)	90-95
C-1 (eq-, free)	95-98
C-1 (ax-, conjugated)	98-103
C-1 (eq-, conjugated)	103-106
C-1 (eq-, conjugated, furane)	106-109
COOH	174-175
C=O	175-180
C_6 (COOH of uronic acid)	186

Applications. Proton and ^{13}C NMR spectroscopy give information on the number and relative proportions of residues, different linkage configurations and positions in homopolymers.

Mannans of yeast constitute highly diverse series of branched homopolymers that have been very profitably investigated by ^{1}H and ^{13}C NMR by Gorin (42). In these polymers, the main chains are constituted by (1→6) linked α-D-mannopyranosyl residues, often substituted at 0-2 by α-linked glycosyl groups. The linkage types in the rest of the side chains vary widely and can consist of 2-0-substituted α- or β-D mannopyranosyl residues or 3-0 substituted α-D-mannopyranosyl residues, which can be arranged in many combinations. The ^{13}C-NMR spectra of D-mannose isolated from bakers' yeast having α-D-linked side chains (fig. 3) is given in Fig. 7.

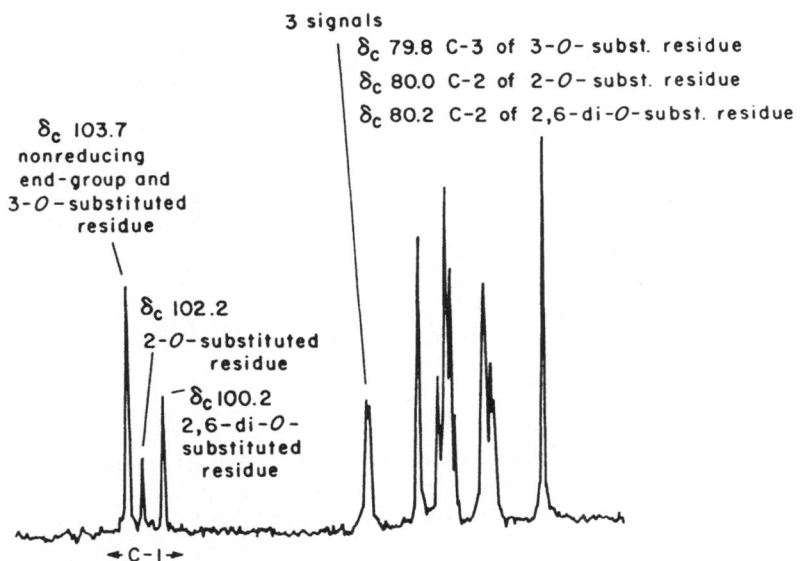

3 signals
δ_c 79.8 C-3 of 3-*O*-subst. residue
δ_c 80.0 C-2 of 2-*O*-subst. residue
δ_c 80.2 C-2 of 2,6-di-*O*-subst. residue

δ_c 103.7 nonreducing end-group and 3-*O*-substituted residue

δ_c 102.2 2-*O*-substituted residue

δ_c 100.2 2,6-di-*O*-substituted residue

◄─C-1─►

Fig. 7.   ^{13}C-NMR spectrum of branched-chain α-D-mannopyrannan from bakers' yeast (solvent, D$_2$O; temperature, 70°; chemical shifts expressed as $\int$ c, relative to external tetramethylsilane (42).

The C-1 signals of non-reducing end groups and internal 2-0-substituted residues are at $\int$103.7 and 102.2, respectively. The signal at $\int$103.7 coincides with the C-1 resonances of 3-0-substituted α-D-mannopyranose. The anomeric carbon C-1 from 2,6-di-0-substituted D mannopyranose is at 100.2. The anomeric carbon C-1 at $\int$101.2 corresponds to C-1

from unsubstituted (1→6) linked α-D-mannopyranosyl residues.
Other characteristic signals at ∫79.8, 80.0, and 80.2 correspond to C-3 of 3-0-substituted, C-2 of 2-0-substituted, and C-
2 of 2,6-di-0-substituted α-D-mannose units, respectively.

Yeast mannans give ^{1}H spectra that are well resolved in
the anomeric region (24,25). Clear chemical shift differences
are observed between the H-1 signals of unsubstituted and 2-0
substituted residues of the main chain, and between the H-1
signals of residues of the side chains. Since there are wide

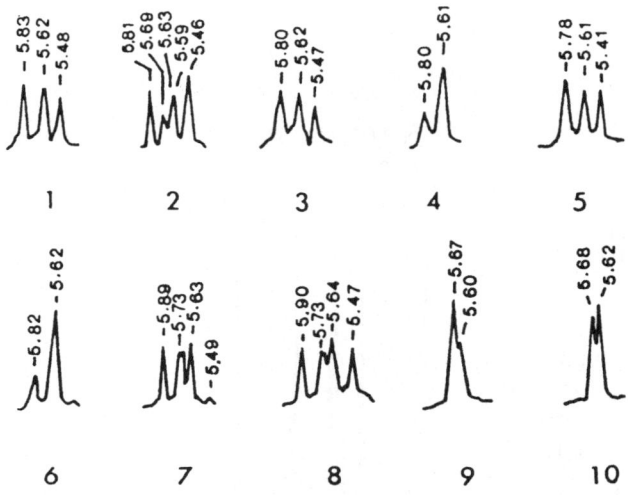

Fig. 8.   Signals due to anomeric protons of mannans
           in D_2O at 100 MH3 from various yeast,
           showing characteristic differences in
           patterns of chemical shift and relative
           intensities (Gorin et al. [24]).

variations as well in the ratios of residue types, the yeast
mannans afford an extensive series of highly distinctive
spectra.

At least carbon 13-NMR spectroscopy is an excellent
method for examining the degree of regularity of sequences in
linear polysaccharides containing different types of linkages,
and/or sugar residues. For example, in the linear α-D-mannan
of Rhodotorula glutinis with an alternating sequence of (1->3)
and (1->4) linkages, Gorin (43) identified 12 resonances.

DISCUSSION

The chemistry and immunology of pathogenic fungi constitute a subject that is likely to receive considerable attention. The need for antigen purification arises from both the
realization by imunologists of the necessity for purified
antigens for their immunoassays and the biochemists' requirements to deal with well-characterized products. Knowledge of
the primary structure of the polysaccharidic antigens is
fundamental to these needs. A good example is given by the
studies of polysaccharidic antigens from <u>Candida</u> <u>albicans</u>
(44). Several groups of investigators have analyzed the
<u>Candida</u> <u>albicans</u> mannan structure. Almost all agree that the
core is composed of $\alpha(1\rightarrow6)$ linked mannose units with side
branches joined to the core by $\alpha(1\rightarrow2)$ linkages. In contrast
to the uniform core, structural variations occur among the
oligomannan branches and these determine immunochemical
specifities. The side chain with mannopyranosyl-units range
from mannobiose to mannohexaose in <u>C.</u> <u>albicans</u> type B, and to
mannoheptaose in serotype A.

The $\alpha(1\rightarrow2)$ linkages predominate but a small proportion
of $\alpha(1\rightarrow3)$ linkages occur in the mannotetraose of serotype A,
and in the mannopentaose and mannohexaose of both serotypes in
ratios of $\alpha(1\rightarrow2)$ linkages to $\alpha(1\rightarrow3)$ linkages in the three
sugars of 2:1, 3:1, and 4:1, respectively. The presence of an α
$(1\rightarrow3)$ linkage represents an additional influence on antigenic
determinance since the inhibitory power of the mannohexaose
containing an $\alpha(1\rightarrow3)$ linkage is greater than the mannoheptaose composed of only $\alpha(1\rightarrow2)$ linkages.

REFERENCES

1. P. A. J. Gorin and J. F. T. Spencer, Structural chemistry
    of fungal polysaccharides <u>in:</u> "Advances in Carbohyddrate
    Chemistry," M. L. Wolfrom and R. Stuart Tipson, ed., Vol.
    23, p. 367, Academic Press, New York (1968).
2. P. A. J. Gorin and E. Barreto-Bergter, The chemistry of
    polysaccharides of fungi and lichens <u>in:</u> "They Polysac-
    charides," G. O. Aspinall, ed., Vol. 2, p. 365, Academic
    Press, New York (1983).
3. Z. Dische, Color reactions of carbohydrates <u>in:</u> "Methods
    in Carbohydrate Chemistry," R. L. Whistler and M. L.
    Wolfrom, ed., Vol. I, p. 477, Academic Press, new York
    (1962).
4. J. Montreuil and G. Spik, Methodes colorimetriques de
    dosage des glucides totaux, Monog. Lab. Chim. Biol. Fac.
    Sci. Lille (1963).
5. C. A. White and J. F. Kennedy, Manual and automated
    spectrophotometric techniques for the detection and assay
    of carbohydrates and related molecules <u>in:</u> "Techniques in
    Carbohydrate Metabolism," H. L. Kornberg, J. C.
    Metcalfe, D. H. Northcote, C. I. Poigson and K. F.
    Tipson, ed., Esevier, Amsterdam (1981).
6. J. Montreuil, G. Spik, B. Fournet and M. T. Tollier,
    Glucides <u>in:</u> "Techniqes d'nalayse et de Controle dans les
    Industries Agro-Alimentaires," B Deymie, J. L. Multon and
    D. Simon, ed., Vol. 4, p. 85, Technique et Documentation,
    Paris (1981).

7. R. R. Selvendran, J. F. March and S. G. Ring, Determination of aldoses and uronic acid content of vegetable fiber, <u>Anal. Biochem.</u>, 96:282 (1979).

8. J. S. Seman, W. E. Moore, R. L. Mitchell and M. A. Miller, Techniques for the determination of pulp constituents by quantitative paper chromatography, <u>Tappi</u>, 37:336 (1954).

9. G. O. Aspinall, E. L. Hirst, E. G. V. Percival and R. G. J. Telfer, Studies on frucosans, part 4; a furctosan from Dactylis glomerata, <u>J. Chem. Soc.</u>, 337 (1953).

10. J. F. Codington, K. B. Linsley and C. Silber, Removal of sialic acids from glycoproteins by chemical methods and determination of sialic acids <u>in</u>: "Methods in Carbohydrate Chemistry," R. L. Whistler and J. N. BeMiller, ed., VII:226, Academic Press, new York (1976).

11. G. D. McGinnis and P. Fang, High-performance liquid chromatography <u>in</u>: "Methods in Carbohydrate Chemistry" R. L. Whistler and J. N. BeMiller ed., VIII:33, Academic Press, New York (1980).

12. B. Fournet, J. Paz Parente, Y. Leroy and J. Montreuil, Analyse des sucres par chromatographie liquide haute pression H.P.L.C., <u>Spectra 2000</u>, 9:28 (1982).

13. G. G. S. Dutton, Applications of gas-liquid chromatography to carbohydrates: Part II <u>in</u>: "Advances in Carbohydrate Chemistry and Biochemistry," S. Tipson and D. Horton ed., 30:9, Academic Press, New York (1974).

14. G. J. Gerwig, J. P. Kamerling, and J. F. G. Vliegenthart, Determination of the absolute configuration of monosaccharides in complex carbohydrates by capillary G.L.C., <u>Carbohydr. Res</u>. 77:1 (1979).

15. J. F. Kennedy, Proteoglycans; Biological and Chemical Aspects in Human Life, Elsevier, Amsterdam, 112 (1979).

16. R. Pecina, G. Bonu, E. Burtscher and O. Bobleter, High-performance liquid chromatographic elution behavior of alcohols, aldehydes, ketones, organic acids and carbohydrates on a strong cation-exchange stationary phase <u>J. Chromatogr</u>. 287:245 (1984).

17. S. I. Hakomori, A rapid permethylation of glycolipid and polysaccaride catalysed by methylsulfinyl carbanion in dimethyl sulfoxide, <u>J. Biochem</u>. (Tokyo), 55:205 (1964).

18. J. Paz Parente, P. Cardon, Y. Leroy, J. Montreuil, B. Fournet and G. Ricart, A convenient method for methylation of glycoprotein glycans in small amounts by using lithium methyl-sulfinyl carbanion, <u>Carbohydr Res</u>. 141:41 (1985).

19. B. Fournet, G. Strecker, Y. Leroy and J. Montreuil, Gas-liquid chromatography and mass spectrometry of methylated and acetylated methyl glycosides. Application to the structural analysis of glycoprotein glycans, <u>Anal. Biochem.</u>, 116:489 (1981).

20. D. Dubourdieu, P. Bidereau-Gayon and B. Fournet, Structure of the extracellular B-D-glucan from <u>Botyrtis cinerea</u>, <u>Carbohydr. Res</u>., 93:294 (1981).

21. O. Samuelson, Partition chromatogrphy on ion-exchange resins, <u>in</u>: "Methodes in Carbohydrate Chemistry," R. L. Whistler and J. N. BeMiller ed., VI:65, Academic Press, New York (1972).

22. R. E. Cherniak, E. Reiss, M. E. Slodki, R. D. Plattner, S. O. Blumer Structure and antigenic activity of the capuslar polysaccharide from <u>Cryptococcus neoformans</u> serotype A., <u>Mol. Immunol.</u>, 17:1025 (1980).

23. Y. C. Lee and C. E. Ballou, Preparation of mannobiose, mannotirose and a new mannotetraose from Saccharaomyces cereviseia Mannan, Biochemistry, 4:257 (1965).

24. P. A. J. Gorin, J. F. T. Spencer and R. J. Magus, Comparison of proton magnetic resonance spectra of cellwall mannans and galactomannans of selected yeasts with their chemical structures, Can. J. Chem., 47:3569 (1969).

25. P. A. J. Gorin, Rationalization of Carbon-13 magnetic resonance spectra of yeast mannans and sructurally related oligosaccharides, Can. J. Chem. 51:2375 (1973).

26. I. J. Goldstein, G. W. Hay, B. A. Lewis and F. Smith, Controlled degradation of polysaccharides by periodate oxidation, reduction and hydrolysis, in: "Methods in Carbohydrate Chemistry," R. L. Whistler and J. N. BeMiller ed., V:361, Academic Press, New York (1965).

27. D. R. Peterson and S. Kirkwood, Studies on the structure and mecanism of an exo-(1->3)-B-D-glucanase from basidiomycete QM 806, Carbohydr. Res., 41:273 (1975).

28. F. W. Parrish, A. S. Perlin and E. T. Reese, Selective enzymolsis of poly-B-D-glucans and the structure of the polymers, Can. J. Chem., 38:2094 (1960).

29. B. S. Valent, A. G. Darvill, M. McNeil, B. K. Robertsen and P. Albersheim, A general and sensitive chemical method for sequencing the glycosyl residues of complex carbohydrates, Carbohydr. Res. 79:165 (1980).

30. S. J. Angyal and K. James, Oxidation of carbohydrates with chromium trioxide in acetic acid, Aust. J. Chem., 23:1209 (1970).

31. J. Hoffman, B. Lindberg and S. Svensson, Determinations of the anomeric configuration of sugar residues in acetylated oligo- and polysaccharides by oxidation with chromium trioxide in acetic acid, Acta Chem. Scand., 26:661 (1972).

32. J. Hoffman and B. Lindberg, Oxidation of acetylated carbohydrates with chromium trioxide in acetic acid, in: "Methods in Carbohydrate Chemistry," R. L. Whistler and J. N. BeMiller ed., VIII:117, Academic Press, New York (1980).

33. F. A. Bovey, High Resolution N.M.R. of macromolecules, Academic Press, New York (1972).

34. J. L. James, in: "Nuclear magnetic resonance in biochemistry," Academic Press, New York (1975).

35. D. W. Jones, Combined applciations of spectroscopy and other techniques, in: "Introduction to the Sepctroscopy of Biological Polymers," Academic Press, New York, 295 (1976).

36. A. S. Perlin and B. Casu, Spectroscopic Methods, in: "The Polysaccharides," G.O. Aspinall ed., 1:133, Academic Press, New York (1982).

37. J. Schaefer and E. O. Stejskal, High-resolution carbon-13 N.M.R. of solid polymers, Top. Carbon-13 N.M.R. Sepctrosc., 3:283 (1979).

38. R. R. Ernst, Sensitivity enhancement in magnetic resonance, in: "Adv. Magn. Reson.," 2:1 (1966).

39. L. D. Hall, Nuclear magnetic resonance, in: "Advances in Carbohydrate Chemistry," M. L. Wolfrom and R. S. Tipson ed., 19:51, Academic Press, New York, (1964).

40. G. Kotowycz and R. U. Lemieux, Nuclear magnetic resonance in carbohydrate chemistry, Chem. Rev., 73:669 (1973).

41. A. S. Perlin, Carbon-13 N.M.R. spectroscopy of carbo-
    hydrates, Int. Rev. Sci. Org. Chem., Ser. Two, 7:1
    (1976).
42. P. A. J. Gorin, Carbon-13 Nuclear magnetic resonance
    spectroscopy of polysaccharices, in: "Advances in
    Carbohydrate Chemistry and Biochemistry," S. Tipson and
    D. Horton ed., 38:13, Academic Press, New York (1981).
43. P. A. J. Gorin, Assignment of signals of carbon-13
    magnetic resonance spectrum of a selected polysaccharide:
    comments on methodlogy, Carbohydr. Res., 39:3 (1975).
44. M. Huppert, Antigens used for measuring immunological
    reactivity in fungi pathogenic from humans and animals;
    Part B. Pathogenicity and detection, D. H. Howard ed.,
    219 Deeker, New York (1983).

CYTOCHEMICAL TECHNIQUES AVAILABLE FOR THE STUDY OF FUNGAL POLYSACCHARIDES

Joseph Schrevel

Laboratoire de Biologie Cellulaire
U.A. CNRS N° 290
Université de Poitiers
86022 Poitiers Cédex, France

The cell wall of fungi consists of homopolysaccharides of glucose (Glc), mannose (Man), N-acetylglucosamine (GlcNAc). The major classes of polysaccharides are: (i) glucans, polymers of glucose containing β-(1 → 3) linkages and some β-(1 → 6)-, or β-(1 → 4)-, or α-(1 → 3) bound units (Rosenberger, 1976); (ii) mannoproteins, mannose polymers covalently linked to peptides (important in yeast, see Ballou, 1976, Cabib et al., 1982); (iii) chitin or chitosan, linear polymers of β-(1 → 4)-GlcNAc and β-(1 → 4)-glucosamine respectively; (iv) galactosamine polymers and polyuronides.

The ultrastructure of most yeast cell walls appears usually with an outer electron dense layer and an inner electron transparent layer in ultrathin-sections of glutaraldehyde/osmium tetroxide fixed material, however five layers were identified in the cell wall of mature blastospores of Candida albicans (Djaczenko and Cassone, 1971) and four in Saccharomyces cerevisiae (Cassone, 1973) upon addition of acrolein and/or Tris (1-aziridinyl) phosphine oxide (TAPO) in fixatives. More intriguingly was the description of eight layers in Candida albicans cell wall using a cytochemical method (Poulain et al., 1978). Such a "layering" structure of the wall is not supported by freeze fracture data where an homogenous fine granular structure is observed with hexagonal arrangements of particles at the plasma membrane level (Moor and Muhlethaler, 1963). The presence of a fibrillar component was demonstrated in the inner area of yeast cell wall by negative staining after specific extraction of the amorphous layer by endo-β-(1 → 3) and exo-β-(1 → 3)-glucanase from Bacillus circulans (Kopecka et al., 1974). The use of lectins in cytochemical studies begins to give accurate results on the carbohydrate distribution in fungal cell walls during the cell cycle (Horisberger and Vonlanthem, 1977; Horisberger et al., 1978; and Tronchin et al., 1981a, 1981b) but also during the cell wall biogenesis (Cabib et al., 1983). The localization of lectin binding sites requires specific markers, both for light microscopy (fluorescent probes, enzymatic markers) or electron microscopy (enzymatic or electron-dense markers). The use of such markers requires the control of technical parameters (sizes of the markers, pre-or postembedding incubation, one step or two-step procedure ...) also encountered in immunocytochemistry. These technical aspects provide the basis for the specificity and the reliability of cytochemical methods and underly the importance of suitable tools in the field of affinity cytochemistry. In this paper, we review cytological methods able to give qualitative and/or quantitative information on carbohydrate components, sugar binding proteins, enzymes involved in the biosynthesis of the cell wall polysaccharides of fungi through its cell cycle; these components could represent of course, potential antigens.

GENERAL PROCEDURES FOR CELL GLYCOCONJUGATE CYTOCHEMISTRY

The major cytochemical methods for cell glycoconjugates use different characteristics and properties of glycans: (i) presence of anionic sites; (ii) oxidation of alcohol to aldehyde groups: (iii) specific binding to lectins (see Schrevel et al., 1981). The antigenicity of some carbohydrate structures will be examined below.

Among methods using cationic substances to detect anionic sites (which are not only restricted to glycoconjugates), the cationized ferritin (Danon et al., 1972) and its fluoresceinylated derivative (King and Preston, 1977) are very useful since they can be used at physiological pH. The application of fluoresceinylated cationized ferritin to Candida albicans showed a bright fluorescence of the germ tube wall in comparison to that of the blastospore cell wall (Tronchin, 1983 and Fig. 1). This labeling could correspond to the phosphodiester bonds of the mannoproteins since it was shown that another cationic dye, Alcian blue, bound to the phosphomannans of the yeast cell wall depending upon the presence of phosphate (Ballou, 1974 and Ballou and Rascke, 1974).

Periodate oxidation is known to cleave the bond between two adjacent hydroxyl groups, but does not oxidize substituted hydroxyl groups: hexapyranose substituted in the C3 hydroxyl group or in both the C2 and the C4 hydroxyl groups, hexosamine substituted on either the 3 and the C4 hydroxyl (see Schrevel et al., 1981). This property is the basis of P.A.S. method. The staining of cell wall of Candida blastospore (see Tronchin et al., 1981a, Fig. 20, Tronchin 1983) and of Conidiobolus obscurus azygospore (Latge et al., 1986) with the periodic acidthiocarbohydrazide-silver proteinate (PA-TCH-Ag or P.A.T.Ag method, Thiery, 1967) an adaptation of the P.A.S. method for electron microscopy, showed that inner parts of the cell wall exhibit little or no reaction; these areas will correspond to β-glucans: the β-(1 → 3) linkages cannot be hydrolyzed by periodic oxidation and the P.A.T.Ag did not give any staining as observed in other β-(1 → 3) linkages or storage or structural polysaccharides (Schrevel, 1970 and Rougier et al., 1973). In contrast, the mannoproteins with their numerous α-(1 → 6) linkages are good candidates for a strong positive staining, as it is observed at the periphery of the cell wall.

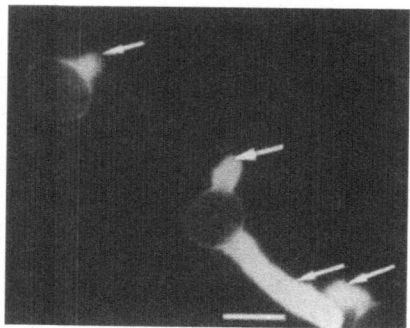

Fig. 1. Candida albicans cell wall labeled with fluoresceinylated cationized ferritin. A bright fluorescence is observed on the germ tube wall (arrows) (G. Tronchin, 1983). Bar represents 5μm.

Lectins, sugar binding proteins (or glycoproteins) of non-immune origin, are classified into a small number of specificity groups according to the most effective monosaccharide inhibiting cell agglutination and/or glycoconjugate precipitation by the lectin (Lis and Sharon, 1986). By using lectins which are specifically inhibited by mannosides (Concanavalin A, Con A), GlcNAc (Wheat Germ Agglutinin, WGA), Gal (Ricinus communis Agglutinin, RcA I; Bandeirea simplicifolia Agglutinin, BsA), the localization of mannoproteins and chitin was clearly defined. Agglutination tests with Con A, cytochemical methods with Con A in a two-step procedure (Con A and HRP; Con A and mannosyl ferritin), demonstrated that mannans are localized in the outer cell wall layer and near the plasma membrane (Cassone et al., 1978 and Tronchin et al., 1979). The contribution of lectin cytochemistry to the localization of chitin is more obvious. This minor component of the cell wall (1%), is the major component of fungal septa (Cabib and Bowers, 1971 and Cabib, 1975). However, its visualization by WGA complexed to colloidal gold (Horisberger and Vanlanthen, 1977; Molano et al., 1980; Roberts et al., 1983; and Hilenski et al., 1986) or with chitobiosyl ferritin (Tronchin et al., 1981b) showed that not only the bud scars but also the periplasmic area and the inner cell wall layer all over the cell were labeled. In contrast to mannoproteins and chitin, the β-glucans did not possess a known specific lectin. Another important result coming from glycoconjugate cytochemistry is the presence of a glycoprotein coat surrounding the entire yeast cell wall (Tronchin et al., 1981a). By using the Con A-HRP indirect method, Tronchin et al. (1981a) visualized a "mucous coat" which could be removed by the proteolytic and reducing treatments described by Cassone et al. (1978). Such a situation offers new approaches for the preparation fo fungal cell surface antigens.

GLYCOSYLATED AFFINITY MARKERS

Much evidence suggests that microorganisms specifically interact with their host cells by a sugar-dependent mechanism through glycoproteins, glycolipids and membrane lectins. Such a situation is usual in enveloped viruses and bacteria. The Sendai virus binds the glycolipids GD la, GT la, GQ lb (Markwell et al., 1981) and, in addition NeuNac α3Gal β3 GalNAc moities of glycophorin (Paulson et al., 1979). Many strains of Escherichia coli contain mannose specific lectin responsible for their binding to cells of the urinary tract (Eshdat and Sharon, 1982; Lis and Sharon, 1986). The adherence of Candida albicans to human epithelial cells is the first step of the host colonization. The cell binding could be modulated by sugars, specially mannose (Douglas et al., 1981; Sandin et al., 1982). Glycosylated affinity markers are useful tools for the visualization of lectin binding sites and for the study of membrane or endogenous lectins.

Some markers are known to possess a natural affinity for specific groups of lectins: horseradish peroxidase (HRP) hemocyanin react with Con A or LcA, glucose oxidase with WGA (see Scghevel et al., 1981). Other markers could be prepared with an induced affinity, for instance cytochemical markers were substituted by specific sugars by coupling the diazonium salts of p-aminophenylglycosides to HRP or to ferritin providing the so-called glycosylated cytochemical markers of G.C.M. (Kieda et al., 1977). The extension of affinity cytochemistry concept allowed to generalize the G.C.M. to neoglycoproteins which are not restricted to protein cytochemical marker (Monsigny et al., 1983). The neoglycoproteins are prepared from p-nitrophenyl glycosides and the coupling occurred between the phenyliso-thiocyanate glycoside and the amino group of bovine serum albumin; they are labeled with fluorescent probes, gold granules. By this procedure, neoglycoproteins usually contain about 25 sugar units per molecule and 3 residues of fluorescein per molecule (Monsigny et al., 1984).

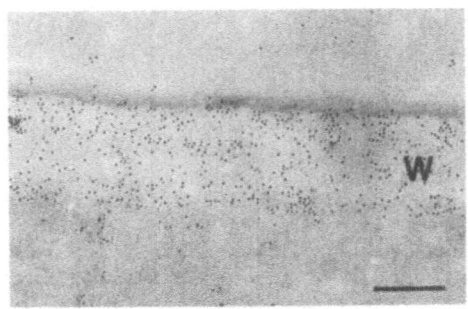

Fig. 2. Labeling of the inner wall of <u>Conidiobolus</u> <u>obscurus</u>
spore on Lowicryl sections with the sequential method
WGA and chitobiosyl ferritin (from Latge et al., 1986).
Bar represents 0.2μm.

The glycosylated cytochemical markers were used to visualize different
lectin-binding sites of various eucaryotic cells (Schrevel et al., 1979)
but also to label Con A and WGA binding sites of fungal cell wall (Tronchin,
1981a and b, Latge et al., 1986 and Fig. 2) The choice of the markers
usually between ferritin (MW 750 to 900,000) and horseradish peroxidase
(MW 40,000) is important since the accessibility of the lectin binding
sites could differ according to G.C.M., in term of steric hindrance.

Membrane or endogenous lectins could be visualized by G.C.M. (Kieda
et al., 1979; Hubert et al., 1985), and more easily with neoglycoproteins
(Ross et al., 1983; Schrevel et al., 1986). The purpose of using neoglycoproteins is to compensate the usually low affinity of a simple sugar by an
avidity effect (Townsend and Stahl, 1981; Monsigny et al., 1983). For
instance, fluorescent neoglycoproteins, label the cell surface of very
small cells such as merozoites from <u>Plasmodium</u>, the causative agent of
malaria, which measure 1.5 μm length (Fig. 3), or organelles such as
nucleus (Seve et al., 1986). By quantitative analysis with flow cytofluorometry, it was demonstrated that lectin-like component of <u>Plasmodium</u>
merozoite primarily binds neoglycoprotein-borne GlcNAc and secondarily
neoglycoprotein-borne α-D-Glc, β-D-Gal and L-Fuc (Schrevel et al., 1986).
The significance of this usual situation in identifying sugar-binding
components by means of neoglygoproteins is that flow cytofluorometry measures
the number of neoglycoprotein molecules bound to cellular components
rather than the actual number of sugar-binding sites for several reasons: i) various
neoglycoproteins can bind different subsites of one type of sugar binding
protein or different types of sugar binding proteins; ii) number of neoglycoproteins bound to their receptors depends on the accessibility of the
receptor, on the number of sugar residues per molecule of carrier protein,
and on the physiological state of the cell.

ENZYME CYTOCHEMISTRY AND GLYCOCONJUGATES

Cell wall glycoconjugates and their metabolism could be partly studied
by enzyme cytochemistry and specific inhibitors. Enzymatic extractions of
fungal cell walls have been reported (Kopecka et al., 1974; Chattaway et al.,
1976; Cassone et al., 1978; Evron and Drewe, 1984). The interpretation
of such extractions for the establishment of cell wall architecture models
have to be taken with care since the polysaccharide scaffolding could be
destroyed by the embedding procedure. The localization of enzymes involved
in cell wall biosynthesis required integrated biochemical-ultrastructural
investigations.

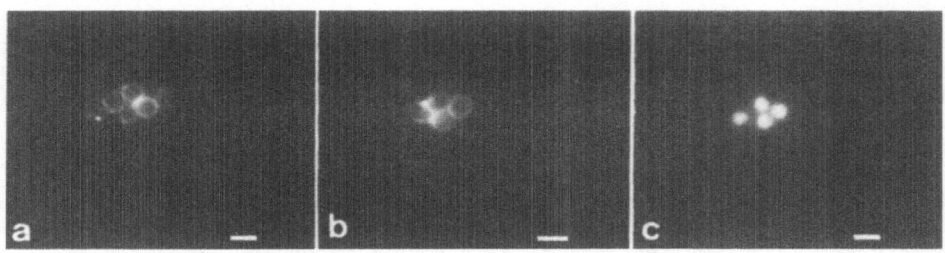

Fig. 3. Free <u>Plasmodium</u> merozoites labeled with
fluoresceinylated GlcNAc-BSA (a and b).
Merozoites as in Fig. b counterstained with
the Hoechst Dye 33258 for DNA visualization
(Fig. c) (from B. El Moudni and M. Philippe,
unpublished data). Bars represent 2 μm.

However, new trends in cytochemistry appear in this complex field.
A new affinity method for the ultrastructural localization of macromolecules
has been developed on the basis of enzyme-substrate interactions, by using
enzyme-gold complexes on tissue thin sections: specific enzymes allow the
direct ultrastructural localization of their corresponding substrates
(Bendayan, 1981, 1984, and 1985). The procedure was able to localize
xylans (Vian et al., 1983), hemicellulose (Ruel and Joseleau, 1984) in
plant cell walls, but also polygalacturonic acids in the cell walls of the
fungus <u>Ascocalyx abietina</u> (Benhamou and Ouellette, 1986). This direct
post-embedding technique resolved some problems of i) accessibility due to
restricted diffusion of marker complexes as in preembedding techniques,
ii) permeabilization of membranes or cell wall components. Efficience of
such a technique in yeast cell wall analysis, for instance by using different
β-glucanases, mannosidases, requires that the substrates are not altered
during the tissue processing.

Fungal cell wall biosynthesis takes place through several pathways:
i) $\beta-(1 \rightarrow 3)$-glucans are synthesized by $\beta-(1 \rightarrow 3)$-glucan synthetase
associated with the plasma membrane (Shematek et al., 1980); ii) chitin
synthetase is also associated with plasma membrane (Duran et al., 1975)
and uniformly distributed all over the cell in its zymogen (latent) form
(Duran et al., 1979); in contrast to glucan synthesis, chitin synthesis
depends upon a proteolytic activation in the primary septum area; iii)
mannoproteins are synthetized similarly to mammalian glycoproteins with
mannosyltransferases in internal and in plasma membranes (Cortat et al.,
1974; Walten-Verstegen et al., 1980) through the usual pathway from endo-
plasmic reticulum to secretory vesicles that fused with plasma membrane
(Novick et al., 1980). In <u>Candida albicans</u>, the distribution of chitin
synthetase in blastospores was similar to that of <u>S</u>. <u>cerevisiae</u>, but a
significant increase of activity was observed in the hyphal forms (Braun
and Calderone, 1978). The main contribution of cytochemical methods was
the demonstration that a new pathway of maturation of glycoconjugates
take place in the plasma membrane without Golgi vectorial process (Cabib
et al., 1983). By labeling the outer surface of <u>Saccharomyces cerevisiae</u>
protoplast membrane with Con A-ferritin and the chitin synthesized from
UDP-GlcNAc by WGA-granules, Cabib et al. (1983) showed that the yeast chitin
synthetase receives GlcNAc residues at the cytoplasmic face of the membrane
and transfers them vectorially to a growing chain of chitin concomitantly
extruded from the protoplast.

Several drugs or antibiotics have been reported to block the polysaccharide cell wall biosynthesis. For example, Polyoxins, antibiotics

from Streptomyces cacaoi (Isono et al., 1967) and especially Polyoxin D, act as powerful inhibitors of chitin synthetase (Hori et al., 1971; Keller and Cabib, 1971); Papulacandrin B and aculeacin A inhibit glucan synthesis (Baguley et al., 1979) and tunicamycin inhibits the synthesis of glyco-proteins in yeast by blocking the transfer of UDP-GlcNAc to dolichol-phosphate (Kuo and Lampin, 1974). By labeling Candida utilis with WGA-gold complexes, Hilenski et al. (1986) demonstrated that polyoxin D did not inhibit the budding in yeast and the lateral growth in mycelium. Primary septa failed to form, no labeling was detected in the cell wall, and the chains were formed by swollen, osmotically sensitive cells. The results were in agreement with the important functon of chitin for the formation of primary septa and for maintenance of structural integrity during morpho-genesis.

ANTI-CARBOHYDRATE ANTIBODIES

The immunodominant structures of yeast are the mannoproteins from the cell wall (Suzuki et al., 1968; Ballou, 1970; Raschke and Ballou, 1971 and 1972). In contrast, yeast β glucans are not antigenic when injected isolated or as cell wall components (Ballou, 1976; Horisberger and Rouvet-Vauthey, 1985; Horisberger et al., 1985). As the rabbit antiserum raised against whole yeast cell wall agglutinates the cells, the identification of immuno-dominant structures in mannans was performed by the inhibitory effects of isolated mannans or mannan acetolysis fragments. The three major haptenic groups identified were: i) mannotetraose unit; ii) GlcNAc-mannotetraose; iii) α-D-Man-phosphorylmannotriose unit (see Ballou et al., 1974). Some antigenic characteristics could be pointed out: i) α -(1 → 3) linked unit is more antigenic than a terminal α -(1 → 2) linked unit (Suzuki et al., 1968; Ballou, 1974); ii) mannosylphosphate group is an important yeast antigen (Raschke and Ballou, 1971). As expected, mannans were localized by homologous anti-mannan antibodies gold complexes at the periphery of the cell wall and near the plasma membrane in Saccharomyces cerevisiae and Candida albicans (Horisberger et al., 1977).

In order to resolve the lack of β-glucans antigenicity, Horisberger and Rouvet-Vauthey (1985) prepared an artificial antigen containing laminariose (3-0-β-D-Glc-D-Glc) as hapten conjugated to a protein carrier (edenstin), and the specificity of the rabbit anti-laminariose antibodies was tested by inhibition of the binding with different sugar derivatives. By comparing the distribution of β-glucans with these antibodies and of α-galacto-mannan by RcA-gold complexes during the fission process of Schizosaccharomyces pombe, it was shown that α-galactomannan was associated with the secondary septa, and β-glucan with the primary septum; β-glucan was detected only on walls generated by fission, and this localization was opposed to that of α-galactomannan. The results suggested that β-glucan is directly implicated in the fission process (Horisberger and Rouvet-Vauthey, 1985). These anti-laminariose antibodies labeled uniformly the inner wall of the ballistospore of Conidiobolus obscurus (Latgé et al., 1986, Fig. 4a)

Some monoclonal antibodies raised against Candida albicans mannoproteins showed a specific labeling of the cell wall layer and also an area near the plasma membrane (Latgé, personal communication, Fig. 4b). The fast develop-ment of methodology using monoclonal antibodies will dramatically change our knowledge of carbohydrate antigens related to the pathogenicity of many fungi as it does in human cancer. Aberrant glycosylation in cancer cell membranes had raised the importance of carbohydrates in tumor-associated antigens for the diagnosis of cancer and precancerous states as well as for therapeutic applications (see Hakomori, 1985; Feizi, 1985). As many tumor associated antigens were carbohydrate structures, they were extensively studies through glycolipid epitopes (Magnani et al., 1982; Hakomori et al.,

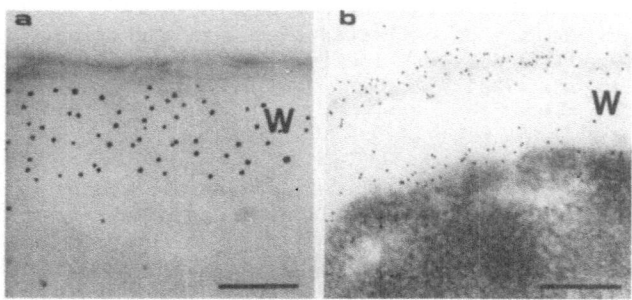

Fig. 4. Labeling of the cell wall of fungi with specific antibodies: a) Immuno-gold labeling of Conidiobolus obscurus spore with antilaminariose antibodies (from Latge et al., 1986). b) Candida albicans cell wall labeled with a monoclonal antibody raised against mannans and visualized by Protein A-gold complexes (from Latge et al., unpublished data). Bars represents 0.2 µm.

1984), since glycolipids can be purified to homogeneity, ther structures can be elucidated, and purified glycolipid retains antigenicity. The same carbohydrate epitopes could be also found in glycoproteins, but oligosaccharide chains in glycoproteins are heterogeneous and difficult to isolate to homogeneity (Hakomori, 1985). It can be expected that monoclonal antibodies raised against fungi cell walls will offer a variety of carbohydrate epitopes, which could be used to select specific strains in different pathogenic species for immunodiagnosis, to target specific antibiotics without damaging the host cells or to visualize specific modifications during the cell cycle of the fungi. For example, a specific monoclonal antibody was used in immunofluroescence to visualize specific ganglioside changes during the chicken neuronal differentiation and synapse formation (Grunwald et al., 1985).

TECHNICAL PARAMETERS IN CYTOCHEMISTRY

Many technical parameters affect the cellular localization of antigens and carbohydrate components.

1. Fixation procedures

Detection of antigens, lectin-binding sites or endogenous lectins requires a good ultrastructural preservation and a minimal loss of binding properties between the ligand and the receptor. Usually, aldehyde fixations do not induce drastic changes in the binding of lectins to cell surfaces, however, comparison between the labeling of unfixed and fixed cells with lectins showed high differences in the distribution of the lectin binding sites on the cell surface (see Schrevel et al., 1981). In immunocytochemistry the fixative parameter is very important since the accessibility to intracellular antigens requires prior permeabilization of the membrane and a good preservation of the native location of each antigen. Immunofluorescence screening of monoclonal antibodies required a panel of fixation procedures, since if only one fixation procedure is used, one can miss very useful antibodies. From detailed measurements on antibody binding to mononuclear phagocyte cell-surface antigens, Leenen et al. (1985) recommended fixation with paraformaldehyde or low concentration of glutaraldehyde for qualitative analysis. In contrast, for quantitative measurements, the fixative effects should be determined for each antigen.

It could be noticed that strong oxidizing agents were reported to restore protein antigenicity on osmicated tissue thin sections, and the antigens could be visualized by the indirect Protein A-gold techinque (Bendayan and Zollinger, 1983).

2. Labeling procedures

The size of markers represents a limiting factor. In preembedding incubation with lectins or antibodies, electron dense markers with a diameter greater than 10 nm (ferritin 12 nm, 10-15 nm for gold granules) are expected to reach cell surface binding sites where the ligand-receptor network is loose enough to allow the diffusion of the markers (Schrevel et al., 1979). In post-embedding incubation, the accessibility of binding sites is dependent on the resin characteristics. In contrast, enzymatic markers with their relative low molecular weight (HRP, 40000; microperoxidase, 1900) usually reach any cell binding sites but give a final, diffuse, electron-dense product preventing qualitative studies.

The incubation with both lectins or antibodies, by using the indirect or two-step procedure, usually gives better labeling than the direct or one-step procedure (see Schrevel et al., 1981; Horisberger, 1985). An enhancement of the protein A-gold immunocytochemical technique was described by introducing two additional steps using an antibody directed against protein A and then, again protein A-gold (Bendayan and Duhr, 1986).

SUGAR SPECIFICITY OF THE GLYCOSYLATED AFFINITY MARKERS AND LECTINS

The phenylthiocarbamyl spacer in neoglycoproteins, may increase the affinity of the sugar as it does with many lectins (Goldstein and Haynes, 1978) but does not hide the sugar specificity as shown by the inefficient labeling with α-Man-bovine serum albumin (Man-BSA) and the efficient labeling of P. chabaudi merozoites by α-Glc-BSA; these two neoglycoproteins only differ by the position of hydroxyl group on the carbon 2 on these sugars (Schrevel et al., 1986). A similar result was observed by the selective binding of Glc-BSA to sugar binding sites in cell nuclei and absence of binding with Man-BSA (Seve et al., 1986).

The nature of the simple sugar hapten used to study the binding of a lectin to its natural ligand does not imply that the ligand contains the sugar. As shown by Monsigny et al. (1980) the WGA binds cell surface N-acetylneuraminic acid and this binding is well inhibited by GlcNAc. Human suppressor T cells selectively bind Rhamnose-BSA (Kieda et al., 1982): no mammalian glycoconjugate is known to contain α-L-Rhamnose, however, the structure of the α-L-Rha is related to that of β-D-Galactose (Monsigny et al., 1983). The binding site of the great majority of known viral, bacterial, plant and animal lectins accomodates a large oligosaccharide (Ashwell and Harford, 1982; Monsigny et al., 1983; Lis and Sharon, 1986). Consequently, a simple sugar borne by a neoglycoprotein is supposed to bind to a limited part of large binding site.

In conclusion, the fungal cell wall architecture is yet poorly understood because: i) the bulk of the components appears to be structurally amorphous; ii) the ratio glucans/mannans is highly variable in fungi; iii) the different "layers" in the cell wall observed in transmission electron microscopy (Djaczenko and Cassone, 1971; Poulain et al., 1978; Farkas, 1979) could reflect modifications due to fixatives or reagents rather than a normal distribution of polysaccharides in a highly cohesive structure for the whole depth of the wall (Cabib et al., 1982; Horisberger and Rouvet-Vanthey, 1985). However, cytochemical methods can now give useful tools for the cellular localization of fungal cell wall polysaccharides: i) anionic sites associated with mannoproteins could be visualized by cationized ferritin; ii) periodate oxidation methods could discriminate β-(1 → 3) glucan area from mannoproteins; iii) lectins could specifically bind mannoproteins (Con A) and chitin (WGA) containing area and could also give information about chitin biosynthesis pathway. New powerful tools are available such as fluorescent or gold labeled neoglyco- proteins for the study of endogenous and/or membrane lectins, or monoclonal antibodies for the identification of strain specific carbohydrate structures. These methods will contribute to give rational fungal cell wall models and also data about the different antigenic fractions isolated by enzymatic or chemical procedures on fungal cells, cell walls or subractions of these structures.

ACKNOWLEDGMENTS

We thank Dr. G. Tronchin (University of Angers), Dr. J.P. Latge (Institut Pasteur, Paris), Dr. J.M. Gallo, Dr. M. Philippe, and B. El Moudni for the critical reading of the manuscript and their kind help in providing original photographs and Miss Decourt for her assistance in preparing the manuscript. Part of the work was supported by grants from INSERM 83-1018, Commission des Communautes Europeennes (TSD 147-F), the Foundation pur la Recherche Medicale and the Foundation Jean Langolis.

REFERENCES

1. G. Ashwell and J. Harford, 1982, Carbohydrate specific receptors of liver, Ann. Rev. Biochem., 51:531-554.
2. C.B. Baguley, G. Rommele, J. Gruner, and W. Wehrli, 1979, Papulacandin B: an inhibitor of glucan synthesis in yeast spheroplasts, Eur. J. Biochem., 97:345-351.
3. C.E. Ballou, 1970, A study of the immunochemistry of three yeast mannans, J. Biol. Chem., 245:1197-1203.
4. C.E. Ballou, 1974, Some aspects of the structure immunochemistry and genetic control of yeast mannans, Adv. Enzymol., 40:239-270.
5. C.E. Ballou, 1976, Structure and biosynthesis of the mannan component of the yeast cell envelope, Adv. Microbiol. Physiol., 14:93-158.
6. C.E. Ballou and W.C. Raschke, 1974, Polymorphism of the somatic antigen of yeast, Science (Wash. D.C.), 184:127-134.
7. M. Bendayan, 1981, Ultrastructural localization of nucleic acids by means of nuclease-gold complexes, J. Histochem. Cytochem., 29:531-541.
8. M. Bendayan, 1984, Enzyme-gold electron microscopic cytochemistry: a new affinity technique for the ultrastructural localization of macromolecules, J. Electron Microsc. Tech., 1:349-372.
9. M. Bendayan, 1985, The enzyme-gold technique: a new cytochemical approach for the ultrastructural localization of macromolecules. in: "Techniques in immunochemistry," G.R. Bullock and P. Petrusz, eds., Academic Press, 3:179-201.
10. M. Bendayan and M.A. Duhr, 1986, Modification of the protein A-gold immunocytochemical technique for the enhancement of its efficiency, J. Histochem. Cytochem., 34:569-575.
11. M. Bendayan and M. Zollinger, 1983, Ultrastructural localization of antigenic sites on osmium-fixed tissues applying the protein A- gold technique, J. Histochem. Cytochem., 31:101-109.
12. N. Benhamou and G.B. Ouellette, 1986, Use of pectinases complexed to colloidal gold for the ultrastructural localization of polygala- cturonic acids in the cell walls of the fungus Ascocalyx abietina, Histochem. J., 18-95-104.
13. P.C. Braun and R.A. Calderone, 1978, Chitin synthesis in Candida albicans comparison of yeast and hyphal forms, J. Bacteriol., 135:1472-1477.

14. E. Cabib, 1975, Molecular aspects of yeast morphogenesis, Ann. Rev. Microbiol., 29:191-214.

15. E. Cabib and B. Bowers, 1971, Chitin and yeast budding. Localization of chitin in yeast bud scars, J. Biol. Chem., 246:152-156.

16. E. Cabib, B. Bowers, and R.L. Roberts, 1983, Vectorial synthesis of a polysaccharide by isolated plasma membranes, Proc. Natl. Acad. Sci. (USA), 80:3318-3321.

17. E. Cabib, R. Roberts, and B. Bowers, 1982, Synthesis of the yeast cell wall and its regulation, Ann. Rev. Biochem., 51:763-793.

18. A. Cassone, 1973, Improved visualization of wall ultrastructure in Saccharomyces cerevisiae, Experientia, 29:1303-1305.

19. A. Cassone, E. Mattia, and L. Boldrini, 1978, Agglutination of blasto-spores of Candida albicans by Concanavalin A and its relationship with the distribution of mannan polymers and the ultrastructure of the cell wall, J. Gen. Microbiol., 105:263-273.

20. F.W. Chattaway, S. Shenolikar, J. O'Reilly, and A.J.E. Barlow, 1976, Changes in the cell surface of the dimorphic forms of Candida albicans by treatment with hydrolytic enzymes., J. Gen. Microbiol., 95:335-347.

21. M. Cortat, Ph. Matile, and F. Kopp, 1973, Intracellular localization of mannan synthetase activity in budding baker's yeast, Biochem. Biophys. Res. Commun., 53:482-489.

22. D. Danon, L. Goldstein, Y. Marikovsky, and E. Skutelsky, 1972, Use of cationized ferritin as a label of negative charges on cell surfaces, J. Ultrastruct. Res., 38:500-510.

23. W. Djaczenko and A. Cassone, 1971, Visualization of new ultrastructural components in the cell wall of Candida albicans with fixatives containing TAPO, J. Cell Biol., 52:186-190.

24. L.J. Douglas, J.G. Houston, and J. Mc Courtie, 1981, Adherence of Candida albicans to human buccal epithelial cells after growth on different carbon sources, FEMS Microbiol. Letters, 12:241-243.

25. A. Duran, B. Bowers, and E. Cabib, 1975, Chitin synthetase zymogen is attached to the yeast plasma membrane, Proc. Natl. Acac. Sci. (USA), 72:3952-3955.

26. A. Duran, E. Cabib, and B. Bowers, 1979, Chitin synthetase distribution on the yeast plasma membranes, Science (Wash.), 203:363-365.

27. Y. Eshdat and N. Sharon, 1982, Escherichia coli surface lectins. Methods Enzymol., 83:386-396.

28. R. Evron and J.A. Drewe, 1984, Demonstration of the polysaccharides in the cell wall of Candida albicans blastospores, using silver methena-mine staining and a sequence of extraction procedures, Mycopathologia, 84:141-149.

29. V. Farkas, 1979, Biosynthesis of cell walls of fungi, Microbiol. Rev., 43:117-144.

30. T. Feizi, 1985, Carbohydrate antigens in human cancer, Cancer Surveys, 4:245-269.

31. I.J. Goldstein and C.E. Haynes, 1978, The lectins: carbohydrate-binding proteins of plants and animals, Adv. carbohydrate Chem. Biochem., 35:128-140.

32. G.B. Grunwald, P. Fredman, J.L. Magnani, D. Trisler, V. Ginsburg, and H. Nirenberg, 1985, Monoclonal antibody 18B8 detects gangliosides associated with neuronal differentiation and synapse formation, Proc. Natl. Acad. Sci. (USA), 82:4008-4012.

33. S.I. Hakomori, 1985, Aberrant glycosylation in cancer cell membranes as focused on glycolipids: overview and perspectives. Cancer Res., 45:2405-2414.

34. S.I. Hakomori, E. Nudelman, S.B. Levery, and R. Kannagi, 1984, Novel fucolipids accumulating in human adenocarcinoma. I glycolipids with di- or trifucosylated type 2 chain, J. Biol. Chem., 259: 4672-4680.

35. L.L. Hilenski, F. Naider, and J.M. Becker, 1986, Polyoxin D inhibits colloidal gold-wheat germ agglutinin labelling of chitin in dimorphic forms of Candida albicans, J. Gen. Microbiol., 132:1441-1451.
36. M. Hori, K. Kakiki, S. Suzuki, and T. Hisato, 1971, Studies on the mode of action of polyoxins. Part III. Relation of polyoxin structure to chitin synthetase inhibition, Agricultural and Biological Chemistry, 35:1280-1291.
37. M. Horisberger, 1985, The gold method as applied to lectin cytochemistry in transmission and scanning electron microscopy, in: "Techniques in immunochemistry," G.R. Bullock and P. Petrusz, eds., Academic Press, 3:155-178.
38. M. Horisberger and M. Rouvet-Vauthey, 1985, Cell wall architecture of the fission yeast Schizosaccharomyces pombe, Experientia, 41:748-750.
39. H. Horisberger, M. Rouvet-Vauthey, V. Richli, and D.R. Farr, 1985, Cell wall architecture of the halophilic yeast Saccharomyces rouxii. An immunocytochemical study, Eur. J. Cell Biol., 37:70-77.
40. M. Horisberger and M. Vonlanthen, 1977, Location of mannan and chitin on thin sections of budding yeasts with gold markers, Arch. Microbiol., 115:1-7.
41. M. Horisberger, M. Vonlanthen, and J. Rosset, 1978, Localization of ⍺-galactomannan and of wheat germ agglutinin receptors in Schizosaccharomyces pombe, Arch. Microbiol., 119:107-111.
42. J. Hubert, A.P. Seve, D. Bouvier, C. Masson, M. Bouteille, and M. Monsigny, 1985, In situ ultrastructural localization of sugar-binding sites in lizard granulosa cell nuclei, Biol. Cell., 55:15-20.
43. K. Isono, J. Nagatsu, K. Kobinata, K. Sasaki, and S. Suzuki, 1967, Studies on polyoxins, antifungal antibiotics. Part V. isolation and characterization of polyoxins C, D, E, F, G, H, and I., Agricultural and Biological Chemistry, 31:190-199.
44. F.A. Keller and E. Cabib, 1971, Chitin and yeast budding. Properties of chitin synthetase from Saccharomyces carlsbergensis, J. Biol. Chem., 246:160-166.
45. C. Kieda, F. Delmotte, and M. Monsigny, 1987, Preparation and properties of glycosylated cytochemical markers, FEBS Letters, 76:257-261.
46. C. Kieda, M. Monsigny, and M.J. Waxdal, 1982, Endogeneous lectins on human peripheral mononuclear leukocytes, in: "Lectins, biology, biochemistry, clinical biochemistry, " Bog-Hansen and Springler, eds., de Gruyter, Berlin, 3:427-433.
47. C. Kieda, A.C. Roche, F. Delmotte, and M. Monsigny, 1979, Lymphocyte membrane lectins. Direct visualization by the use of fluroesceinyl glycosylated cytochemical markers, FEBS Letters. 99:329-332.
48. C.A. King and T.M. Preston, 1977, Fluoresceinylated-cationized ferritin as a membrane probe for anionic sites at the cell surface, FEBS Letters, 73:59-63.
49. M. Kopecka, H.J. Phaff, and G.H. Fleet, 1974, Demonstration of a fibrillar component in the cell wall of the yeast Saccharomyces cerevisiae and its chemical nature, J. Cell Biol., 62:66-76.
50. S.C. Kuo and J.O. Lampen, 1974, Tunicamycin an inhibitor of yeast glycoprotein synthesis, Biochem. Biophys. Res. Commun., 58:287-295.
51. J.P. Latge, G.T. Cole, M. Horisberger, and M.C. Prevost, 1986, Ultrastructure and chemical composition of the ballistospore wall of Conidiobolus obscurus, Exp. Mycol., 10:99-119.
52. P.J.M. Leenen, A.M.A.C. Jansen, and W. Van Ewijk, 1985, Fixation parameters for immunochemistry: the effect of glutaraldehyde or paraformaldehyde fixation on the preservation of mononuclear phagocyte differentiation antigens, in: "Techniques in immunochemistry," G.R. Bullock and P. Petrusz, eds., Academic Press, 3:1-24.
53. H. Lis and N. Sharon, 1986, Lectins as molecules and as tools, Ann. Rev. Biochem., 55:35-67.

54. J.L. Magnani, B. Nilsson, M. Brockhaus, D. Zopf, Z. Steplewski, H. Koprowski, and V. Ginsburg, 1982, A monoclonal antibody-defined antigen associated with gastrointestinal cancer is a ganglioside containing sialylated lacto-N-fucopentose II., J. Biol. Chem., 257:14365-14369.

55. M.A.K. Markwell, L. Svennerholm, and J.C. Paulson, 1981, Specific ganglioside function as host cell receptors Sendai virus. Proc. Natl. Acad. Sci. (USA) 78:5406-5410.

56. J. Molano, B. Bowers, and E. Cabib, 1980, Distribution of chitin in the yeast cell wall. An ultrastructural and chemical study. J. Cell Biol., 85:199-212.

57. M. Monsigny, C. Keida, and A.C. Roche, 1983, Membrane glycoproteins glycolipids and membrane lectins as recognition signals in normal and malignant cells, Biol. Cell., 47:95-110.

58. M. Monsigny, A.C. Roche, and P. Midoux, 1984, Uptake of neoglycoproteins via membrane lectin(s) of L1210 cells evidenced by quantitative flow cytoflurometry and drug targeting, Biol. Cell., 51:187-196.

59. M. Monsigny, A.C. Roche, C.Sene, R. Magnet-Dana, and F. Delmotte, 1980, Sugar-lectin interactions: how does wheat germ agglutinin bind sialoglycoconjugates? Eur. J. Biochem., 101:147-153.

60. H. Moor K. Muhlethaler, 1963, The fine structure in frozen-etched yeast cells, J. Cell Biol., 17:609-628.

61. P. Novick, C. Field, and R. Schekman, 1980, Identification of 23 complementation groups required for post-translational events in the yeast secretory pathway, Cell, 21:205-215.

62. J.C. Paulson, J.E. Sadler, and R.L. Hill, 1979, Restoration of specific myoxovirus receptors to asialoerythrocytes by incorporation of sialic with pure sialyltransferases, J. Biol. Chem., 254:2120-2124.

63. D. Poulain, G. Tronchin, J.F. Dubremetz, and J. Biguet, 1978, Ultrastructure of the cell wall of Candida albicans blastospores: a study of its constitutive layers by the use of a cytochemical technique revealing polysaccharides, Ann. Microbiol. (Inst. Pasteur). 129A:141-153.

64. W.C. Raschke and C.E. Ballou, 1971, Immunochemistry of the phosphomannan of the yeast Kloeckera brevis, Biochemistry, 10:4130-4135.

65. W.C. Raschke and C.E. Ballou, 1972, Characterization of a yeast mannan containing o-N-acetyl-D-glucosamine as an immunochemical determinant. Biochemistry, 11:3807-3816.

66. R.L. Roberts, B. Bowers, M.K. Slater, and E. Cabib, 1983, Chitin synthesis and localization in cell division cycle mutants of Saccharomyces cerevisiae, Mol. Cell. Biol., 3:922-930.

67. R.F. Rosenberger, 1976, The cell wall. In: "The filamentous fungi," J.E. Smith, D.R. Berry, and Arnold Edward, eds., 328-344.

68. P.H. Ross, U. Kolb-Bachofen, Schlepper-Schafer, M. Monsigny, R.J. Stockert, and H. Kolb, 1983, Two galactose specific receptors in the liver with different function. FEBS Letters, 157:253-256.

69. M. Rougier, B. Vian, D. Gallant, and J.C Roland, 1973, Aspects cytochimiques de l'etude ultrastructurale des polysaccharides vegetaux, Annee Biologique, 12:44-76.

70. K. Ruel and J.P. Joseleau, 1984, Use of enzyme-gold complexes for the ultrastructural localization of Hemicellulose in the plant cell wall. Histochemistry, 81:573-580.

71. R.L. Sandin, A.L. Rogers. R.J. Patterson, and E.S. Beneke, 1982, Evidence for mannose-mediated adherence of Candida albicans to human buccal cells in vitro. Infect. Immun., 35:79-85.

72. J. Schrevel, 1970, Etude ultrastructurale et cytochimique de certains polysaccharides de reserve chez les protistes, VII. Congr. Int. Microsc. Electr. (Grenoble), TIII:409-410.

73. J. Schrevel, D. Gros, and M. Monsigny, 1981, Cytochemistry of Cell glycoconjugates, Progr. Hostochem. Cytochem., 14(2):1-269.

74. J. Schrevel, C. Kieda, E. Caigneaux, D. Gros, F. Delmotte, M. Monsigny, 1979, Visualization of cell surface carbohydrates by a general two-step lectin technique: lectins and glycosylated cytochemical markers Biol. Cell., 36:259-266.
75. J. Schrevel, M. Philippe, F. Bernard, M. Monsigny, 1986, Surface Plasmodium sugar-binding components evidenced by fluorescent neo-glycoproteins, Biol. Cell., 56:49-55.
76. A.P. Seve, J. Hubert, D. Bouvier, C. Bourgeois, P. Midoux, A.C. Roche, and M. Monsigny, 1986, Analysis of sugar-binding sites in mammalian cell nuclei by quantitative flow microfluorometry, Proc. Natl. Acad. Sci, (USA), 83:5997-6001.
77. E.M. Shematek, J.A. Braatz, and E. Cabib, 1980, Biosynthesis of the yeast cell wall. I. Preparation and properties of β-(1 — 3) glucan synthetase, J. Biol. Chem., 255:888-894.
78. S. Suzuki, H. Sunayama, and T. Saito, 1968, Antigenic activity of yeasts. I. Analysis of the determinant groups of mannan of Saccharomyces cerevisiae, Jap. J. Microbiol., 12:19-24.
79. J.P. Thiery, 1967, Mise en evidence des polysaccharides sur coupes fines en microscopie electronique, J. Microsc. (Paris), 6:987-1018.
80. G. Tronchin, 1983, Ultrastructure et cytochimie de l'enveloppe cellulaire des levures (Candida albicans), These Universite Lille I, n° 580.
81. G. Tronchin, D. Poulain, and J. Biguet, 1979, Etudes cytochimiques et ultrastructurales de la paroi de Candida albicans. I. Localisation des mannanes par utilisation de Concanavaline A sur coupes ultrafines Arch. Microbiol., 123:245-249.
82. G. Tronchin, D. Poulain, J. Herbaut, and J. Biguet, 1981a, Cytochemical and ultrastructural studies of Candida albicans. II. Evidence for a cell wall coat using concanavalin A, J. Ultrastruct. Res., 75:50-59.
83. G. Tronchin, D. Poulain, J. Herbaut, and J. Biguet, 1981b, Localization of chitin in the cell wall of Candida albicans by means of wheat germ agglutinin. Fluorescence and ultrastructural studies. Eur. J. Cell Biol., 26:121-128.
84. R. Townsend and P.D. Stahl, 1981, Isolation and characterization of a mannose-N-acetylglucosamine/fucose binding protein from rat liver, Biochem. J., 194-209-214.
85. B. Vian, J.M. Brillouet, and B. Satiat-Jeunemaitre, 1983, Ultras-structural visualization of xylans in cell walls of hardwood by means of xylanase-gold complex. Biol. Cell., 49-179-182.
86. G.W. Walten-Verstegen, P. Boer, and E.P. Steyn-Parve, 1980, Lipid-mediated glycosylation of endogenous proteins in isolated plasma membrane of Saccharomyces cerevisiae, J. Bacteriol., 141:342-349.

IMMUNOCHEMISTRY OF FUNGAL ANTIGENS AT THE ELECTRON MICROSCOPY LEVEL

Johannes Müller and R. Kappe

Institute of Parasitology and Mycology
Center of Hygiene, University of Freiburg
Postfach 820, D-7800 Freiburg i. Br.
Federal Republic of Germany

INTRODUCTION

Immunoelectronmicroscopy has proven to be a useful tool for the study of fungal antigenicity at the cellular level. The following paper reviews the different applications of this method and sums up the results so far obtained in our laboratory.

The basic reagents for immunoelectronmicroscopic studies are antibodies of defined specificity labelled with ferritin, peroxidase, or gold. All these conjugates render electron dense complexes. In our experiments we used ferritin-labelled antibodies, since the preparation of stable monomeric antibody-ferritin-conjugates had been improved markedly by our Department of Immunology. Fig. 1 demonstrates the reaction with the direct immuno-ferritin test. Fig. 2 shows the scheme of the indirect immunoferritin test: The fungus is incubated first with an antibody of antifungal specificity and in a second step with a ferritin-labelled antibody directed against the immunoglobulin class and species of the first antibody. In addition anti-bodies which have fixed in vivo to the fungus cell surface can be demonstrated by the direct immunoferritin test using the last incubation step only. The wide variety of defined antibodies against fungal antigens as well as against immunoglobulins has allowed us to enlarge our knowledge of fungal antigens and the humoral immune response to them remarkably. For reviews on this subject see Müller[1, 2].

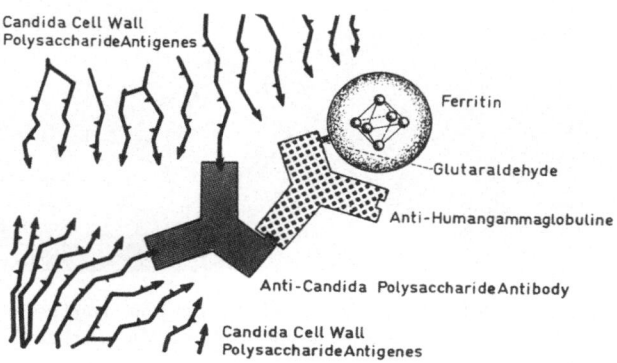

Fig. 1

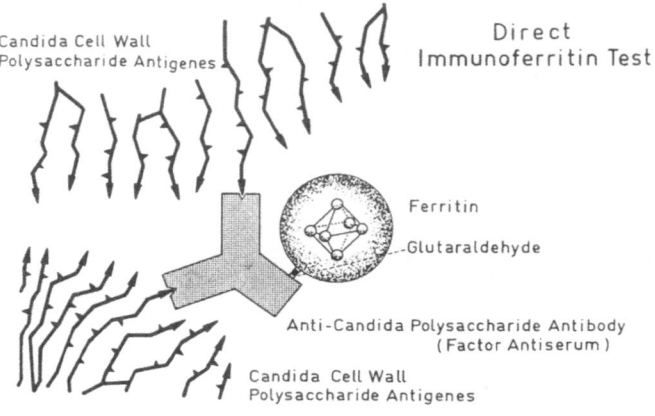

Direct
Immunoferrin Test

Candida Cell Wall
Polysaccharide Antigenes

Ferritin

Glutaraldehyde

Anti-Candida Polysaccharide Antibody
(Factor Antiserum)

Candida Cell Wall
Polysaccharide Antigenes

Fig. 2

DEMONSTRATION OF ANTIGENS IN VITRO

The incubation of cultured Candida albicans cells with a ferritin-labelled hyperimmune serum allows the description of the topography of the agglutination (Fig. 3): In this central site several thousand antibodies cause the agglutination of two cells. In a more tangential section the extreme stretching of the mannan antigens, normally folded down onto the cell wall surface, can be shown (Müller et al.[3], Tokunaga et al.[4], and Berdicevsky et al.[5]).

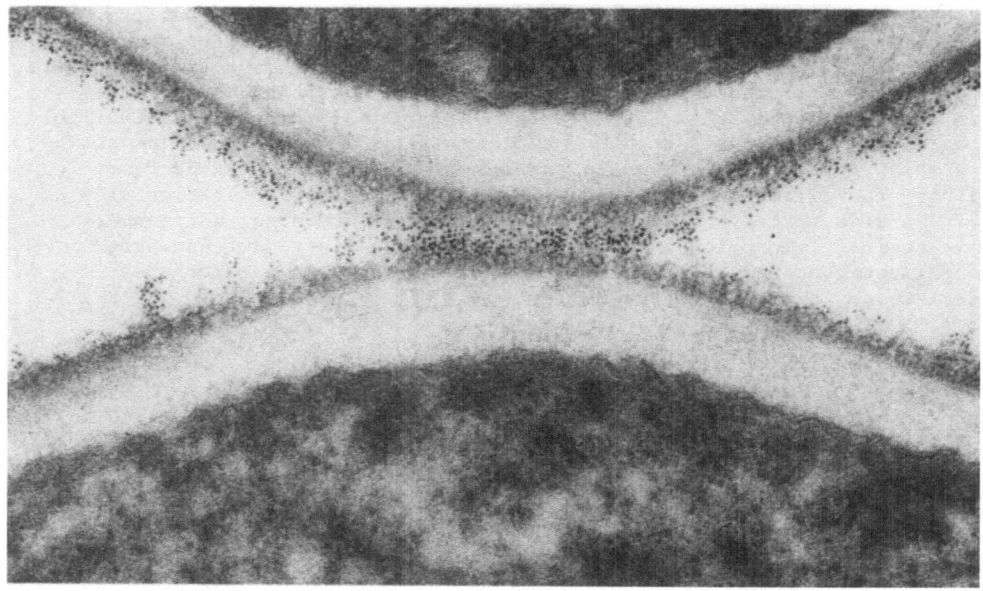

Fig. 3. Demonstration of antigens in vitro by the indirect immunoferritin test: Two cultured Candida albicans cells agglutinated by anti-Candida antibodies of human origin. 100 000 x (Müller et al.[3]).

The use of ferritin labelled monospecific antibodies allows one to study the density of endowment of the cell wall surface with the respective antigens, for example, the antigen 1 (Tsuchiya scheme) on a _Candida_ bud as well as on a _Candida_ filament, or the presence of antigen 6 on the surface of a _Candida_ bud, whereas this antigen is not present on the cell wall surface of _Candida krusei_ – in accordance with Tsuchiya's findings. This method allows the investigation of antigen endowment under a broad variety of environmental conditions in vitro. A more detailed evaluation of such studies leads to the interpretation that the antigens of a single specificity are fixed to the cell wall surface in a bunch-like arrangement.

DEMONSTRATION OF ANTIGENS IN VIVO

The formation of cell wall antigens in vivo differs considerably from the situation in vitro. The fungus cell exposed to the host environment reacts by building up antigen layers which are much thicker and denser than in cultured cells. This is most obvious with fungi in a chronic infection focus. The formation of thick electrontransparent layers can be observed with _Candida_ blastospores, with filaments, as well as with the budding sites of blastospores (Müller[1]). This seems to be a general phenomenon with fungi (Kuttin et al.[6], Drexler et al.[7], and Müller et al.[8]).

The application of labelled monospecific antibodies also permits study of the antigen layers in vivo; their composition might be different from the in vitro status. The most exciting development is in this field, the fact that one can trace the antigens liberated from the fungus cell into the environment and demonstrate their fungal origin, as is shown in Fig. 4 with vaginal scrapings. Several studies of this type have led us

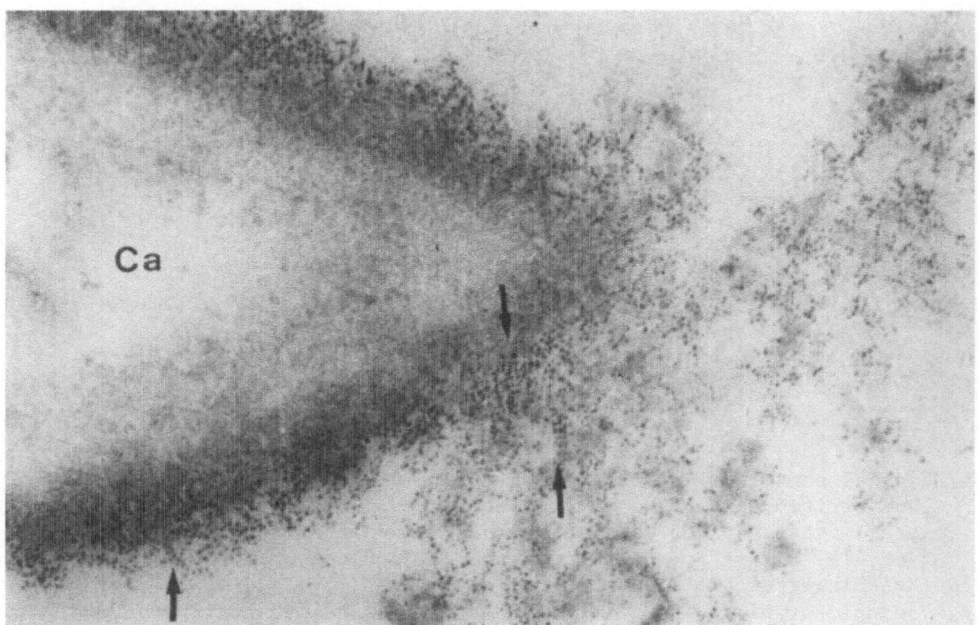

Fig. 4. Vaginal scrapings incubated with ferritin-labelled monospecific anti-_Candida_-1. Antigenic material is released from the exterior cell wall surface into the infection focus and can be identified by its ferritin-marking. The arrows point to mannan filaments. 95 000 x (Melchinger et al.[9]).

to perceive that the liberation of antigens into the infection focus is an important part of at least some sorts of fungal diseases (Melchinger et al.[9], Takamiya et al.[10], and Müller[11]).

REACTIONS OF ANTIGENS WITH ANTIBODIES IN VIVO

CANDIDA: It can be demonstrated by the direct immunoferritin test using ferritin-labelled anti-host-immunoglobulins, that antibodies of host origin contribute to the formation of the thick electron-dense layers surrounding the fungus cell in the infection situation. Mainly IgG and IgA are involved in this process. The ferritin labelling of the substance near the yeast cell proves the release of antigen-antibody complexes into the infection focus (Fig. 5) (Müller et al.[12, 13] and Melchinger et al.[9]). The antigen determinants located on the cell wall surface are never completely saturated with antibodies. Unoccupied antigen sites exist, and this might be due to the obviously rapidly running production of new antigen material deposited

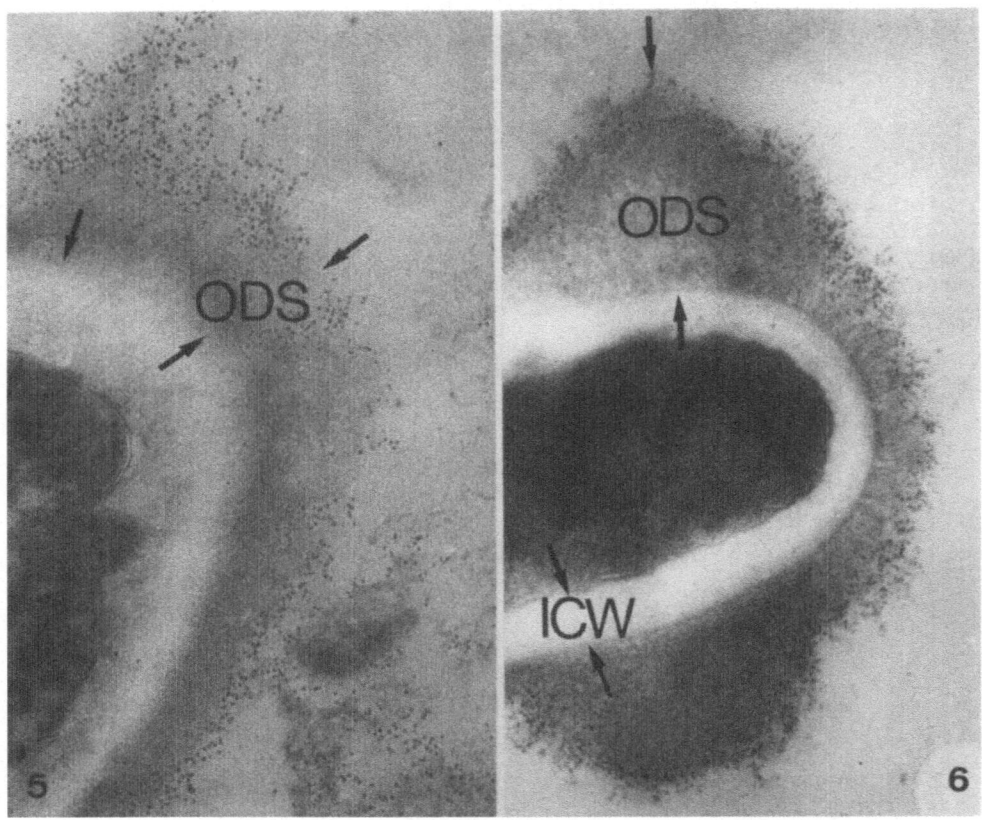

Fig. 5. Vaginal scrapings incubated with ferritin-labelled anti-human IgA. The ferritin-marking of material released from the fungus demonstrates that antigen-antibody complexes besides free antigens (see Fig. 4) are present in the infection focus. 75 000 x (Melchinger et al.[9]).

Fig. 6. Vaginal scrapings incubated with ferritin-labelled anti-human IgA. The huge exterior ODS layer consisting of cell wall mannan and host antibodies forms asteroid protrusions typical for the Splendore-Hoeppli-phenomenon. The filamentous structure of the mannan layer is clearly visible on the right. 10 000 x (Melchinger et al.[9]).

Figs. 5 and 6. ICW = inner electrontransparent cell wall.
             ODS = outer electrondense mannan layer.

on the cell wall surface. These observations allowed us to interpret the so-called "asteroid bodies" or the Splendore-Hoeppli-phenomena with Candida cells as precipitates consisting of antigen and antibody material on the cell wall surface (Fig. 6) (Müller et al.[13] and Melchinger et al.[9]).

The building up of thick antigen layers containing bound antibodies demonstrates the pathological significance of the fungus: Such layers can be seen with yeast cells in the guts of patients with Candida enteritis. Fungi colonizing the gut commensally do not bear an antigen-antibody pre-precipitate on their cell wall surface. This is likewise the case with nonpathogenic fungi: Saccharomyces cerevisiae, ingested in a high quantity by a volunteer, does not form an electron-dense layer of cell wall surface antigens in the gut (Fig. 7), as is the case in culture (Fig. 8). There-fore, the oral uptake of baker's yeast does not falsify the proof of Candida antigen in stool specimens, although both yeast species have some mannan antigens in common (Müller[14] and Kappe and Müller[15]).

CRYPTOCOCCUS: Cryptococcus neoformans possesses a capsule of glucuronoxylo-mannan antigens, the topography of which can be demonstrated by the incuba-tion with a labelled anti-Cryptococcus serum. The capsule of a Crypto-coccus cell which has spent two hours in a non-immunized rabbit shows a spongy texture of antigenic material. The anti-Cryptococcus-ferritin com-plexes offered can occupy all antigenic determinants, including those in the deeper layers of the capsule (Fig. 9) (Kuttin et al.[6, 16]).

The capsule of Cryptococcus cell which has spent two hours in a rabbit immunized against Cryptococcus neoformans reveals a completely different aspect: The capsule is larger in volume, and the capsular material is densely packed. The anti-Cryptococcus ferritin complexes can only react with superficially located antigen determinants (Fig. 10). The dense structure of the Cryptococcus capsules obtained from immunized rabbits is

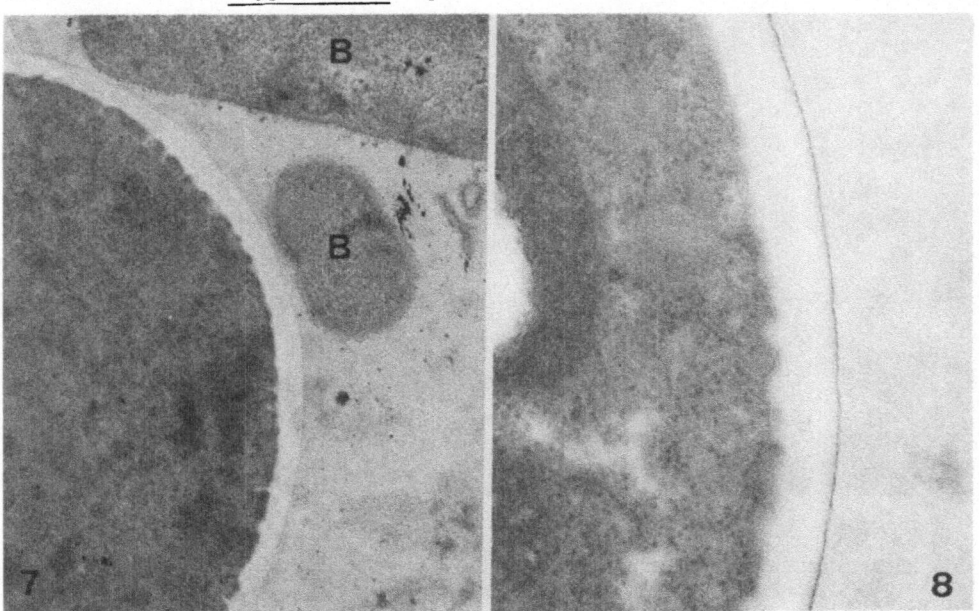

Fig. 7. Saccharomyces cerevisiae (SC) from the gut of a volun-teer who had ingested the yeast cells. The electron-transparent cell wall does not show an electrondense mannan layer on its surface. B = intestinal bacteria. 25000 x.

Fig. 8. Saccharomyces cerevisiae, culture cells for comparison. 50000 x. Figs. 7 and 8 (Kappe et al.[15]).

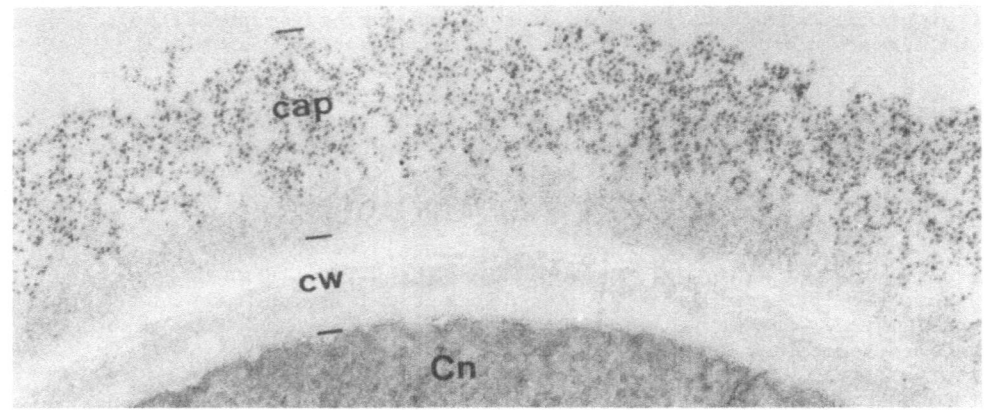

Fig. 9. Cryptococcus neoformans inoculated i.p. into a non-immunized rabbit, reisolated 2 h later and incubated with ferritin-labelled anti-Cryptococcus serum. The capsule is ferritin-labelled down to its deep layers. 75 000 x (Kuttin et al. [6,16]). Cn = Cryptococcus protoplast; cw = cell wall; cap = capsule.

still more evident after incubation with ferritin-conjugated anti-rabbit immunoglobulin (Fig. 11). These features seem to be dependent on host-specific factors: The same experiments with immunized mice render conditions where deeply located capsular antigens are accessible to the anti-Cryptococcus-ferritin complexes.

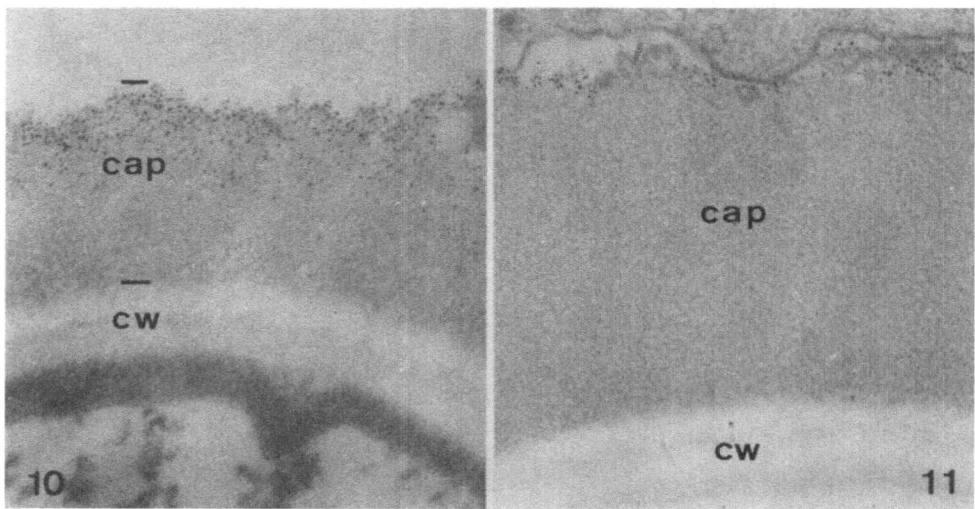

Figs. 10 and 11. Cryptococcus neoformans inoculated i.p. into a rabbit after immunization with ethanol-killed Cryptococcus neoformans cells and reisolated 2 h later.

Fig. 10. Cryptococcus neoformans incubated after reisolation with ferritin-labelled anti-Cryptococcus serum. The dense capsule can react only with its outer parts. 75 000 x.

Fig. 11. Cryptococcus neoformans after reisolation with ferritin-labelled anti-rabbit Ig. The host antibodies can react only on the surface of the capsule due to the dense texture of the capsule under the condition of the immunized host. 75 000 x.
Figs. 10 and 11 (Kuttin et al.[6,16]).

ASPERGILLUS: Asteroid bodies with Aspergillus are a well known phenomenon in light microscopic histopathology. The ultrastructural investigation demonstrates electron-dense extracellular layers of enormous dimension with a membraneous texture, clearly recognizable at higher magnification (Fig. 12, Drexler et al.[7]). It seems plausible that these layers consist of cell wall antigens, of liberated lysosomal enzymes as well as of membrane debris coming from lysed host cells, precipitated altogether on the fungal cell wall surface. It is doubtful that fungal antigen is liberated into the circulation in such a situation.

FONSECAEA: The dematious fungus possesses a cell wall appearing electron-dense in its entire cross-section. The dimension of the cell wall of cells in the infection situation is apparently considerably greater than that in cultured cells. Higher magnification shows a fragile texture of the cell wall with a constant release of cell wall antigens (Müller et al.[8]).

INVOLVEMENT OF COMPLEMENT FACTORS

The application of ferritin-labelled anti-complement factors reveals information on the participation of complement in the host-fungus-interaction. It could be demonstrated with Cryptococcus neoformans in rabbits using anti-rabbit-C_3 that is C_3 is present on the capsule surface (Kuttin et al.[6]). There is no evidence that complement might lead to lysis of the fungal cell as is the case with gramnegative bacteria. But the important question whether complement can be activated when complexed with fungal antigens and antibodies, remains still to be solved.

INVOLVEMENT OF LYSOZYME

Human lysozyme is fixed in low quantities to the Candida cell wall surface (Müller et al.[17]). Additional experimental application of lysozyme in the form of untreated saliva leads to a deformation of the cell wall

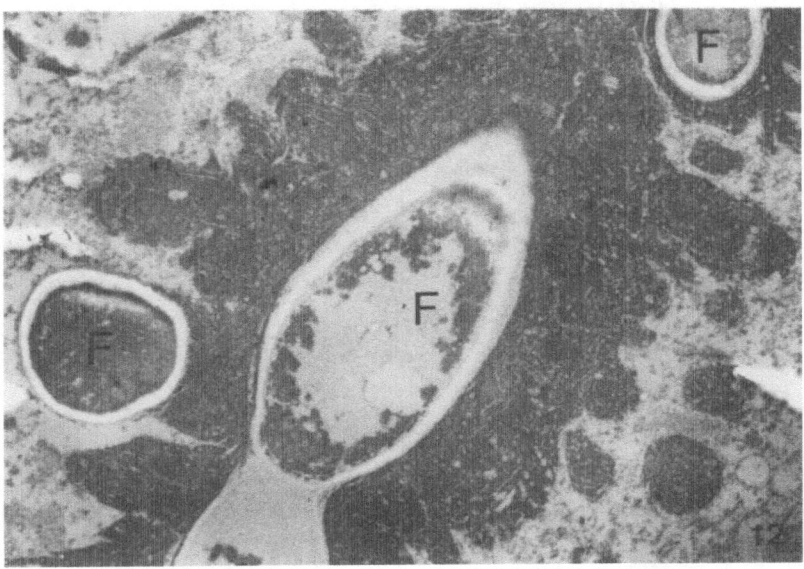

Fig. 12. Asteroid body with an Aspergillus sp. cell in the liver of a leukemia patient. The membranous ultrastructure of the electron-dense layer beyond the electrontransparent cell wall is markedly different from that with Candida asteroid bodies. 14 000 x (Drexler et al.[7]).

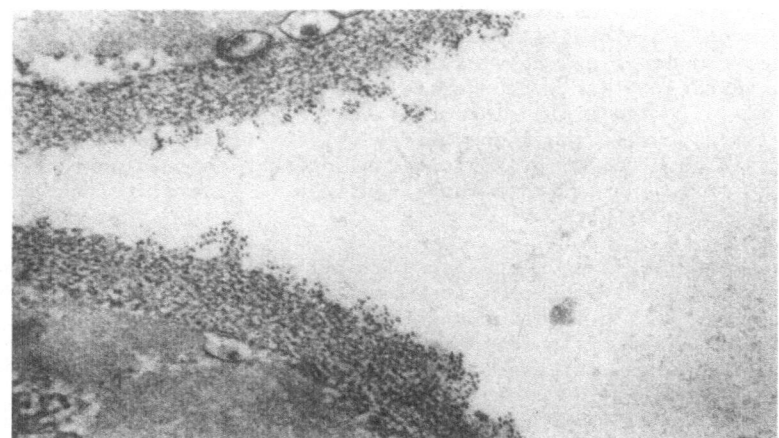

Fig. 13. Autoclaved <u>Candida</u> <u>albicans</u> cells incubated with ferritin-labelled anti-human Ig after pre-incubation with human anti-<u>Candida</u> serum. Note the severe destruction of the cytoplasm, the stratification of the cell wall and the ferritin-marking of the cell wall surface. 75 000 x (Thurner et al.[18], and Thurner[19]).

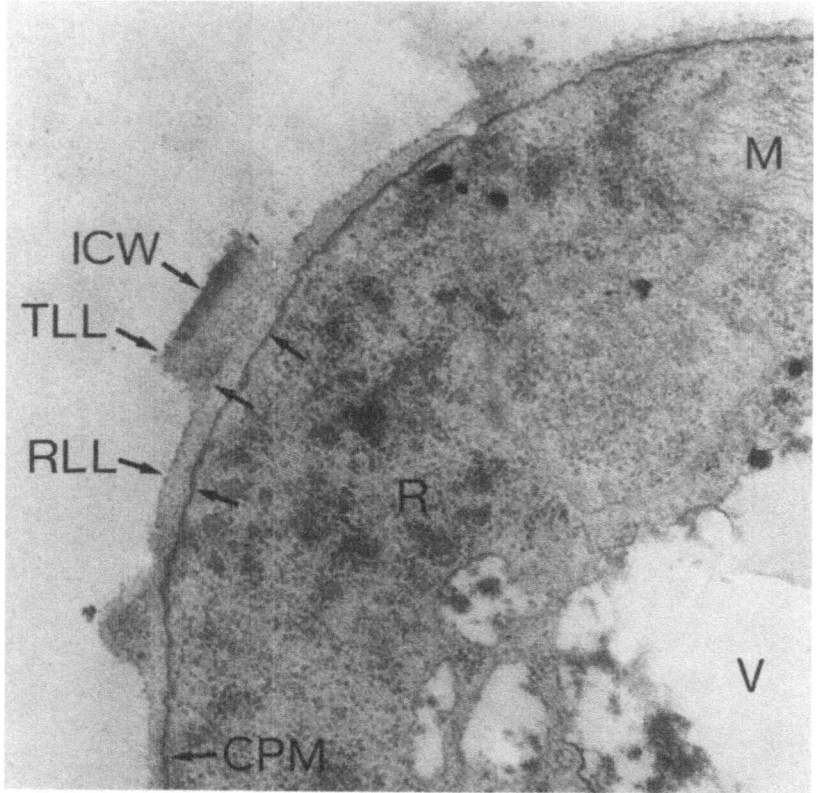

Fig. 14. <u>Candida</u> <u>albicans</u> under digestion by zymolyase. ICW = total cell wall. TLL = tangential lamellar layer which is partially digested. RLL = radial lamellar layer. R = ribosomes. M = mitochondrion. V = vacuole. CPM = cytoplasmic membrane. 75 000 x (Berdicevsky et al.[20]).

surface. Lysozyme fixation obviously favors the release of antigenic
material from the cell wall surface: lysozymal action on fungi does seem
to be restricted to this effect.

DEMONSTRATION OF ANTIGENS AFTER HEAT DEGRADATION

Autoclaving of Candida cells changes tremendously the ultra-structural
aspect of the cell wall which becomes more electron-dense and stratified.
Application of ferritin-labelled Anti-6 reveals that this antigen has disappeared completely from the Candida surface. In contrast, other surface
antigens are still present: incubation with labelled anti-human Ig after
pre-incubation with a human anti-Candida serum proves a remarkable part of
the original antigenicity remains after heating (Fig. 13, Thurner et al[18],
and Thurner[19])

DEMONSTRATION OF ANTIGENS AFTER ENZYMATIC DIGESTION

The Candida cell wall can be digested by zymolyase which is favored
by the presence of certain other enzymes. The outer cell wall layer TLL
(Fig. 14, Berdicevsky et al.[20]), which shows a tangential stratification,
can be digested more easily than an internal cell wall layer RLL with a
radial texture. These partially digested cells have been incubated with
an anti-Candida serum of human origin, which had been absorbed with Candida
cells: All antibodies against superficially located antigens have been
removed. The following incubation with labelled anti-human Ig demonstrates
that in deeper cell wall layers antigens must exist which are not present
on the cell wall surface (Thurner et al.[18], and Thurner[19]).

REFERENCES

1. J. Müller, Immunobiological aspects of Candida mycoses, a review of
   electronmicroscopic studies, Mykosen, Suppl. 1: 289-297 (1978).
2. J. Müller, Electronmicroscopy of Fungi and Fungous Diseases, In:
   Electronmicroscopy in Human Medicine, J.V. Johannessen, ed., McGraw-
   Hill Book Co., New York (1979).
3. J. Müller, H. Takamiya, A. Vogt, and R. Jaeger, Elektronenmikroskopische
   Darstellung von Immunreaktionen an Candida-Zellen: I. In vitro-Inkuba-
   tion von Candida albicans mit Anti-Candida-Humanserum, Mykosen
   19: 295-303 (1976).
4. J. Tokunaga, T. Fujita, A. Hattori, and J. Müller, Scanning electron
   microscopic observation of immunoreactions on the cell surface:
   Analysis of Candida albicans cell wall antigens by the immunoferritin
   method, Proc. 9. Ann. Sem Symp. 301-310, Chicago (1976).
5. I. Berdicevsky, H. Melchinger, and J. Müller, Immunoelectronmicroscopic
   investigations of the cell walls of Saccharomyces uvarum Beijerinck
   (S. carlsbergensis Hansen) and Candida albicans (Robin) Berkhout
   grown in calcium deficient media, Mykosen 28: 77-84 (1985).

6. E.S. Kuttin, R. Jaeger, and J. Müller, Immunelektronenmikroskopische Utersuchungen zur Immunbiologie von Cryptococcus neoformans und Cryptococcus curvatus, Report 19. Meeting of the German speaking Mycological Society, Würzburg (1985).

7. H. Drexler, H. Melchinger, and J. Müller, Asteroid Bodies bei akuter, generalisierter Aspergillose in Organen einer Leukämie-Patientin. Schwerpunkt-Medizin 2: 42-52 (1979).

8. J. Muller, A. Polak, and R. Jaeger, Zur Ultrastruktur der Fonsecaea pedrosoi-Infektion der Maus, Report 17. Meeting of the German Speaking Mycological Society, Luxemburg (1983).

9. H. Melchinger, J. Müller, H. Takamiya, and B. Nold, Immunelektronenmikroskopische Untersuchungen an Asteroid Bodies in Vaginal material von Candida-Kolpitis-Patientinnen, Mykosen 23: 161-182 (1980).

10. H. Takamiya, A. Vogt, S. Batsford, E.S. Kuttin, and J. Müller, Further studies on the immunoelectronmicroscopic localization of poly-saccharide antigens on ultra-thin sections of Candida albicans, Mykosen 28: 17-32 (1985).

11. J. Müller, Mikroökologisch-quantitative Studien über die Sproßpilzflora des Menschen. Immunelektronenmikroskopische Studien uber Pathomechanismen bei Pilzinfektionen. Habilitationsschrift Med. Fak., Freiburg (1978).

12. J. Müller, H. Takamiya, B. Nold, and R. Jaeger, Electronmicroscopic proof of immunoreactions on Candida cells. Evidence of Anti-Candida antibodies in vivo by the direct immunoferritin test, Recent Advances in Medical and Veterinary Mycology, K. Iwata, ed., University of Tokyo Press, Tokyo, 59-65 (1976).

13. J. Müller, H. Takamiya, and R. Jaeger, Elektronenmikroskopische Darstellung von Immunreaktionen an Candida-Zellen. Asteroid Bodies bei Candida albicans im Urin von Nephritiis-Patienten, Sabouraudia 15: 87-93 (1977).

14. J. Müller, Endogene Mykosen und neuere Vorstellungen zu deren Pathomechanismen, Mykosen 24, Suppl. 1: 14-23 (1981).

15. R. Kappe and J. Muller, Cultural and serological examinations after ingestion of baker's yeast by a voluntary subject, Mykosen 30 (1987). In press.

16. E.S. Kuttin, R. Jaeger, and J. Müller, Licht und elektronenmikroskopische Untersuchungen an Peritoneallavagen von nicht-immunisierten und immunisierten, Cryptococcus neoformans-infizierten Versuchstieren. Report 18. Meeting of the German Speaking Mycological Society, Bremen (1984).

17. J. Müller, H. Melchinger, H. Takamiya, C. Douchet, B. Nold, and R. Jaeger, Immunelectronmicroscopic observations on the host-fungus relationships in Candida infections. The role of lysozyme in Candida infections. Proc. 7. Congr. ISHAM, 110-113, Jerusalem (1979).

18. W. Thurner, A. Vogt, J. Müller, and H. Melchinger, Studies on the antigenicity of Candida albicans cell wall constituents with special regard to immunoelectronmicroscopic investigations. Proc. 8. Congr. ISHAM, 216-219, Palmerston North, NZ (1982).

19. W. Thurner, Untersuchungen zur Antigenstruktur der Zellwand von Candida albicans unter besonderer Berücksichtigung der tieflokalisierten Antigene. Inaugural-Dissertation, Medizinische Fakültat, Freiburg (1986).

20. I. Berdicevsky, H. Melchinger, and J. Müller, Calcium deficiency and protoplastization, 9. Congr. Int. Soc. Human and Animal Mycology ISHAM, Atlanta, Georgia, USA (1985).

ENZYME-LINKED IMMUNOSORBENT ASSAY AND RELATED SENSITIVE ASSAYS FOR ANTIGEN AND ANTIBODY DETECTION

Jean-Luc Guesdon[1], Zoilo Pires de Camargo[2],
Edouard Drouhet[2], and Stratis Avrameas[1]

[1]Immunocytochemistry Unit and [2]Mycology Unit
Pasteur Institute, Paris, France

INTRODUCTION

A great deal of effort in the past decade has been directed to the development of rapid, sensitive and accurate methods for detection of antigens of pathogenic microorganisms or the corresponding antibodies. Most of the current rapid tests utilize polyclonal or monoclonal antibodies or antigens labelled with a marker which can be easily quantified with high sensitivity. The labelling of the antigen or the antibody can be achieved with substances such as a radioactive isotope, free radicals, a fluorescent dye, a bacteriophage or an enzyme. In comparison with radioisotope, spin, bacteriophage, and fluorescent dye labelling, the use of enzyme as marker has several interesting features: the reagents are stable for years, possess intrinsic amplification, and there are no problems of health hazards for laboratory personnel or waste disposal when appropriate enzymes and substrates are selected.

The first enzyme immunoassay techniques were independently described by Avrameas and Guilbert, Engvall and Perlmann, and Van Weemen and Schuurs in 1971. Enzyme immunoassays employ antibodies or antigens linked to enzymes in such a way that the immunological and catalytic activity of each moiety is preserved. These assays give reliable and reproducible results and are extremely sensitive when the different parameters of the test procedure are optimized.

Under specific conditions, enzyme immunoassays may be even more sensitive than radioimmunoassays (Harris et al., 1979; Shalev et al., 1980; Imagawa et al., 1982; Labrousse et al., 1982; Johannsson et al., 1986).

ASSAY PRINCIPLES

Enzyme immunoassays are currently classified in two main categories: heterogeneous and homogeneous.

The first category includes all assays which require the separation of free labelled antigen from bound labelled antigen. This separation step is performed after the immunological reaction has been allowed to occur and before enzyme activity is determined. These assays are quite similar to radioimmunassays and need a heterogeneous phase.

In contrast, homogeneous assays obviate the need to perform the separation step. These assays do not have their equivalent in radioimmunoassay, and their applicability is restricted in comparison with the heterogeneous enzyme immunoassay. This method can only be used with a few selected enzymes. However, sensitivity appears to be less than the sensitivity obtained with the heterogeneous immunoassays.

A wide variety of assay formats can be used in heterogeneous enzyme immunoassay techniques (Schuurs and van Weemen, 1977; Guesdon and Avrameas, 1981; Miyai, 1981). They can be broadly divided into competitive and non competitive assays.

Competitive assays

Although competitive assays which detect either antigen or antibody have been described, this type of assay is more commonly used for antigen determination. The more classical competitive assay uses enzyme-labelled antigen and any given bound/free separation method.

During the first step, the antigen to be determined and enzyme-labelled antigen are incubated in various proportions with a limited amount of specific antibody prepared in species A. In a second step, bound and free antigens are separated, usually by using solid phase-attached antibodies directed against the immunoglobulins of species A.

An alternative procedure is to perform the competition step directly on solid phase attached antibody. In this way the time to perform the entire procedure is reduced. The amount of enzyme-labelled antigen linked to the solid phase is finally measured using an appropriate substrate. As more antigen is present in the unknown sample, less labelled antigen will be fixed to the solid phase (Fig. 1).

Although these competitive enzyme immunoassay techniques are specific and easy to perform, they also present several disadvantages. They require purified antigen in relatively large amounts for the preparation of the enzyme-antigen conjugate and they also require contact of the enzyme marked with biological samples such as serum, urine or tissue extract. These fluids may contain proteases, enzyme activators or inhibitors, which may alter the activity of the enzyme used as marker.

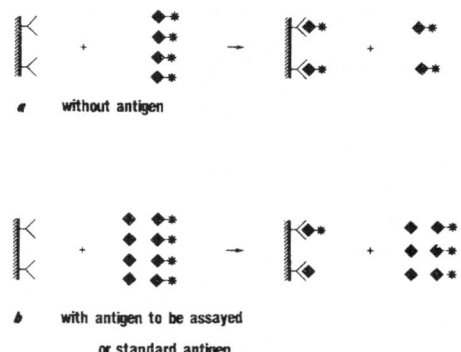

a without antigen

b with antigen to be assayed
or standard antigen

Fig. 1. Competitive enzyme immunoassay involving enzyme labelled antigen for the quantitation of antigen. The antigen to be assayed (◆) and a constant amount of labelled antigen (◆-*) compete for a limiting amount of solid phase-bound antibody.

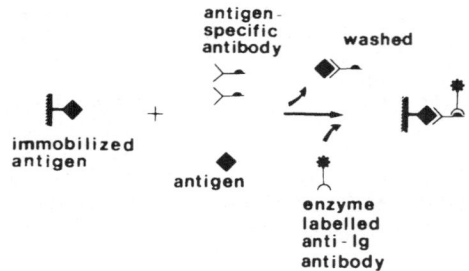

Fig. 2. Competitive enzyme immunoassay involving enzyme
        labelled antibody for the quantitation of antigen.
        The antigen to be assayed and a constant amount of
        immobilized antigen compete for a limiting amount
        of antigen-specific antibody.

Another type of competitive enzyme immunoassay employs enzyme-labelled
antibody and immobilized antigen. This method is analogous to the radio-
immunometric assay. Different forms of this assay are available, and the
two-step assay using enzyme labelled anti-immunoglobulin is most versatile
and convenient (Fig. 2). In this case, antigen bound to the solid phase
is incubated with the antibody solution with or without addition of standard
antigen or samples to be tested. After washing, the antibodies specifically
bound to the solid phase are detected using enzyme-labelled anti-immunoglobulin. In this assay, highly purified antigen is not required and the
same conjugate can be used to quantify several different antigens if the
primary antibodies have been prepared in the same species.

Sandwich procedures

The sandwich procedure is equally suitable to quantify either antigen
or antibody. It makes use of enzyme-labelled antibodies and insolubilized
immune component (i.e., the antibody or the antigen).

In order to quantify an antigen by this procedure, the specific antibody is fixed on a support; then, the sample containing the antigen is
incubated with the immobilized antibody (Fig. 3). After incubation,
substances which are not bound to the solid phase are washed away. The
enzyme-antibody conjugate is then added and attaches to the epitopes
remaining free. After incubation, excess conjugate is washed away and
the amount of enzyme attached to the solid phase is measured. For this
assay, both the immobilized and the labelled antibodies should be used in
excess of the antigen to be measured.

The sandwich technique can be only applied to molecules having at
least two antigenic determinants, and rheumatoid factor is known to be a
potentially interfering factor.

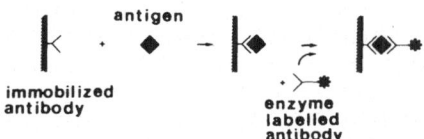

Fig. 3. "Sandwich" enzyme immonoassay for
        measuring antigen.

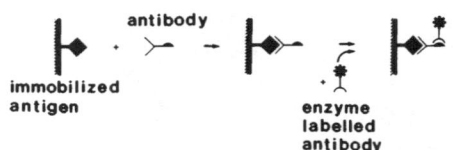

Fig. 4. Indirect enzyme immunoassay involving
enzyme-labelled anti-immunoglobulins
antibodies for the titration of antibodies.

However, the sandwich technique offers several advantages. Since the
assay employs enzyme-labelled antibody, the purification and enzyme labelling
of specific antigen is not necessary. Thus, the same enzyme labelling
procedure and solid phase attachment method can be used for different antibodies. Since the incubation with the test sample is carried out separately
from the incubation with the enzyme-antibody conjugate, the difficulties
encountered with the competitive assays which are due to the presence in
the sample of substances interfering with enzyme activity are not observed
with the sandwich technique. Moreover, this noncompetitive method is
potentially more snesitive and more accurate than the competitive one, and
the useful scale of quantitation is broader. Another advantage of the sandwich procedure when quantifying macromolecular antigen is the possibility
of binding several enzyme labelled antibody molecules to a single antigen
molecule, thus providing an element of amplification.

REAGENTS USED IN SOLID PHASE ENZYME IMMUNOASSAYS

When dealing with enzyme immunoassays one should bear in mind that the
choice of the basic elements required in this type of assay is of great
importance. The characteristics of these elements will influence
the sensitivity, the reproducibility and the practicability of the assay.
In order to obtain accurate and reproducible results in any enzyme immunoassay system, it is necessary to ensure that all reagents and assay conditions (temperature, incubation time, pH...) are optimized. The main elements
involved in heterogeneous enzyme immunoassay are:

1) The antibodies and the coupling procedure to prepare the conjugate.
2) The enzyme label and its method of detection.
3) The solid phase used to prepare the immunoadsorbent required for the
   bound/free separation step.

Antibodies

The sensitivity of the enzyme immunoassay depends greatly on the use of
avid antibodies. High avidity antibodies are usually found in sera of
hyperimmunized animals. The most commonly used schedules of hyperimmunization consist of injecting the Freund's emulsion containing small quantities
of immunogen intradermally at several sites and at several different times
(Vaitukaitis, 1981).

In solid phase enzyme immunoassays, either the globulin fraction
prepared from whole immune serum by salt fractionation, the IgG fraction
obtained by DEAE ion-exchange chromatography, or purified antibodies
isolated by immunoadsorption can be used. Because of the complex nature
of many of the samples, non-specific reactions are a frequent problem. In
order to decrease non-specific adsorption, it is preferable when possible,
to use purified antibodies rather than the globulin or IgG fractions.
Rheumatoid factor which is an anti-globulin antibody that recognizes the

Fc portion of altered immunoglobulins is known to be the cause of false positive results in enzyme immunoassay (Maiolini and Masseyeff, 1975). It is non-species specific. A number of different methods have been used in an attempt to overcome this drawback. One effective solution is to use Fab fragments instead of whole antibodies.

Undoubtedly, improvements in enzyme immunoassays could be made by using monoclonal antibodies with high affinities. Also, higher specificity can be obtained by carefully selecting monoclonal antibodies during the antibody screening and by mixing different selected monoclonal antibodies. Moreover, under certain conditions, the use of monoclonal antibodies allows the simultaneous incubation of conjugate, immunoadsorbent and sample without the risk of a prozone effect (Gupta et al., 1985a).

Conjugate

Selection of the enzyme. In order to obtain a reliable and sufficiently sensitive assay which is not too time consuming, the enzyme chosen should fulfill several criteria, the most important being specific activity and the highest possible turnover rate, and stability during coupling with the antibody and during the performance of the immunoassay.

The choice of the enzyme depends on the nature of the sample to be tested. An enzyme present in the sample or an enzyme dependent on activators or inhibitors present in the sample should not be used as label. In practice, the choice of the enzyme will depend on several factors such as its availability, cost, the speed at which the results should become available, and the number of samples to be determined. When large numbers of tests are carried out simultaneously, a longer enzyme reaction time is preferable, since this reduces timing errors.

Although many enzymes have been suggested as labels for use in solid phase enzyme immunoassay, in practice horseradish peroxidase, calf intestine alkaline phosphatase or, to a lesser extent, E. coli B-D-glactosidase are more employed than others.

Coupling procedures. Many methods exist for covalently coupling haptens, proteins and carbohydrates to proteins. Several reviews describing in detail the various methods for enzyme labelling of antigens and antibodies are available (Wisdom, 1976; Avrameas et al., 1978; O'Sullivan and Marks, 1981; Farr and Nakane, 1981; Ishikawa et al., 1983). The first methods described for coupling enzymes to macromolecules employed the following cross linking agents: cyanuric chloride (Avrameas and Lespinats, 1967), diazonium salts (Avrameas, 1968), water-soluble carbodiimide (Avrameas and Uriel, 1966; Nakane et al., 1966), diisocyanates (Clyne et al., 1973), difluoro dinitrophenyl sulfone (Nakane and Pierce, 1966). They had various disadvantages and none is currently in use.

At present, enzyme-protein conjugates are prepared mainly using sodium-m-periodate (Nakane and Kawoi, 1974), glutaraldehyde (Avrameas, 1969a; Avrameas and Ternynck, 1971), maleimide (Kato et al., 1975), and pyridyl disulfide (Carlson et al., 1978; King et al., 1978). The different coupling procedures have been extensively compared and the products of a number of coupling reactions have been qualitatively and quantitatively studied (Boorsma and Kalsbeek, 1975; Boorsma and Streefkerk, 1976; Boorsma et al., 1976; Avrameas et al., 1978; Ford et al., 1978; Engvall, 1978; Hagenaars et al., 1980; Deelder and de Walter, 1981; Ishikawa et al., 1983).

The covalent binding of enzyme to antigen or antibody is performed using organic compounds and can be regarded as a chemical modification of the enzyme. Such modification often decreases the catalytic activity of the enzyme. To overcome this drawback, several methods involving non-covalent interactions have been proposed. These non-covalent interactions include avidin-biotin (Guesdon et al., 1979) antigen-antibody (Avrameas, 1969b; Mason et al., 1969; Sternberger et al., 1970; Yorde et al., 1976; Butler et al., 1978; Metzger et al., 1981; Ternynck et al., 1983; Guesdon et al., 1983a, 1983b) and lectin-sugar (Guesdon and Avrameas, 1980) inter-actions.

The strong interaction between avidin and biotin has been used in enzyme immumoassays for the quantitation of antigens and antibodies. The avidin-biotin amplified enzyme immunoassay comprises four main steps: incubation of antigen with antibody coated on a solid phase, incubation of immobilized antigen with biotin-labelled antibody, incubation with avidin, incubation with biotin-labelled enzyme. The avidin-biotin system is based on the principle that avidin possesses four binding sites and that the avidin molecule can act as a bridge between two different biotynilated proteins.

Assays have been described which use immunological binding of marker enzymes by means of bridge antibodies (Sternberger et al., 1970). The procedure is based on the principle that an anti-immunoglobulin antibody is bound by one of its combining sites to the antibody directed against the antigen to be tested, the second combining site reacts with the anti-enzyme antibody which then binds the enzyme added during the last step.

A general method which incorporates antigen-specific antibody and anti-enzyme antibody belonging to different species has been described (Guesdon et al., 1983a). This method makes use of chimera antibodies prepared by covalently coupling the antibody specific for the antigen to be quantified with the polyclonal or monoclonal antibody specific for an enzyme. This assay consists of three main steps: incubation of the antigen to be determined containing material on antibody coated solid phase, incubation with the chimera antibody, incubation with the enzyme.

Lectins are able to interact specifically with carbohydrate moieties. Since such moieties are present in certain enzyme molecules. lectin-antibody conjugates can be used in enzyme immunoassays (Guesdon and Avrameas, 1980). The principle of this lectin immunotest, is that specific sites of the lectin can operate as acceptros for glyco-enzymes secondarily added to the system.

The sensitivity of an enzyme immunoassay can generally be enhanced by using one of the biospecific interactions described above.

Substrates

The last step of the enzyme immunoassay is the quantitative determination of the enzyme. Since the basic reaction between antigen and antibody is amplified through the enzyme molecule, the substrate used for enzyme measurement is important. Indeed, the detection limit in a given enzyme immunoassay depends ultimately on the lowest amount of label which can be determined via its activity. The sensitivity of such a determination depends upon the turnover number of the enzyme molecule used and the method employed to detect the product formed during the enzymatic reaction.

High sensitivity has been observed with alkaline phosphatase and β-galactosidase in combination with either a tritiated substrate (Harris et al., 1979) or a substrate which produces a fluorescent compound (Ishikawa and Kato, 1978; Neurath and Strick, 1981; Labrousse et al., 1982; Ali and Ali, 1983).

More recently, Self (1985) has described an enzyme aplification

method for alkaline phosphatase detection. With this method, the first
catalytic step is the dephosphorylation of NADP to produce NAD which cata-
lytically activates a specific redox-cycle involving two enzymes, alcohol
dehydrogenase and dusogirase. This mothod allows the detection of small
amounts of enzyme, as little as 6000 molecules (Johannson et al., 1986).

Horseradish peroxidase has been used with a number of chromogens.
Perhaps the most successful are o-phenylenediamine (OPD) and 2,2'-azino-
di-(3-ethylbenzithiazoline sulfonic acid) ABTS. Chromogen solutions must
be freshly prepared prior to use, as these chromogens are light sensitive
and non-enzymatic conversions occur in the light. Moreover, OPD and ABTS
are mutagenic compounds and must be handled with care. A non-carcinogen
chromogen, 3,3', 5,5' tetramethylbenzidine (TMB), has been proposed by
Bos et al., (1981) to replace OPD and ABTS.

In order to increase the sensitivity of peroxidase detection, light
emission techniques were proposed (Puget et al., 1977; Arakawa et al., 1979;
Whitehead et al., 1983; Thorpe et al., 1984). However, the application of
these techniques to enzyme immunoassays is limited because practical readers
are not yet available.

Solid phases

Heterogeneous enzyme immunoassays make use of antigen or antibody
immobilized on a solid phase and require one or several separation steps
between the liquid and the solid phase during the performance of the assay.

Immobilization of antigen or antibody can be performed either by
passive adsorption or by covalent binding. A large number of solid phases
such as polystyrene, polyvinyl, polypropylene, polycarbonate, aminoalkylsilyl
glass and silicone rubber have been used for the passive adsorption. Among
these, polystyrene has been the most commonly employed in the form of microplates, tubes, balls or rods. Microplates are the most popular, since they
are easy to handle and suitable for total automation (Voller et al., 1976).

The antigen or antibody is passively adsorbed onto the polystyrene
probably as a result of hydrophobic interaction (Cantarero et al., 1980).
The optimal conditions for coating the solid phase (i.e., concentration
of reagent, time of coating, temperature and pH) must be determined by
chequerboard titrations using reference reagents including positive and
negative samples. For most antigen and antibody preparations satisfactory
sensitization is generally achieved with solutions at 1-10 µg/ml in carbon-
ate/bicarbonate buffer pH 9.6. Thus, small amounts of biological material
can be used to coat plastic surfaces. However, wide variation in adsorption properties has been observed among different types of plates (Kricka
et al., 1980) and even among different plates of the same batch. Furthermore, adsorbed protein may undergo denaturation to some extent with loss
of activity (Berkowitz and Webert, 1981) and antigen or antibody can be
detached from the plastic surface. Another disadvantage is that plastic
surfaces have a limited capacity of adsorption.

In order to overcome these drawbacks antigen or antibody can be
covalently bound to the support.

Several solid phases allow the covalent binding of antigen or antibody. These solid supports include polysaccharides and their derivatives
(cellulose, starch, agarose, dextran), vinyl polymers (polyacrylamide),
polyamides (nylon), and inorganic compounds (porous glass, iron oxide powder).

For covalently immobilized antigen or antibody the separation steps
are generally performed by successive centrifugations. The use of magnetic

supports, such as Magnogel (Guesdon and Avrameas, 1976a, 1977) obviates the
need for time consuming multiple centrifugations. Magnogel made of polyacrylamide – agarose beads containing black ferric(ous) oxide was used to
covalently bind various antibodies, complex mixtures of antigens such as
crude extracts of grass pollen (Guesdon et al., 1978) or fungal antigens
(Othman et al., 1978, 1978, 1980).

ERYTHRO-IMMUNOASSAY FOR ANTIBODY AND ANTIGEN TITRATION

Erythro-immunoassay is a solid phase immunoassay which makes use of
chimera antibodies and antigen or antibody coated plates (Guesdon et al.,
1983a). Chimera antibodies are prepared by covalently coupling the antibody specific for the component to be detected with an antibody specific
for erythrocytes. The principle of the use of such chimera antibodies for
antibody titration is illustrated in Fig.5. U or V shaped wells of a polystyrene or polyvinyl chloride microtitration plate are coated with the
antigen. Then, the serum to be titrated is adequately diluted and incubated
in the antigen-coated wells. After washing, the wells are filled with a
solution of anti-IgG antibody coupled to anti-erythrocytes antibody. After
incubation and further washing the plate is filled with a suspension of
erythrocytes. The plate is left undisturbed for several hours and the
degree of erythroadsorption is read with the naked eye.

To measure erythroadsorption in more quantitative terms the erythro-
immunoassay has been modified (Guesdon et al., 1983b). After the addition
of the chimera antibody, incubation and washing, the wells are completely
filled with a large excess of erythrocytes (generally 0.5%). After one
hour's incubation the plate is immersed in a vessel containing isotonic
buffered solution, gently placed bottom up and allowed to float on the
surface. After nonadsorbed erythrocytes have been decanted on the bottom
of the vessel, the plate is gently turned over again and the adsorbance is
read in a ELISA spectrophotometer equipped with a 414 nm filter. In this
way a standard curve can be established.

The erythro-immunoassay is as sensitive as conventional enzyme or
radioimmunoassays. It was successfully used for quantitation of human IgE

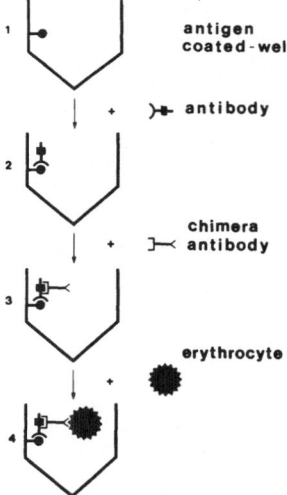

Fig. 5. Erythro-immunoassay for the titration of antibodies.

and for titration of anti- Echinococcus granulosus antibody or anti-DNA
antibody in human sera. An alternative procedure was devised to perform the
erythro-immunoassay without use of chimera antibody. In this procedure
antibody or antigen was covalently coupled to sheep erthrocytes and used in
a sandwich or competitive erythro-immunoassay for detection of human chorionic gonadotropin (Gupta et al., 1985b) or Escherichia coli heat-labile enterotoxin (Germani et al., 1986).

An application of erythro-immunoassay to titrate antibodies specific
for Paracoccidioides brasiliensis will now be described.

APPLICATION OF THE IMMUNOSORBENT ASSAY FOR DIAGNOSIS OF FUNGAL DISEASE

The measurement of antibody in the mycoses may present complex problems
because fungi produce many different antigens and cross reactions between
related   or even dissimilar fungi occur. Purification of a single antigen
from a complex mixture may not provide good sensitivity for disease if antibodies in different sera recognize different antigens of the mixture. Indeed,
false negative results could be obtained. Moreover, antibodies of different
immunoglobulin classes may be present in the same serum thus complicating
the interpretation of the assay results. In spite of these complexities
the measurement of antibodies can provide valuable diagnostic information
and can be of great importance in evaluating the course of disease
(Mackenzie, 1983).

Many different procedures are used to detect specific antibodies against
fungal antigens. The most widely used tests are immunodiffusion, agglutination, immunoelectrophoresis, complement fixation, immunofluroescence and
immunosorbent assays including radio or enzyme immunoassays. Various adaptations of enzyme immunoassay have been reported and compared with
other immunological tests (Hommel et al., 1976; Mathur et al., 1977; Pinon
et al., 1981; Richardson and Warnock, 1983; Hearn et al., 1985).

We developed a magnetic solid phase enzyme immunoassay for quantification of IgG class antibodies to Paracoccidioides brasiliensis in human sera
(Camargo et al., 1984a).

In this assay, somatic and metabolic antigens of Paracoccidioides
brasiliensis were bound to magnetic polyacrylamide agarose beads either
chemically using glutaraldehyde as coupling agent or biospecifically using
a lectin (concanavalin A) as linker. The technical details to perform this
enzyme immunoassay are described in the following.

Determination of immunoglobulin G to Paracoccidioides brasiliensis using
magnetic enzyme linked immunosorbent assay (MELISA)

Materials and methods. Somatic antigen was prepared by mechanical
disruption of the cell wall of Paracoccidioides brasiliensis strain B-339.
Sera were obtained from confirmed Brazilian cases of paracoccidioidomycosis,
or from healthy persons. Sheep anti-human IgG antibodies were isolated by
passage of the whole antiserum on human IgG immobilized on Ultrogel (Guesdon
and Avrameas, 1976b) and coupled to β-D-galactosidase according to Avrameas
et al., (1978) using glutaraldehyde (Serva, France). P. brasiliensis antigen
was bound to magnetic polyacrylamide agarose beads (Magnogel, Industrie
Biologique Francaise, Villeneuve-La-Garenne, France) following the method
previously described (Guesdon and Avrameas, 1977; Guesdon et al., 1978).
Briefly, 5 ml of beads were washed twice with distilled water. The washed
beads were incubated overnight at 37°C in a 6% solution of glutaraldehyde
in 0.1M phosphate buffer, pH 7.4, then extensively washed twice with
distilled water to eliminate excess of glutaraldehyde. The activated beads

were rotated overnight at room temperature with 25 mg of lyophilized antigen (dry weight) in 0.1M phosphate buffer, pH 7.4, previously dialysed overnight against the same buffer. The immunosorbent thus prepared was washed several times with phosphate buffered saline (PBS) until the supernatant had an optical density value at 280 nm less than 0.05. To block andy remaining free aldehyde groups, the beads were suspended in 0.1M phosphate buffer, pH 7.4, containing 0.1M glycine and kept for 3 hours at room temperature. Finally, the immunosorbent was washed, diluted 10-fold in PBS containing 0.2 g Merthiolate and 1 ml Tween 20 per liter and kept at 4°C until use.

In addition, P. brasiliensis antigen was bound to magnetic poly-acrylamide agarose beads using biospecific adsorption. In this case concanavalin A was fixed onto the magnetic beads previously activated with glutaraldehyde. About 1 mg of protein was bound per 1 ml of beads. Poly-saccharides and glycoproteins were then fixed to beads-bound concanavalin A by incubating 5 mg P. brasiliensis antigen and 5 ml beads in a total volume 20 ml PBS for 2 hours at room temperature. Then the beads were washed and kept as described above.

The following procedure was used to detect IgG to P. brasiliensis by magnetic enzyme linked immunosorbent assay. Magnogel beads (0.2 ml) were added to a series of test tubes. The beads were washed three times with PBS containing 0.1% Tween 20 (PBS-Tw) on a magnetic rack. To each tube, 0.5 ml of serum adequately diluted in PBS-Tw containing 3% bovine serum albumin and 0.1% Tween 20 (PBS-Tw-BSA) was added. After 2 hours of rotation at room temperature, the beads were washed three times on a magnetic rack with 4 ml of PBS-Tw. Then, 0.5 ml of galactosidase-labelled sheep anti-human IgG antibodies (5 µg/ml) diluted in PBS-Tw-BSA was added to each tube. The tubes were allowed to rotate at room temperature for 2 hours, then excess conjugate was eliminated by three washes with PBS-Tw. The enzyme activity was then measured by addition of 2.5 ml of substrate solution prepared by dissolving 20 mg of 2-nitro phenylgalactoside in 25 ml of 0.1M phosphate buffer pH 7.0 containing 0.1M 2-mercaptoethanol, 1 mM $MgSO_4$, 0.2 mM $MnSO_4$ and 2 mM EDTA. The enzyme reaction was performed by rotating the tubes horizontally for 1 hour at room temperature. Then, 1 ml of 1M Na_2CO_3 was added to stop the reaction and optical density was measured at 420 nm.

Typical results. In preliminary experiments, metabolic and somatic P. brasiliensis antigens were immobilized on glutaraldehyde-activated beads and the resulting immunosorbents were compared for their value to quantify antibodies. Results obtained showed that the somatic antigen was superior to the metabolic antigen extracted from the culture filtrate (Camargo et al., 1984a). We then decided to use somatic antigen bound to Magnogel either chemically using glutaraldehyde or biospecifically using concanavalin A.

Two typical dose response curves obtained by titrating anti- P. brasiliensis IgG antibodies in serum from proven case of paracoccidioido-mycosis, either with glutaraldehyde-activated Magnogel or concanavalin A-bound Magnogel are shown in Fig. 6. The sensitivity of the enzyme immuno-assay was highest when concanavalin A was used to immobilize P. brasiliensis somatic antigen on Magnogel beads.

The magnetic enzyme linked immunosorbent assay is easy to perform, reproducible and the results are obtained in 5-6 hours. As with other ELISA tests it has the ability to measure the primary interaction between antigen and antibody. Furthermore, with MELISA it was possible to dis-tinguish different antibody levels in sera from paracoccidioidomycosis patients before treatment and patients undergoing antimycotic treatment (Camargo et al., 1986).

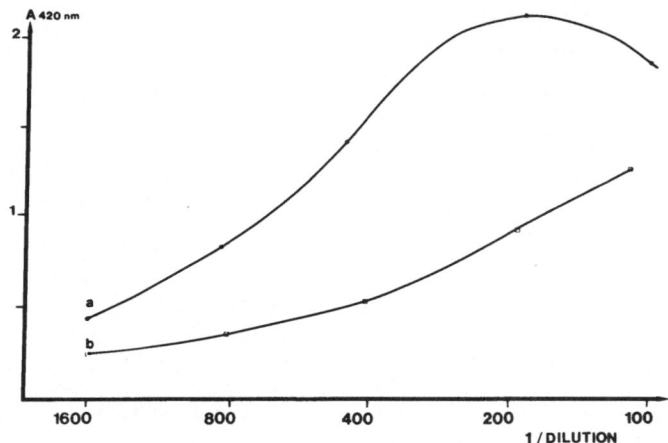

Fig. 6. Quantitation of P. brasiliensis antibodies in a
patient serum. Comparison of the curves obtained
by indirect enzyme immunoassay using antigen
immobilized either on glutaraldehyde activated
Magnogel (b) or on Magnogel-bound concanavalin A (a).

Titration of antibodies to Paracoccidioides brasiliensis by Erythro-immunoassay

Materials and Methods. Lyophilized yeast culture filtrate antigen was
prepared according to Restrepo and Drouhet (1970). Rabbit antiserum against
human immunoglobulins and rabbit antiserum against sheep red blood cells
(anti SRBC) were obtained from Diagnostics Pasteur (Marnes-La-Coquette,
France). Specific anti-human Ig were isolated by affinity chromatography
using human IgG immobilized on glutaraldehyde-activated polyacrylamide –
agarose beads (Guesdon and Avrameas, 1976b). Anti-SRBC antibodies were
isolated from antiserum with a glutaraldehyde copolymer of bovine serum
albumin (BSA) and cell stromate (Avrameas and Ternynck, 1969). Chimera
antibodies were prepared by using the one step glutaraldehyde coupling
procedure (Guesdon et al., 1983a). Two mg of anti-SRBC and 1 mg of anti-
human Ig antibodies were mixed in 1 ml of 0.1M phosphate buffer pH 6.8 and
dialysed overnight at 4°C against the same buffer. Then 0.05 ml of a 1%
glutaraldehyde solution was added. After 2 hours incubation at room temperature, 0.05 ml of 2M glycine solution was added; 2 hours later the mixture
was dialysed overnight at 4°C against PBS and centrifuged (3000g, 30 min).
This chimera antibody preparation was stored at -20°C after addition of 1 ml
glycerol.

Polystyrene plates with U-shaped wells (Greiner, Bischwiller, France)
were coated with 0.1 ml of P. brasiliensis antigen (10 µg/ml) by incubating
2 hours at 37°C and then 18 hours at 4°C. The plates were twice washed
with PBS-Tw. The sera to be tested were absorbed twice with a 50% SRBC
suspension and adequately diluted in PBS-Tw-BSA. Each well received 0.1 ml
of diluted serum (generally from 1/200 to 1/409,600). After incubation for
2 hours at 37°C followed by three washings with PBS-Tw, 0.1 ml of chimera
antibody (25 ng/ml) solution was added to each well. After further incubation for 2 hours at 37°C the plates were washed three times with PBS-Tw and
finally 0.1 ml SRBC suspension (0.1%) was added. The plates were left

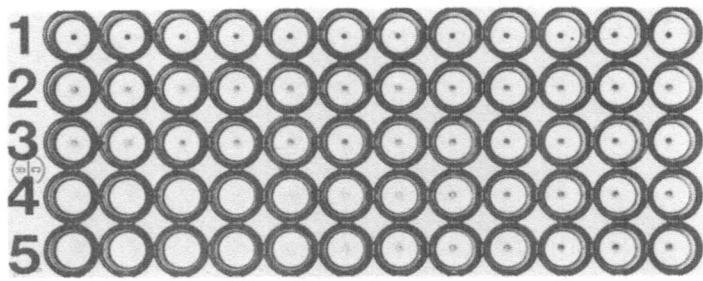

Fig. 7. Titration of P. brasiliensis antibodies in human
sera by erythro-immunoassay employing chimera antibody. The sera were diluted from 1:200 to 1:409, 600).
Row 1 (control), the antigen coated wells received
incubation buffer, chimera antibody, and SRBC suspension.
Rows 2 and 3, titration of a normal serum.
Rows 4 and 5, titration of a patient serum.

undisturbed at room temperature and 3 hours later the results were obtained.
In the positive wells SRBC were uniformly adsorbed on the surface, and the
reaction was considered negative when the SRBC settled, forming a pellet
on the bottom of the well.

Typical result. A representative experiment using one normal and one
patient serum is given in Fig. 7. Erythro-immunoassay is valuable for
detection of antibodies to P. brasiliensis. Good correlation has been
observed between the results obtained by this erythro-immunoassay and those
obtained by the ELISA technique. The findings of similar levels of antibodies
to the P. brasiliensis antigens in ELISA and erythro-immunoassay suggests
that both techniques detect the same kind of antibodies. Such antibodies
may be unrelated to those detected by counterimmunoelectrophoresis, because
sera with high titer in the erythro-immunoassay may have one precipitating
band and sera with lower titer may have five or six precipitating bands
(Camargo et al., 1984a).

Although the antigen utilized for this erythro-immunoassay is not
purified it was found to be useful. It is relatively easy to prepare in
large amounts. Furthermore, it was found that the use of chimera antibody
allowed the development of a sensitive and reliable assay to detect IgG
antibodies to P. brasiliensis in human sera. In particular, the use of
red blood cells as markers makes the procedure very simple and inexpensive:
its detection limit is close to that obtained by conventional enzyme immunoassays (Camargo et al., 1984b; 1984c).

REFERENCES

1. A. Ali and R. Ali, 1983, Enzyme linked immunosorbent assay for anti-DNA
   antibodies using fluorogenic and colorigenic substrates. J. Immunol.
   Methods, 56:341.
2. H. Arakawa, M. Meda, and A. Tsuji, 1979, Chemiluminescence enzyme immunoassay of cortisol using peroxidase as label. Anal. Biochem., 97:248.
3. S. Avrameas and J. Uriel, 1966, Methode de marquage d'antigenes et
   d'anticorps avec des enzymes et son application en immunodiffusion.
   C.R. Acad. Sci. (Paris), 262:2543.
4. S. Avrameas and G. Lespinats, 1967, Enzymes couplees aux proteines.
   Leur utilisation pour la detection des antigenes et des anticorps,
   C.R. Acad. Sci. (Paris), 265:1149.

5. S. Avrameas, 1968, Detection d'anticorps et d'antigenes a l'aide d'enzymes. Bull. Soc. Chim. Biol., 50:1169.

6. S. Avrameas, 1969a, Coupling of enzymes to proteins with glutaraldehyde. Use of the conjugates for the detection of antigens and antibodies, Immunochemistry, 6:43.

7. S. Avrameas, 1969b, Indirect immunoenzyme techniques for the intracellular detection of antigens. Immunochemistry, 6:825.

8. S. Avrameas and T. Ternynck, 1969, The cross linking of proteins with glutaraldehyde and its use for the preparation of immunoadsorbents. Immunochemistry, 6:53.

9. S. Avrameas and B. Guilbert, 1971a, Dosage enzymo-immunologique de proteines a l'aide d'immunoadsorbants et d'antigenes marques aux enzymes. C.R. Acad. Sci. (Paris), 273:2705.

10. S. Avrameas and B. Guilbert, 1971b, A method for quantitative determination of cellular immunoglobulins by enzyme labelled antibodies. Europ. J. Immunol., 1. 394.

11. S. Avrameas and T. Ternynck, 1971, Peroxidase labelled antibody and Fab conjugates with enhanced intracellular penetration. Immunochemistry, 8:1175.

12. S. Avrameas, T. Ternynck, and J.L. Guesdon, 1978, Coupling of enzymes to antibodies and antigens. Scand. J. Immunol., 8 (suppl. 7):7.

13. D.B. Berkowitz and D.W. Webert, 1981, The inactivation of horseradish peroxiadase by a polystyrene surface. J. Immunol. Methods, 47:121.

14. D.M. Boorsma and G.L. Kalsbeek, 1975, A comparative study of horseradish peroxidase conjugates prepared with a one-step and a two-step method. J. Histochem. Cytochem., 23:200.

15. D.M. Boorsma and J.G. Streefkerk, 1976, Peroxidase conjugate chromatography. Siolation of conjugates prepared with glutaraldehyde or periodate using polyacrylamide-agarose gel. J. Histochem. Cytochem., 24:481.

16. D.M. Boorsma J.G. Streefkerk, and N. Kors, 1976, A quantitative comparison of two peroxidase conjugates prepared with glutaraldehyde or periodate and fluorescein conjugate. J. Histochem. Cytochem., 24:1017.

17. E.S. Bos, A.A. Van der Doelen, N. Van Rooy, and A.H.W.M. Schuurs, 1981, 3, 3', 5, 5'-tetramethylbenzidine as an Ames test negative chromogen for horseradish peroxidase in enzyme immunoassay. J. Immunoassay, 2:187.

18. J.E. Butler, P.L. McGibern and P. Swanson, 1978, Amplification of the enzyme-linked immunoadsorbent assay (ELISA) in the detection of class specific antibodies. J. Immunol. Methods, 20:365.

19. Z.P. Camargo, J.L. Guesdon, E. Drouhet, and L. Improvishi, 1984a, Magnetic enzyme-linked immunoadsorbent assay (ELISA) in the detection of class specific antibodies. J. Immunol. Methods, 20:365.

20. Z.P. Camargo, J.L. Guesdon, E. Drouhet, and L. Improvisi, 1984b, Enzyme-linked immunosorbent assay (ELISA) in the paracoccidioidomycosis. Comparison with counterimmunoelectrophoresis and erythro-immunoassay. Mycopathologia, 88:31.

21. Z.P. Camargo, J.L. Guesdon, E. Drouhet, and L. Improvisi, 1984c, Titration of antibodies to Paracoccidioides brasiliensis by erythro-immunoassay (EIA). J. Med. Vet. Mycol., 22:73.

22. Z.P. Camargo, C. Unterkircher, and E. Drouhet, 1986, Comparison between magnetic enzyme-linked immunosorbent assay (MELISA) and complement fixation test (CF) in the diagnosis of paracoccidioidomycosis. J. Med. Vet. Mycol., 22:77.

23. L.A. Cantarero, J.E. Butler, and J.W. Osborne, 1980, The adsorptive characteristics of proteins for polystyrene and their significance in solid phase immunoassays. Anal. Biochem., 105:375.

24. J. Carlsson, H. Drevin, and R. Axen, 1978, Protein thiolation and reversible protein-protein conjugation. Biochem. J., 173:723.

25. D.H. Clyne, S.H. Norris, R.R. Modesto, A.J. Pesce, and V.E. Pollack, 1973, Antibody enzyme conjugates. The preparation of intermolecular

conjugates of horseradish peroxidase and antibody and their use in immunohistology of renal cortex. J. Histochem. Cytochem., 21:233.

26. A.M. Deelder and R. de Wlater, 1981, A comparative study on the preparation of immunoglobulin-galactosidase conjugates. J. Histochem. Cytochem., 29:1273.

27. E. Engvall and P. Perlmann, 1971, Enzyme linked immunoadsorbent assay (ELISA). Quantitative assay of immunoglobulin G. Immunochemistry, 8:871.

28. E. Engvall, 1978, Preparation of enzyme labelled staphylococcal protein A and its use for detection of antibodies. Scand. J. Immunol., 8 (suppl. 7):25.

29. A.G. Farr and P.K. Nakane, 1981, Immunochemistry with enzyme labelled antibodies: A brief review. J. Immunol. Methods, 47:129.

30. D.J. Ford, R. Radin, and A.J. Pesce, 1978, Characterization of glutaraldehyde coupled alkaline phosphatase antibody and lactoperoxidaxe-antibody conjugates. Immunochemistry, 15:237.

31. Y. Germani, E. Begaud, J.L. Guesdon, and J.P. Moreau, 1986 (in press), GM1 Erythroimmunoassay for detection and titration of Escherichia coli heat labile enterotoxin. J. Clin. Microbio. , 24:744.

32. J.L. Guesdon and S. Avrameas, 1976a, Dosages immunoenzymatiques des IgE. Ann. Medicales Nancy, 33.

33. J.L. Guesdon and S. Avrameas, 1976b, Polyacrylamide agarose beads for the preparation of effective immunoabsorbent. J. Immunol. Methods, 11:129.

34. J.L. Guesdon and S. Avrameas, 1977, Magnetic solid phase enzyme immunoassay. Immunochemistry, 14:443.

35. J.L. Guesdon, B. David, and J. Lapeyre, 1978, Magnetic enzyme immunoassay of anti-grass pollen specific IgE in human sera. Clin. Exp. Immunol., 33:430.

36. J.L. Guesdon, T. Ternynck, and S. Avrameas, 1979, The use of avidin-biotin interaction in immunoenzymatic techniques. J. Histochem. Cytochem., 27:1131.

37. J.L. Guesdon and S. Avrameas, 1980, Lectin-immuno-tests: quantitation and titration of antigens and antibodies using lectin antibody conjugates. J. Immunol. Methods, 39-1.

38. J.L. Guesdon and S. Avrameas, 1981, Solid phase enzyme immunoassays, in: "Applied biochemistry and bioengineering", Volume 3, L.B. Wingard, E. Katchalski-Katzir, and L. Goldstein, Eds., Academic Press, New York, 207:232.

39. J.L. Guesdon, F. Naquira Velarde, and S. Avrameas, 1983a, Solid phase immunoassay using chimera antibodies prepared with monoclonal or polyclonal anti-enzyme and anti-erythrocyte antibodies. Annal. Immunol. (Inst. Pasteur), 134C:265.

40. J.L. Guesdon, C. Jouanne, and S. Avrameas, 1983b, Use of hybrid antibody conjugates in enzyme and erythroimmunoassay, in: Immunoenzymatic techniques, Volume 18, S. Avrameas, P. Druet, R. Masseyeff, and G. Feldmann, Eds., Elsevier Science Publishers, Amsterdam, 197.

41. S.K. Gupta, J.L. Guesdon, S. Avrameas, and G.P. Talwar, 1985a, Solid phase sandwich enzyme immunoassays of human chorionic gonadotropin using monoclonal antibodies. J. Immunol. Methods, 83:159.

42. S.K. Gupta, J.L. Guesdon, S. Avrameas, and G.P. Talwar, 1985b, Solid phase competitive and sandwich type erythroimmunoassays for human chorionic gonadotropin. J. Immunol. Methods, 80:177.

43. A.M. Hagenaars, A.J. Kuipers, and J. Negel, 1980, Preparation of enzyme-antibody conjugates. In: Immunoenzymatic assay techniques R. Malvano, Ed., Martinus Nijhoff Publishers, The Hague, 16:27.

44. C.C. Harris, R.H. Yolken, H. Krokan, and I.C. Hsu, 1979, Ultrasensitive enzymatic radio-immunoassay: Application to detection of cholera toxin and rotavirus. Proc. Natl. Acad. Sci. USA, 76:5336.

45. V.M. Hearn, G.C. Donaldson, and M.J.R. Healy, 1985, A method to determine significant levels of immunoglobulin G to Aspergillus fumigatus

antigens in an ELISA system and a comparison with counterimmunoelectro-phoresis and double diffusion techniques. J. Immunoassay, 6:137.

46. M. Hommel, T.K. Truong, and D.E. Bidwell, 1976, Technique immunoenzy-matique (ELISA) appliquee au diagnostic serologique des candidoses et aspergilloses humaines: resultats preliminaires. Nouv. Presse Med., 5:2789.

47. M. Imagawa, S. Yoshitake, E. Ishikawa, Y. Niitsu, I, Urashizaki, R. Kanazawa, S. Tachibana, N, Nakazawa, and H. Ogawa, 1982, Development of a highly sensitive sandwich enzyme immunoassay for human ferritin using affinity purified anti-ferritin labelled with β-D-galactosidase from Escherichia coli. Clin. Chim. Acta., 121:277.

48. E. Ishikawa and K. Kato, 1978, Ultrasensitive enzyme immunoassay, Scand. J. Immunol., 8 (suppl. 7):43.

49. E. Ishikawa, M. Imagawa, S. Hashida, S. Yoshitake, Y. Hamaguchi, and T. Ueno, 1983, Enzyme labelling of antibodies and their fragments for enzyme immunoassay and immunohistochemical staining. J. Immunoassay, 4:209.

50. A. Johannsson, D.H. Ellis, D.L. Bates, A.M. Plumb, and C.J. Stanley, 1986, Enzyme amplification for immunoassays. Detection limit of one hundredth of an attomole. J. Immunol. Methods, 87:7.

51. K. Kato, Y. Hamaguchi, H. Kukui, and E. Ishikawa, 1975, Enzyme-linked immunoassay II. A simple method for synthesis of the rabbit antibody β-D-galactosidase complex and its general applicability. J. Biochem., 78:423.

52. T.P. KIng, Y. Ki, and L. Kochoumian, 1978, Preparation of protein con-jugates via intermolecular disulfide bond. Biochemistry, 17:1499.

53. L.J. Kricka, T.J.N. Carter, S.M. Burt, J.H. Kennedy, R.L. Holder, M.I. Halliday, M.E. Telford, and G.B. Wisdom, 1980, Variability in the adsorption properties of microtitre plates used as solid supports in enzyme immunoassay. Clin. Chem., 26:741.

54. H. Labrousse, J.L. Guesdon, J. Ragimbeau, and S. Avrameas, 1981, Miniaturization of B galactosidase immunoassays using chromogenic and fluorogenic substrates. J. Immunol. Methods, 48:133.

55. S. Lange, H. Nygren, J.E. Brorson, I. Holmberg, P. Larsson, 1981, Diffusion in gel enzyme linked immunosorbent assay (DIG-ELISA) for detection of class specific antibodies to Aspergillus fumigatus and Candida albicans. Acta. Path. Microbiol. Scand., Sect. C, 89:387.

56. D.W.R. Mackenzie, 1983, Serodiagnosis, In: Fungi pathogenic for human and animals. Part B, D.H. Howard, Ed., M. Dekker Inc. Publisher, New York-Basel, 121-218.

57. M.L. McLaren, E.S. Mahgoub, and E. Georgakopoulos, 1978, Preliminary investigation on the use of the enzyme linked immunosorbent assay (ELISA) in the serodiagnosis of mycetoma. Sabouraudia, 16:225.

58. R. Maoilini and R. Masseyeff, 1975, A sandwich method of enzymoimmuno-assay I. Application to rat and human alpha fetoprotein. J. Immunol. Methods, 8:223.

59. T.E. Mason, R.E. Phifer, S.S. Spicer, R.A. Swallow, and R.B. Dreskin, 1969, An immunoglobulin-enzyme bridge method for localizing tissue antigens. J. Histochem. Cytochem., 17:563.

60. S. Mathur, J.M. Goust, E.O. Horger, and H.H. Fudenberg, 1977, Immuno-globulin E anti-Candida antibodies and candidiasis. Infect. Immun. 18:257.

61. K. Miyai, 1981, Enzyme immunoassay system. Schematic representation, In: "Enzyme Immunoassay", E. Ishikawa, T. Kawai, and K. Miyai, Eds., Igaku-Shoin, Tokyo, 123:135.

62. W.J. Metzger, J.E. Butler, P. Swanson, E. Reinders, and H.B. Richerson, 1981, Amplification of the enzyme-linked immunosorbent assay for measuring allergen specific IgE and IgG antibody. Clin. Allergy, 11:523.

63. P.K. Nakane and G.B. Pierce, 1966, Enzyme labelled antibodies. Preparation and application for the localization of antigens. J. Histochem. Cytochem., 14:929.

64. P.K. Nakane, J. Sri Ram, and G.B. Pierce, 1966, Enzyme labelled anti-
    bodies for light and electron microscopic localization of antigens.
    J. Histochem. Cytochem., 14:789.
65. P.K. Nakane and A. Kawaoi, 1974, Peroxidase labelled antibody. A new
    method of conjugate. J. Histochem. Cytochem., 22:1084.
66. A.R. Neurath and N. Strick, 1981, Enzyme linked fluorescence immunoassays
    using β galactosidase and antibodies covalently bount to polystyrene
    plates. J. Virol. Methods, 3:155.
67. T. Othman, J.L. Guesdon, and E. Drouhet, 1978, Dosage enzymoimmunologique
    des anticorps humains anti-Candida albicans en utilisant des perles de
    polyacrylamide agarose magnetiques. Bull. Soc. Fr. Mycol. Med., 7:249.
68. T. Othman, J.L. Guesdon, and E. Drouhet, 1979, Dosage enzymo-immunolo-
    gique des anticorps anti-Cryptococcus neoformans dans les serums
    humains utilisant des perles de polyacrylamide agarose magnetiques.
    Bull. Soc. Fr. Mycol. Med., 8:35.
69. T. Othman and E. Drouhet, 1980, La technique enzymoimmunologique
    utilisant la peroxydase sur frottis de Candida albicans appliquee a
    la recherche des anticorps dans les candidoses. Comparaison avec la
    technique d'immunofluroescence. Bull. Soc. Fr. Mycol. Med., 9:117-120.
70. T. Othman, J.L. Guesdon, and E. Drouhet, 1980, Dosage immunoenzymatique
    des IgE specifiques anti-Candida albicans. Bull. Soc. Fr. Mycol. Med.,
    9:103-110.
71. M.J. O'Sullivan and V. Marks, 1981, Methods for the preparation of enzyme
    antibody conjugates for use in enzyme immunoassay. In: Methods in
    Enzymology, J.L. Langone and H.V. Vunakis, Eds., Academic Press, New
    York, 73:147-166.
72. J.M. Pinon, J.P. Gorse, and G. Dropsy, 1977, Exploration de l'immunite
    humorale dans l'aspergillose: Interet de l'ELIEDA (Enzyme linked
    immuno-electro diffusion assay). Mycopathologia, 60:115.
73. K. Puget, A.M. Michelson, and S. Avrameas, 1977, Light emission tech-
    niques for the microestimation of femtogram levels of peroxidase.
    Application to peroxidase (and other enzymes) coupled antibody-cell
    antigen-cell antigen interactions. Anal. Biochem., 79:447.
74. A. Restrepo and E. Drouhet, 1970, Etude des anticorps precipitants dans
    la blastomycose sud-americaine par l'analyse immunoelectrophoretique
    des antigenes de Paracoccidiodes brasiliensis. Ann. Inst. Pasteur
    (Paris), 119:338.
75. M.C. Richardson and D.W. Warnock, 1983, Enzyme-linked immunosorbent
    assay and its application to the serological diagnosis of fungal
    infection. Sabouraudia, 21:1.
76. A.H.W.M. Schuurs and B.K. Van Weemen, 1977, Enzyme immunoassay. Clin.
    Chim. Acta., 81:1.
77. C.H. Self, 1985, Enzyme amplification. A general method applied to
    provide an immunoassisted assay for placental alkaline phosphatase.
    J. Immunol. Methods, 76:389.
78. A. Shalev, G.H. Greenberg, and P.J. McAlpine, 1980, Detection of
    attograms of antigen by high sensitivity enzyme linked immunosorbent
    assay (HS-ELISA) using a florogenic substrate. J. Immunol. Methods,
    38:125.
79. L.A. Sternberger, P.H. Hardy, J.J. Cuculis, and H.J. Meyer, 1970,
    The unlabelled antibody-enzyme method of immunohistochemistry.
    Preparation and properties of soluble antigen-antibody complex
    (horseradish peroxidase anti-horseradish peroxidase) and its use in
    identification of spirochetes. J. Histochem. Cytochem., 18:315.
80. T. Ternynck, J. Gregoire, and S. Avrameas, 1983, Enzyme anti-enzyme
    monoclonal antibody soluble immune complexes (EMAC): Their use in
    quantitative immunoenzymatic assays. J. Immunol. Methods, 58:109.
81. G.H.G. Thorpe, R. Haggart, L.J. Kricka, and T.P. Whitehead, 1984,
    Enchanced luminescent enzyme immunoassays for rubella antibody,
    immunoglobulin E and digoxin. Biochem. Biophys. Res. Comm., 119:481.

82. J.L. Vaitukaitis, 1981, Production of antisera with small doses of immunogen: multiple intradermal injections. In: Methods in Enzymology, J.L. Langone and H.V. Vunakis, Eds., Academic Press, New York, 73:46-52.

83. B.K. van Weemen and A.H.W.M. Sxhuurs, 1971, Immunoassay using antigen-enzyme conjugates. FEBS Lett., 15:232.

84. A. Voller, D.E. Bidwell, and A. Bartlett, 1976, Enzyme immunoassay in diagnostic medicine: theory and practice. Bull. Wld. Hlth. Org., 53:55.

85. T.P. Whitehead, G.H.G. Thrope, T.J.N. Carter, C. Groucutt, and L.J. Kricka, 1983, Enhanced luminescence procedure for sensitive determination of peroxidase labelled conjugates in immunoassay. Nature, 305:158.

86. G.B. Wisdom, 1976, Enzyme-Immunoassay. Clin. Chem., 22:1243.

87. D.E. Yorde, E.A. Sasse, T.Y. Wang, R.O. Hussa, and J.C. Garancis, 1976, Competitive enzyme linked immunoassay with use of soluble enzyme-antibody immune complexes for labelling. Measurement of human choriogonadotropin. Clin. Chem., 22:1372.

TECHNIQUES AVAILABLE TO PREPARE MONOCLONAL ANTIBODIES DIRECTED TOWARD FUNGAL POLYSACCHARIDE ANTIGENS

Edouard Drouhet and Jean-Paul Latge

Institut Pasteur, Unite de Mycologie, 28 rue du Dr. Roux
75015 Paris, France

Fungal polysaccharides are among the most important antigens involved in humoral immunity reactions developed during the course of human and animal mycoses. These antigens circulate in the body fluids and usually induce the production of specific antibodies. The serodiagnosis of human mycoses is often based on the recognition of these antigens by specific antibodies (Mackenzie, 1983). Reference antibodies are usually produced in experimental animals (rabbit, goat, sheep) and are polyclonal. The characteristic of the polyclonal antibodies is by definition its heterogeneity. Most of the antibodies recognize different antigenic determinants on the same antigen while those directed toward the same determinant generally use different combining sites. This degree of heterogeneity is such that it is quiet impossible to obtain two different sera with identical antibody populations even from the same individual. This difficulty of standardization limits the use of polyclonal antibodies. The ideal reagent would originate from an unique clone of B lymphocytes, according to the clonal selection theory of Burnet : one B lymphocyte - one antibody molecule. Moreover the sensitivity and specificity of the polyclonal systems used to detect antigens are often low : sometimes the number of antigens recognized is limited and cross - reactions between related or even dissimilar fungi are frequent. For example, the galactomannan containing antigen C of Histoplasma capsulatum share antigenic determinants with similar antigens present in Blastomyces dermatitidis, Paracoccidioides brasiliensis and even Alternaria (Azuma et al., 1974). The detection of circulating antigens which is the most current challenge to diagnose a fungal infection in immunocompromised host, is very difficult using polyclonal antibodies because of the very low amount of antigen present in the body fluids.

All these problems may be overcome by the use of monoclonal antibodies (MAb). Introduced by Köhler and Milstein in 1975, these techniques have allowed tremendous progress in parasitology (Handman and Mitchell, 1986) virology (Gerhard and Bachi, 1986) and bacteriology (Bladewell and Winstanley, 1986), but have been scarcely used in medical mycology. The first antifungal MAb has been reported in 1981 (Hall and

Blackstock, 1981) but nowadays, about 20 scientific papers have been published on this topic. Monoclonal anti-carbohydrate antibodies have several advantages over polyclonal antibodies and can be specifically used to : 1) probe antigenic determinants in the analysis of antigen location, organization and availability ; 2) explore the antigenic variability occurring for example during growth or between in vivo and in vitro synthesized antigens ; 3) analyse the structure of the polysaccharidic antigen as an aid or replacement of the chemical method using enzymatic degradation, methylation, gas liquid chromatography and mass spectrometry determination, or lectin affinity ; 4) increase the sensitivity and specificity of tests used for diagnosis of medically important fungi.

This chapter will review the techniques used to prepare monoclonal antibodies directed toward fungal polysaccharidic antigens (Tables 1 and 2) with special emphasis on the protocol used by J.C. Mazié at the Hybridolab of the Pasteur Institute to isolate anti-Candida mannan hybridomas.

I) Immunization protocol (Fig. 1)

Immunization of animals with fungal glycoproteins produces antibodies specific to the protein portion but in general the carbodydrate moiety is not strongly immunogenic and can even be immunosuppressor (Huppert, 1983). The most striking example of the heterogenous antibody reponse is the capsular polysaccharide of Cryptococcus neoformans which at relatively high concentration, can paralyse the cellular and humoral immune defence of the host and in minute amounts can induce the production of specific antibodies (Dromer et al., 1987). The mannan of C.albicans which can regularly provoke the production of specific antibodies is immunosuppressor at high concentration (Drouhet, 1986 ; Durandy et al., 1986). In order to obtain efficient antibody response, carbohydrate may be either mixed with foreign bodies or covalently linked to a proteic carrier such as edestin or bovine serum albumin or sheep erythrocytes (Kozel and Cazin, 1971, 1974, Monsigny et al., 1984, Horisberger and Rousset-Vauthey, 1985). These methods have been used to raise Ab directed to fungal polysaccharides in rabbits. However, using the same method it has been more difficult to obtain good antibody response in mice then in rabbit(Kannagi and Hakomori, 1986). It should also be stressed that the injected antigen must be highly purified and well characterized. In most cases, monoclonal anti-carbohydrate antibodies seem to recognize 2 to 4 sugar residues within a given antigen sequence (Kannagi and Hakomori, 1986) and are directed toward the terminal sugar structure. It is obvious that any contaminating peptide or oligosaccharide present on the antigen may lead to the production of MAb which are not directly related to the antigenic portion. For example, a clone directed toward the carbohydrate antigen in histoplasmin has been obtained after immunization of mice with the major protein of histoplasmin (Reiss et al., 1986). To circumvent the problems of the low immunogenicity of the polysaccharides, whole cells are used to immunize mice and the desired hybridoma clones are identified by their positive reactivity to the carbohydrate antigens. Alive, heat (70°C) or formol-killed cells and crude or partially purified antigen have been used for immunization (Brawner and Cutler, 1984, 1986, Hall and Blackstock, 1981, Barbier et al., 1985, Chardès et al., 1986, Polonelli and Morace, 1986, Reiss et al., 1986). Balb/c or Biozzi HR mice (Biozzi et al., 1975 ; Boumsell

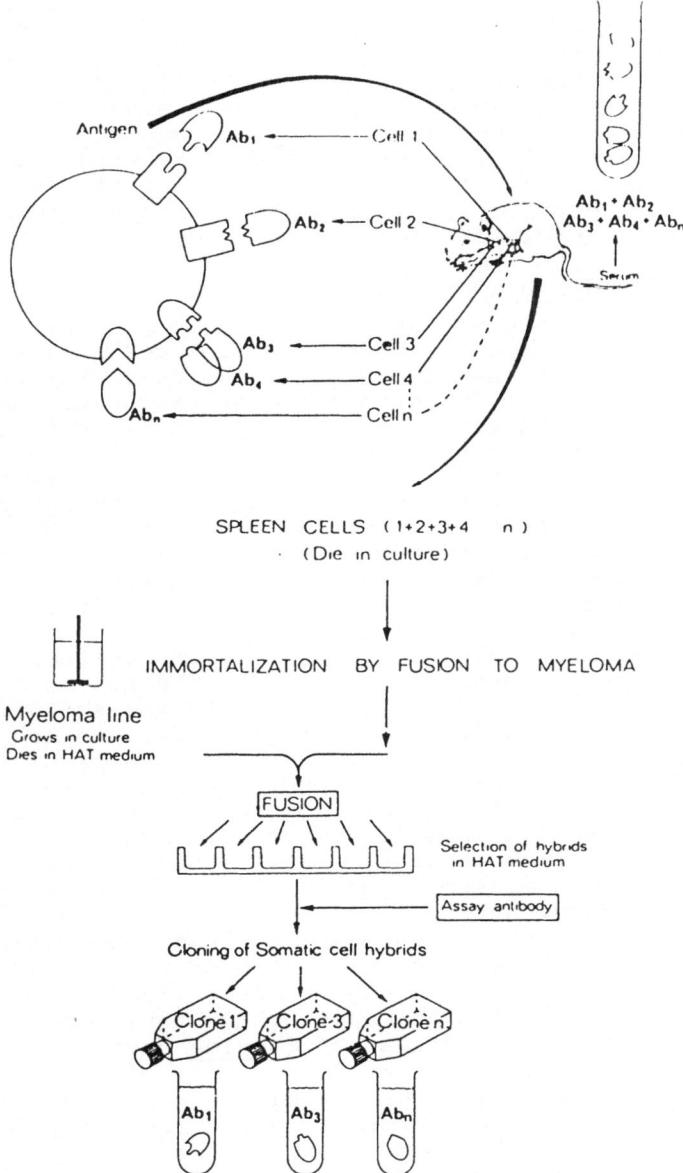

Fig. 1 - Schematic diagram of the basic techniques used for the production
of monoclonal antibodies (from Milstein, 1986)

and Bernard, 1979) have been the animals most successfully used for the production of mAbs directed toward polysacchari- dic antigens. Lou rats have also been employed to study wall cell antigens of <u>Candida</u> specifically produced <u>in vivo</u>. Intra- peritoneal (in presence or absence of Freund adjuvant), and/or intravenous injections have been performed (Tables 1, 2). The concentration of antigen injected, the route of immunization, the timing between injection, the presence of adjuvant and the strain of mice must be determined for each antigen. At the Hybridolab of the Pasteur Institute, the following immuniza- tion protocol was followed in order to select monoclonal antibody specific of the mannan of <u>C.albicans</u>. Whole yeast cells of <u>C.albicans</u> killed by heat (60°C, 1h) were inoculated to Biozzi HR mice. Intraperitoneal injections of 10^7 cells were performed at day 1, 7, 12 and 18. Each mouse received a booster at day 47 and was killed 3 days later.

II) <u>Cell fusion and hybridoma cloning</u>

Cell fusion techniques have been used to produce hybrids between myeloma cells and antibody-producing cells. The hybrid lines derived are permanently adapted to grow in tissue cul- ture and are capable of inducing antibody-producing tumors in mice. The source of B lymphocytes is spleen cells from mice previously immunized with the antigen. The myeloma cells that serve for fusion are mutants that do not secrete antibodies and that lack the enzyme hypoxanthine guanine phosphoribosyl transferase (HGPRT). As this enzyme is involved in the salvage pathway of DNA synthesis, such mutant cells die in the presen- ce of chemical agents that block the main pathway (aminopterin or azaserine) (Goding, 1980, Fazekas et al., 1980, Oi and Herzenberg, 1980, Hurrel, 1981).

Most hybridomas originated from fusion between spleen mice cells and murine myeloma cell line. The two most common Balb/c myeloma lines are SP 2/0-Ag 14 and P3 x63-Ag8.653. The myeloma lines are resistant to either 8-azaguanine or 6- thioguanine or 5-bromo-2'-deoxyuridine. For example P3 x 63- Ag8.653 cells are resistant to 20 μg/ml 8-azaguanine and do not grow in HAT selective medium. Immunologists interested in mouse antigens have also immunized rats with mouse antigen and fused spleen cells from these rats with mouse myelomas. The availability of a rat myeloma line IR 983 F (Bazin, 1982) may make this unnecessary. Hybrid frequencies in rat/mouse fusions can be as good as those generally obtained in mouse/mouse fusions with a myeloma spleen cell ratio of 1 : 1 instead of 10 : 1 generally used in mouse/mouse fusion (Oi and Herzenberg, 1980). Such rat-mouse hybrid have been used to select MAbs specific of antigen of <u>Candida</u> produced <u>in vivo</u> (Hopwood et al., 1986).

Fusion of cell membranes occurs with a very low frequency when cells are brought together, and the frequency can be increased to useful levels by the addition of various substan- ces. Two general types of fusing agents have been used : viruses (Sendai virus in particular) and agents such as lyso- lecithin and polyethylene glycol which affect the membrane in a poorly understood manner. Polyethylene glycol (PEG) is the agent most commonly used.

A successful fusion procedure should bring cells toge- ther, with an optimal frequency of interactions between the two "parent" cell types which allow fusion to occur at a sufficiently rapid rate, and cause minimal damage to the cells. Inevitably an agent which enables membranes to "flow"

Table 1 - Monoclonal antibodies against polysaccharides and cell wall antigens of yeasts

Yeast	Target antigen	Immunization (spleen cells)			Myeloma line (murine)	Number of hybrids and Class of MAb obtained and (studied)	References
		animals used	antigen injected	route duration			
Candida albicans	mannan	Biozzi IIR mice	10^7/ml Y cells heat-killed	i.p. 6 weeks	P3x63-Ag 8.653	5(1) IgM	Munoz et al. (1986)
"	mannan (Ag 5,6)	Balb/c mice	$5 \times 10^7 - 2 \times 10^8$/ml Y cells heat-killed	i.p. 6 weeks	P3x63-8Ag.653	17(2) IgM	Miyakawa et al. (1986)
"	peptidogluco-mannan	Balb/c mice	peptidogluco-mannan		SP 2/0	14(5) IgM	Kerkering and Espinel-Ingroff (1983)
"	cell wall Y	Balb/c mice	Y cells heat-killed	i.p. i.v. 3 weeks	P3x63-Ag8.653	1 IgM	Barbier et al. (1985)
"	cell wall Y	Balb/c mice	10^8/ml Y cells heat-killed	i.p. 6 weeks	P3x63-Ag8.653	3(2) IgM	Brawner and Cutler (1984) Brawner and Cutler (1986)
"	cell wall Y	Biozzi IIR mice	10^5-10^7 Y cells heat-killed	i.p. 6 weeks	P3x63-Ag8.653	31(21)	Chardès et al. (1986)
"	cell wall Y	male Lou, rats	10^5 live Y cells	i.p. 6 weeks	SP 2/0 Ag14	3(1) IgM	Hopwood et al. (1986)
"	surface cell wall Y	Balb/c mice	glycoprotein antigen	i.p. 6 weeks		9(1) IgG	Polonelli and Morace (1986)
Candida tropicalis	mannan	Balb/c mice C57 BL/6 CFW	cell walls	i.p. 6 weeks i.v last week	SP 2/0-Ag 14	41 IgM (4) IgG (1)	Reiss et al. (1984)
Cryptococcus neoformans	glucuronoxylo-mannan (GXM)	Balb/c mice C3H/HeJ	capsular poly-saccharide (CPS) (1 ug)	i.v. 9 days	P3x63-Ag8.653	21 IgG (7) IgM (14)	Dromer et al. (1985)
"	GXM	Balb/c mice	CPS coupled to sheep erythro-cytes	i.v. 1,13,76,114 & 122 th day	SP 2/0-Ag 14	4 IgG_1(4)	Kozel and Eckert (1986)

Y: yeast; i.p.: intraperitoneal; i.v.: intravenous

Table 2 - Monoclonal antibodies against polysaccharides and cell wall antigens of filamentous and systemic dimorphic fungi

Yeast	Target antigen	Immunization (spleen cells)			Myeloma line (murine)	Number of hybrids and Class of MAb obtained and (studied)	References
		animals used	antigen injected	route duration			
MOLDS							
Aspergillus fumigatus	galacto-mannan	Biozzi HR mice	10^7 spores heat-killed	i.p. 6 weeks	P3x63-Ag 8.653	1(1) IgM	Munoz et al. (1986)
SYSTEMIC DIMORPHIC FUNGI							
Histoplasma capsulatum	galacto-mannan(GM) M-protein	Balb/c mice	GM & PM (50mg-25mg)	i.p. 30 days	SP2/0 Ag 14	156 IgM(5)	Reiss et al. (1986)
Coccidioides immitis	spherule and mycelium surface Ag	Balb/c mice	spherules/ endospores HCl inactivated			7(7)	Karu et al. (1985) Kraeger et al. (1986)

together is likely to adversely affect membrane integrity. Most of the methods used for cell fusion are derived from the original PEG technique described by Köhler and Milstein (1975) and Gafre et al. (1977) with minor modifications. The different parameters varying from one procedure to the other are the following : cell ratio, medium used (particularly presence or absence or serum and pH control), conditions for achieving contact, fusing agent and stabilizing additives, conditions for fusion (time, temperature, physical handling) and processing after fusion up to the plating out stage. An additional variable is the operator. It is a common observation that success with fusions increases with experience, even though there is no change in technique detectable to the trained "hybridizer".

The conditions which favor successful fusions are not unique, and there are several variations which seem equally successful. The procedure commonly used at the Hybridolab of the Pasteur Institute is outlined below. The spleens were aseptically removed from the immunized mice teased apart, washed in the HAT (Gibco) culture medium with fetal calf serum (Whitaken, 1972). The cells are recovered by centrifugation at 200xg for 5 min. The spleen cells are resuspended in 10 ml of culture medium with FCS and counts of viable cells are made after Trypan blue exclusion. 10 ml of spleen cells and 10 ml of murine myeloma cells (P3X63-Ag8.653) were mixed (final cell ratio 10:1) and centrifuged at 200xg for 5 min. After removal of the supernatant, 1 ml of warm (37°C) 95 % PEG solution was added to the pellet for 1 min with gentle shaking. After 1 min incubation, the fusion mixture is diluted by slowly adding 20 ml of medium. The cells were centifuged at 200 g for 5 min and resuspended in medium, plated at 2×10^5 myeloma cells/ml in 24 well culture plates and incubated in a 5 % CO_2 incubator at 37°C.

After hybridization the culture plates are usually checked daily for colony growth and aspect of the culture medium (color, pH). Hybridoma were detected macroscopically after 10-20 days of culture. At this time, they are fed by removing 1 ml of spent medium (which can be used for screening) and replacing it slowly with 1 ml medium. This further reduces the concentration of aminopterin and appears to enhance colony growth. If the sample is taken early there is a risk of missing positive hybrids because the antibody concentration in the supernatant is too low. Conversely, it is important to avoid any colony overgrowth as this may result in death of the hybrids. Consequently, tests should be performed before the culture becomes very crowded, but wells giving negative results should be tested further when the medium is yellow. Another problem results from the presence of small amounts of antibody not produced by hybrids, but by surviving spleen cells. This is seen as a weak positive result which disappears when the cultures are fed and retested after a given period of growth.

Selected hybridomas were cloned by the limiting dilution technique (0.5 cell/well in microplate) and injected into a pristane-primed mice for production of ascites. It is essential to clone as early as possible, to ensure that a given culture contains only one cell type and that it is producing only one immunoglobulin specificity. Immediately after hybridization, the fusion products will have approximately 80 chromosomes, and as these cells proceed to divide they will randomly loose some of these chromosomes. By cloning early, it is possible to select those cells which still have the chromosomes coding for antibody production. If cloning is not carried out at this stage, there is a risk that variants, not producing immunoglobulin, will appear and overgrow the culture. In addition, hybridization culture wells often have multiple colonies, which may become disturbed and mixed during the course of feeding. The process of cloning allows the selection of a positive hybrid which is derived from a single cell. Colonies which are composed of a mixture of cell types should be separated into pure populations. Wells containing two or more colonies may still produce only one antibody, if the other hybrids in the well are non secretors. It is nevertheless necessary to clone and isolate the secretor hybrid, as the non secretors may overgrow the culture.

Methods used for measuring antibody activity

The methods employed for screening hybridomas are similar to the one used to study the apparition of specific antibodies during mice immunization. The most adapted method is an indirect enzyme immunoassay (EIA) where the specific antigen is coated on a polystyrene plate or on nitrocellulose paper (Reiss et al., 1986, Dromer et al., 1987). An enzyme anti-mouse immunoglobulin conjugate is used as a second antibody. The binding of the second antibody is measured with a colorimetric reaction. In the case of C. albicans, the MAb B33 was used at a 10^{-4} dilution on a plate coated with mannan antigen at 1 µg/well ; 50 % inhibition was obtained using 0.2 ng/ml.

The antigen coated on the EIA plate should be highly purified. After immunization of a mice with whole cells, numerous clones specific to the fungal species studied can be obtained but very few, if any, will be directed toward the polysaccharidic antigen. The presence of any contaminating peptide or oligosaccharide sequence on the antigen should be monitored because it may lead to the selection of non-specific MAbs.

Since most antigenic polysaccharides have an external position in the cell, indirect immunofluorescence or agglutination tests or EIA with fixed cells coated on the well have also been used to select the MAb reacting positively with the outermost components of the cell (Dromer et al., 1987, Hopwood et al., 1986, Munoz et al., 1986, Reiss et al., 1986). The specificity of the MAb toward the carbohydrate antigen should be evaluated further on.

Isotyping of the Ig is easily performed with rabbit anti-mouse immunoglobulin conjugates. Most of the monoclonal anti-carbohydrate antibodies belong to the IgM class (Table 1 and 2, Kannagi and Hakomori, 1986). Nevertheless, monoclonal antibodies against Cryptococcus neoformans capsular polysaccharides belonging to IgG_1 class have been recently reported (Dromer et al., 1987, Kozel and Eckert, 1986).

Acknowledgements : The authors would specially thank J.C. Mazié for communicating all the technical informations and methods used in his laboratory (Hybridolab, Pasteur Institute).

References

Azuma, I.F., Kanetsuna, Y., Tanaka, Y., Yamamura, Y., and
    Carbonell, J., 1974, Chemical and immunological proper-
    ties of galactomannans obtained from Histoplasma capsu-
    latum, H.duboisi, Paracoccidioides brasiliensis and
    Blastomyces dermatitidis. Mycopathologia, 54:111.
Barbier, E., Sarthou, P., and Le Guern, C., 1985, Obtention
    d'un anticorps monoclonal anti-Candida albicans.
    Bull.Soc.Fr.Mycol.Med., 14:13.
Bazin, H., 1982, Production of rat monoclonal antibodies
    with the Lou rat non secreting IR 983 F myloma cell
    line, pp 615-618, in: Protides of the biological fluids
    29th Colloquium 1981, H. Peeters ed. Pergamon Press,
    Oxford and New York, 615-618.
Biozzi G., Stiffel, C., Mouton, D., and Bouthillier, Y.,
    1975, Selection of lines of mice with high and low
    antibody responses to complex immunogens, pp 123-139,
    in : Immunogenetics and immunodeficiency. B. Benacerraf
    (ed) Academic Press, New York.
Blackwell, C.C., and Winstanley F.P., 1986, Application of
    Monoclonal antibodies in bacteriology pp 114.1-114.4,
    in: Handbook of experimental immunology. Vol. 4.
    Application of immunological methods in biological
    sciences. D.M. Weir (ed). Blackwell Sci.Publ, Oxford.
Boumsell, L., and Bernard, A., 1979, Intérêt des races de
    souris bonnes productrices d'anticorps de Biozzi pour
    la production d'anticorps monoclonaux. C.R.Acad.Sci.
    (Paris) (SérieD),289:723
Brawner, D.J., and Cuttler, J.E., 1984, Variability in
    expression of a cell surface determinant on Candida
    albicans as evidenced by an agglutinating monoclonal
    antibody. Infect.Immun., 43: 966.
Brawner, D.J., and Cuttler, J.E., 1986a. Ultrastructural and
    biochemical studies of two dynamically expressed cell
    surface determinants on Candida albicans. Infect.
    Immun., 51: 327.
Brawner, D.J., and Cuttler, J.E., 1986b, Variability in
    expression of cell surface antigens of Candida antigens
    during morphogenesis. Infect.Immun., 51: 337.
Chardès, T., Piechaczyk, M., Cavaillès, V., Sahli, S.L.,
    Pau, B., and Bastide, J.M., 1986, Production and charac-
    terization of anti-Candida monoclonal antibodies.
    Ann.Immunol. (Inst.Pasteur), 137C:117.
Dromer, F., Salamero, J., Contrepois, A., Carbon, C., Yeni,
    P., 1987, Production, characterization and antibody
    specificity of a mouse monoclonal antibody reactive
    with Cryptococcus neoformans capsular polysaccharide.
    Infect.Imun., 55: 742.
Fazekas de St Groth, S., and Scheidegger, D., 1980, Produc-
    tion of monoclonal antibodies : strategy and tactics,
    J.Immunol.Methods, 35:1.
Galfre, G., Howe, C., Milstin, C., Butcher, G.W., and Howard,
    J.C., 1977, Antibodies to major histocompatibility anti-
    gens produced by hybrid cell lines. Nature (London) 266: 550.
Gerhardt, W., and Bachir, T., 1986. Applications of monoclo-
    nal antibodies in virology, pp 115.1 - 115., in :
    Handbook of experimental immunology, Vol. 4, Applica-
    tions of imunological methods in biological sciences.
    D.M. Weir (ed), Blackwell Sci.Publ. Oxford.

Goding, J.W., 1980, Antibody production by hybridomas, J.Immunol.Methods, 39:285.

Handman, E., and Mitchell G.F., 1986, Monoclonal antibodies in the study of parasites and host-parasites relationships p. 113.1 - 113.4 in : Handbook of experimental immunlogy, Vol.4, Applications of immunological methods in biological sciences D.M. Weir (ed), Blackwell Sci. Publ.Oxford.

Hall, N.K., and Blastock, R., 1981, Production of specific antibody to Cryptococcus neoformans by hybridoma, in vitro. Sabouraudia, 19, 157.

Hardy, R.R., 1986, Purification and characterization of monoclonal antibodies pp. 13.1-13.13 in : Handbook of experimental immunology, Vol 4. Application of immunological methods in biomedical sciences. D.M. Weir (ed), Blackwell Sci.Publ. Oxford.

Hopwood, V., Poulain, D. Fortier, B., Evans, G. and Vernes A., 1986, A monoclonal antibody to a cell wall component of Candida albicans. Infect.Immun, 54: 222.

Horisberger, M. and Rousset-Vauthey, M., 1985. Cell wall architecture of the fission yeast Schizosaccharomyces pombe. Experientia, 41, 748-750.

Huppert, M., 1983, Antigens used for measuring immunological reactivity, pp 121, in: Fungi pathogenic for human and animals, part B. Pathogenicity and detection, Howard D.H. (ed) Marcel Dekker Publisher, New York, Basel.

Hurrell, J.G.R., 1981, Monoclonal hybridoma antibodies : techniques and applications. C.R.C. Press, Boca Raton.Fla.

Kannagi R. and Hakomori, S., 1986, Monoclonal antibodies directed to carbohydrate antigens pp. 117.1 - 117.20 in: Handbook of experimental immunology, vol.4, Application of immunolocal methods in biomedical sciences. D.M. Weir (ed), Blackwell Sci.Publ. Oxford.

Karu, A.E., Gennevois, D.J.P., Koffman, J.W., Kraeger, S.J. and Levine, H.B., 1985, Research and diagnostic applications of monoclonal antibodies to Coccidioides immitis . Abst.Ninth.Intern.Congr.Intern.Soc.Human.Animal.Mycol. (ISHMA) Atlanta (USA) Abst. N°19-3.

Kipps T.J. and Herzenberg L.A. 1986, Schemata for the production of monoclonal antibody-producing hybridomas p. 108.1-108.9, in: Handbook of experimental immunology. Vol.4, Applications of immunological in biological sciences. D. Weir (ed), Blackwell Sci.Publ.Oxford.

Kohler G., and Milstein C., 1975, Continuous cultures of fused cells secreting antibody of predefined specificity. Nature, 256, 495.

Kozel, T.R. and Cazin, J., Jr., 1971, Non-encapsulated variant of Cryptococcus neoformans. I. Virulence studies and characterization of soluble polysaccharide, Infect.Immun., 3, 287.

Kozel, T.R., and Cazin, J., Jr., 1974, Induction of humoral antibody response by soluble polysaccharide of Cryptococcus neoformans, Mycopathol.Mycol.Appl., 54, 21

Kozel,T.R., Oi, V.T. and Herzenberg, L.A., Immunolglobulin-producing hybrid cell lines, in: Selected Methods in Cellular Immunology, Mishell, B.B. and Shiigi, S.M., (Eds)., W.H. Freeman, San Francisco, 1980, 351.

Kozel, T.R., and Eckert, T.E. 1986, Use of monoclonal antibodies for immuchemical analysis of Cryptococcus neoformans capsular polysaccharide, in: "Fungal Antigens" First International Symposium, Paris, Plenum Press, New York.

Mackenzie, D.W.R., 1983, Serodiagnosis, in: Fungi pathogenic for humans and animals. Part B. Pathogenicity and Detection: I Howard D.H. (ed), Marcel Dekker, Publisher, New York, Basel, pp 121.

McKearn, T., Smilek, D., and Fitch F., 1980, Rat-Mouse hybridomas and their application to studies of the major histocompativity complex, in: Monoclonal antibodies hybridomas : a new dimension in biological analysis. McKearn R. (ed) Plenum Press New York, pp 219-232.

Milstein, C., 1986. Overview : monoclonal antibodies. pp 107.1 - 107.12, in: Handbook of experimental immunology. Vol.4. Application of immunological methods in biological sciences. D.M. Weir (ed) Blackwell Sci.Publ. Oxford.

Miyakawa, Y., Kagaya, K., Fukazawa, Y., and Soe, G., 1986. Production and characterization of agglutinating monoclonal antibodies against predominant antigenic factors for Candida albicans. J.Clin.Microbiol., 23: 881.

Monsigny, M., Roche, A-C., and Midoux, P. 1984. Uptake of neoglycoprotein via membrane lectin (s) of L1210 cells evidenced by quantitative flow cytofluorimetry and drug targeting. Biol.Cell., 51:187.

Munoz, C., Mazié, J.C., Delga, J.M., Dupont, B., Latgé, J.P. and Drouhet, E., 1986. Monoclonal antibodies anti-Candida albicans mannan and anti-Aspergilus fumigatus galactomannan, in: Fungal antigens. First International Symposium. Paris. Plenum Press. New York.

Polonelli, L. and Morace, G. 1986. Specific and common antigenic determinants of Candida albicans isolates detected by monoclonal antibody. J. Clin. Microbiol., 23:366.

Reiss, E., Knowles, J.B., Bragg, S.L and Kaufman, L. 1986. Monoclonal antibodies against the M protein and carbohydrate antigens of histoplasmin characterized by the enzyme linked immunoelectrotransfer blot method., Infect. Immun., 53:540.

Strockbine, N.A. Largen, M.T., and Buckley, H.R. 1984. Production and characterization of three monoclonal antibodies to Candida albicans proteins, Infect.Immunity, 43:1012.

Whitaker, A.M. Tissue and Cell Culture, Williams and Wilkins, Baltimore, 1972.

FUNGAL EXOANTIGENS

Leo Kaufman and Paul Standard

Division of Mycotic Diseases
Center for Infectious Diseases
Centers for Disease Control
Public Health Service
U.S. Department of Health and Human Services

INTRODUCTION

Exoantigens are valuable for the immunoidentification of fungal pathogens and for resolving taxonomic problems. Most fungi produce unique antigens that allow specific identification. Exoantigens are soluble immunogenic macromolecules produced by fungi early in their development. These antigens are readily detected in culture filtrates and aqueous extracts of slant cultures. The exoantigen test depends on interaction between concentrated or unconcentrated antigens produced by fungi in culture and homologous precipitins. Any precipitate formed is readily checked for fusion with preselected specific reference precipitates in counter-immunoelectrophoretic or immunodiffusion tests to establish the identity of the fungus producing the antigen(s). Lines of identity are diagnostic. Specific identification of fungi thus may be accomplished within 2 to 5 days of receipt of mature cultures. Conventional identification of dimorphic and other pathogenic fungi has been known to take weeks or months.

Fungi that produce specific antigens cam be immunologically identified in their typical, atypical, or non-sporulating states. Mould-form antigens are excellent markers, and the exoantigen technique obviates the need for time-consuming, temperature-dependent conversion or in vivo inoculation studies. Exoantigens can be detected even in contaminated and non-viable fungi. Potential exposure to biohazardous fungi is reduced by eliminating the need for in vit conversion and the culture manipulations used in animal inoculations Using exoantigen tests eliminates costs arising from time-consuming and laborious morphologic, physiologic, and cultural studies. In addition, rapid exoantigen identification of a pathogen benefits the patient by permitting administration of appropriate therapy soon after receipt of a mature isolate.

The earliest attempt to identify fungi by their soluble antigens was made by Manych and Sourek in 1966. They used a single-diffusion-tube precipitin test and inoculated cultures of Blastomyces dermatitidis, Coccidioides immitis, Histoplasma capsulatum, and Paracoccidioides brasiliensis on the surface of serum agar. They were able to detect precipitin bands as the antigens produced by the proliferating fungi were precipitated. The test, however, took up to 21 days to complete. In addition, the bands could not be

specifically identified because of the absence of control antigens. These
studies, however, laid the ground-work for the rapid and practical procedures
that evolved 10 years later.

Extensive serologic studies carried out by a variety of investigators
(Kaufman et al., 1983a) revealed that many patients with blastomycosis,
coccidioidomycosis, histoplasmosis, and paracoccidioidomycosis react to
infections by producing diagnostically specific precipitins. These findings
led us to speculate that selected precipitins could be used in agar gel
immunodiffusion (ID) tests to specifically precipitate cell-free extracts of
pathogenic fungi for rapid identification. Antigenic analyses revealed that
B. dermatitidis, C. immitis, H. capsulatum, and P. brasiliensis shared
antigens not only among themselves but also with a variety of saprophytic
fungi. Each of these pathogens, however, elaborated soluble antigens that
permitted their specific identification. These specific antigens are
designated A for B. dermatitidis; HS, HL and F for C. immitis; H and M for H.
capsulatum; and 1, 2, and 3 for P. brasiliensis. In a series of studies that
began in 1976, we developed and perfected micro-ID procedures, or as we
prefer to call them, exoantigen tests, for many fungi of medical importance
(Standard and Kaufman, 1976).

METHODS

The exoantigen test involves reacting soluble antigens of an unknown
fungus in parallel with selected specific antigens from a known fungus
against reference antiserum specific to the latter. Antigens may be derived
from supernatants of merthiolate treated, 3-day-old, shaken brain-heart
infusion broth cultures or merthiolate extracts of 7 to 10 day-old Sabouraud
dextrose agar (SDA) slant cultures with at least 15 x 30 mm of growth
(Standard et al. 1985).

A brief, direct, microscopic examination of the unidentified fungal
culture to determine whether conidia or conidiophores are produced is first
required. This preliminary examination facilitates the selection of
appropriate exoantigen test reagents.

Shaken broth cultures are recommended as a source of exoantigens whenever
a false-negative result is suspected from the slant extraction procedure.
After 24 hr of treatment with merthiolate, 5 ml of the culture supernatant or
slant extract is transferred with a 9 in. Pasteur pipette to an Amicon
Minicon Macrosolute B-15 concentrator. Some extracts, for example those from
suspected cultures of B. dermatitidis and H. capsulatum, should be
concentrated 25 to 50 times, whereas extracts from cultures suspected of
being C. immitis are tested either unconcentrated or concentrated 5 times.
The unknown antigens are then tested in parallel with the control antigens
and homologous antiserum by micro-ID using only 20 ul of reagents (Standard
et al., 1985).

Charged ID plates are incubated at 25°C for 24 hr. The unknowns are
then examined for formation of lines of identity with the reference
precipitates.

Since fungal pathogens may produce shared as well as specific antigens,
the currently used exoantigen tests incorporate reference antigens and
polyclonal antibodies rich in a specific antibody. The only antigen
considered diagnostic is one that forms a distinct line of identity with the
reference precipitate. Unrelated or partially related lines are not
diagnostic.

Although false positive reactions have not been encountered, false negative reactions can occur. They may be attributable to technical error, absence of specific homologous antibody in the reference antiserum, or inadequate antigen concentration due to insufficient growth or improper growth conditions. Tests should be repeated when mycological or clinical judgment suggests that a negative reaction was erroneous. We also recommend that cultures that are negative by the exoantigen test be re-examined by conventional morphologic tests to make certain that they are not known pathogens.

The exoantigen test has been evaluated extensively with Centers for Disease Control (CDC) reference reagents and commercial reagents. It is extremely sensitive and specific for systemic fungal pathogens. Positive results are equivalent to cultural isolation and identification by conventional tests.

Specific antibodies needed for exoantigen tests have been produced by few techniques. High quality reference antiserum can be produced by immunizing rabbits with purified or unpurified culture filtrates or with electrophoretic or immunodiffusion precipitin arcs. Precipitin arcs are highly recommended as immunogens, since they elicit antiserum that is less cross-reactive than that produced by other techniques. To date no monoclonal antibodies that react specifically against established diagnostic exoantigens have been produced.

Acceptable antiserum may be obtained within 3 weeks by injecting rabbits intramuscularly with culture-filtrate antigens mixed with equal volumes of Freunds' incomplete adjuvant, followed by intravenous injections of the antigen alone (Standard et al., 1985).

Antiserum to arc antigens is prepared by injecting mixtures of equal quantities of emulsified arcs and Freunds' complete adjuvant subcutaneously in the dorsal area of a rabbit. Multiple injections are given about a month apart. Serum samples are checked for specific precipitins approximately 3 weeks after the second injection (Standard et al., 1985).

SAFETY MEASURES

To ensure that laboratory workers do not become infected, we recommend that all fungal cultures be handled in a biological safety hood and that broth cultures and extracts be treated with a preservative and/or be filter-sterilized. The preservative chosen should render the culture non-viable without destroying its exoantigens. Although 0.2% formaldehyde can effectively kill the mycelial forms of B. dermatitidis, C. immitis, and H. capsulatum, it unfortunately also inactivates the B. dermatitidis A and C. immitis HL antigens. Solutions of 0.02% merthiolate consistently kill these fungi in broth culture but not in slant extracts. We routinely use 0.02% merthiolate to treat broth cultures and slant extracts for 24 h, and we have noted that viable elements, on rare occasions, can survive in the merthiolate-treated extract. Merthiolate does not deleteriously affect the exoantigens. For those workers who prefer to work with sterile slant extracts, we recommend sterilization of the merthiolate treated extracts by passage through a 0.45-um membrane filter (Standard and Kaufman, 1982).

DIMORPHC FUNGI

Exoantigen tests are extremely valuable in identifying dimorphic pathogens, particularly those that form atypical cultures or are difficult to convert from one form to the other.

B. dermatitidis causes blastomycosis in North America (except Mexico), Africa, the Middle East, and India. At 25°C, the fungus grows as a white-to-tan mould on SDA. Microscopically, it may demonstrate round-to-oval conidia ranging in size from 3 to 12 um. However, many pathogenic and saprophytic fungi, such as the Chrysosporium spp. and atypical isolates of H. capsulatum, can bear similar conidia. Proper identification of B. dermatitidis requires conversion to its characteristic tissue form on culture media or in vivo. The exoantigen test is a rapid and reliable alternative for these procedures. Using precipitin A-positive anti-B. dermatitidis serum, we identified a specific exoantigen homologous to the A antibody in slant extracts of 56 isolates of B. dermatitidis (Kaufman and Standard, 1978). Although extracts of Chrysosporium spp., C. immitis, and H. capsulatum may cross-react with non-A antibodies in anti-B. dermatitidis serum, in no instance do they produce an A precipitate.

More recently, we evaluated the value of antibody to B. dermatitidis A- precipitin arcs for identifying B. dermatitidis isolates from the United States, Canada, Africa, India, and Israel. With the exception of some African isolates of B. dermatitidis, the specific A antigen was found in isolates of B. dermatitidis from all parts of the world (Kaufman et al., 1983b). Our studies also revealed that all isolates of B. dermatitidis shared an antigen designated "K." Thus, it is apparent that at least two serotypes of the fungus exist, one that possesses the A antigen and one that does not. The serotypic differences among isolates of this species appear to be associated with their geographic distribution and morphologic differences. The A-deficient serotype appears to be prevalent only in Africa. Antigen K is frequently produced earlier than the A antigen and, although it is also produced by Chrysosporium parvum var. parvum and C. parvum var. crescens, it may be useful for screening cultures of suspected B. dermatitidis.

Carmichael (1952) considers the genus Blastomyces to be a synonym of Chrysosporium. Studies by Sekhon et al. (1986a) indicate that the Chrysosporium species produce antigens that are shared by the morphologically similar mould forms of B. dermatitidis and H. capsulatum. Stronger antigenic relationships, however, were noted with B. dermatitidis than with H. capsulatum. These antigenic analyses support Carmichael's treatment of B. dermatitidis as a Chrysosporium species.

A definitive diagnosis of coccidioidomycosis depends on isolation of the mould form of C. immitis, which typically produces alternate, barrel-shaped arthroconidia, and its successful conversion to the parasitic or endosporulating spherule form. Identifying this pathogen is not always easy since several members of the family Gymnoascaceae, notably species of Arachniotus, Auxarthron, Uncinocarpus, and species belonging to the form-genus Malbranchea, produce superficially similar arthroconidia. Furthermore, atypical, non-sporulating, and pigmented forms of C. immitis can also be encountered. The exoantigen test is completely sensitive and specific for identifying C. immitis in its typical or atypical mould forms. Slant extracts are excellent sources of diagnostic exoantigens (Kaufman et al., 1983a). These exoantigens may be either heat-stable (HS) (remaining stable after boiling for 10 min) or heat labile (HL or F) (losing activity after heating at 60°C). Detection of any one of these antigens identifies the pathogen.

The HS antigen is usually the earliest of the three diagnostic antigens to appear. Reference reagents used for the immunologic identification of C. immitis must contain the HS antigen and preferably a combination of the three antigens. Our experience indicates that false-negative results ensue when the reagents used do not meet this standard. The C. immitis heat-stable tube precipitinogen (TP), used for detecting early coccidioidomycosis IgM antibodies in humans, is of limited value for identifying C. immitis isolates in cultures because this antigen is also common to certain Gymnoascaceous saprophytes, such as Arachniotus, Auxarthron, and Malbranchea species (Kaufman et al., 1985). Taxonomic relatedness among these arthroconidia-forming fungi is reflected by the TP antigen, whereas the HS, HL, and F antigens are found only in the pathogen.

The occurrence of specific antigens can help separate fungi from morphologically similar non-pathogenic species and justify their taxonomic position. The genus Histoplasma contains only one species, H. capsulatum, which has three varieties, var. capsulatum, var. duboisii, and var. farciminosum. The mycelial forms of two of these varieties (var. capsulatum and var. duboisii) are characterized by the production of tuberculate macroconidia. Detection of these macroconidia, however, can only suggest an H. capsulatum variety since many saprophytic fungi in the genera Arthroderma, Chrysosporium, Corynascus, Renispora, and Sepedonium have mycelial forms that grossly and microscopically resemble those of H. capsulatum. To further complicate identification, atypical non-sporulating and red-pigmented isolates of H. capsulatum have been encountered (Morris et al., 1986). However, cultures of all three varieties of H. capsulatum, whether typical, atypical, or pigmented, produce H and M antigens, whereas morphologically similar non-pathogenic fungi do not (Kaufman et al., 1983a). The antigenic and morphologic similarities exhibited by the three varieties of H. capsulatum are the basis for their classification. Renispora flavissima, which bears tuberculate macroconidia, was originally thought to be H. capsulatum; however, negative H and M exoantigen tests indicated that this fungus was distinct. It was subsequently described and classified as a new species (Sigler et al., 1979).

Our immunodiffusion studies do not support recent reports (McGowan and Buckley, 1985; Restrepo and Moncada, 1974) that the M antigen of H. capsulatum is identical to P. brasiliensis antigen 3 (Standard and Kaufman, 1980).

HYALINE NON-DIMORPHIC FUNGI

Identification of aspergilli involves the use of standardized media and slide cultures, a time-consuming process. In some instances, non-sporulating isolates, such as the "albino-type" of Aspergillus fumigatus, are encountered, which require even more extensive studies. Antisera prepared against extracts to A. fumigatus, A. flavus, A. nidulans, A. niger, and A. terreus demonstrated variable intra-generic cross-reactions. These were readily eliminated by adsorptions with antigens of selected heterologous Aspergillus spp. (Sekhon et al., 1986b). Accordingly, group-specific antisera were developed for specifically identifying the Aspergillus species of medical importance. Exoantigens proved reliable for identifying the sterile albino-type isolates of A. fumigatus. To date, exoantigen analyses have supported the traditional classification of Aspergillus into groups of species based on color of the colonies and morphologic characteristics.

BASIDIOBOLUS AND CONIDIOBOLUS SPP.

Within the Order Entomophthorales are two genera, Basidiobolus and Conidiobolus, with pathogenic species that cause subcutaneous zygomycosis. Exoantigen studies have not yet been carried out on the Conidiobolus spp. Some investigators have resorted to exoantigen analyses in the hope of resolving taxonomic problems among the Basidiobolus spp. Polonelli and Morace (1984), using polysaccharide and concentrated soluble filtrate antigens, found that B. haptosporus, B. meristosporus, B. microsporus, and B.

ranarum were closely related. However, distinct antigens were recognized with the use of adsorbed antiserum, which permitted antigenic separation of the four species. More recently, Yangco et al. (1986) described two heat-stable exoantigens common to B. haptosporus and B. ranarum, designated N and Y. B. meristosporus, B. microsporus, C. coronatus, and C. incongruus shared only the N antigen. No other heterologous zygomycete of the order Mucorales nor dimorphic fungus studied produce either antigen. The data suggest that B. haptosporus and B. ranarum are antigenically similar to each other but distinct from B. meristosporus and B. microsporus. The sharing of the N antigen by the Basidiobolus and Conidiobolus species implies a taxonomic relationship between the two genera and supports their taxonomic classification in the order Entomophthorales.

DEMATIACEOUS FUNGI

The dematiaceous fungi responsible for phaeohyphomycosis are often difficult to identify because they are polymorphic. Exophiala jeanselmei and Wangiella dermatitidis are common agents of phaeohyphomycosis. Because of their polymorphic nature and the superficial similarities of their conidiogenous cells, they may be difficult to identify and to differentiate from one another as well as from other Exophiala spp. such as E. moniliae, E. spinifera, and E. werneckii.

McGinnis and Padhye (1977) demonstrated that Phialophora jeanselmei and P. gougerotii are morphologically similar and should be considered a single species in the genus Exophiala, E. jeanselmei. Exoantigen studies have supported this conclusion. Kaufman et al. (1980) showed that E. jeanselmei has three serotypes. Isolates that cause mycetomas were identified as E. jeanselmei serotype 1, whereas those that caused phaeohyphomycosis and were originally identified as P. gougerotii belonged to serotypes 1, 2, or 3 of E. jeanselmei. As a result of these studies, exoantigen reagents have been developed for specifically identifying 6-day-old cultures of E. jeanselmei and W. dermatitidis. The antigenic distinction noted between these fungi and the diagnostic value of the exoantigens have been confirmed by Espinel-Ingroff et al. (1984).

REFERENCES

Carmichael, J.W. 1952. Chrysosporium and some other aleuriosporic hyphomycetes. Can. J. Bot. 40:1137-1173.
Espinel-Ingroff, A., Shadomy, S., Kerkering, T.M., Shadomy, H.J. 1984. Exoantigen test for differentiation of Exophiala jeanselmei and Wangiella dermatitidis isolates from other dematiaceous fungi. J. Clin. Microbiol. 20:23-27.
Kaufman, L., Standard, P.G. 1978. Immuno-identification of cultures of fungi pathogenic to man. Curr. Microbiol. 1:135-140.
Kaufman, L., Standard, P.G., Huppert, M., Pappagianis, D. 1985. Comparison and diagnostic value of the coccidioidin heat-stable (HS and tube precipitin) antigens in immunodiffusion. J. Clin. Microbiol. 22:515-518.
Kaufman, L., Standard, P., Padhye, A.A. 1980. Serologic relationship among isolates of Exophiala jeanselmei (Phialophora jeanselmei, P. gougerotii) and Wangiella dermatitidis. Proc. of the Fifth Internat. Conf. on the Mycoses, Scientific Publication No. 396, Pan Amer. Health Organ., Washington, D.C. p. 252-258.
Kaufman, L., Standard, P.G., Padhye, A.A. 1983a. Exoantigen tests for the immunoidentification of fungal cultures. Mycopathologia. 82:3-12.
Kaufman, L., Standard, P.G., Weeks, R.J., Padhye, A.A. 1983b. Detection

of two Blastomyces dermatitidis serotypes by exoantigen analysis. J.
Clin. Microbiol. 18:110-114.

Manych, J., Sourek, J. 1966. Diagnostic possibilities of utilizing
precipitation in agar for the identification of Histoplasma capsulatum,
Coccidioides immitis, Blastomyces dermatitidis, and Paracoccidioides
brasiliensis. J. Hyg. Epidemiol. Microbiol. Immunol. 10:74-84

McGinnis, M.G., Padhye, A.A. 1977. Exophiala jeanselmei, a new
combination for Phialophora jeanselmei. Mycotaxon. 5:341-352.

McGowan, K.L., Buckley, H.R. 1985. Preparation and use of cytoplasmic
antigens for the serodiagnosis of paracoccidioidomycosis. J. Clin.
Microbiol. 22:39-43.

Morris, P.R., Terreni, A.A., DiSalvo, A.F. 1986. Red pigmented
Histoplasma capsulatum - an unusual variant. J. Med. Vet. Myco.
24:229-231.

Polonelli, L., Morace, G. 1984. Rapid immunoidentification of
pathogenic fungi. Proceedings of the European Symposium on New Horizons
in Microbiology. Edited by A. Sanna and G. Morace. Elsevier Science
Publishers, N.Y. p. 203-219.

Restrepo, A., Moncada, L.H. 1974. Characterization of the precipitin
bands detected in immunodiffusion test for paracoccidioidomycosis. Appl.
Microbiol. 28:138-144.

Sekhon, A.S., Standard, P.G., Kaufman, L., Garg, A.K. 1986a.
Reliability of exoantigens for differentiating Blastomyces dermatitidis
and Histoplasma capsulatum from Chrysosporium and Geomyces species.
Diagn. Microbiol. Infect. Dis. 4:215-221.

Sekhon, A.S., Standard, P.G., Kaufman, L., Garg, A.K., Cifuentes, P.
1986b. Grouping of Aspergillus species with exoantigens. Diagnostic
Immunol. 4:112-116.

Sigler, L., Gaur, P.K., Lichtwart, R.W., Carmichael, J.W. 1979.
Renispora flavissima, a new gymnoascaceous fungus with tuberculate
chrysosporium conidia. Mycotaxon. 10:133-141.

Standard, P.G., Kaufman, L. 1976. Specific immunological test for
rapid identification of members of the Genus Histoplasma. J. Clin.
Microbiol. 3:191-199.

Standard, P.G., Kaufman, L. 1980. A rapid and specific method for the
immunological identification of mycelial form cultures of aracoccidioides
brasiliensis. Curr. Microbiol. 4:297-300.

Standard, P.G., Kaufman, L. 1982. Safety considerations in handling
exoantigen extracts from pathogenic fungi. J. Clin. Microbiol.
15:663-667.

Standard, P.G., Kaufman, L., Whaley, S.D. 1985. Rapid identification
of pathogenic isolates by immunodiffusion. CDC Lab Manual, Centers for
Disease Control, Public Health Service, U.S. Dept. of Health and Human
Services, Atlanta, GA 30333.

Yangco, B.G., Nettlow, A., Okafor, J.I., Park, J., Strake, D.T. 1986.
Comparative antigenic studies of species of Basidiobolus and other
medically important fungi. J. Clin. Microbiol. 23:679-682.

Invertebrate
Pathogens

FUNGAL ELICITORS OF INVERTEBRATE CELL DEFENSE SYSTEM

Drion Boucias[1] and Jean-Paul Latgé[2]

[1]Department of Entomology, University of Florida
Gainesville, FL 32611
[2]Unité de Mycologie, Institut Pasteur
76724 Paris Cedex 15, France

INTRODUCTION

Arthropods, lacking the true immunoglobulin and T-type lymphocyte system characteristic of vertebrates, possess an array of defense systems which protect them against fungal attack. Whether common defense mechanisms are possessed by the arthropod group is unclear. The defensive needs of this diverse group of organisms are extremely variable. To date, the immune responses of a relatively few species has been examined in detail. Both cellular and/or humoral (non-cellular, inducible antimicrobial substances) responses have been reported to be elicited in arthropods when challenged with pathogenic organisms. In light of the available data, considering such responses to be primitive precursors to the vertebrate recognition systems is speculative. However, the defense systems possessed by arthropods have been sufficient to insure the continued survival of this ancient and diverse group of animals.

Potentially, host defense mechanisms may operate against the invading fungus at any stage of the infection process. The general infection sequence of arthropod mycopathogens involves a coordinated differentiation of the fungus. Initial events include the attachment of the spore propagule to the outer surface of the epicuticle. Spores, receiving the proper chemical stimuli from the host epicuticle, will produce germ tubes which orientate to and penetrate both the cuticle and underlying epithelium. For infection to occur the fungi must penetrate the exoskeleton of arthropods. The outer cuticle, a composite layer of chitin, protein, and lipid, is considered to be the primary barrier to fungal invasion. Pathogenic fungi possess the enzymatic and mechanical processes needed for entry through the cuticle layers. Additionally, they are capable of tolerating the various antimycotic components (phenolics, fatty acids, protease inhibitors, etc.) associated with the cuticle. Penetrant hyphae upon reaching the hemocoel (open circulatory system) produce vegetative cells which multiply rapidly, depleting the nutrient stores present in the hemolymph. In many cases, the fungal elements present in the hemocoel will be yeast-like hyphal bodies or in certain cases wall-less protoplasts. Eventually these vegetative cells (usually at host death) will give rise to elongate mycelial forms which will then exit the cuticle and produce externally borne conidiophores.

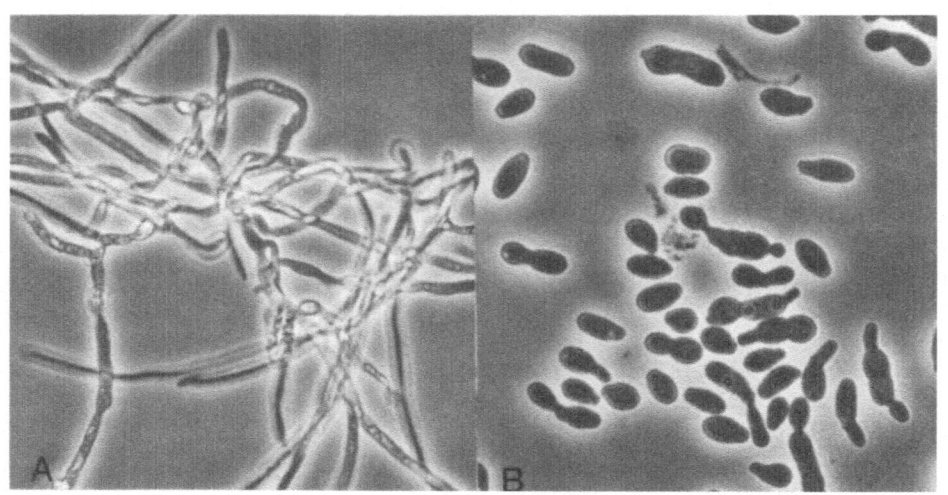

Fig. 1. Light micrographs of the mycelial stage (A) and
yeast-like hyphal body stage (B) of the insect
pathogenic fungus Nomuraea rileyi.

The dimorphic life style possessed by many invertebrate pathogenic
fungi is believed to be required for successful colonization of hosts. For
example, attenuation of various strains of the deuteromycete, Nomuraea
rileyi, is associated with the loss in ability to produce the yeast-like
hyphal body stage under in vitro conditions (Fig. 1). Such attenuated
isolates in vitro grow as mycelium without producing hyphal bodies
(Boucias, unpublished data). Unlike the zoopathogenic fungi which can be
induced to produce yeast cells at 37°C (Cole and Nozawa, 1981), the
invertebrate pathogenic fungi, Entomophaga aulicae and Nomuraea rileyi do
not respond to temperature change. However, similar to zoopathogenic fungi,
dimorphism in certain entomopathogenic fungi can be regulated by CO_2 and
nutrient alternation. For example, Verticillium lecanii blastospores are
induced to be produced by high CO_2 concentrations (Hall and Latgé, 1980).
Growth of N. rileyi on media supplemented with yeast extract and/or fetal
calf serum is characterized by the production of hyphal bodies from germinating conidia. Germination of conidia or transfer of hyphal bodies to a
salt-dextrose-amino acid media results in the production of mycelium (Morrow,
1986). Cultures of E. aulicae in media containing dextrose, yeast extract,
salt(s) and fetal calf serum results in the production of protoplasts. If
the salt component is replaced by a neutral osmotic stabilizer such as
sucrose, which is not metabolized by this fungus, these protoplasts will
convert to a walled mycelium (Beauvais and Latgé, 1987a). The hyphal body
and protoplast conversion to the mycelial form will result in detectable
changes in the carbohydrate composition of the cell wall. The mycelial
form of N. rileyi produce high levels of exogenous β (1,3), (1,6), branched
glucans (Latgé et al., 1987) and does contain surface carbohydrates which
are receptive to galactose-specific lectins (Pendland and Boucias, 1986a).
In E. aulicae and E. musca β (1,3) glucans, the major wall component of
mycelium are absent on the surface of protoplasts. However, the chemical
modifications related to dimorphism have not been stdied in invertebrate
fungi as much as in medical fungi where, for example, dimorphism has been
associated to changes in the conformation (α or β) and composition of the
wall polysaccharides (Cole and Nozawa, 1981).

The fate of mycopathogens which breach the cuticle of resistant host
invertebrates is not well understood. Two principle host defense mechanisms,

cellular and/or humoral, have been reported to protect invertebrates from fungal infections. Whether or not these two mechanisms can operate independently and be effective against invading fungi is uncertain. Recent work on the prophenoloxidase cascade (Soderhall and Smith, 1986) suggests that many of the humoral components (opsonins, protease inhibitors, phenoloxidase) are degranulation products of hemocytes which may be released in response to the presence of cell wall components. However, the majority of studies on arthropods have been carried out under simulated conditions with investigators either challenging hosts with an injection of fungal elements, studying host-fungal interaction under in vitro conditions (hemocyte monolayers), or using artificial substrates (non-fungal, and/or non-arthropod pathogens) as elicitors. In light of the relative small size of many of the invertebrates examined, the defence response obtained in experiments involving injection of fungal elements could appear similar to host wound repair system versus a specific antimicrobial reaction. In certain cases, wounding with a needle has been demonstrated to elicit the release of opsonic material into the hemolymph (Komano et al., 1980). Similarly, the bleeding process used to collect blood cells for "in vitro" studies can be expected to elicit massive degranulation of hemocytes and subsequent self-activation of the prophenoloxidase system (Soderhall and Smith, 1986). In only a few cases have efficient bleeding methods, utilizing osmotically balanced anticoagulant buffers, been developed for hemolymph collection and subsequent isolation (Mead et al., 1986).

Historically, the immune system of arthropods has been reported to lack the "specificity" and "immunologic memory" which characterize the vertebrate system. However, several studies have clearly demonstrated that many arthropod species (mainly Crustacea) possess discriminative clearance mechanisms (Renwrantz, 1983). In general, primary challenge with either live or dead fungal cells has failed to confer a secondary immune response in arthropod systems. However, certain other antigens (bacteria, soluble proteins) have been shown to elicit immune responses which do confer protection against a secondary challenge. For example, several insect species when challenged with bacteria will produce an array of antibacterial response proteins (ARs) (Faye et al., 1975; Hultmark et al., 1983; Boman et al., 1986). The titer of these ARs is retained over time conferring protection to a second challenge (DeVerno et al., 1984; Azambuja et al., 1986; Robertson and Postlethwait, 1986). The production and release of these ARs by the fat body is believed to be induced by peptidoglycans which are released from phagocytised bacterial cells (Dunn et al., 1985). The immune response of the American cockroach to water-soluble antigens such as honeybee and rattlesnake venoms has been shown to possess specificity and immunologic memory (Karp and Rheins 1980; Rheins et al., 1980; Hreins and Karp 1984; Karp, 1985). In fact, the inducible humoral factor produced by these insects to venom challenge has been shown to act like an antibody-like molecule, forming precipitin bands (Rheins and Karp, 1982). Furthermore, the overall response to venom challenge has been correlated with age, with younger adult roaches having a higher degree of immunocompetency than older adults (Rheins and Karp, 1985).

When arthropods are challenged with intact fungal elements (spores, hyphae) an immediate and readily observable cellular response usually occurs, resulting in the attachment of hemocytes to the fungal cell wall surface. Whether or not released fungal metabolites (exocellular enzymes, mycotoxins, etc.) can induce the production of specific. antimycotic components in a fashion similar to the ARs proteins is unclear. It is known that challenge by fungal cells can stimulate increased titers of agglutinins (opsonins), enzymes (prophenoloxidase), and protease inhibitors in the serum of various arthropod hosts (Soderhall and Smith, 1986). However, elicitors other than fungi (i.e., inert substances, cuticle wounding, non-fungal microbes, etc.) may also induce arthropods to produce a similar complex of potential antimicrobial substances.

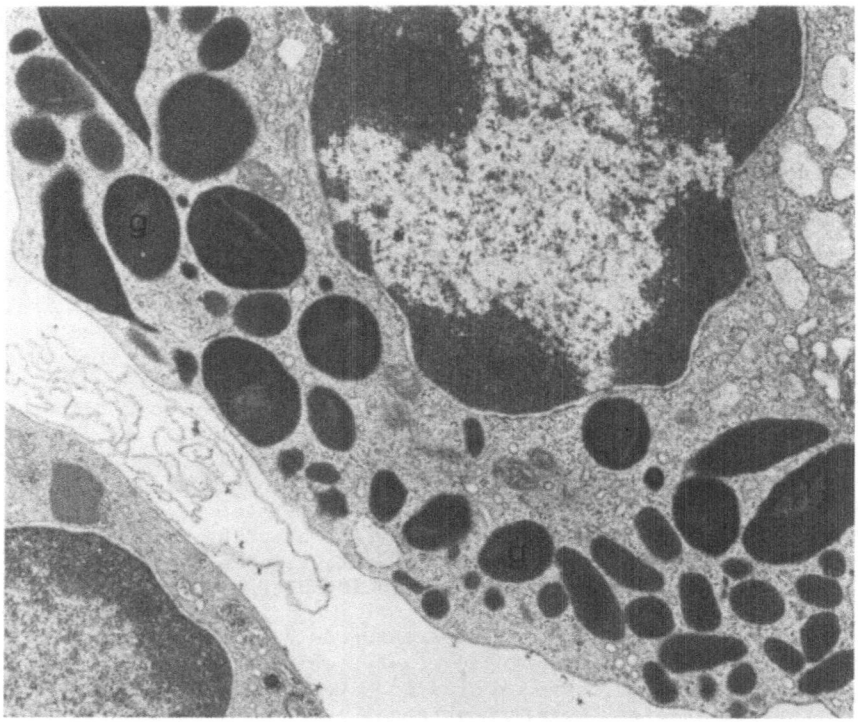

Fig. 2. Electron micrograph of an arthopod granulocyte of the
crayfish <u>Pacifastacus</u> <u>leniusculus</u>. Note the numerous
granules (g) throughout the cytoplasm of this cell.

GENERAL HEMOCYTE RESPONSE AGAINST FUNGI

The hemolymph of invertebrate blood, while containing an array of
morphological discrete blood cell classes, possesses two blood cell types,
the granulocyte and the plasmatocyte which are involved in recognition of
fungal cells (Fig. 2); the names used for these two cells depending on the
arthropod species and on the author nomenclature. The cellular response
to mycopathogens results in phagocytosis, phagocytic encapsulation, or
nodulation (Rowley and Ratcliffe, 1978; Ratcliffe and Rowley, 1975; Gotz,
1986; Vey et al., 1975). The overall kinetics of this cellylar response
in arthropods is very rapid, with initial adherence of cells to foreign
bodies occurring within minutes of their entry within the hemocoel. Small-
sized foreign bodies such as bacteria, yeast cells, or inanimate material
if present in relatively low concentration may be taken up and digested
by phagocytic hemocytes (mostly plasmatocytes) (Ratcliffe and Walters,
1983; Mullainadhan et al., 1984; Brehelin and Hoffman, 1980). Arthropods
challenged with high numbers of bacteria, yeast cells, or organisms larger
than blood cells, such as mycelia or nematodes, will respond by producing
multi-hemocytic granulomas comprised of several layers of more or less
degranulated hemocytes and foreign material aggregates (Nappi, 1975;
Ratcliffe and Gagen, 1976; Ratcliffe and Walters, 1983) (Fig. 3). Initial
contact between hemocytes and foreign material may be mediated by a direct
interaction between host cells receptors and surface components on the
foreign body. Although possessing an open circulatory system, arthropods
do possess constricted sinuses which probably enhance the chances of direct
contact between hemocytes and foreign material. Alternatively the presence
of foreign material may induce the hemocytes and/or other host tissues to
release opsonizing substances which coat foreign material.

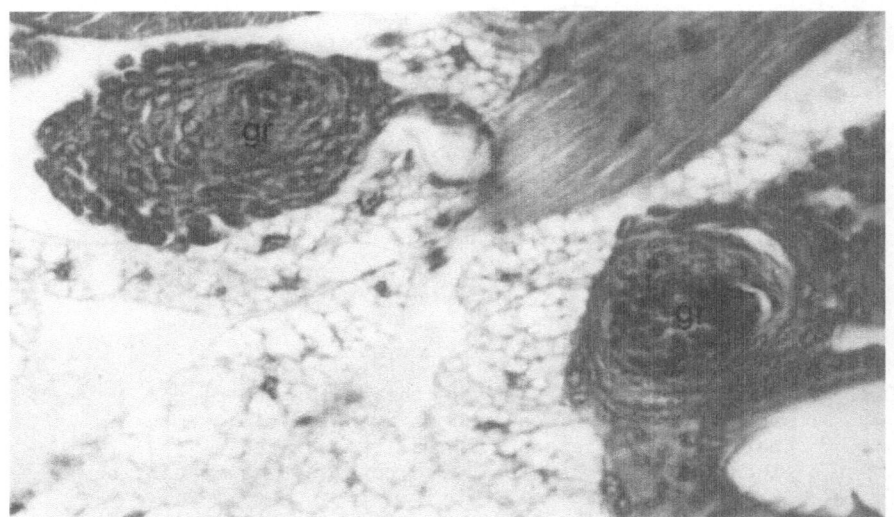

Fig. 3. Granuloma (gr) formed around conidia of the insect
pathogen <u>Cordyceps</u> <u>militaris</u> by hemocytes of <u>Galleria</u>
<u>mellonella</u>.

The selective cellular response to "nonself" by arthropods suggests
recognition of the chemical surface by hemocytes. In arthropods, lectin-
type molecules having specificity to various carbohydrates have been
implicated as mediating the initial recognition response. To date, a
variety of constitutive and humoral agglutinins have been detected in the
serum and/or associated with cell membranes of arthropod hemocytes (Vasta
and Marchalonis, 1985). Many of these lectin-type molecules possess hetero-
agglutinin activity as reflected by their ability to agglutinate a wide
array of vertebrate erythrocytes. These in turn are inhibited by a spectrum
of different carbohydrate and non-carbohydrate molecules (Hartman et al.,
1978; Komano et al., 1980; Yeaton, 1981; Jurenka et al., 1982; Cenini, 1983;
Vasta and Cohen, 1984; Ceri, 1984; Pendland and Boucias, 1985). For example,
the well-characterized humoral lectin, limulin, present in the hemolymph
of horseshoe crabs (Marchalonis and Edelman, 1968) is comprised of multiple
subunits possessing distinct binding affinities (Shimizu et al., 1977;
Robey and Lui, 1981).

In arthropods, the presence of heteroagglutinins having complex binding
capabilities has suggested to many researchers that these molecules are
playing a role in recognition. However, Yeaton (1982) points out that such
molecules, in addition to playing an important role in the immune response
(agglutination, opsonization, etc.), could be also playing a role in the
host developmental processes (sugar and cation transport, mitogenicity,
cell integration, etc.). Several studies have demonstrated that humoral
lectins from arthropods are capable of binding to cell wall surfaces of
various microorganisms. For example, the lectins from lobsters actively
agglutinate non-pathogenic microbes (Cornick and Stewart, 1968). Similarly,
the agglutinins detected in the insects <u>Rhodnius</u> <u>prolixus</u>, <u>Periplaneta</u>
<u>americana</u>, and <u>Melanoplus</u> <u>sanguipes</u> have been found to react with different
protozoan parasites (Pereira et al., 1981; Lackie, 1981; Jurenka et al.,
1982). Recently, a naturally-occurring galactose-binding agglutinin
extracted from the beet armyworm <u>Spodoptera</u> <u>frugiperda</u> has been shown to
bind the cell walls of the non-pathogenic fungus, <u>Paecilomyces</u> <u>farinosus</u>,
which is rich in galactose residues (Pendland and Boucias, 1986b). However,
adsorption assays and rhodamine-labeled lectin studies demonstrated that

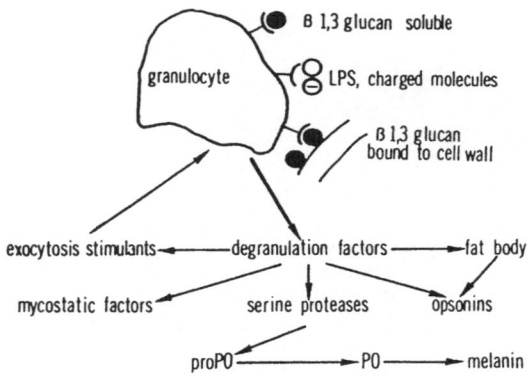

Fig. 4. Schematic representation of the activation of the
prophenoloxidase localized in granulocytes. Note
that both soluble and cell wall associated   (1,3)
glucan molecules are believed to elicit degranulation.

this lectin does not bind to the vegetative cells of the entomopathogenic
fungus <u>Nomuraea</u> <u>rileyi</u>.

The agglutinins which recognize foreign cells could be acting as
mycostatic agents as do certain plant seed lectins (Brambl and Gade, 1985),
as activators for lytic responses (Komano and Natori, 1985), or as opsonins
enhancing their recognition by host phagocytes or phagocytic tissues
(Renwrantz, 1983). As discussed by Vasta and Marchalonis (1985), the
opsonic role that arthropod lectins play remains a subject of controversy.
Pretreatment of foreign bodies with agglutinins or with hemolymph containing
agglutinins has in certain cases resulted in enhanced recognition and
removal by hemocytes (Renwrantz, 1983). However, in other cases humoral
opsonins have been shown not to be necessary for effective recognition
(Yeaton, 1981). The presence of a full complement of cell membrane-bound
opsonins may in part be responsible for this discrepancy. Amirante and
Mazzali (1978) have shown that FITC antibodies against the humoral lectin
of the cockroach will bind to its hemocytes. More recently, Nappi and
Silvers (1984) have reported discrete differences in the cell surface
carbohydrate composion between non-responsive and immunocmpetant hemocyte
populations of <u>Drosophila</u> <u>melanogaster</u>. Whether the immunocompetent cells
are more receptive to the humoral lectin described by Ceri (1984) remains
to be examined.

In recent years several research groups have proposed that the
prophenoloxidase (proPO) activating system is responsible for "nonself"
recognition in arthropods (Soderhall and Smith, 1986). Pioneer research on
the crayfish, <u>Astacus</u> <u>astacus</u> (Soderhall, 1982a; Soderhall and Smith, 1986)
and on the silkworm <u>Bombyx</u> <u>mori</u> (Ashida et al., 1983) has provided valuable
insight into the complexity of this recognition system. This system involves
a cascade reaction comprised of a series of enzymes present as zymogens
which are sequentially activated by divalent cation-dependent proteolysis
(Sonderhall, 1986). The cascade is initiated when the highly sensitive
granulocytes, exocytose and release degranulation factors into the serum
(Fig. 4). Exocytosis of granulocytes will occur immediately when arthropods
are bled or when challenged with proper elicitor molecule. The proPO
activating enzyme, a Ca^{+2} requiring serine protease located in the granulocytes (Fig. 5), will cause limited proteolysis of the proPO resulting
in its conversion to phenoloxidase. Alternatively, the proPO may be
activated by a second serine protease which is triggered by low calcium
concentrations (Soderhall and Smith, 1986). These proteases are believed to
cleave a small peptide from the proPO (Ashida et al., 1974; Soderhall and

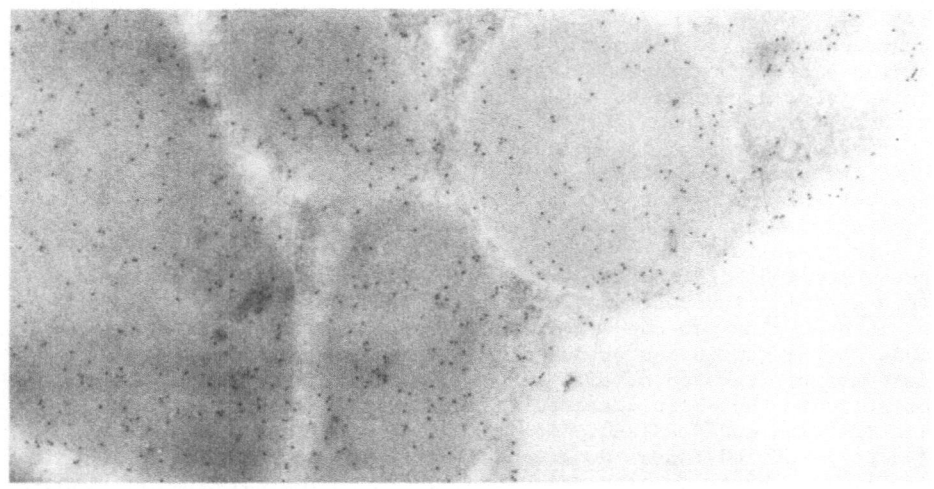

Fig. 5. Cytochemical localization of a serine protease in
the granules of a granulocyte of a crayfish using
an antiserine protease antiserum labeled with a
Protein A-5 nm gold complex.

Smith, 1986). In an inactive form, proPO is located in the same granules
where the serine protease has been visualized. After release and activation phenoloxidase is very "sticky" and has been shown to bind tightly
to foreign surfaces, including the surface of fungal cell walls (Soderhall
et al., 1979; Soderhall, 1981). Phenoloxidase acting on various phenolics
substrates will generate quinones and melanin, which are mycostatic
(Soderhall and Ajaxon, 1982). In addition to the phenoloxidase, activation
of this cascade is believed to result in both the release of other factors
responsible for both eliciting additional granulocytes to exocytosis and
the production of cell adhesive proteins. Activated phenoloxidase alone
or in combination with these cell-adhesive proteins has been shown to act
as an opsonin enhancing phagocytosis of bacterial cells and stimulating
the recruitment and adhesion of plasmatocytes (containing little or no
phenoloxidase) which form multicellular sheets around the coated fungal
cell (Soderhall et al., 1984; Gunnarsson and Lackie, 1985; Ratcliffe et al.,
1984; Smith et al., 1984).

The mass elicitation of the proPO cascade in host invertebrates is
prevented by compartmentalizing the proPO activation to the surface or at
the proximity of the fungal cell wall. A series of serum protease inhibitors
including scavenging macroglobulin type molecules and a series of serine
protease inhibitors are believed to play important roles in regulating this
cascade (Soderhall 1986; Hall and Soderhall, 1982; Sugurmaran et al., 1985).
In addition, several investigators have speculated that the high levels of
protease inhibitors detected in various invertebrate hosts are important
in the defense against invading microorganisms (Kucera, 1982, 1984; Eguchi,
1982; Eguchi and Kanabe, 1982; Eguchi et al., 1982; Sadaki, 1978; Sasaki
and Kobayashi, 1984; Hall and Soderhall, 1982, 1983; Boucias and Pendland,
1986). Mycopathogens are known to utilize a complex of hydrolases for
both the penetration and digestion of host tissues as well as for their
own growth processes. Inhibitors of such activity may abort the infection
process and confer resistance to hosts possessing protease inhibitor
activity. Many arthropods possess a variable number of protease inhibitors
active against a spectrum of different hydrolases. In the silkworm a series
of inhibitors (7,000-60,000 daltons) having either trypsin or chymotrypsin

inhibitor activities have been isolated and found to be active against fungal proteases (Sasaki, 1978; Eguchi et al., 1982). Comparisons among different races of B. mori demonstrated a general polymorphism in protease inhibitor activities (Eguchi et al., 1984). Correlation between the presence of these inhibitors and the susceptibility of the different B. mori races has not yet been reported. Low molecular weight protease inhibitors detected in the wax moth Galleria mellonella have been shown to be highly active against the toxin proteases produced by the mycopathogen Metarhizium anisopliae (Kucera 1982, 1984). Hall and Soderhall (1982, 1983) have also detected a protease in both the blood cells and cuticle of the crayfish A. astacus which inhibits the protease from the crustacean mycopathogen Aphanomyces astaci. Interestingly, this inhibitor was active against peptidases hydrolyzing MeO-Suc-Arg-Pro-Try-PNa, one of the few substrates susceptible to the peptidases produced by mycopathogens A. astaci and Beauveria bassiana (Persson et al., 1984). Recently, high levels of protease inhibitor activity have been detected in the caterpillar Anticarsia gemmatalis (Boucias and Pendland, 1986). Challenging this host with vegetative cells of the fungus Nomuraea rileyi results in increased inhibitor levels (Table 1). Under in vitro conditions, extracted inhibitor fractions inhibited germination and subsequent germ tube elongation of Nomuraea rileyi but not the vegetative reproduction of the hyphal bodies of this pathogenic fungus.

As discussed by Soderhall (1981) and Soderhall and Smith (1986) the proPO cascade is in many respects similar to the well-studied clotting mechanisms of Limulus and other Crustaceans. The clotting system, involving the activation of coagulogen to coagulin by a serine protease can be elicited by either the presence of Ca^{+2}, bacterial endotoxin and/or (1,3) glucans (Kakinuma et al., 1981). Limulus, unlike the crayfish A. astacus, lacks the terminal enzyme phenoloxidase and does not elicit a melanization reaction (Armstrong, 1985). A large molecular weight macroblobulin type inhibitor has been shown to regulate the clotting process in Limulus (Armstrong and Quigley, 1985; Quigley and Armstrong, 1985; Armstrong et al., 1985). Additionally, a family of acid-stable, low molecular weight, site-specific inhibitors have been detected in Limulus which are active against both serine and metal proteases. Both classes of protease inhibitors are released by hemocytes which have been induced to undergo exocytosis (Armstrong, 1985). Comparisons between the arthropod proPO and the

Table 1. Inhibitor of different proteolytic enzymes by native and N. rileyi challenged A. gemmatalis hemolymph samples.

Enzyme	Relative units/ul hemolymph[a]	
	Control	Challenged
Trypsin	1.2	1.9
Chymotrypsin	1.5	1.6
Aspergillus protease	0.2	1.6
Rhizopus protease	5.6	10.0
Protease K	1.2	0.9
Pepsin	6.3	10.4
Subtilisin	1.1	0.7

[a]One inhibitor unit is equivalent to a decrease in the absorbance of 0.01 of inhibitor-enzyme substrate mixtures. Substrates used were either azocoll or hide azure blue.

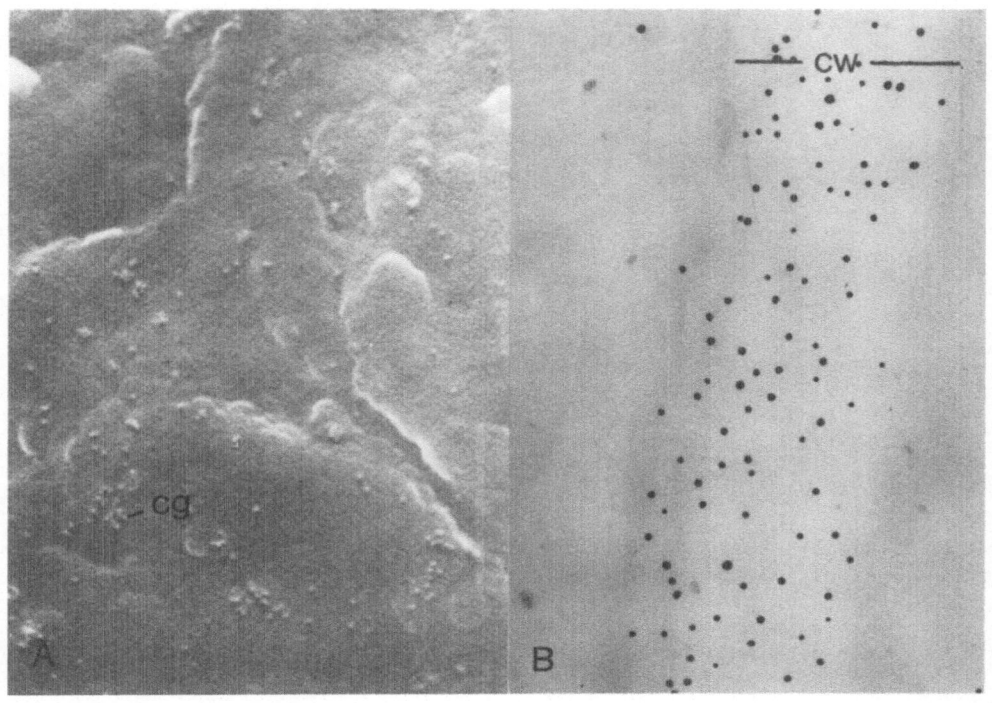

Fig. 6. Cytochemical localization of β (1,3) glucans on the
surface (6A) and within the cell wall (cw, 6B) of the
Entomophtorales. Anti-laminaribiose-IgG in combination
with anti-IgG-colloidal gold 45 nm (A) or Protein
A-colloidal gold 12 nm (B) was used to localize the
(1,3) glucan molecules.

vertebrate complement system have also shown certain similarities
(Soderhall and Smith, 1983; Soderhall and Smith, 1986). Whether or not
this cascade is the universal recognition system operating in arthropods
is unclear. Only a few species of Crustacea and Insects have been examined
to date. It is known, as evidenced by previous research on certain
arthropods, that the complete cascade is not required for cellular recognition. For example, in the black cell mutant of <u>Drosophila</u> <u>melanogaster</u>,
the phenoloxidase-containing crystal cells are absent and are replaced by
a mutant form of these cells. Despite the lack of functioning crystal
cells and the absence of phenoloxidase in the cell-free hemolymph, these
mutants are able to effectively encapsulate foreign bodies, producing
amelonotic capsules (Rizki and Rizki, 1984).

ACTIVATION OF THE ARTHROPOD CELL DEFENSE REACTION

The cellular defense response of arthropods can be elicited by minute
quantities of β (1,3) glucan molecules which are ubiquitous to the fungal
cell wall (Smith and Soderhall, 1983; Soderhall and Unestam, 1978; Soderhall,
1982b). Both exocellular (soluble) and wall bound β (1,) glucan molecules
have been found to effectively activate the degranulation process. For
example, the mycelium of the Entomophthorales which is very effective in
inducing a cellular response, produces a cell wall containing high levels
(>50%) of linear β (1,3) glucan (Latge and Beauvais, 1987). Cytohistochemical
localization of β (1,3) glucan using anti-laminaribiose serum IgG coupled with
protein-A colloidal gold has revealed the presence of this polysaccharide on the
surface as well as within the cell wall (Latge, unpublished) (Fig. 6).

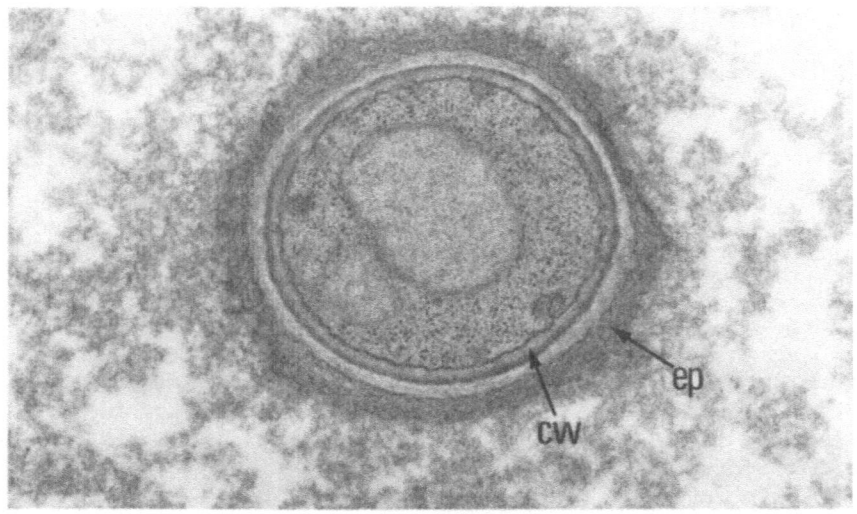

Fig. 7. Electron micrograph of mycelium of <u>Nomuraea</u> <u>rileyi</u>
after intrahemocoelic injection into <u>Galleria</u> <u>mellonella</u>.
Note the association of degranulation products with the
exocellular polysaccharide material (ep) which surrounds
the cell wall (cw) of the mycelium.

Alternatively the mycelium of the deuteromycete <u>Nomuraea</u> <u>rileyi</u>, unlike
the hyphal body stage, produces large quantities of exocellular polysaccharide composed of a branched   (1,3) (1,6) glucan complex (Latge et al.,
1987). Micrographs of mycelia which have been injected into lepidopterous
host larvae have revealed that the degranulation products are released and
are bound to this extracellular polysaccharide material (Fig. 7). Various
other carbohydrates such as   (1,4) glucans, chitin and   (1,6) glucans
which are also important wall constituents do not elicit a cellular defense
response (Soderhall and Unestam, 1978). However, other microbial wall
components, such as the lipopolysaccharides (LPS) associated with Gram-negative
bacteria are highly effective in eliciting this cell defense reaction.
Recent research has indicated the existence of separate receptors for
(1,3) glucan and LPS activation (Ashida et al., 1986). The presence of
  (1,3) glucan or LPS elicitors will initiate the activation of a proPO
activating serine protease present as a zymogen in the granulocytes
(Ashida et al., 1974; Ashida and Dohke, 1980; Soderhall and Smith, 1983;
Leonard et al., 1985). Potentially other proteases, such as those exocellular proteases released by the fungus during penetration and growth
may also serve as activators to the proPO cascade. In addition to these
neutral molecules charged radicals can also stimulate the cascade in a
non-specific way as demonstrated with DEAE resins (Lackie, 1986).

INHIBITION OF THE HOST DEFENSE REACTION

A variety of pathogens and parasites are capable of overcoming the
cellular and/or humoral defense mechanisms of arthropods. Current understanding of the mechanisms responsible for the successful in vivo development of mycopathogens in arthropods is limited. As mentioned previously,
many of the mycopathogens will replicate in the yeast-like hyphal body form
in the host hemocoel. Recent studies on the Deuteromycete, <u>Nomuraea</u> <u>rileyi</u>
have shown that this stage, unlike its mycelial form does not elicit a strong
cellular response and is virulent when injected into the hemocoel of

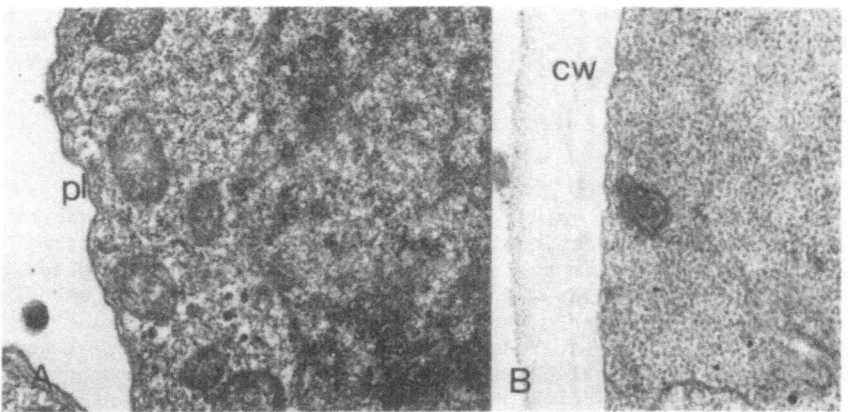

Fig. 8. Electron micrograph of the protoplast (A) and
walled hypha (B) of the Entomophthorale
Entomophaga aulicae. Note the lack of a cell
wall (cw) associated with the plasmalemma (pl)
of the protoplast.

lepidopterous larvae (Boucias, unpublished). Injections of mycelial
fragments elicits an immediate cellular response resulting in the formation
of melanized capsules. None of insect larvae injected with mycelium succumb
to mycosis. These results indicate that the hyphal body stage either does
not release high levels of the (1,3) glucans and/or does not contain
the receptor complex suitable for opsonin or hemocyte attachment. Alternatively, the hyphal body stage may be producing inhibitors such as mycotoxins to the host defense response (Vey, 1986). Work by Huxham and Lackie
(1986) using an in vitro insect hemocyte monolayer system has demonstrated
that the fungus, Metarhizium anisopliae produce metabolites which suppress
the proPO cascade associated with granulocytes. In most cases, mycopathogens
have been shown to be resistant to the encapsulation reaction. For example,
the histological study by Gotz and Vey (1974) showed that the fungus,
Beauveria bassiana actively penetrated and elicited the humoral encapsulation
response of host Chironomus larvae. This fungus was not inhibited by the
deposition of melanin, and continued to grow and eventually kill the host.
Certain members of the Entomophthorales, produce a wall-less protoplast
stage (Fig. 8) within the host hemocoel which undergoes growth and division
without eliciting a defense reaction (Dunphy and Nolan, 1980; 1981; 1982).
Injection of walled hyphal bodies of these "protoplast" strains into their
respective hosts elicited the multicellular encapsulation reaction (Latge
et al., 1986). Chemical analyses of the hyphal body wall of the protoplast
producing species Entomophaga aulicae demonstrated the presence of unbranched
(1,3) glucans (50-60% wall dry weight) and chitin (Latge and Beauvais,
1986). Analysis of the (1,3) glucan synthetase system in protoplasts
versus hyphal bodies have demonstrated that glucan synthetase is present in
both forms. The constituents in the host hemolymph are able to block the
"activation" of this enzyme in protoplasts and the relative activity of this
enzyme is much lower in protoplasts than in the walled form (Beauvais and
Latge, 1987b). Consequently, no (1,3) glucan is synthesized and is
present on the surface of the protoplast membrane (Latge, unpublished).
Formaldehyde killed protoplasts did not induce cell reactions after intrahemocoelic injections indicating that, contrary to deuteromycetes, Entomophthorale toxins or metabolites are not involved in the inhibition of
granuloma formation by protoplasts. Moreover, using an anti-protoplast antiserum, it was demonstrated that insect serum proteins do not stick to protoplast membranes resulting in their masking to the arthropod defense system.

The inability of protoplasts to elicit proPO activation in its respective host insects seems to result only from the lack of β (1,3) glucans on the fungal surface.

SUMMARY

Arthropods lacking the typical immunoglobulin and T-type lymphocyte recognition system characteristic of vertebrates, possess an effective cellular defense system. This system is comprised of a complex cascade of biomolecules including a variety of opsonins, enzymes (proteases, phenoloxidase) and enzyme inhibitors. At present only a few components of this defense system have been isolated and characterized. Very little if anything is known about the interaction among fungal antigens (elicitors) and components of the host cell defense reaction. The cellular defense of arthropods can be readily activated in a non-specific fashion by soluble and cell-bound β (1,3) glucan molecules associated with invading fungi. The lack of such molecules in certain stages of fungi causing systemic mycosis has been documented in several species (E. aulicae, N. rileyi) and is believed to be vital to their successful development in various insect hosts. Further characterization of other fungal antigens (elicitors) such as exocellular proteases which may be specific are needed to explain the variable host response of arthropods to fungal attack.

REFERENCES

1. G.A. Amirante and F.G. Mazzeli, 1978, Synthesis and localization of hemoagglutinins in hemocytes of the cockroach Leucophaea maderae, Dev. Comp. Immunol., 2:735.
2. P.B. Armstrong and J.P. Quigley, 1985, Proteinase inhibitory activity released from the horseshoe crab blood cell during exocytosis, Biochemica et Biophysica Acta 827:453-459.
3. P.B. Armstrong, 1985, Adhesion and motility of the blood cells of Limulus, in: "Blood Cells of Marine Invertebrates: Experimental Systems in Cell Biology and Comparative Physiology", Alan R. Liss, Inc.
4. P.B. Armstrong, M.T. Rossener, and J.P. Quigley, 1985, An 2-macroglobulinlike activity in the blood of chelicerate and mandibulate arthropods, J. Exp. Zool., 236:109.
5. M. Ashida, K. Dohke, and E. Ohnishi, 1974, Activation of prephenoloxidase III. Release of a peptide from prephenoloxidase by the activating enzyme, Biochem. Biophy. Res. Comm., 57:1089-1095.
6. M. Ashida and K. Dohke, 1980, Activation of pro-phenoloxidase by the activating enzyme of the silkworm, Bombyx mori, Insect Biochem. 10:37-47.
7. M. Ashida, Y. Ishizaki, and H. Iwahana, 1983, Activation of prophenoloxidase by bacterial cell walls or (1,3) glucans in plasma of the silkworm Bombyx mori, Biochem. Biophy. Res. Comm., 113:562-568.
8. M. Ashida, M. Ochiai, and H. Yoshida, 1986, (1,3) glucan receptor and peptidoglycan receptor within prophenoloxidase activating system in insects. In: Molecular Aspects of Invertebrate Immunology, ISDCJ Conference, Berlin, 20-21.
9. P. Azambuja, C.C. Freitas, and E.S. Garcia, 1986, Evidence and partial characterization of an inducible antibacterial factor in the haemolymph of Rhodnuis prolixus, J. Insect Physiol., 32:807-812.
10. A. Beauvais and J.P. Latgé, 1987a, A simple medium to grow Entomophthoralean protoplasts, J. Invertebr. Pathol. (submitted).

11. A. Beauvais and J.P. Latgé, 1987b, Glucan synthetases of protoplastic entomophatorales, Exp. Mycol. (submitted).

12. D.G. Boucias and J.C. Pendland, 1986, Detection of protease inhibitors in the hemolymph of resistant Anticarsia gemmatalis which are inhibitory to the entomopathogenic fungus, Nomuraea rileyi, Experientia, (in press).

13. H.G. Boman, I. Faye, P.V. Hofsten, K. Kockum, J.Y. Lee, K.G. Xanthopoulos, H. Bennish, A. Engstrom, B.R. Merrifield, and D. Andrau, 1986, Antibacterial immune proteins in insects: a review of some current perspectives, in "Immunity in Invertebrates, M. Brehelin, ed., Springer-Verlag, NY.

14. R. Brambl and W. Gade, 1985, Plant seed lectins disrupt growth of germinating fungal spores, Phsiol. Plant, 64:402.

15. M. Brehelin and J.A. Hoffman, 1980, Phagocytosis of inert particles in Locusta migratoria and Galleria mellonella: study of ultrastructure and clearance, J. Insect Physiol., 23:103-111.

16. P. Cenini, 1983, Comparative studies of haemagglutinins and haemolysins in an annelid and a primitive crustacean, Dev. Comp. Immunol., 7:637-640.

17. H.Ceri, 1984, Lectin activity in adult and larval Drosophila melanogaster, Insect Biochem., 14:547-549.

18. G.T. Cole and Y. Nozawa, 1981, Dimorphism in: "Biology of Conidial Fungi, Vol. 1, G.T. Cole and B. Kendrick, eds., Academic Press, NY, 97-133.

19. J.W. Cornick and J.E. Stewart, 1968, Interaction of the pathogen Gaffikya homari with natural defense mechanisms of Homarus americanus, J. Fish. Res. Board, Can., 25:695-709

20. P.J. DeVerno, J.S. Chadwick, W.P. Aston, and G.B. Dunphy, 1984, The in vitro generation of an antibacterial activity from the fat body and hemolymph of non-immunized larvae of Galleria mellonella, Dev. Comp. Immunol., 8:537-546.

21. P.E. Dunn, Dai, Wei, M.R. Kanost, and C. Geng, 1985, Soluble peptidoglycan fragment stimulate antibacterial protein synthesis by fat body from larvae of Manduca sexta, Dev. Comp. Immunol., 9:559-568.

22. G.B. Dunphy and R.A. Nolan, 1980, Response of Eastern hemlock looper hemocytes to selected stages of Entomophthora egressa and other foreign particles, J. Invertebr. Pathol., 36:71-84.

23. G.B. Dunphy and R.A. Nolan, 1981, A study of the surface proteins of Entomophthora egressa protoplasts and of larval spruce budworm hemocytes, J. Invertebr. Pathol., 38:352-361.

24. G.B. Dunphy and R.A. Nolan, 1982, Cellular immune responses of spruce budworm larvae to Entomophthora egressa protoplasts and other test particles, J. Invertebr. Pathol., 39:81-92.

25. M. Eguchi, 1982, Inhibition of the fungal protease by haemolymph protease inhibitors of the silkworm Bombyx mori L. (Lepidoptera: Bombycidae), Appl. Ent. Zool., 17(4):589-590.

26. M. Eguchi and M. Kanbe, 1982, Changes in haemolymph protease inhibitors during metamorphosis of the silkworm Bombyx mori L. (Lepidoptera; Bombycidae), Appl. Ent. Zool., 17(2):179-187.

27. M. Eguchi, I. Haneda, and A. Iwamoto, 1982, Properties of protease inhibitors form the haemolymph of silkworms, Bombyx mori, Antheraea pernyi and philosamia Cynthia Ricini, Comp. Biochem. Physiol., 71B:569.

28. M. Eguchi, K. Ueda, and M. Yamashita, 1984, Genetic variants of protease inhibitors against fungal protease and chymotrypsin from hemolymph of the silkworm, Bombyx mori, Biochem. Genet., 22:1093-1102.

29. I. Faye, A. Pye, T. Rasmuson, H.G. Boman, and I.A. Boman, 1975, II. Simultaneous induction of antibacterial activity and selective synthesis of some hemolymph proteins in diapausing pupae of Hyalophora cercropia and Samia cynthia, Infect. Immunol., 12: 1426-1438.

30. P. Gotz and A. Vey, 1974, Humoral encapsulation in Diptera (Insecta):

defence reactions of <u>Chironomus</u> <u>larvae</u> against fungi, Parasitology, 68:192–205.

31. P. Gotz, 1986, Encapsulation in arthropods, <u>in</u>: "Immunity in Invertebrates", M. Brehelin, ed., Springer-Verlag, NY.

32. S.G.S. Gunnarsson and A.M. Lackie, 1985, Hemocytic aggregation in <u>Schistocerca</u> <u>gregaria</u> and <u>Periplaneta</u> <u>americana</u> as a response to injected substances of microbial origin, <u>J</u>. <u>Invertebr</u>. <u>Pathol</u>., 46:312–319.

33. R.A. Hall and J.P. Latgé, 1980, Etude de quelques facteurs stimulant la formation in vitro de blastospores de <u>Verticillium</u> <u>lecanii</u> (Zimm) C.R. Viegas, <u>Acad</u>. <u>Sci</u>., Paris, 291D, 75–78.

34. R.A. Hall and K. Soderhall, 1982, Purification and properties of a protease inhibitor from crayfish hemolymph, <u>J</u>. <u>Invertebr</u>. <u>Pathol</u>., 39:29–37.

35. R.A. Hall and K. Soderhall, 1983, Isolation and properties of a protease inhibitor in crayfish (<u>Astacus</u> <u>astacus</u>) cuticle, <u>Comp</u>. <u>Biochem</u>. <u>Physiol</u>., 76B:699–702.

36. K.D. Hapner, 1983, Haemagglutinin activity in the haemolymph of individual acridiae (grasshopper) specimens, 1983, <u>J</u>. <u>Insect</u> <u>Physiol</u>., 29:101–106.

37. A.L. Hartman, P.A. Campbell, and C.A. Abel, 1978, An improved method for the isolation of lobster lectins, <u>Develop</u>. <u>Comp</u>. <u>Immunol</u>., 2:617–625.

38. D. Hultmark, A. Engstrom, K. Andersson, H. Steiner, H. Bennich, and H.G. Boman, 1983, Insect immunity. Attacins, a family of antibacterial proteins from <u>Hyalophora</u> <u>cercropia</u>, EMBO 2:571–576.

39. I.M. Huxham and A.M. Lackie, 1986, A simple visual method for assessing the activation and inhibition of phenoloxidase production by insect haemocytes in vitro, <u>J</u>. <u>Immunol</u>. <u>Methods</u> (in press).

40. R. Jurenka, K. Manfredi, and K.D. Hapner, 1982, Haemagglutinin activity in Acrididae (grasshopper) haemolymph, <u>J</u>. <u>Insect</u>. <u>Physiol</u>., 28:177–181.

41. A. Kakinuma, T. Asano, H. Torii, and Y. Sugino, 1981, Gelation of <u>Limulus</u> amoebocyte lysate by an antitumor (1,3)-β-D-glucan, <u>Biochem</u>. <u>Biophy</u>. <u>Res</u>. <u>Comm</u>., 101:434–439.

42. R.D.Karp and L.A. Rheins, 1980, Induction of specific humoral immunity to soluble proteins in the American cockroach (<u>Periplaneta</u> <u>americana</u>) II. Nature of the secondary response, <u>Dev</u>. <u>Comp</u>. <u>Immunol</u>., 4:629–639.

43. R.D. Karp, 1985, Preliminary characterization of the inducible humoral factor in the American cockroach (<u>Periplaneta</u> <u>americana</u>), <u>Dev</u>. <u>Comp</u>. <u>Immunol</u>., 9:569–575.

44. H. Komano, D. Mizuno, and S. Natori, 1980, Purification of lectin induced in the hemolymph of <u>Sarcophaga</u> <u>peregrina</u> larvae on injury, <u>J</u>. <u>Biol</u>. <u>Chem</u>., 255:2919–2924.

45. H. Komano and S. Natori, 1985, Participation of <u>Sarcophaga</u> <u>peregrina</u> humoral lectin in the lysis of sheep red blood cells injected into the abdominal cavity of larvae, <u>Dev</u>. <u>Comp</u>. <u>Immunol</u>., 9:31–40.

46. M. Kucera, 1982, Inhibition of the toxic proteases from <u>Metarhizium</u> <u>anisopliae</u> by extracts of <u>Galleria</u> <u>mellonella</u> larvae, <u>J</u>. <u>Invertebr</u>. <u>Pathol</u>., 40:299–300.

47. M. Kucera, 1984, Partial purification and properties of <u>Galleria</u> <u>mellonella</u> larvae proteolytic inhibitors acting on <u>Metarhizium</u> <u>anisopliae</u> toxin protease, <u>J</u>. <u>Invertebr</u>. <u>Pathol</u>., 43:190–196.

48. A. Lackie, 1986, Immune mechanisms in Invertebrate Vectors, Oxford University Press, Oxford, 300 pp.

49. A. Lackie, 1981, Humoral mechanisms in the immune response of insects to larvae of <u>Hymenoplepis</u> <u>diminuta</u> (Cestoda), <u>Parasit</u>. <u>Immunol</u>., 3:201.

50. J.P. Latgé, A. Beauvais, and A. Vey, 1986, Wall synthesis in the entomophthorales and its role in the immune reaction of infected insects, <u>in</u>: Molecular Aspects of Invertebrate Immunology, ISDCJ, Conference, Berlin, 92–93.

51. J.P. Latgé and A. Beauvais, 1987, Wall composition of protoplastic entomophthorales, J. Invertebr. Pathol. (in press).

52. J.P. Latgé, D.G. Boucias, and B. Fournet, 1987, Structure of the extracellular polysaccharide produced by the fungus Nomuraea rileyi, Carbohydrate Res. (in press).

53. C. Leonard, K. Soderhall, and N.A. Ratcliffe, 1985, Studies of prophenoloxidase and protease activity of Blaberus craniifer haemocytes, Insect Biochem., 15:803-810.

54. J.J. Marchalonis and G.M. Edelman, 1968, Isolation and characterization of a natural hemagglutinin from Limulus polyphermus, J. Mol. Biol., 32:453-465.

55. G.P. Mead, N.A. Ratcliffe, and L.R. Renwrantz, 1986, The separation of insect haemocyte types on percoll gradients; methodology and problems, J. Insect Physiol., 32:167-177.

56. B. Morrow, 1986, In vitro growth of the dimorphic fungal entomopathogen Nomuraea rileyi (Farlow) Samson, emphasizing protoplast production and cell wall analysis, Masters Thesis, University of Florida, Gainesville, FL, USA, 99 pp.

57. P. Mullainadhan, M.H. Ravindranath, R.E. Wright, and E.L. Cooper, 1984, Crustacean defense strategies I. Molecular weight dependent clearance of dyes in the mud crab Scylla serrata (Forskal) (Portunidae: brachyura), Dev. Comp. Immunol., 8:41-50.

58. A.J. Nappi, 1975, Parasite encapsulation in insects, in: "Invertebrate Immunology", K. Maramorsch, ed., Academic Press Inc., NY.

59. A.J. Nappi and M. Silvers, 1984, Cell surface changes associated with cellular immune reactions in Drosophila, Science, 225:1166-1168.

60. J.C. Pendland and D.G. Boucias, 1985, Hemagglutinin activity in the hemolymph of Anticarsia gemmatalis larvae infected with the fungus Nomuraea rileyi, Dev. Comp. Immunol., 9:21-30.

61. J.C. Pendland and D.G. Boucias, 1986a, Lectin binding characteristics of several entomogenous hyphomycetes: Possible relationship to insect hemagglutinins, Mycologia., 78:818-824.

62. J.C. Pendland and D.G. Boucias, 1986b, Characteristics of a galactose-binding hemagglutinin (lectin) from hemolymph of Spodoptera exigua larvae, Dev. Comp. Immunol., 9:21-30.

63. M.E.A. Pereira, A.F.B. Andrade, and J.M.C. Ribeiro, 1981, Lectins of distinct specificity in Rhodnius prolixus interact selectively with Trypanosoma cruzi, Science, 211:597.

64. M. Persson, L. Hall, and K. Soderhall, 1984, Comparison of peptidase activities in some fungi pathogenic to arthropods, J. Invertebr. Pathol., 44:342-348.

65. J.P. Quigley and P.B. Armstrong, 1985, A homologue of α-2-macroglobulin purified from the hemolymph of the horseshoe crab Limulus polyphemus J. Biol. Chem., 260:12715-12719.

66. N.A. Ratcliffe and A.F. Rowley, 1975, Cellular defense reactions of insect hemocytes in vitro: phagocytosis in a new suspension culture system, J. Invertebr. Pathol., 26:225-233.

67. J.A. Ratcliffe and S.J. Gagen, 1976, Cellular defense reactions of insect hemocytes in vivo: nodule formation and development in Galleria mellonella and Pieris brassicae larvae, J. Invertebr. Pathol., 28:373-382.

68. N.A. Ratcliffe and J.B Walters, 1983, Studies on the in vivo cellular reactions of insects: clearance of pathogenic and non-pathogenic bacteria in Galleria mellonella larvae, J. Insect. Physiol, 29: 407-415.

69. N.A. Ratcliffe, C. Leonard, and A.F. Rowley, 1984, Prophenoloxidase activation: nonself recognition and cell cooperation in insect immunity, Science, 226:557-559.

70. L. Renwrantz, 1983, Involvement of agglutinins (lectins) in invertebrate defense reactions: The immuno-biological importance of carbohydrate-specific binding molecules, Dev. Comp. Immunol., 7:603-608.

71. L.A. Rheins, R.D. Karp, and A. Butz, 1980, Induction of specific humoral immunity to soluble proteins in the American cockroach (Periplaneta americana), Dev. Comp. Immunol., 4:447-458.

72. L.A. Rheins and R.D. Karp, 1982, An inducible humoral factor in the American cockroach (Periplaneta americana): Precipitin activity that is sensitive to a proteolytic enzyme, J. Invertebr. Pathol., 40:190-196.

73. L.A. Rheins and R.D. Karp, 1984, The humoral immune response in the American cockroach Periplaneta americana: reactivity to a defined antigen from honeybee venom, phospholipase A_2, Dev. Comp. Immunol., 8:791-801.

74. L.A. Rheins and R.A. Karp, 1985, Ontogeny of the invertebrate humoral immune response: studies on various developmental stages of the American cockroach (Periplaneta americana), Dev. Comp. Immunol., 9:395-406.

75. T.M. Rizki and R.M. Rizki, 1984, The cellular defense system of Drosophila melanogaster, in: "Insect Ultrastructure Vol. 2", R.C. King and H. Akai, Eds., Plenum, NY.

76. M. Robertson and J.H. Postlethwait, 1986, The humoral antibacterial response of Drosophila adults, Dev. Comp. Immunol., 10:167-179.

77. F.A. Robey and T-Y Lui, 1981, Limulin: A C-reactive protein from Limilus polyphemus, J. Biol. Chem., 256:969

78. A.F. Rowley and N.A. Ratcliffe, 1976, The granular cells of Galleria mellonella during clotting and phagocytic reactions in vitro, Tissue and Cell, 8(3):437-446.

79. T. Sasaki, 1978, Chymotrypsin inhibitors from hemolymph of the silkworm, Bombyx mori, J. Biochem., 83:367-376.

80. T. Sasaki and K. Kobayashi, 1984, Isolation of two novel proteinase inhibitors from hemolymph of silkworm larva, Bombyx mori. Comparison with human serum proteinase inhibitors, J. Biochem., 95:1009-1017.

81. S. Shimizu and M. Niwa, 1977, Lectins in the hemolymph of the Japanese horseshoe crab, Tachypleus tridentatus, Biochem., Biophys. Acta 500, FL.

82. V.J. Smith and K. Soderhall, 1983, β-(1,3)-glucan activation of crustacean hemocytes in vitro and in vivo, Biol. Bull., 164:299-314.

83. V.J. Smith, K. Soderhall, and M. Hamilton, 1984, β-(1,3)-glucan induced cellular defence reactions in the shore crab, Carcinus maenas, Comp. Biochem. Physiol., 77A:635-639.

84. K. Soderhall and T. Unestam, 1978, Activation of serum prophenoloxidase in arthropod immunity. The specificity of cell wall glucan activation and activation by purified fungal glycoproteins of crayfish phenol-oxidase, Can. J. Microbiol., 25:406-414.

85. K. Soderhall, L. Hall, T. Unestam, and L. Nyhlen, 1979, Attachment of phenoloxidase to fungal cell walls in arthropod immunity, J. Invertebr. Pathol., 34:285-294.

86. K. Soderhall, 1981, Fungal cell wall β-(1,3)-glucans induce clotting and phenoloxidase attachment to foreign surfaces of crayfish hemocyte lysate, Dev. Comp. Immunol., 5:565-573.

87. K. Soderhall and J. Ajaxon, 1982, Effect of quinones and melanin on mycelial growth of Aphanomyces spp. and extracellular protease of Aphanomyces astaci, a parasite on crayfish, J. Invertebr. Pathol., 39:105-109.

88. K. Soderhall, 1982a, Prophenoloxidase activating system and melanization a recognition mechanisms of arthropods? A review, Dev. Com. Immunol., 6:601-611.

89. K. Soderhall, 1982b, β-(1,3)-glucan enhancement of protease activity in crayfish hemocyte lysate, Comp. Biochem. Physiol., 74B:221-224.

90. K. Soderhall, 1986, The cellular immune system in crustaceans. In: "Fundamental and Applied Aspects of Invertebrate Pathology, eds., R.A. Samson, J.M. Vlak, and D. Peters, 4th Int. Symp., Invertebr. Pathol., Veldhoven, The Nethrlands, 417-420.

91. K. Soderhall and V.J. Smith, 1983, The prophenoloxidase activating system – a complement like pathway in arthropods? In: "Infection Processes of Fungi", D.W. Roberts and J.R. Aist, eds., Rockefeller Foundation, Bellagio Conference 1983.

92. K. Soderhall, A. Vey, and M. Ramstedt, 1984, Hemocyte lysate enhancement of fungal spore encapsulation by crayfish hemocytes, Dev. Comp. Immunol., 8:23–29.

93. K. Soderhall and V.J. Smith, 1986, The prophenoloxidase activating system: the biochemistry of its activation and role in arthropod cellular immunity, with special reference to crustaceans, in: "Immunity in Invertebrates", M. Brehelin ed., Springer-Verlag, NY.

94. M. Sugumaran, S.J. Saul, and N. Ramesh, 1985, Endogenous protease inhibitors prevent undesired activation of prophenolase in insect hemolymph, Biochem. Biophy. Res. Comm., 132:1124–1129.

95. G.R. Vasta and E. Cohen, 1984, Carbohydrate specificities of Birgus latro (coconut crab) serum lectins, Dev. Comp. Immunol., 8:197–202.

96. G.R. Vasta and J.J. Marchalonis, 1985, Humoral and cell membrane-associated lectins from invertebrates and lower chordates: Specificity, molecular characterization and their structural relationships with putative recognition molecules from vertebrates, Dev. Comp. Immunol., 9:531–539.

97. A. Vey, M. Bouletreau, J.M. Quiot, and C. Vago, 1975, Etude in vitro en microcinematographie des reactions cellulaires d'invertebres vis-a-vis d'agents bacteriens et cryptogamiques, Entomophage 20: 337–351.

98. A. Vey, 1986, Immunosuppressive effect of toxins: action of destruxins on the multicellular defence reaction of insects, Berlin ICCDI Conference.

99. R.W. Yeaton, 1981, Invertebrate lectins: II. Diversity of specificity, biological synthesis and function in recognition, Dev. Comp. Immunol., 5:535–545.

100. R.W. Yeaton, 1982, Are invertebrate lectins primordial receptors, in: "Developmental Immunology: Clinical Problems and Aging", Academic Press, NY.

Candida

IMMUNOCHEMISTRY OF <u>CANDIDA ALBICANS</u> ANTIGENS

Sharon Lynn Salhi and Jean-Marie Bastide

Unite de Recherche en Immunologie
Faculte de Pharmacie
Montpellier, France

INTRODUCTION

Immunochemical studies on <u>Candida</u> species are aimed at classifying and identifying these yeasts for taxonomic and diagnostic purposes. Although the members of the genus <u>Candida</u> have a complex array of cell wall and cytoplasmic antigens, species-specific immunodeterminants have been identified by various immunochemical techniques.

SPECIES-SPECIFIC CELL WALL ANTIGENS

Mannan, which is the generic name for mannose-containing glycoproteins, is one of the major components of the yeast cell wall, and it is also the immunodominant structure on the yeast cell (see review articles by Ballou[2] and Sentandreu et al.[9]). De Repentigny and Reiss[18] described the relationship between the general structure of yeast mannan and its antigenic specificity in the following terms. "The general plan of yeast mannan is a linear $\alpha(1, 6)$ backbone heavily substituted with $\alpha(1, 2)$ and $\alpha(1, 3)$-linked oligomannoside chains. The length of side chains, location of 1, 3 bonds, and substitution with phosphodiesters determine antigenic specificity. Two domains recognized in yeast mannans are the inner core and the outer chain[3]. The inner core structure is a common feature of yeast, plant, and mammalian glycoproteins. It consists of structural protein linked to core mannan. The polymannosyl outer-chain region distinguishes yeast mannans from other glycoproteins. The outer chain is a trimer to pentamer of a repeat unit consisting of 15 mannose residues[3]. This region is of primary concern because it contains the immunodominant epitopes."

Using absorbed or monospecific antisera in slide agglutination tests, Tsuchiya et al.[24] determined the surface antigenic patterns of medically important yeasts and proposed antigenic formulas for their serological classification. The comprehensive studies of Tsuchiya and coworkers and studies of other investigators on the antigenic structures of a large number of yeasts have recently been reviewed[23].

Table 1 lists the surface antigenic formulas for the three serotypes of C.albicans as presented by Tsuchiya et al.[25]. Hasenclever and Mitchell[12] described serotypes A and B in 1961; Fukazawa, Shinoda, and Tsuchiya[10] made reference to serotype C in 1968. There are five antigens common to the

Table 1. Surface Antigenic Compositon of
the Serotypes of *Candida albicans*.

Serotype	Antigenic Formula[a]
C. albicans A	1,2,3,4,5,6,(13b)[b]
C. albicans B	1,2,3,4,5,(7)[b],13b
C. albicans C	1,2,3,4,5,6,7

[a] According to Tsuchiya et al.[25]
[b] Not always detected.

three serotypes. The type differences are based on three antigens: serotype A has antigen 6 and occasionally antigen 13b, serotype B antigen 13b
and occasionally antigen 7, serotype C antigens 6 and 7.

Combined chemical and hapten inhibition studies show that the serotype
specificity of C.albicans resides in the sequence of the mannohexaose side
chains of mannan[18]. The immunodominant hapten of serotype A is a straight
chain 1, 2-linked oligosaccharide with a 1, 3 terminal; type B lacks a
terminal 1, 3 linkage but has a single C-1, C-2, C-3 branch point and an
additional internal 1, 3 linkage[18]. Suzuki and Fukazawa[21] presented evidence
suggesting that antigen 6 may be bound via a β (1 → 6) linkage to C-6 of
mannosyl residues in the oligosaccharide side chain of serotype A mannan.

The surface antigenic composition of C.albicans is compared to that
of other medically important Candida species in Table 2. Inspection of these
antigenic patterns presented by Fukazawa et al.[10] reveals that C.albicans,
C.tropicalis, and C.stellatoidea are very closely related. In their published table, Fukazawa et al. did not indicate the serotype of the C.albicans
strain tested; they stated in the text, however, that they only used serotype
C strains, which prevail in Japan. Given this information, comparison of the
antigenic profiles in Tables 1 and 2 shows that C.tropicalis is closely
related to C.albicans serotype A and C.stellatoidea to C.albicans serotype B.

A comment on the antigenic formulas for C.albicans serotypes proposed
by Tsuchiya and coworkers concerns the schemes for these strains presented
in their recent review[23] where serotype A is listed as having the following
antigenic formula: 1,2,3,4,5,6,7; the authors gave no explanation for
including antigen 7 in this formula for C.albicans A surface factors.

Table 2. Surface Antigenic Composition of Six
Medically Important Species of *Candida*.

Species	Antigenic Formula[a]
C. albicans[b]	1,2,3,4,5,6,7
C. tropicalis	1,2,3,4,5,6
C. stellatoidea	1,2,3,4,5,10,32
C. guillermondii	1,2,3,4,9
C. krusei	1,2,5,11
C. pseudotropicalis	1,8,10,28,31

[a] According to Fukazawa et al.[10]
[b] The authors indicated in their publication
that they only used the type C strain.

Table 3. Surface Antigenic Composition of the
Medically Important Species of *Candida*.

Species	Antigenic Formula[a]												
C. albicans A	1	2	3	4	5	6	7						
C. albicans B	1	2			5					10	11	12	13
C. tropicalis	1	2	3	4	5						11		
C. stellatoïdea	1	2			5				9	10	11	12	
C. guillermondii	1	2	3		5				9	10	11		
C. krusei	1					6					11		
C. parapsilosis	1	2						8		10	11	12	13
C. pseudotropicalis	1						7						

[a] According to Sweet and Kaufman [21]

Sweet and Kaufman [22] also developed a slide agglutination test using
monospecific antisera applicable to a rapid and accurate identification
of medically important species of *Candida*. Their scheme is shown in Table
3. The authors underscored the fact that their coding system is not the
same as that of Tsuchiya and colleagues. Sweet and Kaufman's scheme also
points out the cell wall antigenic similarity between C.albicans A and
C.tropicalis and between C.albicans B and C.stellatoidea.

Miyakawa et al.[15] drew attention to the fact that cross-absorption
for the preparation of specific antisera is often accompanied by a concomitant decrease in antibody titers because of the existence of common
antigens among the *Candida* species. In an attempt to overcome this problem these investigators prepared two monoclonal antibodies (MAbs) directed against mannan antigens of C.albicans serotype A. The specificity
of each MAb was determined by slide agglutination tests for cross-reactivity
patterns with the homologous and other strains of *Candida*. MAb CA4-2
appears to be directed against antigen 6 because this MAb agglutinates
C.albicans A, C.tropicalis, and T.glabrata (which also has antigen 6). MAb
CA5-4 is designated as specific for a new antigen related to antigen 5
because it agglutinates both C.albicans A and B and C.tropicalis, but not
C.krusei or C.parapsilosis. Consequently, Miyakawa et al.[12] proposed
improved antigenic formulas (Table 4) based in part on the two new MAbs
for identifying medically important *Candida* species. The authors deleted
factor 13b, which has been shown to be variable in agglutinating C.albicans
serotype B strains for the following reasons: (i) a high-titered new factor
serum (designated 5b) specific for all the members of C.albicans has been
obtained as MAb and (ii) C.stellatoidea has been reduced to synonomy with
C.albicans, in spite of the latter's ability to assimilate sucrose[15]. This
recent move was based on DNA complementarity studies[14].

Other investigators have also produced anti-Candida MAbs to study the
expression of haptens on the surface of *Candida* species[5,9,16,17]. Chardes
et al.[9] produced 21 MAbs against C.albicans serotype A cells. These MAbs
were selected by an indirect ELISA technique using whole yeast cells bound
to the microtiter plates. Five of the MAbs (CA3-7) show restricted specificity. They are directed against C.albicans serotype A, but not serotype
B, and recognize C.tropicalis and T.glabrata (CA6 and 7) or C.guilliermondii
(CA3, 4, 5). From the surface antigenic formulas presented by Miyakawa
et al.[15], MAbs CA6 and 7 might be directed against antigen 6 and MAbs
CA3, 4, 5 against antigens 6 and 9. The sixteen other MAbs cross-react
with several *Candida* species. These findings showing that MAbs against
C.albicans appear to recognize more than one antigen might be explained by
cross-reactivity between common epitopes [8,20].

Table 4. Revised Surface Antigenic Composition of
Medically Important Species of <u>Candida</u>.

Species	Formula[a] Standard[b]	Revised[c]
C. albicans A	1,4,5,6	1,4,5b,6
C. albicans B	1,4,5,13b	1,4,5b
C. tropicalis	1,4,5,6	1,4,5b,6
C. guillermondii	1,4,9	1,4,9
C. krusei	1,(5)[d],11	1,11
C. parapsilosis	1,(5)[d],13,13b	1,13
C. pseudotropicalis	1,8	1,8

[a] According to Miyakawa et al.[15]
[b] Based on slide agglutination tests with factor sera.
[c] Based on slide agglutination tests with two monoclonal antibodies and factor sera.
[d] Occasionally negative, depending on the lot of antiserum.

Another explanation for the difficulty encountered in identifying specific epitopes on the cell surface of <u>Candida</u> species may lie in the fact that cell wall antigenic expression is a dynamic process as shown by Brawner and Cutler[5]. These workers produced an agglutinating MAb against <u>C.albicans</u> that also reacts with <u>C.tropicalis</u>, <u>C.albicans</u> var. <u>stellatoidea</u> and <u>T.glabrata</u>. The immunodeterminant appears to vary in concentration on different strains of <u>C.albicans</u> (Table 5). Its expression is also dependent upon nutrition and growth phase.

Polonelli and Morace[16] evaluated the potential use of MAbs for serotyping <u>C.albicans</u>; they produced a MAb against a <u>C.albicans</u> antigen extract, called an exoantigen. Immunoblot analysis of exoantigens from yeast isolates with the MAb indicated that all <u>C.albicans</u> strains have a species-specific determinant of 12,000 daltons. The MAb reacts with other antigens in extracts from several strains of <u>C.albicans</u> and cross-reacts with exoantigens from other <u>Candida</u> species (Table 6). The authors characterized five different presumptive types of <u>C.albicans</u>. One of these types has a 20,000 dalton antigen that is also present in the <u>C.tropicalis</u> exoantigen, but the 12,000 dalton component common to all <u>C.albicans</u> strains tested is not present in the <u>C.tropicalis</u> exoantigen. These results suggest that <u>C. albicans</u> <u>C.tropicalis</u> can be differentiated on the basis of their exoantigens.

Table 5. Monoclonal Agglutinin Reactivity
with Various <u>Candida</u> species.

Species	No.of Strains	Agglutination Reaction
C. albicans	42	3-4+
	15	2+
	19	1+
C. albicans var. stellatoidea	1	+
C. tropicalis	7	2-4+
C. parapsilosis	1	-
C. krusei	1	-

[a] Data from Brawner and Cutler[9].

Table 6. Immunoblot Analysis of Yeast Isolates with
an anti-<u>Candida albicans</u> Monoclonal Antibody[a].

Species	Strain	Mol Wt (10^3) of Bands Recognized
C. albicans	CDC B385	12,20,35
	CBS 5983	12,54
	CBS 103	12
	CBS 9931	12,50
	CDC B612	12,35
C. tropicalis	CBS 94	20,44,60
C. guilliermondii	CBS 566	25,45,78
C. krusei	CBS 573	25,48,78
C. parapsilosis	CBS 604	37,66
C. pseudotropicalis	CBS 607	25,45,70

[a] Data from Polonelli and Morace [16]

SPECIES-SPECIFIC CYTOPLASMIC ANTIGENS

The complexity of <u>C.albicans</u> cytoplasmic antigens was shown by Axelsen[1] who detected 78 individual <u>C.albicans</u> antigenic fractions by crossed immunoelectrophoresis (CIE). To detect such a large number of immunoprecipitates, Axelsen ran several electrophoreses with different ratios between the amount of antigen and antibody. The results of Axelsen were confirmed by Lebecq et al.[13] who detected 77 <u>C.albicans</u> serotype A somatic antigens; all these precipitates were detected on the same immuno-plate using a novel antibody overlay technique for CIE.

The CIE technique can be used to identify species-specific antigens in yeast extracts. Guinet and Gabriel[11] performed CIE of a <u>C.tropicalis</u> cytoplasmic extract with anti-<u>C.tropicalis</u> antiserum in the upper part of the gel and an anti-<u>C.albicans</u> antiserum in an intermediate gel; they identified five antigens that might be specific for <u>C.tropicalis</u>.

Biguet et al.[4] found cytoplasmic antigens specific for several <u>Candida</u> species by immunoelectrophoresis. Their results, summarized in part in Table 7, show that the <u>C.albicans</u> strain of undetermined serotype that was studied is antigenically very similar to <u>C.stellatoidea</u> and that <u>C.tropicalis</u> differs from this <u>C.albicans</u> strain by at least six cytoplasmic antigens.

Bruneau and Guinet[6] used crossed immunoaffinoelectrophoresis to identify and characterize glycoproteins in yeast cytoplasmic extracts. On

Table 7. Somatic Antigenic Composition
of Five Candida species.

Species	Antigenic Formula[a]
C. albicans	1-15
C. tropicalis	1-7,17-22
C. stellatoïdea	1-12,15, 16
C. krusei	2,3,30-38
C. pseudotropicalis	1,30,39-48

[a] According to Biguet et al. [4]

the basis of the percentage of antigens having affinity for ligands such as free concanavalin A, concanavalin A-Sepharose, and wheat-germ lectin-Sepharose, C.tropicalis was surprisingly found to be closely related to C.albicans serotype B and a C.stellatoidea strain to C.albicans serotype A. The authors stressed that these results must be confirmed by a study on a larger number of isolates.

Burnie et al.[7] prepared anti-C.albicans antisera against yeast phase and mycelial phase pressates and immunoblotted these antisera against C.albicans in the yeast and mycelial phases and the yeast phase of various species of Candida. Cross-reactivity was greatest against C.parapsilosis. The authors identified a yeast-specific mannoprotein with a molecular mass of 49,000 daltons. An important finding of this study is that no mycelial-specific antigens could be identifed. The authors stated that the absence of such an antigen argues against any need to use the mycelial phase as the basis of a serological test for invasive candidiasis.

It emerges from the studies reviewed herein that the combined use of highly sensitive immunochemical techniques and judiciously selected mono-clonal antibodies should make it possible to identify and characterize immunodominant haptens specific for Candida albicans and related species.

REFERENCES

1. N.H. Axelsen, Quantitative immunoelectrophoretic methods as tools for a polyvalent approach to standardization in the immunochemistry of Candida albicans, Infect. Immun. 7:949 (1973).
2. C.E. Ballou, Some aspects of the structure, immunochemistry, and genetic control of yeast mannans, Adv. Enzymol. 40:239 (1974).
3. C.E. Ballou, Structure and biosynthesis of the mannan component of the yeast cell envelope, Adv. Microb. Physiol. 14:93 (1976).
4. J. Biguet, P. Tran Van Ky, and S. Andrieu, Etude immunoelectrophoretique et immunochimique comparee des antigenes de quelques levures du genre Candida (C.albicans, C.pseudotropicalis, C.macedoniensis), Mycopathologia 17:239 (1962).
5. D.L. Brawner and J.E. Cutler, Variability in expression of a cell surface determinant on Candida albicans as evidenced by an agglutinating mono-clonal antibody, Infect. Immun. 43:966 (1984).
6. S.M. Bruneau and R.M.F. Guinet, Characterization of Candida antigens by crossed-immunoaffinoelectrophoresis, J. Med. Vet. Mycol. 23:107 (1985).
7. J.P. Burnie, R. Matthews, A. Fox, and S. Tabagchali, Use of immuno-blotting to identify antigenic differences between the yeast and mycelial phases of Candida albicans, J. Clin. Pathol. 38:701 (1985).
8. J.P. Burnie, R. Matthews, D. Featherbe, and S. Tabaqchali, 47 kD antigen of Candida albicans, Lancet i:1155 (1985).
9. T. Chardes, M. Piechaczyk, V. Cavailles, S.L. Salhi, B. Pau, and J.-M. Bastide, Production and partial characterization of anti-Candida monoclonal antibodies, Ann. Inst. Pasteur/Immunol. 137 C;117 (1986).
10. Y. Fukazawa, T. Shinoda, and T. Tsuchiya, Response and specificity of antibodies for Candida albicans, J. Bacteriol. 95:754 (1968).
11. R. Guinet and S. Gabriel, Comparaisons par immunoelectrophoreses quantitatives des antigenes de C.albicans groupes A et B et de C. tropicalis, Bull. Soc. Fr. Mycol. Med. 9:241 (1980).
12. H.F. Hasenclever and W.O. Mitchell, Antigenic studies of Candida. I. Observation of two antigenic groups in Candida albicans, J. Bacteriol. 82:570 (1961).
13. J.C. Lebecq, S.L. Salhi, J.-M. Bastide, A novel antibody overlay techni-que for two-dimensional immunoelectrophoresis, J. Immunol. Methods 66:219 (1984).

14. S.A. Meyer, D.G. Ahearn, and D. Yarrow, Discussion of the genera belong-
    ing to the imperfect yeasts: genus Candida, in: "The Yeasts, a Taxonomic
    Study", 3rd. edn., N.J.W. Kreger-van Rij, ed., Elsevier Science Pub-
    lishers B.V., Amsterdam (1984).
15. Y. Miyakawa, K. Kagaya, Y. Fukazawa, and G. Soe, Production and
    characterization of agglutinating monoclonal antibodies against pre-
    dominant antigenic factors for Candida albicans. J. Clin. Microbiol.
    23:881 (1986).
16. L. Polonelli and G. Morace, Specific and common antigenic determinants
    of Candida albicans isolates detected by monoclonal antibody, J. Clin.
    Microbiol. 23:366 (1986).
17. E. Reiss, M. Huppert, and R. Cherniak, Characterization of protein and
    mannan polysaccharide antigens of yeast, moulds, and Actinomycetes,
    in: "Current topics in Medical Mycology," Vol. 1, M.R. McGinnis, ed.,
    Springer-Verlag, New York (1985).
18. L. de Repentigny and E. Reiss, Current trends in immunodiagnosis of
    candidiasis and aspergillosis, Rev. Infect. Dis. 6:301 (1984).
19. R. Sentandreu, E. Herrero, and J.P. Martinez-Garcia, Biogenesis of the
    yeast cell wall, in: "Subcellular Biochemistry", Vol. 10, D.B. Roodyn,
    ed., Plenum Press, New York (1984).
20. N.A. Strockbine, M.T. Largen, and H.R. Buckley, Production and charac-
    terization of three monoclonal antibodies to Candida albicans proteins,
    Infect. Immun. 43:1012 (1984).
21. M. Suzuki and Y. Fukazawa, Immunochemical characterization of Candida
    albicans cell wall antigens: Specific determinant of Candida albicans
    serotype A mannan, Microbiol. Immunol. 26:387 (1982).
22. C.E. Sweet and L. Kaufman, Application of agglutinins for the rapid and
    accurate identification of medically important Candida species, Appl.
    Microbiol. 19:830 (1970).
23. T. Tsuchiya, M. Taguchi, Y. Fukazawa, and T. Shinoda, Serological
    characterization of yeasts as an aid in identification and classifica-
    tion, in: "Methods in Microbiology", Vol. 16, T. Bergan, ed., Academic
    Press, London (1984).
24. T. Tsuchiya, Y. Fukazawa, and S. Kawakita, Significance of serological
    studies on yeasts, Mycopathol. Mycol. Appl. 26:1 (1965).
25. T. Tsuchiya, Y. Fukazawa, M. Taguchi, T. Nakase, and T. Shinoda, Sero-
    logic aspects of yeast classification, Mycopathol. Mycol. Appl.
    53:77 (1974).

CYTOLOGICAL, CHEMICAL, AND SEROLOGICAL IMMUNODETECTION OF A CANDIDA ALBICANS ANTIGEN

Daniel Poulain*, Gérard Strecker**, Jean-Francois Dubremetz*, Bernard Fortier*, Roselyne Rousseau***, and Jan Van Cutsem****

*Inserm U42-369, rue J. Guesde 59650 Villeneuve d'Ascq;
**Laboratoire de Chimie Biologique and U.A. CNRS 217, USTL 59655 Villeneuve d'Ascq;
***U.A. CNRS 409, IRCL 1 Place Verdun 59000 Lille, France;
****Janssen Pharmaceutica 20, Turnhoutseweg 20, Turnhoutseweg, Beerse, Belgium

Candida albicans is a common human harmless commensal yeast until factors favouring pathogenic transformation lead to often severe protean syndromes(18). The mechanisms of the pathogenicity of this highly versatile micro-organism(15) are unknown. Studies on antigenic variability might allow the identification of antigenic markers specific to the pathogenic phase of growth (for a review see Ref.24). A monoclonal antibody (Mab) recognizing polysaccharide antigen has been generated(11) and has been coupled to colloidal gold particles (8,25). The same probe has been used in cytological and immunochemical studies as well as for diagnostic tests. The Mab recognized antigens in the cytoplasm, which are expressed on various glycoproteins of C.albicans. The latter probably contain Man (α1-3) Man which are excreted from the cell. Their presence in animal and human sera is indicative of Candida infection.

A rat IgM Mab has recently been isolated following an experimental infection by a cloned strain of C.albicans; preliminary studies have shown that the Mab recognises an antigen carried by C.albicans, C.tropicalis, C.glabrata, and C.guilliermondii(11). Expression of the antigen varies according to the strains of a given species and to the cells of a given strain in vitro as well as in vivo. Complex and variable expression of surface antigens of C.albicans has already been reported by us and others using both polyclonal(24,27) and monoclonal (4,5) antibodies. A possible explanation for the transitory expression of superficial antigens is that they are excreted discontinuously. In order to further investigate the precise cellular processing of the antigen, the puried Mab (3) has been coupled to colloidal gold (8)as described by Roth et al. (25)and used in reactions on ultrathin sections of C.albicans embedded in Lowicryl K4M (Fig. 1). The epitope recognized is detected in small quantities in the cytoplasm and in larger amounts in the region of the plasma membrane and the cell wall surface. This peripheral location does not correspond to all mannoprotein layers (6,22) it is reminiscent of acid phosphatase for example (29). These observations suggest a process of secretion from the cell membrane through the cell wall.

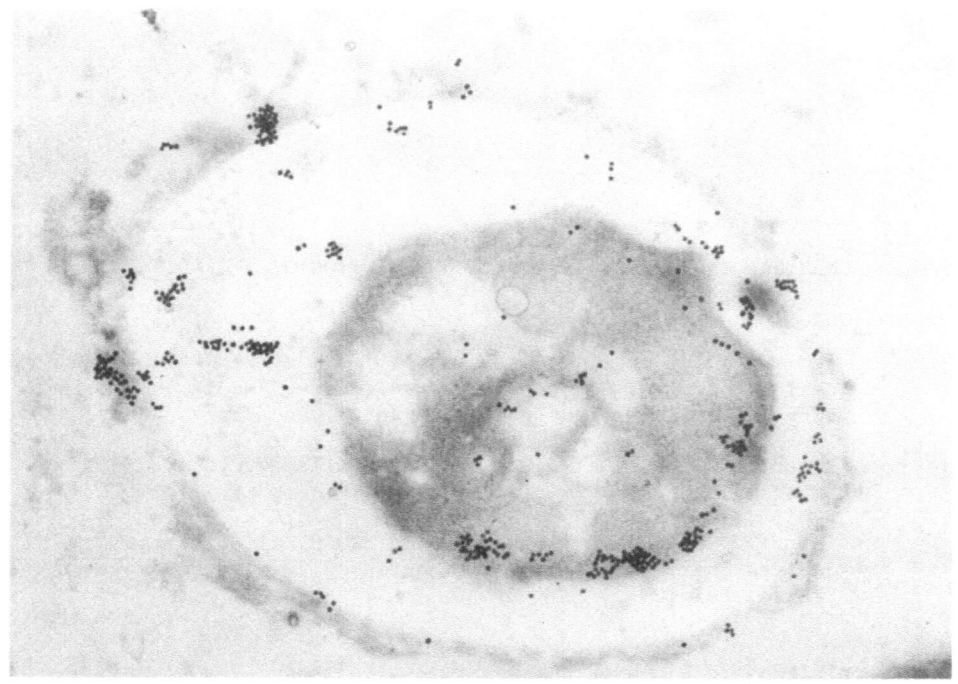

Fig.1.Ultrastructural localization of the antigen in C.albicans parasitic form. Kidney of a rabbit infected by intravenous injection of C.albicans, clone VW.32 (11). Cells were fixed by a mixture of 4% formaldehyde and 0.5% glutaraldehyde in PBS, treated with 50 mM NH_4Cl and embedded in Lowicryl K4M (11). Ultrathin sections were incubated with the colloidal gold labeled Mab.

Faint labelling is observed in the cytoplasm and higher reactivity is found in the plasma membrane region, expecially where the cell wall is cross sectioned. On the left the section is more tangential and the distribution of gold particles reveals the excretion pathways of the corresponding antigen through the entire cell wall. On the cell surface, label is associated with the floccular layer located at the interface of host and parasite. Gold particles are also seen at some distance from the fungal cell, within host tissues. This direct reaction produced a highly specific labelling as shown by absence of background, and all the controls performed on in vitro and in vivo cells. The absence of IgM-gold agglutinates was also checked, so that the intense accumulations observed here did correspond to greater amounts of antigen.

The Mabs coupled to colloidal gold has also been used to study the presence of the antigen in various cell wall extracts by the technique of immunodots on nitrocellulose membranes. The immunodots were visualized by Immunogold silver staining (16) (IGSS dots). The antigenic extracts investigated were prepared from C.albicans germ tubes obtained in RPMI medium , by autoclaving the germ tubes as described by Peat et al. (20) for the removal of cell wall mannans, by treating the germ tubes with ethylenediamine as described by Korn and Northfold (12) or, by grinding with a Braun homogenizer (9). An example of the results using Peat's extract is shown in Figure 2a.

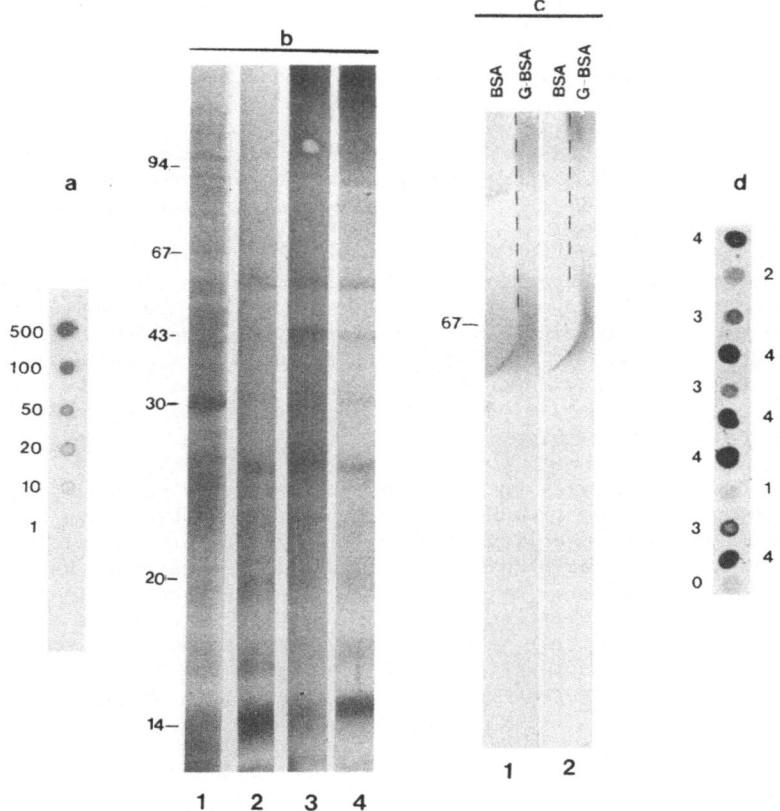

Fig.2.(a) Example of dot reaction observed with _in vitro_ extracts of
C.albicans cells. One ul samples of serial dilutions of a cell wall
extract according to Peat et al.(20) were deposited. The strips were
processed according to Peat et al. (16), after saturation with 5% BSA
in Tris pH 8.2 buffer and transfer to a solution of gold-labeled antibody.
Reactions were visualized with the silver enhancement kit from Janssen
Pharmaceutica (Beerse, Belgium). The sensitivity of the reaction is close
to that of 1 $\mu g/ml^{-1}$ which amounts to 1 ng of unpurified antigen.

    (b) C.albicans germ tube antigen prepared by Braun homogenization
(9) (1 volume germ tubes, 6 volumes sample buffer (14) LCO_2 cooling).
Electrophoresis 14h in 12% polyacrylamide gel with a current of 8 mA;
electrotransfer (28) was performed at 60 V for 6h. b1, stained with
Coomassie blue; b2, IGSS with Mab colloidal gold; b3, stained with indirect
method (Mab followed by anti-rat μ chain peroxidase conjugate-Nordic-); b4,
silver stained with Concanavalin A coupled to colloidal gold (IBF, Paris,
as described for Mab). Up to 10 transferred bands are stained by gold
conjugated Mab (b2), these results are confirmed by the indirect reaction,
(b3), stained with Con A-gold confirms the glycoprotein nature of the
molecules recognized and demonstrate that all the glycoproteins revealed
carry the epitope (b4).

    (c) Immunoblot of BSA Man (α 1-3) Man (β 1-4) GlcNAc (G-BSA) and BSA
deposited side by side with an excess of BSA in order to have a control on
the same strip (same technical conditions as above) c1, Coomassie blue

staining; c2, Mab-gold silver staining. The band of Man (α 1-3) Man (β 1-4) GlcNAc extends over a wide range of molecular weights due to the variable number of mannosidic glycans linked by molecules of BSA. There is a strong labelling of the synthetic glycoprotein whereas the carrier protein does not react at all.

(d) Example of dot IGSS reactions observed with guinea pig sera. The results appreciated with the naked eye and scored according to a scale from 0 to 4. A reaction is considered positive if it gives a score greater than 2.

The sensitivity in detection of ethylendiamine fractions A, B, and C were between 1 μg ml^{-1} to > 1 mg ml^{-1}. These results are consistent with the presence of the epitope within the matricial water soluble cell wall polysaccharides (12,20,22). Ground up cell extracts were also electrophoresed on 12% polyacrylamide gels (14). After transfer to nitrocellulose (28), the blots were treated with the Mab coupled to colloidal gold and revealed by the silver staining method.

Up to 10 bands carrying the epitope were observed, whose glycoprotein nature was demonstrated by treatment of a similar immunoblot with Concanavalin A (Con A) coupled to colloidal gold (Fig. 2b). These results confirm the observations of earlier electrophoretic studies suggesting that the eiptope is carried on several molecules (4). The presence of repetitive antigenic polysaccharide moieties carried on both polysaccharide constituents (19) and on glycoproteins (26) is one of the characteristics of yeasts (24), since the antigen has been located at early biosynthetic stage in the cytoplasm it is not surprising that numerous carriers of the epitope can be revealed.

The results of immunoblotting with the Mab-gold have been confirmed by an indirect reaction with anti-rat IgM chain (Biosoft, Paris) labeled with peroxidase (Fig. 2b). Immunoprecipitation, acetolysis, fractionation and methylation of oligosaccharides are in progress to determine the chemical structure of the epitope. In order to relate conversely a chemotype (2) with this serotype an attempt was made with the Man (α 1-3) Man structure known to be immunodominant in yeasts (17). Man (α 1-3) Man (β 1-4) GlcNAc was prepared from the urine of a patient with mannosidosis. It was coupled to bovine serum albumin (BSA) according to the method of Gray (10) with a sugar/protein molar ratio of 11/1. Simultaneous blot Mab-IGSS analysis of this structure and of its protein carrier revealed a strong reactivity bounded to the glycosylated BSA (Fig. 2c). Controls performed with hen ovalbumin, known to contain high levels of terminal glycans of various mannosidic types were totally negative despite a high reactivity with Con A-gold.

Cytological investigations suggested that the antigen is secreted from the cell, (Fig. 1). The sensitive IGSS dot assay was, therefore, used to look for the presence of antigen in the serum of animals and patients with systemic candidiasis. An experimental infection model was set up as described by Fransen et al. (7). Male Parbright guinea pigs free of C.albicans in their gastro-intestinal tract were infected by intravenous inoculation of 8 x 10^3 colony forming units of C.albicans strain ATCC 44858 per g. body weight. Forty sera from control animals and 100 sera from infected animals were tested blind by the IGSS method (amongst these 4 were infected for 1-2 days, 9 for 5-7 days, and 87 for 15-17 days.)

Histopathology and culture studies were simultaneously performed as controls. Serum was taken from each animal, numbered randomly and frozen. An example of dot IGSS reactions performed with these sera is shown in Fig. 3d.

Fig. 3. Evolution of antigenemia and the corresponding humoral response in a patient (Service de Chirurgie adulte Ouest C.H.U. de Lille) presenting with an infection caused by C.albicans. This patient had undergone surgery for neoplasm of the gastrim antrum. Four days later, the patient became befrile and a positive blood culture yielding C.albicans was obtained 11 days after surgery. The first serum A1 was taken 5 days after hemoculture, A2 was taken 22 days after. IGSS dot reaction (bottom of the figure) and co-counterimmunoelectophoresis on cellulose acetate (21) were performed on each serum sample. For co-counterimmunoelectophoresis the two sera were placed on either side of a deposit of purified Mab (400 µg ml^{-1}) and run against a somatic antigen of strain VW32, the strip was stained by Schiff's reagent to reveal polysaccharide antigens. The transient occurrence of antigen detected by the Mab precedes the synthesis of antibodies of same specificity.

The results obtained, in relation to the presence and duration of infection, show that the detection of antigen in these animals is highly significant since the sensitivity, specificity, and predictive values (13) of the text were 94%, 95%, and 98% respectively, if only sera taken from animals 1-2 days and 15-17 days post infection are included. If results obtained from serum samples taken from animals 5-7 days post infection are also included, the sensitivity of the test falls to 85%. After the initial circulation of the infecting organism, the pathogen lodges in the tissues (7) and the amount of detectable antigen may decrease. The presence of this circulating antigen in the sera of humans with candidiasis has also been studied.

The efficacy of a test for diagnosis of human candidiasis is more difficult to assess. For this reason, the IGSS dot method has been evaluated with well documented observations of human candidiasis, each patient being its own control. We report here an example of such simple bioclinical observation concerning one patient who developed a C.albicans infection after surgery. The patient showed positive IGSS dot reactions with serum taken during infection (first serum) whereas for serum taken more than 10 days later (serum 2), when the patient was recovering, the same reaction was negative (Fig. 3).

Co-counterimmunoelectrophoresis on cellulose acetate (21) was also performed on the serum samples together with the Mab which had previously been shown to precipitate when run against a somatic extract of C.albicans (11). After staining with Schiff's reagent the strips showed the absence of detectable antibodies in the first serum whereas a reaction of identity was observed between the Mab and the second serum demonstrating the production of antibodies specific to the same polysaccharide antigen. At this time IGSS dot for antigen detection was negative.

This study demonstrates the existence of a fungal antigen carried on several glycoproteins which is discontinuously secreted and circulates during systemic candidiasis. The originality of this detection technique lies in the use of the same probe through a panel of immunogold methods, some of which are directly applicable in routine diagnostic laboratories.

REFERENCES

1. B. Armbuster, E. Carleman, R. Chiovetti, R. Garavito, J. Hobot, E. Kellenberger, and W. Villiger, Resin development for electron microscopy and an analysis of embedding at low temperature, J. Microsc. (Paris) 126:123 (1982).
2. C. Ballou and W. Raschke, Polymorphism of the somatic antigen of yeast, Science 184:127 (1974).
3. H. Bazin, L.M. Xhurdebise, G. Brutonboy, A.M. Lebacq, L. De Clercq, and F. Cormont, Rat monoclonal antibodies. I Rapid purification from in vitro culture supernatants, J. Immunol. Methods 66:261 (1984).
4. D. Brawner and J. Cutler, Variability in expression of cell surface antigens of Candida albicans during morphogenesis, Infect. Immun. 51:337 (1986).
5. D. Brawner and J. Cutler, Variability in expression of a cell surface determinant on Candida albicans as evidenced by an agglutinating monoclonal antibody, Infect Immun. 43:966 (1984).
6. A. Cassone, E. Mattia, and M. Boldrini, Agglutination of blastopores of Candida albicans by Concanavalin A and its relationship with the distribution of mannan polymers and the ultrastructure of the cell wall, J. Gen. Microbiol. 105:263 (1978).
7. J. Fransen, J. Van Custen, R. Vandesteene, and P.A. Janssen, Histopathology of experimental systemic candidosis in guinea pigs. Sabouraudia 22:455 (1984).
8. G. Frens, Controlled nucleation for the regulation of particle size in monodisperse gold solutions, Nature Phys. Sci. 241:20 (1973).
9. S. Gabriel-Bruneau and R. Guinet, Antigenic relationships among some Candida species studied by crossed-line immunoelectrophoresis taxonomic significance, Int. J. Syst. Bact. 34:227 (1984).
10. G. Gray, The direct coupling of oligosaccharides to proteins and derivatized gels, Arch. Biochem. Bioph. 163:246 (1974).
11. V. Hopwood, D. Poulain, B. Fortier, G. Evans, and A. Vernes, A monoclonal antibody to a cell wall component of Candida albicans, Infect. Immun. (in the press).
12. E.D. Korn and D.H. Northcote, Physical and chemical properties of polysaccharides and glycoproteins of the yeast cell wall, Biochem. J. 75:12 (1960).
13. P. Kozinn, C. Taschdjian, P. Goldberg, W. Protzman, D. Mackenzie, J. Remington, and S. Anderson, Efficiency of serologic tests in the diagnosis of systemic candidiasis, Am. J. Clin. Pathol. 70:893 (1978).
14. U.K. Laemmli, Cleavage of structural proteins during the assembly of the head of bacteriophage T4, Nature 227:680 (1970).
15. B. Lutsky, J. Buffo, and D. Soll, High frequency switching of colony morphology in Candida albicans, Science 230:666 (1985).
16. M. Moeremans, G. Daneels, A. Vand Dijck, G. Langanger, and J. De Mey, Sensitive visualization of the antigen antibody reaction in dot and blot immuneoverlay assays with immunogold and immunogold/silver staining, J. Immunol. Methods 74:353 (1984).
17. H. Nakajima, N. Itoh, T. Kawasaki, and I. Yamashina, Reaction of anti-mannan antibodies with oligomannosides and glycopeptides, J. Biochem. 85:209 (1979).
18. F.C. Odds,"Candida and Candidosis," Leicester University Press, Leicester (1979).
19. A. Parodi, Biosynthetic mechanisms for cell envelope polysaccharides, in "Yeast cell envelopes: Biochemistry, Biophysics, and Ultrastructure", W. Arnold, ed., C.R.C. Press Inc. Boca Raton, FL., U.S.A. 2:47 (1981).
20. S. Peat, W.J. Whelan, and T.E. Edwards, Polysaccharides of baker's yeast. Part IV Mannan, J. Chem. Soc. 1:29 (1961).
21. D. Poulain and J.M. Pinon, Diagnosis of systemic candidiasis: development of co-counterimmunoelectrophoresis, Eur. J. Clin. Microbiol. 5:420 (1986).

22. D. Poulain, G. Tronchin, S. Jouvert, J. Herbaut, and J. Biguet, Architecture parietale de Candida albicans, Localisation de composants chimiques et antigeniques, Ann. Microbiol. (Inst. Pasteur) 132A:219 (1981).

23. D. Poulain, J. Fruit, L. Fournier, E. Dei-Cas, and A. Vernes, Diagnosis of systemic Candidiasis: Application of Co-counterimmunoelectro-phoresis, Eur. J. Clin. Microbiol. 5:427 (1986).

24. D. Poulain, V. Hopwood, and A. Vernes, Antigenic variability of Candida albicans, C.R.C. Crit. Rev. Microbiaol. 12:223 (1985).

25. J. Roth and M. Binder, Colloidal gold, ferritin and peroxidase as markers for electron microscopic double labeling lectin technique, J. Histochem. Cytochem. 26:1 (1978).

26. F. Shonholzer, A. Schweingruber, H. Trachel, and E. Schweingruber, Intracellular maturation and secretion of acid phosphatase of Saccharomyces cerevisiae, Eur. J. Biochem. 147:273 (1985).

27. E. Smail and J. Jones, Demonstration and solubilization of antigens expressed primarily on the surfaces of Candida albicans germ tubes, Infect. Immun. 45:74 (1984).

28. H. Towbin, T. Steahlin, and J. Gordon, Electrophoretic transfer of proteins from polyacrylamide gels to nitrocellulose sheets: procedure and some applications, Proc. Nat. Acad. Sci., U.S.A. 76:4350 (1979).

CANDIDA ANTIGENS AND THEIR ROLE IN ADHERENCE MECHANISMS

Raymond Robert, Jean-Marcel Senet, Guy Tronchin, and
Abdelhamid Bouali

Laboratory of Immunology, Parasitology, and Mycology
Pharmacy Department 16
bd Daviers, 49100 Angers
France

Fungal adherence has grown into one of the most active, if not the most exciting, area of study in the field of mycological ecology. Shortly after birth, the mucosal surfaces of the upper respiratory and gastrointestinal tracts of man are colonized by Candida albicans. Infection by the candida species can be envisioned as a stepwise process in which the fungus must first adhere to a tissue surface. Failure to adhere would result in their being swept away in the fluids which constantly bathe the tissue surface. The precise role of adherence in tissue colonization and in the development of candidosis has been difficult to define. This is primarily because interactions between Candida and various tissue surfaces are physicochemically complex and that specificity is thus difficult to assess in a single model system. However, one can identify interactions between the fungus and synthetic smooth surfaces which involve opposite charge attractions and/or surface hydrophobic affinity with a specific molecular ligand (or adhesin as some authors prefer to call it). The isolation and identification of the adhesive molecular structures on both surfaces is likely to suggest new approaches to the control of candidosis. One obvious approach would be to try to block the adherence of Candida by the application of ligand or receptor material once the substances or their analogues have been identified.

Much work has been done to improve the understanding of such mechanisms. Thus far, most of these studies have been performed in vitro. This approach makes it easy to assess the effects of various conditions of incubation or different substances upon the adherence process. However, the disadvantage is that many of the physiological barriers of the intact tissue surface are bypassed. Thus, the data obtained from in vitro experiments need to be interpreted with caution as to their in vivo significance. Furthermore, the diversity and the dimorphism of Candida albicans implies modifications at the surface level leading to great complexity of an experimental model.

This paper describes the main attachment mechanisms involved at the molecular level and reviews current work on Candida albicans attachment to host tissues (1).
The attachment mechanisms may be divided into two main processes:
Non specific adherence mechanisms essentially due to net surface charges and/or hydrophobic forces.

Specific adherence mechanisms which imply complementary steric configuration of molecules called receptors and/or ligands.

These mechanisms allow a close interaction between the two partners we are interested in: Candida and host. The greater the steric complementarity, the closer the reciprocal sites: this can exert powerful attraction mainly due to electrostatic, London, van der Waals forces and hydrogen bonds.

The linking molecules may be components of either surface. Alternately, it could also be imagined that indirect links may be mediated by a bivalent ligand making a bridge between the receptors. That bridge might be a microorganisms (2, 3) or a soluble protein like fibronectin (4). Non-specific and specific mechanisms act in synergy and are largely dependant on the physico-chemical parameters of the surrounding medium including ionic strength, pH and protein concentration.

ADHERENCE OF CANDIDA TO SYNTHETIC SURFACES

As has been demonstrated by several authors (5, 6, 7, 8, 9, 10) Candida can adhere to the surface of many plastic prosthetic devices. It seems possible to correlate this attachment to forces due to surface charges and/or hydrophobic forces. The net negative surface potential of individual cells and the ultrastructural delineation of their superficial mannoprotein coat has been demonstrated with the use of ruthenium red (7, 9). Surface fixation of cationized ferritin molecules has let Tronchin to similar results (11) (Fig. 1). Following the approach of Watt et al. (12) who measured the surface electrostatic potential of Neisseria gonorrhoeae, who have compared adhesion of Candida albicans to DEAE and CM sephadex over a 5 to 9 pH range (Fig. 2). The results obtained confirm that yeasts are endowed with a net negative surface potential.

Adherence also seems to be related to extracellular metabolic products induced by previous incubation of the fungus in a sugar-containing medium (6, 7). The amount of polysaccharide at the cell surface could be related to an increase of the net negative surface charge and thus to the enhancement of adherence mechanisms to plastic surfaces. Klotz et al. (8) recently demonstrated that the hydrophobicity of the C. albicans cells, determined by a water-hydrocarbon two phase assay, correlated with the tendency of years to adhere to polystyrene. These results are in agreement with the

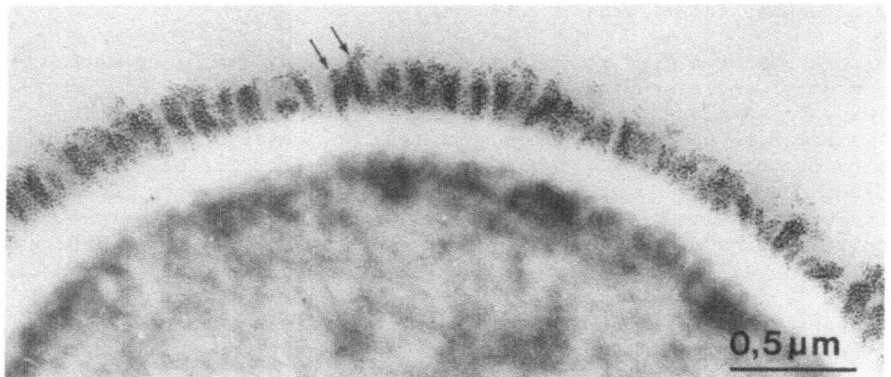

Fig. 1. Cationized ferritin binding to anionic groups on
C. albicans blastoconidia

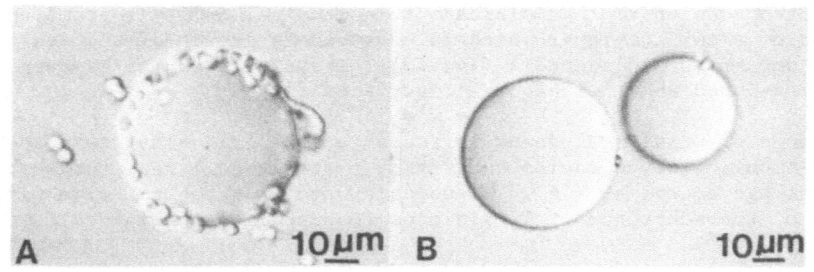

Fig. 2. Adherence of blastoconidia to positively charged DEAE
microspheres(A) and lack of adherence to negative charged
CM microspheres (B).

findings of Minagi et al. (13) on the adherence of Candida albicans and
Candida tropicalis to denture material.

In vitro experiments are incapable of unravelling what happens in
vivo in view of the multiplicity of events which play a role in physio-
pathological phenomena, particularly the dynamic cascade due to the numerous surrounding factors activated by introduction of the plastic devices.
Several authors (5, 10) have shown the early deposit of many biological
products such as fibrin, platelets and leucocytes simultaneously with the
fungal adherence.

However, one can conclude from the afore-mentioned observations that
the polysaccharide antigens play a major role in adherence mechanisms of
Candida to synthetic prosthetic devices such as dental prostheses, intra-
vascular and urinary catheters and prosthetic cardiac valves.

SPECIFIC ADHERENCE MECHANISMS

Candida species adhere readily to a variety of cells among which buccal
and vaginal mucosal cells have been more extensively studied in view of the
natural ecology of the fungus (Fig. 3). Experiments are sometimes difficult
to compare because of the different conditions in which Candida has been
harvested (stationary or logarithmic phase) and used (heat killed or viable)
for experiments. In addition, the host tissue cells can be of different
origin and diversely prepared. Some of several results reported on the
attachment process are described below.

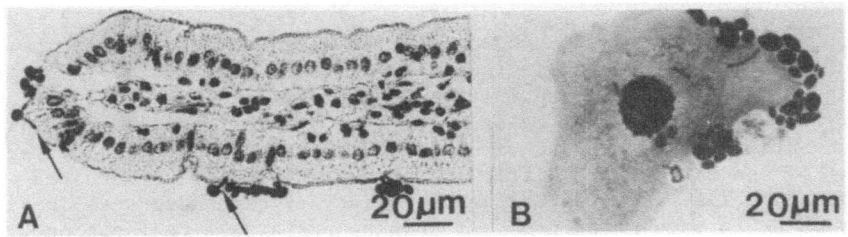

Fig. 3. C. albicans adhering to (A) intestinal (B) vaginal
epithelial cells.

In vitro adherence of C. albicans to epithelial cells is promoted by production of a surface layer which is enhanced by growth of the yeast in different sugar containing media (14, 15). These results may be compared to those mentioned above with acrylic surfaces (7).

Adherence of Candida albicans to buccal epithelial cells seems to be the same in children and adults suggesting a stable cell receptor system that is not age dependant (16). These data were in agreement with Davidson's work (17) on the adherence of C. albicans to buccal epithelial cells of neonates.

Sandin et al. (18) have noted that germinated yeasts adhere to buccal cells more effectively than non germinated yeasts. A strong correlation between germination and increased adherence of C. albicans to human buccal epithelial cells has been reported by other workers (19, 20, 21). Sobel et al. (22) studied a germ tube negative mutant and found it to be relatively avirulent.

In order to obtain clear proof of differences in behaviour between the different forms of C. albicans, Anderson and Odds (23) have promoted a standardized cellular ATP measurement method to test the ability of yeast cells, germ tubes and hyphae to adhere to vaginal cells. In terms of surface area of fungi attached per epithelial cell the hyphae and germ tubes attached significantly better than the yeast forms.

King et al. (24) examined the in vitro adherence capabilities of seven Candida species. Microscopic evaluation indicated that Candida albicans adheres to vaginal and buccal epithelial cells to a significantly greater degree that the other species tested. C. tropicalis and C. stellatoidea demonstrated moderate adherence capabilities while C. parapsilosis adhered only to a slight degree. Other species failed to interact with isolated mucosal cells. These findings suggest that there is a relationship between the adherence capabilities and the frequency with which the species colonize mucosal surfaces.

Segal et al. (25) found a correlative relationship between adherence of Candida albicans to human vaginal epithelial cells in vitro and candidal vaginitis.

These results, which demonstrate the greater adherence of hyphae versus yeast and of virulent species versus less virulent species, are likely to have some bearing on our own work on the localization and distribution of a fibrinogen binding factor in Candida species which will be discussed later.

As mentioned above, most of the studies have been made on epithelial buccal or vaginal cells. Nevertheless, several investigators have focused on other cell types which might play an important role in adherence and colonization.

The two main cell types which have been investigated are polymorpho-nuclear cells (26), and endothelial cells (27, 28, 29, 30, 31, 32). Attachment of Candida has been explored in vivo in the perfused rabbit kidney and in vitro with human endothelial cells. In vitro studies revealed good correlation between adherence and virulence of the Candida species tested (32). Most of the results are difficult to compare with those obtained with epithelial cells. In view of the difficulty of extending such in vitro model systems to in vivo mechanisms, it appears preferable to focus on the adherence mechanism at the molecular level.

Kinetic, cytochemical, and ultrastructural studies of Candida albicans during adherence to human buccal epithelial cells (33) have produced clear

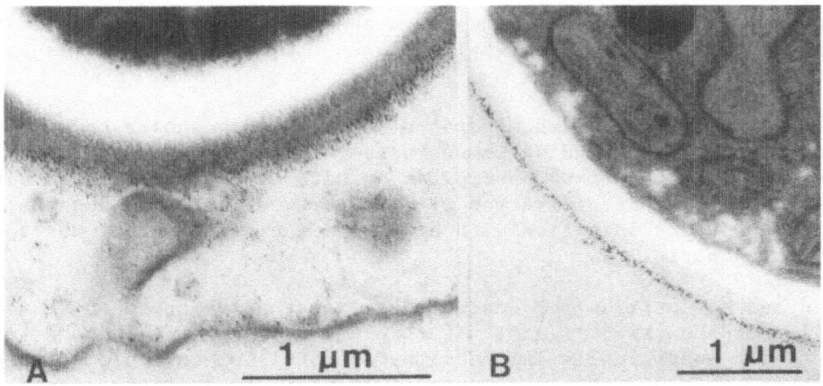

Fig. 4. Ultrastructural aspects of the cell wall coat of
blastoconidia adhering to epithelial cells (A) compared to blastoconidia harvested from Sabouraud's
medium (B) (Concanavalin A - Mannosyl-ferritin).

evidence for modifications of the cell wall coat (Fig. 4). Attachment of
the fungus to epithelial cells appears to involve spatial rearrangement of
the cell wall surface. In particular, adhering yeast developed a fibro-
granular surface layer. Fibrillar structure distributed on the cell coat
appeared to be involved in the attachment mechanisms (Fig. 5). This work
is in agreement with that of Douglas et al. (7, 34) and suggests an active
role for the fungus in promoting adherence to cell surfaces.

In order to identify the specific molecules or ligands involved in the
attachment mechanism, many approaches have been utilized among which competitive inhibition or controlled degradation process are the most commonly
used.

Sandin et al. (18, 35) found that Concanavalin A, a lectin that
recognizes mannose and glucose, inhibited adherence of pretreated yeasts
to buccal cells and reciprocally inhibited adherence of pretreated buccal
cells to non pretreated yeasts. Only mannose-recognizing lectins play such
a role. It was found that D-methyl mannopyranoside in the incubation medium
during the assay also inhibited adherence, whereas other sugars did not.

Segal et al. (36) have demonstrated that binding of C. albicans to
human vaginal cells is inhibited by chitin, its hydrolysate derivative and
N-acetylglucosamine. Inhibition was also obtained by glucosamine and
mannosamine. Similar experiments were performed in vivo (37) to block
attachment and prevent vaginal infections by pretreating mice with chitin

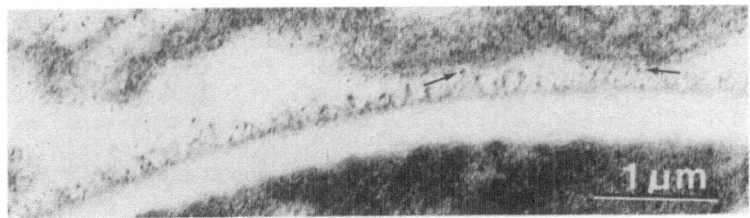

Fig. 5. Attachment of blastoconidia via external granules
to the surface of a buccal cell.

soluble extract, N-acetyl glucosamine and mannan. The data revealed that
only pretreatment with chitin derivative blocks attachments of yeasts to
vaginal mucosal surfaces.

Rotrosen et al. (32) tested adherence to cultured vascular endothelial
cells and noted that anti-Candida immune serum blocks attachment; this
activity was abolished by immunoprecipitation with C. albicans mannan but
was unaffected by immunoprecipitation with S. cerevisiae mannan but was
unaffected by immunoprecipitation with S. cerevisiae mannan or by absorption
with particulate chitin.

Doublas and Mc Courtie (38) described an alternative approach to the
identification of the yeast ligand. It involved treating C. albicans with
tunicamycin which specifically inhibits the synthesis of mannoprotein and
not that of the other major wall components (glucan and chitin). Their
results suggest that yeast mannoprotein is of primary importance in the
adherence of C. albicans to buccal epithelial cells.

A different approach complementary to competitive inhibition studies
is that of the controlled degradation of the yeast surface by means of
enzymes or chemical substances. Epithelial adherence of intact Candida
is diminished after incubation with proteolytic enzymes (19, 39). In a
similar manner Lee et al. (40) studied the papain release of low molecular
weight glycoproteins which competitively block attachment of intact Candida
to vaginal epithelial cells. The variety of methods used makes it very
difficult to suggest a precise attachment mechanism and a specific receptor/
ligand system has yet to be demonstrated. However, most of the evidence
supports the involvement of mannan residues or mannoproteins as has been
recently demonstrated by Mc Courtie and Douglas (41). Rather than a yeast
adhesin, Skerl et al. (4) have proposed a cellular adhesin. Fibronectin,
a major glycoprotein of mammalian cells which has been tested for its
capability of binding Candida species. Blocking experiments using competitive inhibition or enzyme pretreatment of the yeast led the authors
to propose a possible role of fibronectin in the adherence of C. albicans
to buccal and vaginal epithelial cells.

Fibrin Platelet Matrix (FPM) is an interesting model for the demonstration and analysis of Candida attachment. This in vitro system has the
advantage of having good correlation to the in vivo process and has been
explored by Maish and Calderone (42).

The main results obtained by the use of rediolabelled Candida may be
summarized as follows: Heat and formalinized Candida cells do not adhere
to the clot as well as viable cells. Enzyme (pronase, chymotrypsin, trypsin)
pretreatment of C. albicans reduces adherence. Specific anti-Candida anti-
bidies can inhibit attachment to the clot whereas normal gamma globulin does
not. Capability of adherence of different Candida species appears to be
related to virulence. These findings are similar to the afore-mentioned
results on cell attachment.

In order to identify the structure involved in such adherence mechanisms, Maish et al. (43) fixed a soluble extract from a cell wall preparation
of Candida albicans to sheep erythrocytes (SRBC). Conjugated SRBC readily
attached to a fibrin platelet matrix, and different enzyme treated extracts
were conjugated with SRBC and attachment to fibrin platelet matrix was
measured. The results confirmed studies cited above suggesting that intact
polysaccharide moiety is essential for adherence.

In vivo experiments confirmed the capability of Candida albicans to
fix fibrin platelet deposits (10, 44) and to participate in the pathogenesis
of endocarditis. Calderone (45), using an adherence negative mutant of

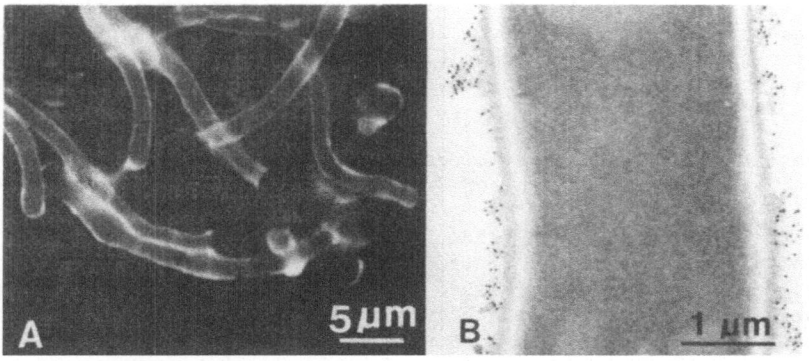

Fig. 6. Fibrinogen binding to C. albicans hyphae revealed
by immunofluorescence (A) and gold labelling (B).

Candida albicans demonstrated that low in vitro adherence is correlated
with low virulence in vivo. This work illustrated that adherence is undoubtedly one of the most important steps of the pathogenic process of
candidosis. Robert (unpublished) has found by intravenous injection of
Candida albicans yeasts (10^8 CFU) in mice that the fungal elements readily
adhere 1h after the yeast injection to an artificial abcess provoked on the
skin. Another in vivo experiment has been done with C. albicans yeasts
injected in mice and rabbits. Autopsy materials after 3 and 7 days revealed
that fungal elements trapped in the kidneys were surrounded by fibrin.

As the fibrin platelet matrix (FPM) appears to be a good in vitro and
in vivo model system to study Candida adherence, Bouali et al. postulated
the existence of an adhesin at the surface level of the fungus capable of
binding to the FPM. In order to confirm this hypothesis, the ability of
the fungus to fix soluble fibrinogen in vitro was investigated by means of
indirect immuno fluorescence and transmission electron microscopy (46).
Figure 6 shows strong fluorescence due to fibronogen on germ tube surfaces
and hyphae contrasting with the non fluorescence of the yeast form. Electron
microscopy with gold labelling confirms these observations and shows that
fibrinogen is deposited in patches on viable fungus and regularily spread
on formaldehyde killed germ tubes.

The binding was determined using^{125} labelled fibrinogen. Compared to
the yeast form, the relative binding was eight times greater for germ tubes
and twelve times greater for mycelium (BOUALI et al., unpublished). Fibrinogen binding on various Candida species showed a parallel between the amount

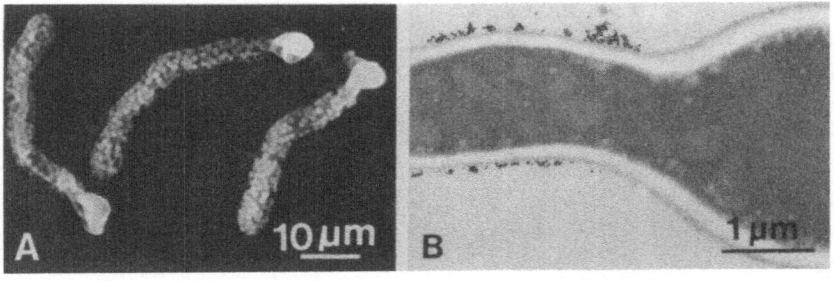

Fig. 7. Fibrinogen coated microspheres (A) and fibrinogen coated
gold particles (B) adhering to C. albicans germ tubes.

of fixed fibrinogen and virulence in terms of frequency with which the species is encountered in pathology. For example, C. albicans hyphae fixed 3, 5 times more fibrinogen than C. pseudotropicalis hyphae. In the same way, twice as much fibrinogen was bound to freshly isolated C. albicans strains than to strains from culture collections.

In order to elucidate at the molecular level this adhesin-like feature, competitive inhibition and controlled degradation processes were used.

Pretreatment of yeasts with E and D fibrinogen fragments showed a higher affinity for the latter. Pretreatment of yeasts with trypsin, mannosidase, mercaptoethanol diminished fibrinogen fixation whereas pronase or chitinase did not. Inhibition achieved by preincubation of the fibrinogen before testing of adhesion to yeast showed that dialyzed C. albicans culture filtrate inhibited the binding more strongly than C. albicans mannan. In contrast, mannose and several other hexoses or chitin did not.

The interaction of the fungus with polymerized fibrin was considered to be a closer in vivo model of adherence and was tested using fibrin coated microspheres and gold particles in scanning and transmission electron microscopy (Fig. 7). These results cnfirmed the above observations and revealed that the fibrinogen binding sites appeared to be associated with the flocculent surface layer, increased during growth of germ tubes.

Incubation of C. albicans germ tubes with freshly washed free platelets showed strong binding to the tubes (Fig. 8). This bindig was inhibited by preincubation of the germ tubes with fibrin (Robert, unpublished). It is interesting to compare these findings to results obtained by Skerl et al. (44) and Nosal et al. (47) who demonstrated the role of a cationic protein that stimulated germination of yeast cells and a platelet aggregation mechanism mediated by cell wall fragments.

These findings, related to the Fibrin Platelet Matrix and more specifically to the fibrinogen cell wall adherence of germinated yeast and mycelium have prompted Robert et al. to undertake further investigations on specific fibrinogen receptors at the surface of Candida albicans. We have isolated a so-called fibrinogen binding factor by the following process: Fibrinogen coated column (ultrogel) was embedded with a crude dialyzed culture filtrate of C. albicans. After several washings with phosphate buffer (pH 5) this factor was eluted by phosphate buffer (pH 9), demonstrating that fixation is Ph dependent.

Eluted material revealed a 67 kD native mannoprotein that has a strong affinity for fibrinogen. This fibrinogen binding factor is being studied for its in vitro and in vivo role and in physiopathology of candidosis (48).

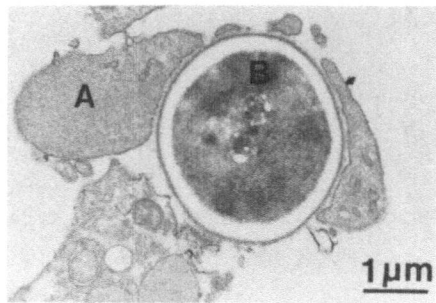

Fig. 8. Platelets (A) adhering to C. albicans hyphae (B).

Among investigations needed to characterize molecules involved in adherence mechanisms, the following seem to be the most promising:
- Monoclonal antibody studies of the precise biosynthetic locus, surface localization and fate of such molecules.
- Inhibition assays with small molecules such as amino sugars, oligopeptides or degradation products of receptors. It has already been demonstrated that fibronectin and fibrinogen share the same peptide amino-acid sequence (Arg-Gly-Asp) capable of adhering to cell receptors.
- Affinity constants of ligand/receptor sites would be useful in the understanding of their in vivo functions.
- DNA cloning of the responsible genes coding for the receptor.

REFERENCES

1. D. Rotrosen, R.A. Calderone, and J.E. Edwards, Jr., Adherence of Candida species to host tissues and plastic surfaces. Rev. Infect. Dis., 8, 1:73 - 85 (1986).
2. A. Centeno, C.P. Davis, M.S. Cohen, and M.M. Warren, Modulation of Candida albicans attachment to human epithelial cells by bacteria and carbohydrates. Infect. Immun., 39:1354 - 60 (1983).
3. M.J. Kennedy and P.A. Voltz, Ecology of Candida albicans gut colonization: Inhibition of Candida adhesion, colonization, and dissemination from the gastrointestinal tract by bacterial antagonism. Infect. Immun., 49:654 - 663 (1985).
4. K.G. Skerl and R.A. Calderone, In vitro binding of Candida albicans yeast cells to human fibronectin. Can. J. Microbiol., 30:221 - 227 (1984).
5. D. Rotrosen, T.R. Gibson, J.E. Edwards, Jr., Adherence of Candida species to intravenous catheters. J. Infect. Dis., 147, 3:594 (1983).
6. L.P. Samaranayake and T.W. Mc Farlane, An in vitro study of the adherence of Candida albicans to acrylic surfaces. Archs. Oral. Biol., 25:603 - 609 (1980).
7. L.J. Douglas and J. Mc Courtie, Adherence of Candida albicans to denture acrylic as affected by changes in cell-surface composition. In: "Current developments in yeast research". Ed. G.G. Steward and I. Russel, Toronto, Pergamon Press, 375 - 381 (1981).
8. S.A. Klotz, D.J. Drutz, and J.E. Zajic, Factors governing adherence of Candida species to plastic surfaces. Infect. Immun., 50, 1:97 - 101 (1985).
9. J. Mc Courtie and L.J. Douglas, Relationship between cell surface composition of Candida albicans and adherence to acrylic after growth on different carbon sources. Infect. Immun., 32, 3:1234 - 1241 (1981).
10. R.A. Calderone, M.F. Rotondo, and M.A. Dande, Candida albicans endocarditis: Ultrastructural studies of vegetation formation. Infect. Immun., 20, 1:279 - 289 (1978).
11. G. Tronchin, Ultrastructure et cytochimie de l'enveloppe cellulaire des levures (Candida albicans). These de Doctorat d'Etat. Universite de Lille 1, France (1983).
12. P.J. Watt and M.E. Ward, Adherence of Neisseria gonorrhoeae and other Neisseria species to mammalian cells. In: "E.H. Beachey, ed., Bacterial adherence", London, Chapman and Hall, 251 - 88 (1980).
13. S. Minagi, Y. Miyake, K. Inagaki, H. Tsuru, and H. Suginaka, Hydrophobic interaction in Candida albicans and Candida tropicalis adherence to various denture base resin materials. Infect. Immun., 47:11 - 4 (1985).
14. L.J. Douglas, J.G. Houston, and J. Mc Courtie, Adherence of Candida albicans to human buccal epithelial cells after growth on different carbon sources. FEMS Microbiol. Let., 12:241 - 243 (1981).
15. L.P. Samaranayake and T.W. Mc Farlane, The adhesion of the yeast Candida albicans to epithelial cells of human origin in vitro. Archs. Oral. Biol., 26:815 - 820 (1981).

16. F. Cox, Adherence of Candida albicans to buccal epithelial cells in Children and adults. J. Lab. Clin. Med., 102:960 - 972 (1983).
17. S. Davidson, M. Brish, and F. Rubinstein, Adherence of Candida albicans to buccal epithelial cells of neonates. Mycopathologia, 85:171 - 173 (1984).
18. R.L. Sandin, A.L. Rogers, R.J. Patterson, and E.S. Beneke, Evidence for mannose-mediated adherence of Candida albicans to human buccal cells in vitro. Infect. Immun., 35, 1:79 - 85 (1982).
19. J.D Sobel, G. Muller, P.G. Myers, D. Kaye, and M.E. Levison, Adherence of Candida albicans to human vaginal and buccal epithelial cells. J. Infect. Dis., 143:76 - 82 (1981).
20. L.H. Kimura and N.H. Pearsall, Adherence of Candida albicans to human buccal epithelial cells. Infect. Immun., 21:64 - 8 (1978).
21. L.H. Kimura and N.N. Pearsall, Relationship between germination of Candida albicans and increased adherence to human buccal epithelial cells. Infect. Immun., 28, 2:464 - 468 (1980).
22. J.D. Sobel, G. Muller, and H.R. Buckeley, Critical role of germ tube formation in the pathogenesis of Candida vaginitis. Infect. Immun., 44:576 - 80 (1984).
23. M.L. Anderson and F.C. Odds, Adherence of Candida albicans to vaginal epithelia: Significance of Morphological form and effect of ketoconazole. Mykosen, 28, 11:531 - 540 (1985).
24. R.D. King, J.C. Lee, and A.L. Morris, Adherence of Candida albicans and other Candida species to mucosal epithelial cells. Infect. Immun., 27, 2:667 - 674 (1980).
25. E. Segal, A. Soroka, and A. Schechter, Correlative relationship between adherence of Candida albicans to human vaginal epithelial cells in vitro and Candida vaginitis. J. Med. Vet. Mycol., 22:191 - 200 (1984).
26. R.D. Diamond and R. Krzesicki, Mechanisms of attachment of neutrophils to Candida albicans pseudohyphae in the absence of serum, and of subsequent damage to pseudohyphae by microbicidal processes of neutro-in vitro. J. Clin. Invest., 61:360 - 369 (1978).
27. W.H. Johnston and H. Latta, Acute focal glomerulonephritis in the rabbit induced by injection of Saccharomyces cerevisiae: An electron microscopic study. Lab. Invest., 26:741 0 54 (1972).
28. W.H. Johnston and H. Latta, Acute hematogenous pyelonephritis induced in rabbit with Saccharomyces cerevisiae: An electron microscopic study. Lab. Invest., 29:495 - 505 (1973).
29. J.L. Barnes, R.W. Osgood, J.C. Lee, and R.D. Stein, Host-parasite interactions in the pathogenesis of experimental renal candidiasis. Lab. Invest., 49:460 - 7 (1983).
30. D. Rotrosen, J.E. Edwards, J. Moore, J. Adler, A.M. Cohen, and I. Green. Pathogenesis of hematogenous candidiasis: Adherence of Candida species to vascular endothelium and mechanisms of endothelial cell invasion. Clin. Res., 31:374 A (1983).
31. S.A. Klotz, D.J. Drutz, J.L. Harrison, and M. Huppert, Adherence and penetration of vascular endothelium by Candida yeasts. Infect. Immun., 42:374 - 84 (1983).
32. D. Rotrosen, J.E. Edwards, Jr., T.R. Gibson, J.C. Moore, A.H. Cohen, and I. Green, Adherence of Candida to cultured vascular endothelial cells: Mechanisms of attachment and endothelial cell penetration. J. Infect. Dis., 152, 6:1264 - 1274 (1985).
33. G. Tronchin, D. Poulain, A. Vernes, Cytochemical and ultrastructural studies of Candida albicans. III Evidence for modifications of the cell wall coat during adherence to human buccal epithelial cells. Arch. Microbiol., 139:221 - 224 (1984).
34. L.J. Douglas, J.G. Houston, and J. Mc Courtie, Adherence of Candida albicans to human buccal epithelial cells after growth on different carbon sources. FEMS Microbiol. Let., 12:241 - 243 (1981).
35. R.L. Sandin and A.L. Rogers, Inhibition of adherence of Candida albicans to human epithelial cells. Mycopathologia, 77:23 - 26 (1982).

36. E. Segal, N. Lehrer, and I. Ofek, Adherence of Candida albicans to human vaginal epithelial cells: Inhibition by amino sugars. Expl. Cell. Biol., 50:13 - 17 (1982).

37. N. Lehrer, E. Segal, and L. Barr-Nea, In vitro and in vivo adherence of Candida albicans to mucosal surfaces. Ann. Microbiol., 134 B, 2:293 - 306 (1983).

38. L.J. Douglas and J. Mc Courtie, Effect of tunicamycin treatment on the adherence of Candida albicans to human buccal epithelial cells. FEMS Microbiol. Let., 16:199 - 202 (1983).

39. J.C. Lee and R.D. King, Characterization of Candida albicans adherence to human vaginal epithelial cells in vitro. Infect. Immun., 41:1024 - 30 (1983).

40. J.C. Lee and R.D. King, Adherence mechanisms of Candida albicans. In: D. Schlesinger, Ed., Washington, DC: American Society for Microbiology, 269 - 72 (1983).

41. J. Mc Courtie, and L.J. Douglas, Extracellular polymer of Candida albicans: Isolation, analysis and role in adhesion, J. Gen. Microbiol., 131:495 - 503 (1985).

42. P.A. Maisch and R.A. Calderone, Adherence of Candida albicans to a Fibrin-Platelet Matrix formed in vitro. Infect. Immun., 27, 2:650 - 656 (1980).

43. P.A. Maisch and R.A. Calderone, Role of surface mannan in the adherence of Candida albicans to Fibrin-Platelet clots formed in vitro. Infect. Immun., 32, 1:92 - 97 (1981).

44. K.G. Skerl, R.A. Calderone, and T. Sreevalsan, Platelet interactions with Candida albicans. Infect. Immun., 34, 3:938 - 943 (1981).

45. R.A. Calderone, R.L. Cihlar, D.D.S. Lee, K. Hoberg, and W.M. Scheld, Yeast adhesion in the pathogenesis of endocarditis due to Candida albicans: Studies with adherence-negative mutants. J. Infect. Dis., 152, 4:710 - 715 (1985).

46. A. Bouali, R. Robert, G. Tronchin, and J.M. Senet, Binding of human fibrinogen to Candida albicans in vitro: A preliminary study. J. Med. Vet. Mycol., 24:345 - 348 (1986).

47. R. Nosal and Z. Menyhardtova, The effect of glycoprotein from Candida albicans on functions of fat platelets. Toxicon, 14:313 - 318 (1976).

48. R. Robert, A. Bouali, G. Tronchin, and J.M. Senet, Le facteur fixant le Fibrinogene (F.F.F.) de Candida albicans a-t-il un role dans la physiopathologie des candidoses? Ann. Biol. Clin., (in press).

ANTIGEN AND METABOLITE DETECTION IN INVASIVE CANDIDIASIS

Louis de Repentigny

Department of Microbiology and Immunology
Sainte-Justine Hospital and University of Montreal
Montreal, Quebec, Canada   H3T 1C5

INTRODUCTION

Invasive candidiasis is the most frequent opportunistic fungal infection in immunocompromised patients, especially those with acute leukemia. Rapid diagnosis is of great importance because early treatment may be successful in resolving this potentially fatal infection. Unfortunately, diagnosis of invasive candidiasis remains difficult. A premortem diagnosis, made early enough for treatment, occurs in only 15-40% of patients.[1] Although fungal surveillance cultures may help predict invasive candidiasis,[2] blood cultures are negative in 56% of patients with autopsy-proven disseminated infection.[3] Detection of antibodies to the Candida species by a number of immunoassay formats has been used for a number of years,[4] but unfortunately the predictive value of these methods is unsatisfactory in profoundly immunosuppressed patients who are at greatest risk of invasive candidiasis. Insufficient sensitivity may be due to inability of the host to produce an antibody response, or to an inadequate period of time for agglutinins or precipitins to reach detectable levels. In addition to low sensitivity, these antibody tests have a limited ability to discriminate between colonization and deep tissue invasion and are thus not sufficiently specific. Because of these limitations, attempts have been made to identify and purify more specific antigens for antibody tests.[4] Most recently, a refined immunoblot analysis of the serological response in invasive candidiasis has shown the presence of antibodies specific to a 45- to 54-kilodalton major immunodominant cytoplasmic protein antigen.[5-11] In addition, patients who produced antibody to this antigen had an improved prognosis in invasive candidiasis.[9]

While such approaches are desirable, different strategies have been experimented in profoundly immunosuppressed patients in whom infection may not elicit a measurable antibody response. One such approach currently under investigation involves the detection of circulating antigens or metabolites of the Candida species.[4] In candidiasis, methods have been devised for the detection of the cell wall polysaccharide antigen mannan in serum by monoclonal enzyme immunoassay (EIA),[12] radioimmunoassay (RIA),[13,14] or latex agglutination,[15,16] of a 48-kilodalton cytoplasmic protein antigen by monoclonal EIA,[7,8,17] and of a heat-labile glycoprotein antigen by latex agglutination.[15,16,18,19] Commercial kits are under evaluation for the latter two methods.[15-19]

Despite sensitivities in the low ng/ml range, current immunoassays for candidal antigenemia detect at best 70% of patients with disease.[4] Antigenemia may be transient because of the formation and rapid elimination of immune complexes,[20,21] and improved detection may result mainly from repeated blood samplings rather than from the development of more sensitive immunoassays.[16,19] In addition to the detection of circulating antigens, the candidal monosaccharide metabolites arabinitol and mannose are quantitated in serum by gas-liquid chromatography and have been proposed as marker substances for invasive candidiasis.[4] In this paper, a survey will be made of our current knowledge of antigen and metabolite detection in invasive candidiasis.

ANTIGEN DETECTION IN INVASIVE CANDIDIASIS

Mannan

Much attention has been directed to the detection of the polysaccharide mannan, which is the main antigenic component of the cell wall of the Candida species.

Structure, methods of preparation, production of antisera. The general plan of yeast mannan is a linear α (1,6) backbone heavily substituted with α (1,2)- and α (1,3)-linked oligomannoside chains. The length of side chains, location of 1,3 bonds, and substitution with phosphodiesters determine antigenic specificity. Yeast mannan contains two domains, the inner core and the outer chain.[22,23] The inner core structure is a common property of yeast, plant, and mammalian glycoproteins. It consists of structural protein linked to core mannan. The polysaccharide moiety is linked via a diacetylchitobiose bridge by an N-glycosidic bond to an asparaginyl residue in the protein part of the molecule.[23] The polymannosyl outer chain region distinguishes yeast mannans from other glycoproteins and contains the immunodominant epitopes.

C. albicans occurs in two serotypes, A and B, and serotype specificity resides in mannan. The fine structure of mannan from *C. albicans* serotype A has been elucidated by Suzuki and Fukazawa.[24] They proposed a tree-like structure composed of a protein trunk with a linear branch of alternating 1,6-mannose residues two to three units long, flanking 1,2-linked mannobiose. The specific determinant of *C. albicans* serotype A mannan resides in terminal mannohexaose sub-branches consisting in one terminal α (1,3) linkage in addition to four α (1,2) linkages and might be bound via the β (1,6) linkage, to C-6 of the mannose residues of the oligosaccharide side chains. Serotype B is more complex, lacking a terminal 1,3 linkage but having instead a single C-1, C-2, C-3 branch point and an additional internal 1,3 linkage. Using polyclonal adsorbed factor-specific antisera,[25-27] these same researchers also demonstrated that *C. albicans* serotype A possesses antigenic factor 6 while serotype B has factor 13b instead of 6, and that antigenic factor 6 is important for the identification of *C. albicans* serotype A (Table 1). The mannans of *C. albicans* serotype A and *C. tropicalis* are closely related, and Candida stellatoidea is now considered by many yeast taxonomists to be synonymous with *C. albicans*.[28]

Recently, monoclonal antibodies have proved very useful as immunoprobes of the antigenic determinants of the mannans of the Candida species. Based on reactivity with several species of Candida and with Torulopsis glabrata, cross-adsorption experiments with polyclonal factor sera, and competitive binding with polyclonal factor sera in an indirect immunofluorescence test, Miyakawa et al.[27] characterized one monoclonal

Table 1. Antigenic Formulas of Candida albicans
and Related Species[a]

Strain	Antigenic factors
C. albicans serotype A	1, 4, 5, 6
C. albicans serotype B	1, 4, 5, 13b
C. tropicalis	1, 4, 5, 6
C. guilliermondii	1, 4, 9
C. krusei	1, 5, 11
C. parapsilosis	1, 5, 13, 13b
C. pseudotropicalis	1, 8
T. glabrata	1, 4, 6, 34

[a]Adapted from references 24-26.

antibody against C. albicans that corresponds to polyclonal factor serum 6, while another is closely related to polyclonal factor serum 5. Brawner and Cutler[29,30] also produced two agglutinating monoclonal antibodies against cell surface determinants on C. albicans, one of which may correspond to polyclonal factor serum 6. One determinant was continually expressed on mother cells and germ tubes throughout germination, while another was initially lost from the mother cell surface and preferentially expressed on hyphae during the first four hours of germination, later to reappear on the mother cell by 20 h.[31] Interestingly, the antigens that were reactive with these monoclonal antibodies were associated with a flocculent surface layer,[30] and the sloughing of these carbohydrate layers may correlate with the mannan antigenemia observed in invasive candidiasis.

In our own studies,[12] mice were immunized with C. tropicalis cell walls, and antibodies against mannan were detected in 3 of 9 BALB/c mice, 4 of 11 C57B1/6 mice, and in 4 of 8 CFW mice. Forty-one clones secreting anti-mannan monoclonal antibodies were obtained, and four clones were selected for propagation (Table 2). These included one IgM and one IgG monoclonal antibody that reacted with mannans of C. albicans serotypes A and B and of C. tropicalis and two IgM monoclonal antibodies specific for an epitope only in the mannans of C. albicans serotype A and C. tropicalis. Two monoclonal antibodies, CB6 (C. tropicalis and C. albicans A specific) and AC3 (C. tropicalis and C. albicans A and B specific) functioned in place of polyclonal antisera in the serotyping of C. albicans by immunofluorescence. Competitive inhibition between monoclonal antibody CB6 and polyclonal factor sera 1, 4, and 6 indicated that it recognizes antigenic factor 6. Similarly, monoclonal antibody AC3 appears to correspond to polyclonal factor serum 4.

In order to reliably detect mannan in the serum of patients with invasive candidiasis, the antibodies used as reagents (monoclonal or polyclonal) should recognize antigenic factors which are shared among the mannans of the species most commonly involved, C. albicans serotypes A and B and C. tropicalis.

The preparation of Candida mannan needs to be considered since the purified antigen is used as a positive control in immunoassays for mannan antigenemia. Traditionally, autoclaving in neutral citrate buffer by the method of Peat et al.[32] has been most widely used to obtain crude mannan, which can be purified by precipitation as its

Table 2. Monoclonal antibodies against <u>Candida tropicalis</u> mannan[a]

| MAbs | Class | Mannans recognized | | |
		C. albicans 20A (serotype A)	C. albicans 526B (serotype B)	C. tropicalis
AC3	IgM	+	+	+
DC5	IgG	+	+	+
BC4	IgM	+	−	+
CB6	IgM	+	−	+

[a]Adapted from reference 12.

insoluble copper complex,[32-35] or as a borate-cetyltrimethylammonium bromide complex.[36,37]

Different immunogens and immunization schedules can be utilized to produce antisera against mannan. Heat-killed <u>C. albicans</u> cells injected intravenously, or subcutaneously in complete or incomplete Freund's adjuvant, have been most commonly used,[14-16,38-42] and the efficiency of this immunogen correlates well with the presence of a flocculent cell surface layer of mannan as observed by Brawner and Cutler.[30] Other immunogens include a <u>C. albicans</u> homogenate,[13] a carbohydrate extract,[43] and cell walls.[20] The intravenous immunization of rabbits with cell walls[20] allows the purification of high-titered anti-mannan IgG.

<u>Immunoassays for detection of serum mannan.</u> The presence of mannan in serum was first described by Axelsen and Kirkpatrick in 1973 in a patient with chronic mucocutaneous candidiasis.[44] Since 1973, when serum mannan was detected by two-dimensional crossed rocket immuno-electrophoresis, tests of increased sensitivity and simplicity have been devised, evolving through hemagglutination inhibition, radioimmunoassay, inhibition EIA, sandwich EIA, and latex agglutination (reviewed in 4). Two factors complicate the detection of mannan in serum: first, it circulates in the range that is below the level of detection of most conventional serologic tests (ng/ml); and second, even in immuno-suppressed patients mannan circulates in the form of soluble immune complexes that must be dissociated before mannan can be reliably detected.[20,21] The presence of immune complexes has been directly demonstrated in the serum of patients with invasive candidiasis or chronic mucocutaneous candidiasis, using four screening techniques (direct nephelometry, Clq binding assay, PEG-IgG assay, and PEG-C4 assay).[21] In addition, specific <u>Candida</u> antigen and anti-<u>Candida</u> antibody were detected in the isolated immune complexes. Clinically, the dissociation of the complexes can be accomplished by briefly boiling serum in a diluent containing Na_2 EDTA to inhibit coagulation.[20] The dissociated serum extract is recovered by centrifugation and then added to microtiter plates coated with rabbit anti-<u>C. albicans</u> cell walls IgG. After incubation at 4°C and washing, indicator antibodies labeled with peroxidase detect mannan in the clinical range of 1-100 ng/ml, with absorbance directly proportional to concentration of mannan. This sandwich EIA requires three hours to perform (Fig. 1). Heat extraction of immune complexes has also been used by Weiner et al.,[13] Scheld et al.,[40] and Lew et al.[42] Antigen-antibody complexes can also be disso-ciated by incubating sera with protease for 1 h at 37°C.[15,16] Optimal detection of serum mannan in the clinical laboratory ultimately depends

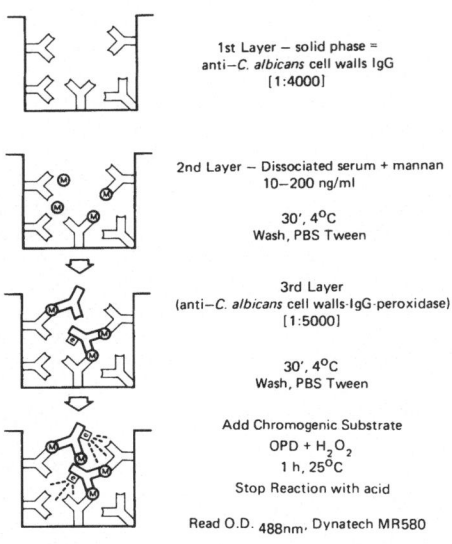

1st Layer — solid phase =
anti—*C. albicans* cell walls IgG
[1:4000]

2nd Layer — Dissociated serum + mannan
10–200 ng/ml

30', 4°C
Wash, PBS Tween

3rd Layer
(anti—*C. albicans* cell walls-IgG-peroxidase)
[1:5000]

30', 4°C
Wash, PBS Tween

Add Chromogenic Substrate
OPD + H_2O_2
1 h, 25°C
Stop Reaction with acid

Read O.D. 488nm, Dynatech MR580

Calculate mannan ng/ml w.r.t. standard curve

Fig. 1. Double-antibody sandwich EIA format.

on the dissociation of immune complexes and a simple, sensitive immunoassay format such as the double-antibody sandwich EIA or latex agglutination that also provide adequate shelf-life of reagents. It remains to
be determined whether more sensitive immunoassay formats, such as amplification of the sandwich EIA using avidin and biotin,[45] will improve the
detection of serum mannan in patients with invasive candidiasis. If the
presence of mannan in serum is transient, greater sensitivity to detect
antigen may not necessarily correlate with improved ability to detect
disease.

Recently, monoclonal antibodies were produced against C. tropicalis
mannan[12] (Table 2). An IgM monoclonal antibody, CB6, was an effective
substitute for rabbit antibodies in the double-antibody sandwich EIA to
detect antigenemia produced in rabbits infected with C. albicans serotype A or C. tropicalis. It could function either as the peroxidase-
conjugated indicator antibody or as the capture antibody. These
monoclonal antibodies avoid the need to repeatedly raise rabbit antisera
and are the source of reproducible, well-characterized reagents.

Evaluation of serum mannan detection in experimentally infected
animals. Animal models of invasive candidiasis are useful tools for the
evaluation of immunoassays for serum mannan. They provide a reproducible, controlled infection and the ability to obtain serum samples at
precise time intervals. In addition, the magnitude of antigenemia can
be quantitatively related to the extent of infection in deep organs. It
is also possible to compare the magnitude of antigenemia in animals
infected with C. albicans serotypes A or B or C. tropicalis.

Rats,[42] mice,[14] and rabbits[13,39,40,46,47] have been used in
experimental models, and rabbits are especially convenient because of

the ease in handling and the substantial volumes of serum which can be obtained from repeated bleedings of the same animal (reviewed in 4). Immunosuppression with cortisone acetate, although not essential to produce invasive candidiasis, potentiates the infection. Most investigators have used a single intravenous inoculation of candidal blastoconidia to infect the animals, although in one report the polyethylene catheter-induced experimental aortic valve endocarditis model was used.[40]

We evaluated[46,47] the double-antibody sandwich enzyme immunoassay for serum mannan in rabbits immunosuppressed with cortisone acetate and infected intravenously with 1.0×10^7 blastoconidia of C. albicans 3181A (serotype A). Serum mannan, negative before infection, peaked in all twelve infected animals 4 days after infection (mean, 18 ng/ml) and decreased thereafter (Fig. 2). Mannan antibodies, also negative before infection, began to rise 2 days after infection and coincided with decreasing serum mannan. All 12 animals showed antibodies to mannan. Control animals had no detectable serum mannan or antibodies to mannan.

These experiments illustrate several important concepts. First, serum mannan reproducibly appears in detectable concentrations (ng/ml) in experimental candidiasis. Second, after an initial rise, serum mannan decreases rapidly with the appearance of antibodies to mannan, probably by the formation and rapid elimination of immune complexes. Although immune complexes were not measured in these rabbit experiments, they have been detected in humans with invasive candidiasis or chronic mucocutaneous candidiasis.[21] Transient appearance of serum mannan is also observed in humans.[16] It would thus appear that the sensitivity of serum mannan for invasive candidiasis may eventually be limited by the way in which the antigen is handled by the host, which stands in contrast to the cryptococcal capsular polysaccharide which can be detected for several weeks in patient sera.[48] Thirdly, the infecting dose of candidal blastoconidia which was used reproducibly infected all animals, while allowing them to survive sufficiently long to observe the kinetics of appearance and disappearance of antigen.

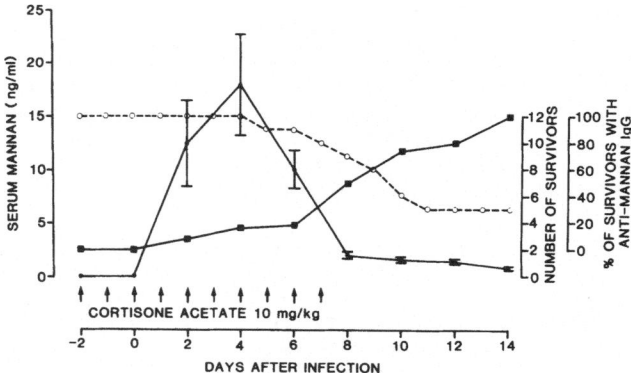

Fig. 2. Twelve rabbits infected intravenously with C. albicans 3181A on day 0. Symbols: o, number of survivors; ●, mean mannan in serum ± standard error of the mean; ■, percent of survivors with anti-mannan immunoglobulin G. Adapted from reference 47.

Table 3. Relationship Between CFU/g of Kidney and the Concentration of Serum Mannan Measured by Double-Antibody Sandwich EIA[a]

Rabbit no.	CFU/g of kidney[b]	Concentration of serum mannan[c] (ng/ml ± SD)
1	1.0×10^7	48 ± 8.7
2	1.2×10^7	77 ± 4.6
3	1.9×10^7	91 ± 10.5
4	5.0×10^7	128 ± 15.4

[a]Adapted from reference 12.
[b]Fresh weight; mean of right and left kidneys.
[c]Concentration four days after infection with respect to a standard curve of C. tropicalis mannan in serum. All data are triplicates and the mean concentrations from three daily runs (nine replicates).

Other experiments demonstrated that mannan is a quantitative marker and that its concentration is proportional to the number of candidal cells both in vivo and in vitro. When C. albicans 3181A was grown in yeast nitrogen base supplemented with 3 g of glucose per liter, mannan increased in parallel with viable count and optical density in this culture medium[47] (Fig. 3). Similarly, when rabbits were immunosuppressed with cortisone and infected intravenously with C. albicans 3181A, there was a direct relationship between CFU/g of kidney and the concentration of serum mannan[12] (Table 3).

Invasive candidiasis is most often caused by C. albicans serotypes A or B or C. tropicalis, and the optimal immunoassay should detect all three mannans. Using polyclonal anti-cell walls IgG as capture and indicator antibodies, the double-antibody sandwich EIA detects mannan from C. albicans serotype A or C. tropicalis.[12] These findings are in agreement with the identical antigenic formulas of the mannans of these two species, which share factors 1, 4, 5 and 6. Monoclonal antibody

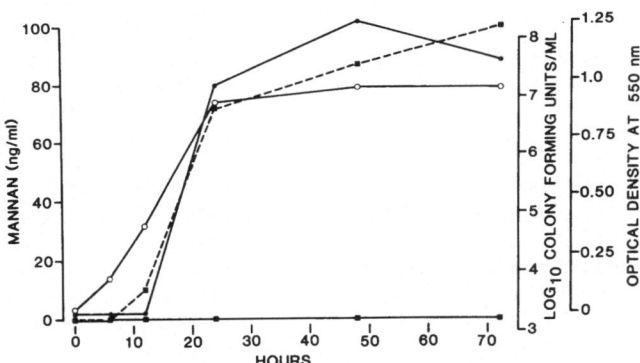

Fig. 3. In vitro production of mannan in yeast nitrogen base supplemented with 3 g of glucose per liter. Symbols: o, $\log_{10}$ CFU/ml; •, optical density at 550 nm; ■, mannan in uninoculated (—) and inoculated (---) media. Adapted from reference 47.

CB6, which binds to an epitope which is probably factor 6, can substitute as indicator antibody for the detection of these two mannans in experimentally infected rabbits.[12] This observation supports the view that the mannans of C. albicans serotype A and C. tropicalis are very closely related if not identical and can be detected by the same double-antibody sandwich EIA using polyclonal or monoclonal antibodies.

Antibodies against cell walls of C. albicans serotype A fail to detect mannan in sera of rabbits infected with C. albicans serotype B. Detection of serotype B mannan appears to be a special case and requires the use of antisera produced in rabbits immunized with whole blastoconidia heated and killed at 60°C.[49] C. albicans serotypes A and B share antigenic factors 1, 4 and 5; C. albicans serotype B additionally contains factor 13b (Table 1). It is hypothesized that antigens that are released into the plasma during infection with C. albicans serotype B are depleted in the common factors 1, 4, and 5 and enriched in factor 13b or in an as yet unrecognized factor,[49] which are not detected by the serotype A-specific reagents.

Evaluation of serum mannan detection in patients with invasive candidiasis. A number of factors need to be considered in evaluating published clinical studies on the detection of serum mannan: i) the study design may be either prospective, with sera drawn at predetermined intervals from a given number of patients at risk, or retrospective, using a limited number of frozen sera from patients known to have had the disease in the recent past. The number of sera available per patient is usually greater in a prospective study, and if antigenemia is transient this will tend to produce a falsely high estimate of sensitivity for disease, especially if the test is ultimately ordered on clinical suspicion of candidiasis. In a prospective study by Kahn and Jones,[16] the sensitivity of a latex agglutination test for serum mannan was much greater (78%) when several sera were obtained prospectively, as compared to sera collected only at the time candidiasis was clinically suspected (22%). These authors also showed that the presence of mannan in serum is often transient,[16] and optimal diagnostic sensitivity would require at least weekly serum samplings. ii) The proximity of sera to the time of diagnosis or death can influence diagnostic sensitivity, especially if antigenemia is not sustained but transient. This time interval has varied greatly and often complicates the comparison of published studies. iii) The case definition of invasive candidiasis is a critical aspect which deserves close attention. Clearly, the most well-defined study population would be composed of patients at risk, with or without invasive candidiasis at autopsy. This case definition has several advantages. There is no doubt about the diagnosis, either in patients with invasive candidiasis or in control patients. The extent of dissemination can also be ascertained. However, it is very difficult to assemble a study population based on these criteria who also have a sufficient number of sera in the weeks preceding death. There may also be a bias towards selecting the more fulminant forms of the disease with greater candidal organ content and antigenemia. At the opposite extreme, some case definitions of invasive candidiasis have relied on repeated culture of Candida species at mucosal sites of patients at risk, and the diagnosis of disease in these studies is in grave doubt. A more realistic approach is to stratify patients according to the probability of disease.[16] iv) Patient populations studied have varied with regard to underlying disease, and the degree of immunosuppression of these populations may influence the production of antibodies to mannan and consequently the clearance of circulating antigen. v) In most clinical studies the serotype of C. albicans was not determined and the ability to detect serum mannan in invasive candidiasis caused by C. albicans serotype B could not be ascertained.

vi) Immunoassay formats and antisera have also varied greatly, and these factors can have a profound influence on the outcome of clinical studies. vii) The dissociation of immune complexes increases the sensitivity of immunoassays for serum mannan, and thus the overall sensitivity of the test for disease. For example, Bailey et al.[15] detected mannan by latex agglutination in 17 of 21 patients with disseminated candidiasis when sera were treated with protease and heat, while antigen was found in only 3 of the 21 patients when this step was omitted.

Despite these limitations in comparing published studies, the following conclusions can be drawn about the value of serum mannan detection for the diagnosis of invasive candidiasis.[4] Available methods, including latex agglutination[15,16] and EIA,[50] have moderate sensitivity (50-70%) but high specificity (>99%) for invasive candidiasis. For example, our retrospective evaluation of the double-antibody sandwich EIA in cancer patients with or without invasive candidiasis at autopsy showed a sensitivity of 65% and a specificity of 100% using 1 ng/ml as the cutoff value.[50] Further gains in sensitivity for detecting disease may be obtained from more frequent blood samplings[16] because of the transient nature of antigenemia. Available methods (EIA, latex agglutination) are sufficiently simple to be performed in most diagnostic laboratories, but the widespread use of assays for serum mannan awaits the development of a reliable commercial kit.

Cytoplasmic Proteins

Cytoplasmic protein antigens also circulate in the serum of patients with invasive candidiasis.[17,51,52] For example, Araj et al.[51] detected a C. albicans cytoplasmic antigen that appears to be a heat stable protein in sera of patients with cancer. Likewise, Stevens et al.[52] described a radioimmunoassay for the detection of an antigen that may be a cytoplasmic protein and which does not cross-react with mannan. Finally, the major 48-kilodalton cytoplasmic protein antigen,[5,6] which appears promising as a specific reagent to detect antibody in invasive candidiasis,[7-9] has been detected by a monoclonal enzyme immunoassay in sera of mice with experimental invasive candidiasis.[17] The value of these methods awaits further clinical trials.

Heat-Labile Glycoprotein

The tests described thus far rely on the prior selection of a specific antigen as the candidal component most likely to be detected in systemic disease. A different approach was used by Gentry et al.[18] in which a circulating heat-labile antigen was detected using latex sensitized with serum from rabbits immunized with whole heat-killed C. albicans blastoconidia. The circulating antigen was sensitive to heat (56° for 30 min), pronase, 2-mercaptoethanol, and sodium periodate, which suggested that it might be a glycoprotein; its definite characterization as a component of the fungus was not reported. The sensitivity of the antigen to heat and the inability of sensitized latex to agglutinate mannan[18] suggest that the circulating antigen is other than mannan.

The latex agglutination test developed by Gentry et al. has been commercialized as the Cand-Tec Candida Detection System.[18] It has the advantages of being commercially available and of being easy and rapid (10 min) to perform in almost any diagnostic laboratory. However, kits do not contain control latex sensitized with normal rabbit serum, and thus false-positive results due to rheumatoid factor[19] are not excluded. Furthermore, published clinical evaluations[15,16,18,19,53] of the Cand-Tec test suggest that a titer of 1:8 is usually diagnostic of invasive disease while a titer of 1:4 occurs more commonly in invasive disease but is also found in colonized patients. The test appears insensitive (19-71%)[15,16,18,19,53] using a titer of 1:8 which excludes most false-positive results, and this lack of sensitivity is especially frequent on single serum specimens.[16,19] For example, Burnie et al.[19] found that in ten patients who had invasive candidiasis and an antigen titer of 1:8 or greater, only 28 out of 108 sera (26%) were positive at this level. Likewise, Kahn and Jones[16] found that the sensitivity of the Cand-Tec test was 48% in a panel of 355 sera sequentially collected from 75 leukemic patients, while antigenemia was detected in only 19% of patients with invasive candidiasis in a second panel of 364 sera collected from 150 patients with a variety of underlying diseases only at the time candidiasis was suspected. Despite the lack of sensitivity of the Cand-Tec procedure, it seems highly specific for invasive candidiasis at a titer of 1:8. Furthermore, the method detects antigenemia in patients infected with C. albicans serotypes A and B,[19] C. tropicalis[15,16,18] and C. parapsilosis.[18,19] The Cand-Tec test was compared to a latex agglutination method for serum mannan in two studies,[15,16] and the mannan test was found to have greater sensitivity and equivalent specificity for invasive candidiasis.

DETECTION OF METABOLITES IN INVASIVE CANDIDIASIS

In addition to the detection of antigens, much attention has been directed to the quantitation of the candidal sugar metabolites, arabinitol and mannose, in serum by gas-liquid chromatography.[4] The initial descriptions of the presence of these two substances in sera of patients with candidiasis were made independently.

Arabinitol

In 1979, Kiehn et al.[54] found increased serum concentrations of a sugar, arabinitol, in 15 of 20 patients with invasive candidiasis. It was also observed that arabinitol concentrations were higher in patients with renal failure than in those with normal renal function. These investigators went on to show that arabinitol is produced in vitro by C. albicans, C. tropicalis, C. parapsilosis, and Candida pseudotropicalis, but not by Candida krusei, Torulopsis glabrata, or Cryptococcus neoformans.[55] The mean concentration of arabinitol produced by C. tropicalis (1.6 μg/ml) was much lower than that of C. albicans (14.1 μg/ml), which may be of concern because the former species has become the leading etiologic agent of invasive candidiasis.[2] In later studies[56,57] these same investigators determined that arabinitol is cleared by the kidneys at the same rate as creatinine and that serum concentrations in renal failure can be adjusted by calculating the ratio of arabinitol to creatinine; they also showed that the appearance of arabinitol in rats with experimental invasive candidiasis is proportional to renal colony counts. A retrospective clinical evaluation of arabinitol quantitation in 25 patients with invasive candidiasis was done, and the sensitivity and the specificity of the ratio of arabinitol to creatinine were 64% and 96%, respectively.[58] Increased concentrations of arabinitol in invasive candidiasis were also reported by Roboz et al.[59] and Eng et al.,[60] but no adjustment was made for renal function. Thus in these latter studies, increased arabinitol concentrations in patients with invasive candidiasis may have resulted from renal failure rather than from the fungal disease.

Our own work has not confirmed the value of serum arabinitol as a diagnostic marker for invasive candidiasis.[47,50] Significant increases in arabinitol and the arabinitol/creatinine ratio were found in 12/12 and 8/12 experimentally infected rabbits, respectively, but also in all 6 uninfected animals receiving cortisone.[47] Similarly, the sensitivities and specificities were 26 and 87% for arabinitol, and 13 and 93% for the arabinitol/creatinine ratio, in cancer patients with autopsy-proven invasive candidiasis.[50] The diagnostic value of arabinitol and of the ratio of arabinitol to creatinine remains unclear and further studies are needed. Technically, immunoassays for circulating antigen, especially latex agglutination, are easier to perform than gas-liquid chromatography in the diagnostic laboratory and do not require special equipment.

Mannose

Miller et al.[61] first reported increased concentrations of serum mannose in six patients with candidemia and two of four patients with invasive candidiasis. This mannose was presumably derived from mannan. Different approaches have been used to assay mannose (reviewed in 4). First, either trimethylsilyl (TMS) or per-O-acetylated aldononitrile (PAAN) derivatives of this sugar were prepared. The disadvantage of the TMS derivatization procedure is that one sample component yields two anomeric peaks for mannose.[61] More conveniently, PAAN derivatives that produce a single peak per sample component were later used. The second consideration applies to the hydrolysis of serum before derivatization. Since mannan is often present in the serum of patients with invasive candidiasis, one can measure exclusively free mannose,[62,63] or the mannan can be hydrolyzed prior to derivatization.[64] In the latter case, both free mannose and the mannose generated by the mannan hydrolysis will be measured. The approach involving hydrolysis has the potential for yielding higher mannose concentrations in these patients, but this must be weighed against the possibility that hydrolysis may split mannose away from glycoproteins in normal serum.

Published studies suggested that elevated concentrations of mannose can be found in the majority of patients or laboratory animals with invasive candidiasis. However, increased concentrations of free mannose were also found in uninfected patients with diabetic ketoacidosis,[63] perhaps a reflection of the disturbance in carbohydrate metabolism.

We have been unable to confirm the value of serum mannose as a diagnostic marker for invasive candidiasis.[47,50] Significant increases in serum mannose were found in 12/12 experimentally infected rabbits, but also in all 6 uninfected animals receiving cortisone.[47] Likewise, the sensitivity and specificity were 39% and 87% in cancer patients with invasive candidiasis.[50] As for serum arabinitol, further clinical studies are needed to resolve these conflicting results.

CONCLUSION

Antigen and metabolite detection have been closely studied in experimental and human invasive candidiasis. The cell wall polysaccharide antigen mannan can be detected by latex agglutination or monoclonal EIA, and these methods have moderate sensitivity but high specificity for invasive candidiasis. Further gains in sensitivity for disease may be limited by the transient nature of mannan antigenemia. The widespread use of tests for serum mannan awaits the development of a reliable commercial kit. The commercial Cand-Tec latex agglutination test detects a heat-labile antigen, but this marker appears less sensitive than serum mannan for invasive candidiasis. Active areas for research are the development of an immunoassay for detection of the

major cytoplasmic protein antigen and application of monoclonal antibodies. The value of gas-liquid chromatography for serum arabinitol and
mannose remains controversial and awaits further clinical studies.

ACKNOWLEDGMENTS

The author sincerely thanks Errol Reiss and Serge Montplaisir for
their continued collaboration, and Sylvie Tassé for assistance in the
preparation of the manuscript.

REFERENCES

1. Edwards JE, Lehrer RI, Stiehm ER, Fischer TJ, Young LS. Severe
    candidal infections: clinical perspective, immune defense mecha-
    nisms, and current concepts of therapy. Ann Intern Med 1978;89:91-
    106
2. Sandford GR, Merz WG, Wingard JR, Charache P, Saral R. The value of
    fungal surveillance cultures as predictors of systemic fungal
    infections. J Infect Dis 1980;142:503-9
3. Myerowitz RL, Pazin GJ, Allen CM. Disseminated candidiasis: changes
    in incidence, underlying diseases, and pathology. Am J Clin Pathol
    1977;68:29-38
4. de Repentigny L, Reiss E. Current trends in immunodiagnosis of
    candidiasis and aspergillosis. Rev Infect Dis 1984;6:301-12
5. Jones JM. Quantitation of antibody against cell wall mannan and a
    major cytoplasmic antigen of Candida in rabbits, mice, and humans.
    Infect Immun 1980;30:78-89
6. Greenfield RA, Jones JM. Purification and characterization of a
    major cytoplasmic antigen of Candida albicans. Infect Immun
    1981;34:469-77
7. Strockbine NA, Largen MT, Zweibel SM, Buckley HR. Identification
    and molecular weight characterization of antigens from Candida
    albicans that are recognized by human sera. Infect Immun
    1984;43:715-21
8. Strockbine NA, Largen MT, Buckley HR. Production and
    characterization of three monoclonal antibodies to Candida albicans
    proteins. Infect Immun 1984;43:1012-8
9. Matthews RC, Burnie JP, Tabaqchali S. Immunoblot analysis of the
    serological response in systemic candidosis. Lancet 1984;ii:1415-8
10. Au-Young J, Troy FA, Goldstein E. Serologic analysis of
    antigen-specific reactivity in patients with systemic candidiasis.
    Diagn Microbiol Infect Dis 1985;3:419-32
11. Manning-Zweerink M, Maloney CS, Mitchell TG, Weston H. Immunoblot
    analyses of Candida albicans - associated antigens and antibodies
    in human sera. J Clin Microbiol 1986;23:46-52
12. Reiss E, de Repentigny L, Kuykendall RJ, Carter AW, Galindo R,
    Auger P, Bragg SL, Kaufman L. Monoclonal antibodies against Candida
    tropicalis mannan: antigen detection by enzyme immunoassay and
    immunofluorescence. J Clin Microbiol 1986;24:796-802
13. Weiner MH, Coats-Stephen M. Immunodiagnosis of systemic
    candidiasis: mannan antigenemia detected by radioimmunoassay in
    experimental and human infections. J Infect Dis 1979;140:989-93
14. Poor AH, Cutler JE. Partially purified antibodies used in a
    solid-phase radioimmunoassay for detecting candidal antigenemia. J
    Clin Microbiol 1979;9:362-8
15. Bailey JW, Sada E, Brass C, Bennett JE. Diagnosis of systemic
    candidiasis by latex agglutination for serum antigen. J Clin
    Microbiol 1985;21:749-52

16. Kahn FW, Jones JM. Latex agglutination tests for detection of _Candida_ antigens in sera of patients with invasive candidiasis. J Infect Dis 1986;153:579-85

17. Eng M, Walsh T, Maret M, Lockatell V, Johnson D, Rosenstein R, Buckley H. Evaluation of a monoclonal ELISA for detection of invasive candidiasis in mice. Prog Abstr Ann Meeting Am Soc Microbiol 1986;F31:402

18. Gentry LO, Wilkinson ID, Lea AS, Price MF. Latex agglutination test for detection of _Candida_ antigen in patients with disseminated disease. Eur J Clin Microbiol 1983;2:122-8

19. Burnie JP, Williams JD. Evaluation of the Ramco latex agglutination test in the early diagnosis of systemic candidiasis. Eur J Clin Microbiol 1985;4:98-101

20. Reiss E, Stockman L, Kuykendall RJ, Smith SJ. Dissociation of mannan-serum complexes and detection of _Candida albicans_ mannan by enzyme immunoassay variations. Clin Chem 1982;28:306-10

21. Burges G, Holley HP, Virella G. Circulating immune complexes in patients with _Candida albicans_ infections. Clin Exp Immunol 1983;53:165-74

22. Ballou C. Structure and biosynthesis of the mannan component of the yeast cell envelope. Adv Microb Physiol 1976;14:93-158

23. Farkas V. Biosynthesis of cell walls of fungi. Microbiol Rev 1979;43:117-44

24. Suzuki M, Fukazawa Y. Immunochemical characterization of _Candida albicans_ cell wall antigens: specific determinant of _Candida albicans_ serotype A mannan. Microbiol Immunol 1982;26:387-402

25. Tsuchiya T, Fukazawa Y, Taguchi M, Nakase M, Shinoda T. Serologic aspects on yeast classification. Mycopathologia 1974;53:77-91

26. Shinoda T, Kaufman L, Padhye AA. Comparative evaluation of the Iatron serological Candida check kit and the API 20C kit for identification of medically important _Candida_ species. J Clin Microbiol 1981;13:513-8

27. Miyakawa Y, Kagaya K, Fukazawa Y, Soe G. Production and characterization of agglutinating monoclonal antibodies against predominant antigenic factors for _Candida albicans_. J Clin Microbiol 1986;23:881-6

28. Meyer SA, Ahearn DG, Yarrow D. Discussion of the genera belonging to the imperfect yeasts; genus _Candida_, p 585-844. In NJW Kreger-van Rij (ed.), The Yeasts. Elsevier Science Publishers B.V., Amsterdam, 1984

29. Brawner DL, Cutler JE. Variability in expression of a cell surface determinant on _Candida albicans_ as evidenced by an agglutinating monoclonal antibody. Infect Immun 1984;43:966-72

30. Brawner DL, Cutler JE. Ultrastructural and biochemical studies of two dynamically expressed cell surface determinants on _Candida albicans_. Infect Immun 1986;51:327-36

31. Brawner DL, Cutler JE. Variability in expression of cell surface antigens of _Candida albicans_ during morphogenesis. Infect Immun 1986;51:337-43

32. Peat S, Whelan WJ, Edwards TE. Polysaccharides of baker's yeast. Part IV. Mannan. J Chem Soc (London) 1961;1:29-34

33. Gorin PAJ, Spencer JFT. Galactomannans of _Trichosporon fermentans_ and other yeasts-proton magnetic resonance and chemical studies. Can J Chem 1968;46:2299-2304

34. Kocourek J, Ballou CE. Method for fingerprinting yeast cell wall mannans. J Bacteriol 1969;100:1175-81

35. Gorin PAJ, Spencer JFT. Proton magnetic resonance spectroscopy - an aid in identification and chemotaxonomy of yeasts. Adv Appl Microbiol 1970;13:25-89

36. Lloyd KO. Isolation, characterization, and partial structure of peptido galactomannans from the yeast form Cladosporium werneckii. Biochem 1970;9:3446-3470

37. Nakajima T, Ballou CE. Characterization of the carbohydrate fragments obtained from Saccharomyces cerevisiae mannan by alkaline degradation. J Biol Chem 1974;249:7679-84

38. Segal E, Berg RA, Pizzo PA, Bennett JE. Detection of Candida antigen in sera of patients with candidiasis by an enzyme-linked immunosorbent assay - inhibition technique. J Clin Microbiol 1979;10:116-8

39. Harding SA, Brody JP, Normansell DE. Antigenemia detected by enzyme-linked immunosorbent assay in rabbits with systemic candidiasis. J Lab Clin Med 1980;95:959-66

40. Scheld WM, Brown RS Jr, Harding SA, Sande MA. Detection of circulating antigen in experimental Candida albicans endocarditis by an enzyme-linked immunosorbent assay. J Clin Microbiol 1980;12:679-83

41. Meunier-Carpentier F, Armstrong D. Candida antigenemia, as detected by passive hemagglutination inhibition, in patients with disseminated candidiasis or Candida colonization. J Clin Microbiol 1981;13:10-4

42. Lew MA, Siber GR, Donahue DM, Maiorca F. Enhanced detection with an enzyme-linked immunosorbent assay of Candida mannan in antibody-containing serum after heat extraction. J Infect Dis 1982;145:45-56

43. Kerkering TM, Espinel-Ingroff A, Shadomy S. Detection of Candida antigenemia by counterimmunoelectrophoresis in patients with invasive candidiasis. J Infect Dis 1979;140:659-64

44. Axelsen NH, Kirkpatrick CH. Simultaneous characterization of free Candida antigens and Candida precipitins in a patient's serum by means of crossed immunoelectrophoresis with intermediate gel. J Immunol Methods 1973;2:245-9

45. Yolken RH. Enzyme immunoassays for the detection of infectious antigens in body fluids: current limitations and future prospects. Rev Infect Dis 1982;4:35-68

46. de Repentigny L, Kuykendall RJ, Reiss E. Simultaneous determination of arabinitol and mannose by gas-liquid chromatography in experimental candidiasis. J Clin Microbiol 1983;17:1166-9

47. de Repentigny L, Kuykendall RJ, Chandler FW, Broderson JR, Reiss E. Comparison of serum mannan, arabinitol, and mannose in experimental disseminated candidiasis. J Clin Microbiol 1984;19:804-12

48. Prevost E, Newell R. Commercial cryptococcal latex kit: clinical evaluation in a medical center hospital. J Clin Microbiol 1978;8:529-33

49. Reiss E, Kuykendall RJ, Kaufman L. Antigenemia in rabbits infected with Candida albicans serotype B: detection by enzyme immunoassay and preliminary characterization of the antigen. J Med Vet Mycol 1986;24:259-69

50. de Repentigny L, Marr LD, Keller JW, Carter AW, Kuykendall RJ, Kaufman L, Reiss E. Comparison of enzyme immunoassay and gas-liquid chromatography for the rapid diagnosis of invasive candidiasis in cancer patients. J Clin Microbiol 1985;21:972-9

51. Araj GF, Hopfer RL, Chesnut S, Fainstein V, Bodey GP. Diagnostic value of the enzyme-linked immunosorbent assay for detection of Candida albicans cytoplasmic antigen in sera of cancer patients. J Clin Microbiol 1982;16:46-52

52. Stevens P, Huang S, Young LS, Berdischewsky M. Detection of Candida antigenemia in human invasive candidiasis by a new solid phase radioimmunoassay. Infection 1980;8:S334-8

53. Fung JC, Donta ST, Tilton RC. Candida detection system (Cand-Tec) to differentiate between Candida albicans colonization and disease. J Clin Microbiol 1986;24:542-7

54. Kiehn TE, Bernard EM, Gold JWM, Armstrong D. Candidiasis: detection by gas-liquid chromatography of D-arabinitol, a fungal metabolite, in human serum. Science 1979;206:577-580

55. Bernard EM, Christiansen KJ, Tsang S-F, Kiehn TE, Armstrong D. Rate of arabinitol production of pathogenic yeast species. J Clin Microbiol 1981;14:189-94

56. Wong B, Bernard EM, Gold JWM, Fong D, Silber A, Armstrong D. Increased arabinitol levels in experimental candidiasis in rats: arabinitol appearance rate, arabinitol/creatinine ratios, and severity of infection. J Infect Dis 1982;146:346-52

57. Wong B, Bernard EM, Gold JWM, Fong D, Armstrong D. The arabinitol appearance rate in laboratory animals and humans: estimation from the arabinitol/creatinine ratio and relevance to the diagnosis of candidiasis. J Infect Dis 1982;146:353-9

58. Gold JWM, Wong B, Bernard EM, Kiehn TE, Armstrong D. Serum arabinitol concentrations and arabinitol/creatinine ratios in invasive candidiasis. J Infect Dis 1983;147:504-13

59. Roboz J, Suzuki R, Holland JF. Quantification of arabinitol in serum by selected ion monitoring as a diagnostic technique in invasive candidiasis. J Clin Microbiol 1980;12:594-602

60. Eng RHK, Chmel H, Buse M. Serum levels of arabinitol in the detection of invasive candidiasis in animals and humans. J Infect Dis 1981;143:677-83

61. Miller GG, Witwer MW, Braude AI, Davis CE. Rapid identification of Candida albicans septicemia in man by gas-liquid chromatography. J Clin Invest 1974;54:1235-40

62. Monson TP, Wilkinson KP. D-mannose in human serum, measured as its aldononitrile acetate derivative. Clin Chem 1979;25:1384-7

63. Monson TP, Wilkinson KP. Mannose in body fluids as an indicator of invasive candidiasis. J Clin Microbiol 1981;14:557-62

64. Marier RL, Milligan E, Fan Y-D. Elevated mannose levels detected by gas-liquid chromatography in hydrolysates of serum from rats and humans with candidiasis. J Clin Microbiol 1982;16:123-8

MANNAN ANTIGEN OF CANDIDA ALBICANS AND CELLULAR IMMUNE RESPONSES IN

VITRO AND IN VIVO

Anne Durandy*, Alain Fischer*, Edouard Drouhet**, and
Claude Griscelli

*Inserm U 132 Hôpital Necker-Enfants Malades,
Paris, France
**Institut Pasteur, Paris, France

Candida albicans (CA) is an ubiquitous pathogen that can cause severe
infection in immunocompromised subjects. By studying the types of immuno-
deficiencies (ID) which are associated with CA infections, it is possible
to delineate the effectors of immunity against CA. Invasive CA infections
are observed in neutropenic subjects, probably because phagocytic cells are
necessary to kill CA yeasts. Secondly, mucocutaneous CA infections are
found in patients with various types of cellular ID, indicating that T
cell sensitization and T cell response (probably through the activation
of macrophage lymphokines) are essential components of host defense against
CA. Chronic mucocutaneous candidosis (CMCC) is an exciting model for
studying the immune response towards CA, because this disease is characterized by an elective cellular ID towards CA. We have therefore attempted to
analyse the cause of this ID. For this purpose, we have set up different
in vitro assays of cellular and humoral immune response to CA.

I - T cell response to mannan

We have studied the T cell response to the polysaccharide antigen
mannan which is extracted from the cell wall of CA. It has been shown
to be the major immunogen in CA-infected subjects (1). It is a 8,000 MW
polysaccharide with a polymannose repetitive structure.

Mannan was purified using a slight modification of the method of
Summers et al. (13), and the preparation obtained was weakly contaminated
by protein (< 4%) and by traces of glucan. After coupling to a non
immunogenic carrier, such as methylated bovine serum albumin (MBSA) (15)
or latex beads (10) mannan was able to induce a T cell response in sensitized
normal donors, at doses ranging from 10 to 100 µg/ml (14). Sensitized donors
were detected by the presence of serum antimannan antibodies (Ab) and by a
positive cutaneous hypersensitivity to CA antigens. The majority of normal
adults are spontaneously sensitized, after obvious or latent infection. The
T cell proliferation, evaluated by the number of recovered T blast cells
and the incorporation of ^{3}H-thymidine, peaked at day 5 was shown to
be specific. In fact, peripheral blood lymphocytes (PBL) obtained from
newborns or from unsensitized adult subjects were unable to be stimulated
by mannan, whatever the dose of antigen used and the duration of the
culture (Table 1).

Table I. T cell response to mannan.[+]

Cultures stimulated with	Number of experiments performed	Proliferative response	
		cpm (X10^{-3})	Number of recovered cells (X10^{-3})
Sensitized donors			
Mannan-MBSA (10 µg/ml)	7	60 ± 16	545 ± 116
Mannan-MBSA (50 µg/ml)	35	90 ± 21	707 ± 135
Mannan-MBSA (100 µg/ml)	7	48 ± 15	682 ± 123
Mannan-MBSA (500 µg/ml)	3	4 ± 3	110 ± 90
Mannan (40 µg/ml)	6	7 ± 3	277 ± 80
MBSA (10 µg/ml)	6	5 ± 2	170 ± 103
Unsensitized donors			
Mannan-MBSA (50 µg/ml)	5	2 ± 1	185 ± 72
Mannan-MBSA (50 µg/ml)	3	3 ± 1	130 ± 95

Cultures were stimulated either with mannan absorbed on MBSA at a ratio 4:1, mannan alone or MBSA alone in flat-bottom microwells. Proliferative response was evaluated at day 5 by ^{3}H-thymidine incorporation (cpm x 10^{-3}) and by the number of recovered cells (x 10^{-3}) for 1 x 10^{6} input cells. Results are expressed as the mean ± 1 SD of the responses obtained with PBL from different donors.

Genetic restriction necessary for T cell proliferation was investigated by the addition to cultures of monoclonal antibodies (MAb) specific for HLA molecules. As shown on Fig. 1, MAb directed to HLA-DQ exerted a strong suppressive effect on the mannan-induced T cell proliferation, whereas anti HLA-DR or DP had no detectable effect (13).

Co-culture experiments mixing monocytes and purified T lymphocytes from HLA-DR and DQ typed subjects confirmed the necessity of

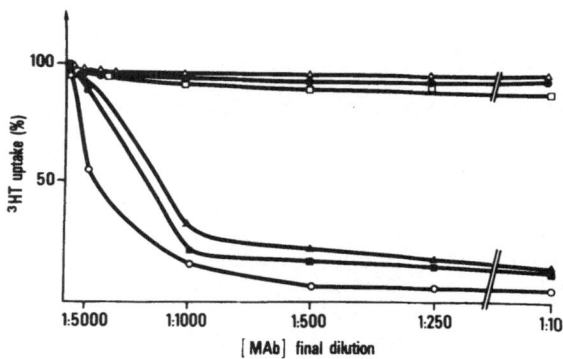

Fig. 1. Effect of different HLA-class II MAb on the in vitro mannan-induced T cell proliferation.

PBL from a HLA DR 2/2 DQ w_1/w_1 sensitized donor were cultured for 5 days in presence of mannan MBSA (50µg/ml). The results of one experiment out of 3 are shown.

anti HLA-DR:	CA 206 and		CA 135
anti HLA-DQ:	SDR 1	BT 3-4 and	SG 465
anti-HLA-DP:	B7/21		

(reprinted from Human Immunol., 1986, 16:114)

HLA-DQ for T cell proliferation. Similar results were obtained using mannan-specific, IL 2-dependent T cell clones. We also observed that a 2 h pulse of monocytes with antigen was sufficient to allow a T cell proliferation. Addition of HLA-DQ MAb during the monocyte pulse was able to inhibit the T cell proliferation. These results demonstrate that HLA-DQ molecules are required for presentation of mannan antigen to T lymphocytes.

It is striking to note that the genetic restriction linked to HLA-DQ molecules was found in all 12 sensitized subjects tested. This HLA-DQ linked restriction is also required for the proliferation of T cell clones specific for synthetic peptide of influenza virus hemagglutinin (5) and for the T cell proliferation stimulated by Candida antigens (11). There has yet been no indication that relatively simple antigenic structures such as mannan may always be presented to T lymphocytes in association with one given type of HLA-class II molecules. If the constant association of the in vitro human response to mannan with an HLA-DQ governed restriction is confirmed on a large panel of subjects, this would suggest that HLA-DQ molecule (s) -but neither HLA-DR nor HLA-DP- can interact with mannan in order to activate specific T lymphocytes.

Whether the role of HLA-DQ molecule (s) is related to the type of antigen remains an unanswered question, which will require the use of several distinct systems to be resolved. In the mouse, it has been shown that poly Glu^{40}-Ala^{60} (2) and poly Glu^{51}, Lys^{34}, Tyr^{15} (12) are preferentially recognized in association with either IA or IE molecules respectively. It will be worthwhile to study whether structurally simple antigens such as polysaccharides- are consistently recognized in humans in association with a unique type of HLA-class II molecules.

II - B cell response to mannan

Besides the T cell proliferation, mannan also triggered B lymphocytes to produce specific Ab in 7 day cultures. The detection of anti-mannan Ab was achieved using ^{3}H-mannan. Mannan was tritiated by L. Pichat (Centre d'Energie Atomique, France) by catalyzed exchange in solution with tritium as described by Evans (7). Analysis of labelled antigen revealed that the radioactivity was bound solely to mannose with a high specific activity (1.5 mCi/mg). A radioautography technique allowed us to directly visualize cells containing intracytoplasmic anti-mannan antibodies (AMACC). Approximately 5-10% of recovered blast cells were detected as AMACC (Fig. 2).

The B cell nature of the AMACC was confirmed by coupling autoradiography technique and membrane immunofluorescence. In fact, these cells bore surface immunoglobulins (IgM or IgG) and were not labelled by MAb directed against monocytes or T lymphocyte membrane antigens. The specificity of the mannan binding onto AMACC was confirmed by the complete absence of labelled cells in cultures stimulated with mitogens (such as pokeweed mitogen) or with antigens other than mannan (tetanus toxoid influenza virus and Toxoplasma gondii). Moreover, AMACC obtained in mannan-stimulated cultures did not bind another tritiated polysaccharide, the pneumococcal SIII (4).

Double labelling with intracytoplasmic immunofluorescence and autoradiography allowed the characterization of the isotype of the produced Ab. Both IgM and IgG (more precisely IgG1 and IgG2) Ab were detected. The evidence for specific anti-mannan Ab production was confirmed using an enzyme-linked immunoassay that detected the presence of antimannan Ab in mannan stimulated culture supernatants.

Purification of cell preparations by the E. rosette technique (E(+) = T lymphocytes, E (-) = B lymphocytes + monocytes) revealed the T cell dependency of B cell response to mannan. Co-culture experiments mixing

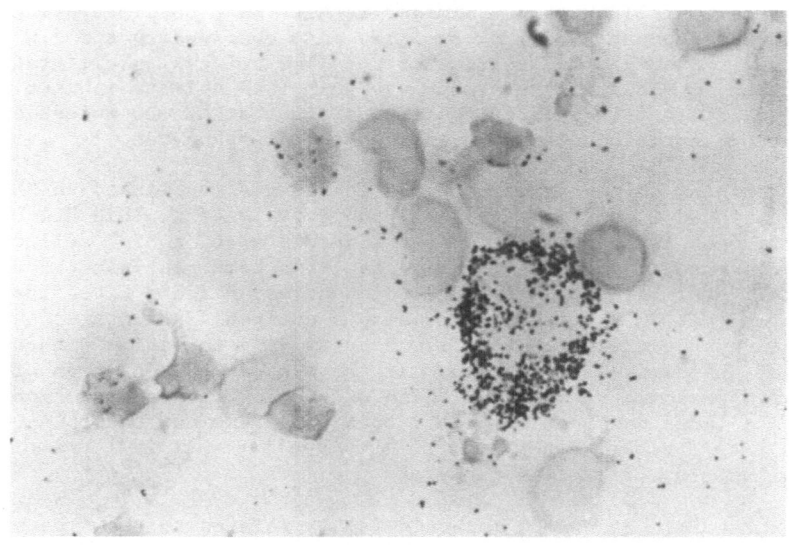

Fig. 2. Anti-mannan antibody containing cells.

AMACC were detected at the end of a seven day mannan-MBSA (50 ug/ml) stimulated culture by autoradiography using ^{3}H mannan and an eight day exposure in emulsion.
(reprinted from J. Clin. Invest., 1983, 71:1602)

E (+) and E (-) cell populations from different donors confirmed the previous observation of the genetic restriction between monocytes and T lymphocytes. However, they revealed that, although the B cell response to mannan was T cell-dependent, there was no genetic restriction between T lymphocytes and B lymphocytes, as shown in Table II.

Table II. Cellular requirements for T cell dependency of antimannan antibody production in mannan-MBSA-stimulated cultures[+].

Cultured cells	Number of Experiments performed	Antibody containing cells (X10^{-3})	Supernatant antibody concentrations	
			IgM (U/ml)	IgG
Unseparated PBL	10	12.3 ± 5.4	16.9 ± 10.1	23.7 ± 12.1
E(+)	10	<0.1	1.2 ± 1.7	1.9 ± 1.5
E(-)	10	<0.1	2.3 ± 0.9	1.4 ± 2.1
E(+) + autologous E(-)	10	14.0 ± 8.9	18.5 ± 6.1	20.5 ± 11.8
E(+) + allogeneic E(-)	10	<0.1	1.8 ± 1.5	0.8 ± 0.5
E(+) + autologous monocytes + allogeneic E(-)	6	6.8 ± 4.1	ND	ND
E(+) + autologous monocytes	6	<0.1	ND	ND
E(+) + semi-identical E(-)	4	11.2 ± 3.5	ND	ND

Experiments were performed with total or separated PBL from normal volunteers and E(-) cells isolated from either unrelated HLA-different or related HLA-semi-identical (mother or daughter) controls.
Cultures were stimulated with mannan-MBSA (50 ug/ml) for 7 d in flat-bottom microwells. Results of antibody containing cells are given for 10^6 initially cultured cells and results of supernatant antibodies are given in units per milliliter as compared with a reference serum.

Table III. Helper activity of mannan-induced T cell
supernatants for anti-mannan antibody production.(+)

T-cell supernatants (concentration %)	anti-mannan antibody containing cell generation by E(-) cells
-	100
5	400
10	2,400
20	7,100

T lymphocytes + autologous monocytes were cultured for 48 hrs with
mannan-MBSA (50 μg/ml). Supernatants were then added to 7 day mannan-
stimulated cultures of E(-) cells. Results of one experiment out of 3
is herein shown. Similar results were observed with allogeneic E(-)
cells.

T lymphocytes could be replaced by supernatants of T lymphocytes +
monocytes stimulated 48 h with mannan (Table III), indicating that there was
no cognate interaction between T lymphocytes and B lymphocytes for anti-
mannan Ab production. T cells exerted helper activity through the release
of lymphokines to B cells directly preactivated by mannan itself through
cross-linking of antigen receptors (membrane immunoglobulins).

T lymphocytes + autologous monocytes were cultured for 48 h with mannan-
MBSA (50 μg/ml). Supernatants were then added to 7 day mannan-stimulated
cultures of E(-) cells. Results of one experiment out of three is shown.
Similar results were observed with allogeneic E(-) cells.

Taking advantage of the knowledge of the in vitro T and B cell responses
to mannan of CA, we were able to study the abnormal immune response to CA
in CMCC.

III - The antigen-specific immunodeficiency in CMCC

CMCC is characterized by the persistence and the recurrence of CA
infections of the skin and of the mucosae. Besides an elevated in vivo
production of anti-CA antibodies, especially anti-mannan, cutaneous delayed
type hypersensitivity to CA is constantly absent during the infectious phase
of the disease · (6). In contrast, in vivo and in vitro lymphocyte responses
to unrelated antigens are normal in most patients, suggesting an immune
deficiency specific to CA. This selective cellular immunodeficiency is
found during active infection, while it disappears in most patients with
the healing of candidosis (Table IV).

Table IV. Immunodeficiency in CMCC

- Selective in vitro and in vivo T-cell unresponsiveness towards
  Candica antigens

- Present in infectious phase of the disease

- Normal or increased antibody production to Candida albicans
  antigens, especially mannan

- Detection in a large group of patients of a serum inhibitory
  activity specifically blocking T-cell responses to Candida.

In the serum of many patients, a strong inhibitory activity of the
CA antigen induced T cell proliferation has been observed (13, 14 ;
Figure 4). This suppressive effect was only observed during the acute
phase of the disease and disappeared after treatment.

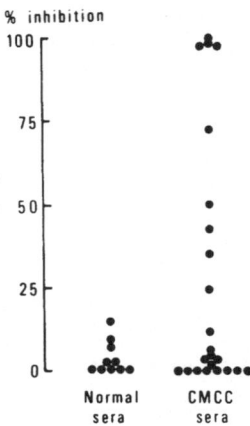

Fig. 3. Inhibitory activity in the serum of CMCC patients.

Effect of serum from 23 CMCC patients and 10 normal individuals on proliferative response of normal CA-sensitized lymphocytes to CA antigens.
(reprinted from J. Clin. Invest., 1978, 62:1005)

In the serum of many patients, a strong inhibitory activity of the CA antigen induced T cell proliferation has been observed (8, 14; Fig. 3). This suppressive effect was only observed during the acute phase of the disease and disappeared after treatment.

The suppression was specific to CA antigens since it did not generally affect T cell proliferation induced by mitogens (such as ConA) or unrelated antigens, such as PPD, as shown in Fig. 4.

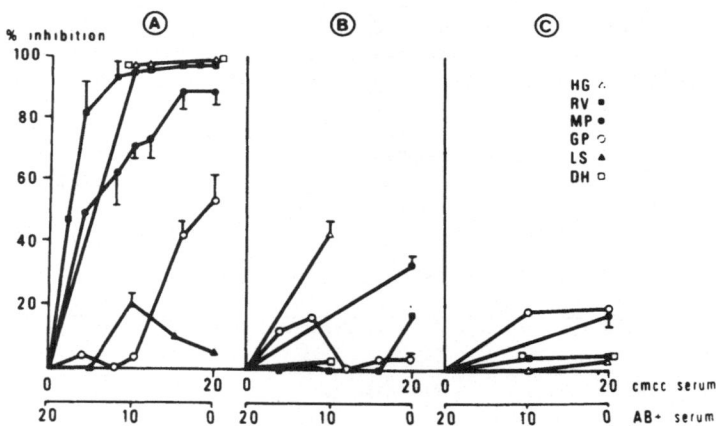

Fig. 4. CA antigen specific inhibitory activity in
the serum of CMCC patients.

Percentage of inhibition induced by various concentrations of CMCC patients' sera and tested on CA (A), PPD (B) and ConA (C) induced normal lymphocyte proliferation. Results are given as the mean (± of the percentage of inhibition in two experiments with lymphocytes of different normal individuals.
(reprinted from J. Clin. Invest., 1983, 71:1602)

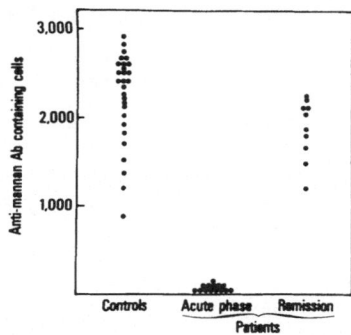

Fig. 5. Absence of anti-mannan antibody production in patients
in acute phase of CMCC.

Biochemical analysis of the serum inhibitory activity revealed that
the suppressive effect was found in the polysaccharide fraction, and was
ascribed to mannan itself. These results suggest that mannan circulates
in serum of patients with CMCC in the acute phase and is responsible for
the observed specific inhibitory activity. One can deduce from these
results that there should be an immunosuppressive concentration of mannan
in infected tissues resulting in a local unresponsiveness to CA.

Beyond this serum inhibitory activity, we also observed a cellular
suppressor activity mediated by T lymphocytes. Indeed, we found an
unresponsiveness fo T lymphocytes to mannan antigen, in CMCC patients
during the acute phase of the disease, leading to an absence of in vitro
anti-mannan antibody production (Fig. 5).

The lack of T cell proliferation was secondary to the presence of T
suppressor (TS) lymphocytes (Fig. 6).

The TS lymphocytes observed in patients with CMCC in acute phase had
the following characteristics: they belonged both to CD8(+) and CD8(-)
lymphocytes, and they were shown to be radiosensitive and indomethacin
resistant. They were specific to CA antigens since they did not affect

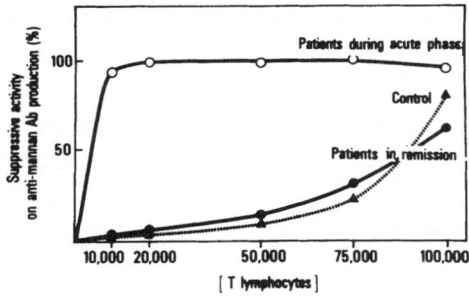

Fig. 6. Evidence for a T suppressor lymphocyte activity in patients
in acute phase of CMCC.

T lymphocytes from patients or from normal subjects were added to 200 x 10³
normal PBL stimulated with mannan-MBSA (50 g/ml). Results are expressed
as percentage of inhibition of the anti-mannan AB production. One representative experiment out of five is shown.

Table V. Induction of TS activity by excess of mannan[+]

Preincubation of normal subjects'cells		Suppressive activity on in vitro production (%)	
		of anti-mannan antibody	of PWM-induced Ig
PBL			
Mannan	100 µg/ml	81 ± 23	< 10
T lymphocytes			
Mannan	100 µg/ml	< 10	< 10
T lymphocytes + autologous monocytes			
Mannan	100 µg/ml	88 ± 14	< 10
T-lymphocytes + HLA-mismatched monocytes			
Mannan	100 µg/ml	< 10	< 10
Monocytes			
Mannan	100 µg/ml	< 10	–

Preincubated cells (PBL, T lymphocytes, or monocytes) were added to autologous PBL stimulated with mannan-MBSA (50 µg/ml) or pokeweed antigen (10 µl/ml) for 7 days in a ratio of 1:2. Results are expressed as the percentage of inhibition on the in vitro anti-mannan Ab production or Ig production (means of 5 experiments).

cell proliferation induced by mitogens or unrelated antigens. They acted as well on HLA identical or HLA-incompatible target cells, giving evidence of the absence of genetic restriction for the effector phase of suppression.

IV - Immunosuppressive activity of mannan

In order to better understand the generation of such TS lymphocytes in CMCC patients, we tried to reproduce the phenomenon by incubating normal subjects' PBL in presence of an excess of mannan (100 µg/ml). Such preincubation resulted in a strong suppressive effect on T and B cell responses to mannan. T lymphocytes were responsible for the suppressive activity but their triggering occurred only in presence of autologous monocytes (Table V).

These results demonstrate that normal subjects' T lymphocytes can be activated to exert a suppressive activity by mannan. The characteristics of the in vitro mannan-induced TS cells were shown to be very comparable to the in vivo induced TS cells found in CMCC patients (Table VI).

Table VI. Characteristics of the mannan induced suppression.

Mannan-induced suppression observed	
In CMCC patients	With control cells
CD8(+) and CD8(-) lymphocytes	CD8(+) and CD8(-) lymphocytes
Radiosensitive	Radiosensitive
Indomethacin resistant	Indomethacin resistant
Specific to CA	Specific to CA
Resistant to exogenous 112	Resistant to exogenous 112
–	HLA restricted (induction phase)
Not HLA restricted (effector phase)	Not HLA restricted (effector phase)
Active on T cell proliferation and on antibody production	Active mainly on antibody production

Table VII. Suppression by mannan of monocyte-dependent T
lymphocyte proliferation to CA antigens[+].

Cultured cells obtained from normal CMA-responders	Expt.1 (cpm x 10^{-3})	Expt.2 (cpm x 10^{-3})
Monocytes	255 ± 36	387 ± 82
CA-pulsed monocytes	344 ± 41	311 ± 50
T lymphocytes	172 ± 24	143 ± 14
T lymphocytes + CA	423 ± 28	277 ± 31
Monocytes + T lymphocytes	417 ± 32	515 ± 28
CA-pulsed monocytes + T lymphocytes	26,536 ± 743	12,730 ± 603
Mannan-pulsed monocytes + T lymphocytes	463 ± 12	398 ± 27
CA and mannan-pulsed monocytes + T lymphocytes	10,852 ± 447	3,445 ± 208
CA and mannan-pulsed monocytes + T lymphocytes + CA-pulsed monocytes	24,260 ± 926	8,235 ± 511

The results of two representative experiments are given as the means
of triplicate cultures ± 1 SD. Monocytes were pulsed with 100 µg/ml
mannan.

Mannan could exert its inhibitory effect through a secondary mechanism.
We have indeed shown that an excess of mannan added to a pulse of normal
monocytes with CA antigen inhibits the presentation of antigen to T lymphocytes (Table VII).

This effect is not due to the induction of suppressor monocytes because
the mixing of mannan-incubated and non-incubated monocytes resulted in
normal T cell responses to CA antigens. The precise mechanisms by which
mannan may impair either its own antigen presentation or the presentation
of unrelated CA antigens remains unknown.

It might well be that in vivo the immunosuppressive effect of mannan
is related to both mechanisms.

Persistence of mannan in CMCC

Although we have given some evidence for the immunosuppressive effect
of mannan, we are not able so far to understand why patients with CMCC
are more susceptible to this immunosuppressive effect. Indeed, as shown
in Table VIII, lymphocytes from CMCC subjects obtained in time of remission
are more susceptible to inhibition by mannan than normal lymphocytes (9).

Table VIII. Increased sensitivity to inhibition of Candida-
induced lymphocyte proliferation in three CMCC patients[+].

Mannan Concentration (ug/ml)	Proliferation responses to CA (cpm)			
	Control (mean of 10 subjects)	Patient 1	Patient 2	Patient 3
0	32,800 ± 11,800	18,700	30,500	19,500
10	35,400 ± 13,200 (0 %)	7,400 (60.2 %)	20,000 (34.4 %)	10,500 (46.3 %)
100	15,400 ± 8,400 (53.1 %)	2,500 (86.7 %)	6,700 (78.2 %)	4,800 (75.6 %)
1.000	2,700 ± 1,100 (91.8 %)	500 (97.5 %)	900 (97.0 %)	1,300 (93.5 %)

2×10^5 PBL were cultured for 5 days in presence of CA antigens.
Results are the difference between ^{3}H-thymidine uptake in CA-
stimulated and unstimulated cultures. In brackets, percentage of
suppression.

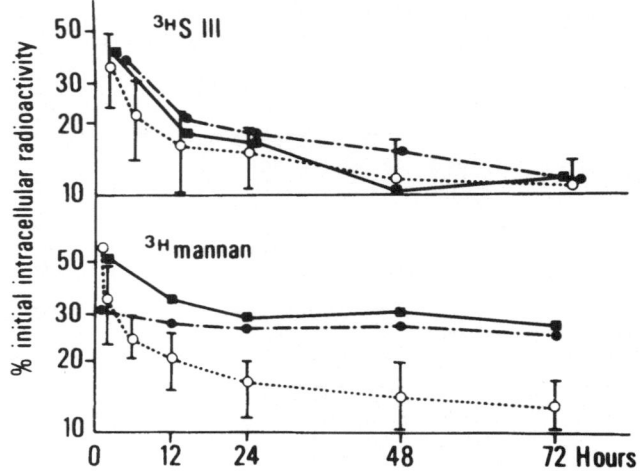

Fig. 7. Uptake and catabolism of labelled mannan of SIII
pneumococcal polysaccharide by control or patient
monocytes.

These data suggest that the defect in CMCC may be related to an
abnormal persistence of mannan in susceptible patients. In order to test
this hypothesis, we compared the uptake and catabolism of ^{3}H-mannan and
[3]H-pneumococcal SIII by patients' and controls' monocytes. As shown in
Fig. 7, we found in two out of six patients a defective catabolism of
mannan (but not of SIII) following a normal uptake. This abnormality was
not due to a dilution of ^{3}H-mannan in cold mannan since these assays were
performed in remission phase (9).

This finding may thus explain, at least in some patients, why mannan
persists and thereby exerts a strong immunosuppressive activity. One cannot,
however, tell whether the abnormality in mannan catabolism is the primary
defect or is secondary to another phenomenon.

As shown in Fig. 8, one can propose that an excess of free mannan
secondary to a defective handling by monocytes, at least in some patients,
can induce specific T cell unresponsiveness in two ways: 1) induction of
suppressor T cells; 2) inhibition of antigen presentation.

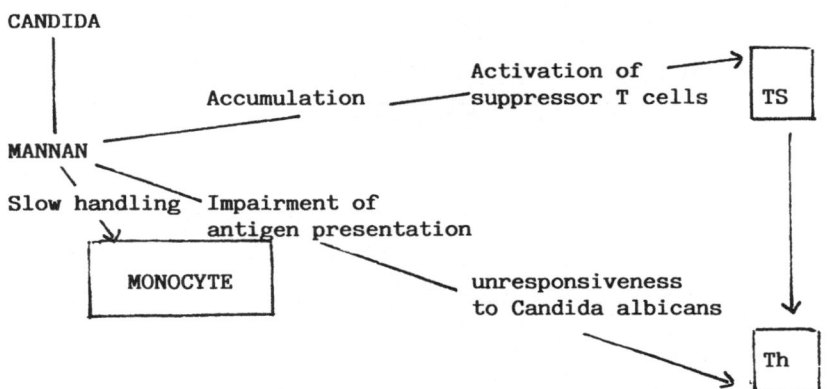

Fig. 8. Proposed mechanisms of the specific immune
deficiency observed in CMCC.

In this hypothesis, it remains to be understood why the specific B cell response is normal. As mentioned in this review the antibody response to mannan, the main immunogen of CA, is T-cell dependent. However, we have shown that T cells may act as amplifiers of the response rather than inducers. One can easily envisage that the B cell response to mannan is directly triggered by mannan and that non-specific factors (Il 1) may be sufficient in vivo for production of anti-mannan antibodies.

ACKNOWLEDGMENT

Reprints from Human Immunology, Journal of Clinical Investigation and Clinical Experimental Immunology were reproduced with Editor's copyright permission.

REFERENCES

1. N.H. Axelsen and C.H. Kirkpatrick, 1973. Simultaneous characterization of free Candida antigens and Candida precipitins in a patient's serum by means of crossed immunoelectrophoresis with intermediate gel. J. Immunol. Methods, 2:245.
2. C.N. Baxevants, D. Vernet, Z.A. Nagy, P.H. Maurer, and J. Klein, 1980. Genetic control of T cell proliferative responses to poly (Glu40 Ala60) and poly(Glu51 Lys34 Tyr15): Subregion a specific inhibition of the responses with monoclonal anti-Ia antibodies. Immunogenetics, 11: 617
3. A. Durandy, A. Fischer, D. Charron, and C. Griscelli, 1986. Restriction of the in vitro anti-mannan antibody response by HLA-DQ molecules. Hum. Immunol., 16:114.
4. A. Durandy, A. Fischer, and C. Griscelli, 1983. Specific in vitro anti-mannan rich antigen of Candida albicans antibody production by sensitized human blood lymphocytes. J. Clin. Invest., 71:1602.
5. D.D. Eckels and R.J. Hartzman, 1982. Characterization of human T lymphocyte clones (TLCS) specific for HLA region gene products. Immunogenetics, 16:117.
6. J.E. Edwards, R.S. Lehrer, C.R. Stiehm, T.J. Fischer, and C.J. Yount, 1978. Severe candidal infections, clinical perspective, immune defence mechanisms and current concept of therapy. Ann. Med., 85:51.
7. E.A. Evans, H.C. Sheppard, J.C. Turner, and D.C. Warrel, 1974. A new approach to specific labelling of organic compounds with tritium gas. J. Labelled Comp., 10:569.
8. A. Fischer, J.J. Ballet, and C. Griscelli, 1978. Specific inhibition of in vitro Candida-induced lymphocyte proliferation by polysaccharidic antigens present in the serum of patients with chronic mucocutaneous candidiasis. J. Clin. Invest., 62:1005.
9. A. Fischer, L. Pichat, M. Audinot, and C. Griscelli, 1982. Defective handling of mannan by monocytes in patients with chronic mucocutaneous candidiasis resulting in a specific cellular unresponsiveness. Clin. Exp. Immunol., 47: 653.
10. S.M.M. Gettner and D.W.R. Mackenzie, 1981. Responses of human peripheral lymphocytes to soluble and insoluble antigens of Candida albicans. J. Med. Microb., 14:333.
11. T.A. Gonwa, L.J. Pickler, H.V. Raff, S.M. Goyert, J. Silver, and J.D. Stobo, 1983. Antigen-presenting capabilities of human monocytes correlates with their expression of HLA-DS, an Ia determinant distinct from HLA-DQ. J. Immunol., 170:706.
12. E.A. Lerner, L.A. Matts, C.A.J. Janeway, P.P. Jones, R.M. Schwartz, and D.B. Murphy, 1980. Monoclonal antibody against an Ia gene product. J. Exp. Med., 152:1085.

13. D.F. Summers, A.P. Grollman, and H.F. Hasenclever, 1964. Polysaccharide antigens of Candida cell wall. J. Immunol., 92:491.

14. J.S. Twomey, C.C. Wadell, S. Frantz, R. O'Reilley, P. L'Esperance, and R.A. Good, 1974. Chronic muco-cutaneous candidosis with macrophage dysfunction, a plasmic inhibitor and co-existent aplastic anemia. J. Lab. Clin. Med., 85:968.

15. E. Wiener and A. Bandieri, 1974. Differences in antigen handling by peritoneal macrophages from the Biozzi high and low responder lines of mice. Eur. J. Immunol., 4:457.

ANTIGENICITY OF <u>CANDIDA</u> <u>ALBICANS</u> CELL WALL AFTER ENZYMATIC
DIGESTION

I. Berdicevsky* and J. Muller**

*Technician Faculty of Medicine, Microbiology,
Haifa, Israel
**Inst. of Parasitology and Mycology, Center of
Hygiene, Univ. of Freiburg, Postfach 820, D-7800
Freiburg, Germany FR

Studies of the cell wall of <u>Candida</u> <u>albicans</u> may provide
significant data on the virulence mechanism(s) of this medically important fungus. The arrangement of wall components has
a marked effect on the activity of various lytic enzymes used
for protoplast production. Since the cell wall plays such an
important role in antigenicity, we used selective enzymatic
digestion as a method for examining the immunoreactivity of
different wall components. The purpose of this study was to
follow the cell wall degradation by various enzymes and the
changes in antigenicity during wall digestion.

<u>Candida</u> <u>albicans</u> was submitted to digestion by zymolyase,
chitinase, lysozyme, or all three together. At various intervals, aliquots were taken and prepared for electron- microscopic (EM) observation, or pretreated with anti-human anti-
<u>Candida</u> anti-serum labeled with immunoferritin followed by EM
observations. No differences in morphology and labelling
between the control cells and those exposed to chitinase or
lysozyme were found even after 120 min of treatment. Yeasts
treated with zymolyase for this same period completely converted into protoplasts. The addition of chitinase or lysozyme to zymolyase shortened the time necessary for protoplast
formation from 120 min to 30 min.

Since the mannan-protein layer contains the major antigenicity sites, we focused on possible morphological changes in
this layer and carefully examined the distribution of immunoferritin label during various stages of cell wall enzymatic
degradation. Control cells and lysozyme- or chitinase-treated
cells were heavily labelled with immunoferritin whereas those
treated with zymolyase in addition to the two other enzymes
caused total digestion of the cell wall after short periods of
time (30 min) and no immunoferritin labelling was subsequently
observed.

DIFFERENTIAL EXPRESSION OF SECRETORY PROTEINASE BY <u>CANDIDA</u> <u>ALBICANS</u> AND <u>C</u>. <u>PARAPSILOSIS</u> INVADING HUMAN MUCOSA

M. Borg and R. Rüchel

Institute of Hygiene, University of Göttingen
Kreuzbergring 57, D-3400 Gottingen, FRG

Opportunistic yeasts of the oral microbial flora can cause infections like thrush or denture stomatitis. The most relevant fungal species in such loci is <u>Candida</u> <u>albicans</u>, followed by <u>C</u>. <u>tropicalis</u> and to a lesser extent by <u>C</u>. <u>parapsilosis</u>. Among the various yeast-like fungi found in clinical specimens, these three species are distinguished by secretion of acid proteinase <u>in</u> <u>vitro</u> (R. Ruchel et. al., 1986, <u>Infect</u> <u>Immun</u>., 53:411). The <u>in</u> <u>vitro</u> activity does not reflect the large difference of virulence among the three proteolytic <u>Candida</u> species. Previously we have shown that avirulent <u>C</u>. <u>parapsilosis</u> do not express proteinase during infection of cultured phagocytes (R. Rüchel et al., 1986, <u>Infect</u>. <u>Immun</u>., 53:411). In the present investigation we tried to answer the question whether such differential expression of proteinase might account for the different virulence of the three fungal species in the oral cavity.

Samples of non-keratinized oral mucosa were taken by punch biopsy (diameter 2 mm) from the cheek of healthy volunteers. The samples were attached to cover slips using a fibrin adhesive and were kept in medium 1640-199 + 10% fetal calf serum + 0.5 mg/ml ampicillin. The mucosal samples were infected with $5x10^5$ yeasts harvested in the late exponential phase from glucose peptone broth. After incubation up to 7h (37^oC, 6% CO_2), samples were submitted to immunoscanning electron microscopy for detection of proteinase antigen (M. Borg, Z. Naturforsch, 1985, 40c:539). Concomitantly samples of the culture medium were drawn in the course of incubation for monitoring of soluble acid proteolytic activity (modified hemoglobin assay at pH 3.5).

Blastospores of <u>C</u>. <u>albicans</u> and <u>C</u>. <u>tropicalis</u> quickly adhered to the mucosal surface. After 4h the first germ tubes were detected invading the epithelium. At this stage of infection, proteinase antigen was abundant on the surface of blastospores and germ tubes of <u>C</u>. <u>albicans</u> serotype A (CBS 2730), and <u>C</u>. <u>tropicalis</u> (isolate 293/80). Among <u>C</u>. <u>albicans</u> serotype B (ATCC 48867, ATCC 36802) proteinase antigen was abundant only on blastospores and was hardly detectable on filamentous cells. This pattern may reflect the noted difference of virulence between the serotypes in the oral cavity (M.V. Martin and D.J. Lamb, 1982, J. Clin. Pathol., 35:888).

Blastospores of C. parapsilosis (isolate 265/80) showed little adherence and filamentous cells were scarce. Proteinase antigen was not detectable on either growth form, thus confirming our previous findings in infected cultured phagocytes and mice (R. Rüchel et al., 1986, Infect. Immun., 53:411).

The results correlated with the development of soluble proteolytic activity in the culture medium. With C. albicans and C. tropicalis a continuous rise of enzymatic activity was detectable during incubation. With C. parapsilosis, a plateau of low activity established early in the incubation. Comparison with non-infected mucosal samples suggested that this fraction of activity is due to cathepsin-D leaking from the mucosal tissue. In vitro most isolates of C. albicans, C. tropicalis and C. parapsilosis are strongly proteolytic. However, during infection of oral mucosa in tissue culture only C. albicans serotype A and C. tropicalis expressed the proteinase antigen on the surface of all cell types. C. albicans serotype B expressed the antigen on blastospores only, and C. parapsilosis cells did not express proteinase antigen at all. These results reflect the comparatively low virulence of C. parapsilosis and C. albicans serotype B in the oral cavity. The presence of acid proteolytic activity in the culture medium suggests that the surface antigen of C. albicans and C. tropicalis represents an active enzyme which may play a role during invasion.

PRODUCTION OF ANTI-CANDIDA MONOCLONAL ANTIBODIES:

IMMUNOPURIFICATION OF C. ALBICANS CELL WALL ANTIGENS

T. Chardès, J.C. Lebecq, B. Pau and J. Bastide

Unité de Recherche en Immunologie
Faculté de Pharmacie
Montpellier, France

Our main objective is the immunopurification and characterization of C. albicans cell wall antigens. By an indirect ELISA technique, we screened 4870 hybridoma culture supernatants in search of anti-C. albicans monoclonal antibody (MAb)-secreting hybridomas. Two hundred and thirty-nine of these hybridomas were found to secrete anti-Candida MAbs. All these MAbs recognized C. albicans and C. tropicalis. Twenty-one of the MAbs were selected for further study (Chardes et al. Ann. Inst. Pasteur/-Immunol. 137c:117-125, 1986). By immunoblotting and immunoaffinity chromatography, we attempted to partially characterize and isolate the antigens recognized by MAbs CA2 and CA4; the former MAb reacts with C. albicans serotypes A and B and C. tropicalis, and the latter only reacts with C. albicans serotype A and C. tropicalis. The results of immunoblotting experiments showed that both these MAbs reacted with antigens having a molecular weight greater than 100,000. We isolated the antigens recognized by MAbs CA2 and CA4 by passage of a C. albicans A crude cell extract through immunoaffinity columns containing each MAb bound to Sepharose. The antigenic fractions recognized by CA2 and CA4 were eluted with a pH 4.3 phosphate-citrate buffer. The fractions contained 95% and 80% carbohydrate, respectively. We estimated the purification factor of antigen CA2 by inhibition of the indirect ELISA reaction between antigen CA2 and MAb CA2 using various inhibitors: antigen CA2, the crude cell extract, C. albicans mannan (Figure 1). The ratio of the IC_{50} values for the crude cell extract and antigen CA2 indicated a 555-fold purification. Given the carbohydrate nature of this antigen and the fact that the C. albicans mannan preparation inhibited the reaction between antigen CA2 and MAb CA2, we conclude that antigen CA2 is a mannoprotein. The techniques developed in this study may be applied to a molecular, structural approach to the classification of yeasts and the eventual identification of circulating antigens in patients with invasive candidiasis.

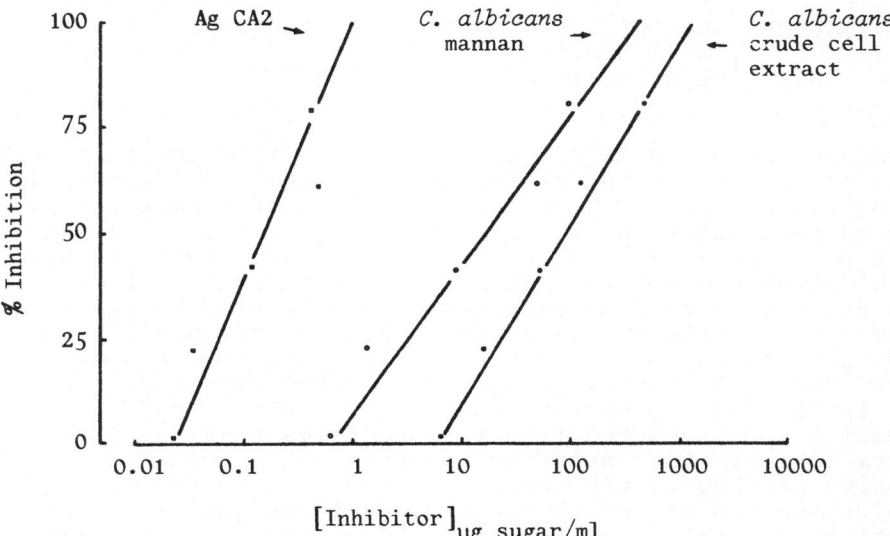

Figure 1. Inhibition of the reaction between
MAb CA2 and Ag CA2 by various antigenic preparations

CANDIDA ANTIGEN AND ARABINITOL LEVELS IN THE SERA OF PATIENTS
WITH PROVEN OR PROBABLE INVASIVE CANDIDOSIS

A.G. Deacon[1] and M.D. Richardson[2]

Department of Bacteriology and Immunology[1]
and Medical Mycology Unit[2],
Western Infirmary, Glasgow G11 6NT

Two approaches that have gained considerable interest over
the last few years for the diagnosis of systemic candidosis in
immunocompromised patients are the detection of arabinitol by
gas-liquid chromatography (g.l.c.) and Candida cell wall mannan
by latex particle agglutination. Since January, 1986, we have
been evaluating a new latex particle reagent capable of detecting 25 ng of mannan/ml of serum. We attempted to correlate the results from the latex tests with those obtained on
the same patients by g.l.c. for arabinitol and the semi-quan-
titative detection of anti-Candida precipitating antibodies by
counter immunoelectrophoresis (CIE).

Results are presented from retrospective testing of 40
sera from 11 patients, including 6 proven and 5 probable cases
of invasive candidosis that were previously collected during
the evaluation and use of the arabinitol technique, and from a
prospective examination of sera obtained since January, 1986.
In the following months, 89 sera from 31 adults and 22 children
were examined by all three techniques. In addition, antigen
and CIE estimations were performed on a further 164 sera from
158 patients.

None of the 11 patients from the retrospective series,
including the 6 proven cases, yielded a positive antigen reaction. All 11 had at least one elevated arabinitol result in
the sera tested and 6 of the 11 had positive CIE results. In
the prospective series, 11 of the 89 patients had positive
antigen tests but none had raised arabinitol levels and none
were independently proven to have systemic infection. However,
when the sera from 158 patients with suspected systemic disease
were tested by latex agglutination and CIE only 3 patients had
detectable antigen in serum and were subsequently documented as
having deep-seated infection. One of these patients was positive for both antigen and antibody.

Possible reasons for this lack of correlation may include:
poor sensitivity of either arabinitol or latex tests in relation to the tissue biomass of Candida, the rapid clearance of
mannan from the circulation, and late or insufficient sampling
relative to the onset of symptoms of systemic infection.

MANNAN DETECTION BY ELISA-INHIBITION IN DISSEMINATED CANDIDIASIS:

EXPERIMENTAL STUDY IN RABBITS

J. Garcia de Lomas*, C. Morales*, M.A. Grau*,
J.M. Delga**, E. Drouhet**, B. Dupont** and
L. Improvisi**

*Department of Microbiology, University Hospital,
Medical School, Valencia - 46010, Spain
**Pasteur Institute, Unit of Mycology, Paris, France

Introduction

The diagnosis of disseminated candidiasis is an important problem and much attention has been directed to serodiagnosis. Unfortunately, the presence or absence of antibody has not been sufficiently predictive of disease. During the last 10 years, methods for antigen detection in serum have been evaluated in patients and in experimentally infected animals. We selected the ELISA-inhibition assay because of its sensitivity and simplicity and evaluated the method in rabbits with disseminated candidiasis.

Material and methods

The ELISA-inhibition test was performed as described by Segal et al (1979, J. Clin. Microbiol. 10:116) on sera and urine from rabbits inoculated intravenously with 10^7 blastoconidia of Candida albicans strain 3153A. Reference mannan was prepared by the method of Peat et al (1961, J. Chem. Soc. 1:29). Antisera against C. albicans serotype A, C. albicans serotype B and C. tropicalis were raised in rabbits inoculated subcutaneously weekly with a yeast lysate. The ELISA inhibition test was carried out using polystyrene plates coated with 10 g of mannan. Anti-C. albicans serum was diluted 1:2,000 as a positive control. Standard curves were obtained with concentrations of mannan from 0, 5, 10 to 1,280 g/ml diluted in PBS.

Serum samples diluted 1:4 and urine (undiluted) were boiled at 100ºC to dissociate immune complexes and the supernatants were incubated with anti-C. albicans serum for 2 h at 37ºC. The mean concentration detected in serum and in urine taken from control, uninoculated rabbits was 19.8 and 23.9 ng/ml, respectively.

Blood cultures and assays for serum mannan were done daily until death and urine obtained at post mortem was examined for Candida. Organs were examined for histopathology and organ homogenates were cultured.

Results

Macroscopic granulomas were observed in the kidneys of most of the animals, and all animals had cultural evidence of invasive candidiasis (Table 1).

The concentration range of mannan in vesical urine was 248-8372 ng/ml, and the concentration was independent of the presence of viable yeast cells in urine (Table 2).

The concentration range of mannan in serum was 20-9310 ng/ml, but it was always greater than 40 ng/ml after 48 hours post-inoculation (Table 3).

Mannan was detected independent of the coating mannans and specific sera used (Table 4). High correlation coefficients (R) between mannan concentrations and optical densities detected by ELISA readings were obtained. A prospective study to quantify mannan in sera and urine from patients is being done to relate test results with clinical findings.

TABLE 1. NECROPSY STUDY
(MACROSCOP LESIONS/ORAGN CULTURES)

RABBIT NO.	HOUR OF DEATH POSTINOCULATION	MACROSCOPIC LESIONS	ORGAN CULTURES
1	168	K . L	K.L
2	96	K . L	K.L.S.Lu.B.H
3	120	K.LuB.H.C	K.L.S.Lu.B.H.C
4	96	K.Lu.B	K.L.S.Lu.B.H
5	72	K.Lu.B.H	K.Lu.B.H.C
6	120	K.Lu.B.H	K.L.S.Lu.B.H.C
7	72	K . Lu	K.L.S.Lu.B.H
8	72	K	K.L.S.Lu.B.H
9	72	-	K.S.Lu.B.H
10	72	-	K.S.Lu.B.H
11	72	-	K.L.S.Lu.B.H
12	96	K.Lu	K.L.S.Lu.B.H
13	96	K	K.S.Lu.B
14	96	K	K.B
15	96	K.Lu.C	K.S.Lu.B.H

K:KIDNEY; L:LIVER; S:SPLEEN; Lu:LUNG; B:BRAIN; H:HEART; C:COLON

TABLE 2. INTRAVESICAL URING SPECIMEN
(POSTMORTEM)

RABBIT NO.	MANNAN (NG/ML)	CULTURE
1	ND	ND
2	ND	ND
3	533	NEGATIVE
4	989	POSITIVE
5	2220	NEGATIVE
6	336	POSITIVE
7	2052	POSITIVE
8	ND	ND
9	356	POSITIVE
10	248	NEGATIVE
11	638	NEGATIVE
12	ND	ND
13	8372	POSITIVE
14	440	POSITIVE
15	352	POSITIVE

ND : NOT DONE

TABLE 3. QUANTIFICATION OF MANNAN IN SERUM (NG/ML)

RABBIT NO.	HOURS POSTINOCULATION						
	24	48	72	96	120	144	168
1	90	140	170	90	190	130	180
2	52	40	130				
3	30	40	180	140			
4	52	100	190	210			
5	20	50	110				
6	180	210	500	880	990		
7	-	260	772				
8	270	930	2200				
9	80	100					
10	470	910					
11	260	-					
12	50	170	140	300			
13	2760	8970	3630				
14	1640	9310	6800				
15	1840	7240	1840				

TABLE 4. COMPARISON OF RESULTS WITH DIFFERENT COATING MANNANS. SPECIFIC SERA AND MANNANS DETECTED

COATING	SERUM ANTI-..	MANNAN DETECTED	R
C.ALBICANS A	C.TROPICALIS	C.ALBICANS A	0.938
C.ALBICANS B		C.ALBICANS B	0.976
C.TROPICALIS		C.TROPICALIS	0.988
C.GUILLIERMONDII		C.GUILLIERMONDII	0.984
C.ALBICANS A	C.ALBICANS A	C.ALBICANS A	0.991
C.ALBICANS B		C.ALBICANS B	0.986
C.TROPICALIS		C.TROPICALIS	0.936
C.GUILLIERMONDII		C.GUILLIERMONDII	0.927
C.ALBICANS A	C.ALBICANS B	C.ALBICANS A	0.968
C.ALBICANS B		C.ALBICANS B	0.987
C.TROPICALIS		C.TROPICALIS	0.973
C.GUILLIERMONDII		C.GUILLIERMONDII	0.964

EVALUATION OF THE CANDIDATE SUPER LATEX AGGLUTINATION TEST FOR THE DIAGNOSIS OF VAGINAL CANDIDOSIS

V. Hopwood, D.W. Warnock, J.D. Milne, T. Crowley, C.T. Horrocks and P.K. Taylor

Departments of Microbiology and Genito-Urinary Medicine, Bristol Royal Infirmary, Bristol BS2 8HW, U.K.

We have compared a new, commercial slide latex agglutination (SLA) test for the detection of Candida antigens with conventional methods for the diagnosis of vaginal candidosis.

263 non-pregnant women who had not received antifungal treatment during the preceding month were investigated. 23 women with vulvitis and/or vaginitis, in whom culture was positive (C. albicans, 22; C. glabrata, 1) were defined as having vaginal candidosis. 40 women with no clinical signs in whom culture was positive were regarded as having commensal carriage.

The SLA test was positive for 15 of the 23 women (65.2%) with vaginal candidosis. 6 of the 8 antigen negative women had counts of $<1 \times 10^3$ cfu per ml. 5 of the 7 women with commensal carriage who were antigen positive had counts of $\geq 1 \times 10^3$. The incidence of a positive SLA test increased in proportion to the amount of yeast isolated.

Table 1. Analysis of three methods for the diagnosis of vaginal candidosis

	Smear	Culture	SLA
Sensitivity	69.6%	100%	65.2%
Specificity	95.4%	83.3%	96.7%
Predictive value positive	59.3%	36.5%	65.2%
Predictive value negative	97.0%	100%	96.7%
Efficiency	93.2%	84.8%	93.9%

The results indicate that SLA is a useful alternative to conventional methods of diagnosis, particularly in a clinical setting.

AN ATTEMPT TO OBTAIN CANDIDA SKIN ANTIGENS FOR THE DIAGNOSIS OF SYSTEMIC CANDIDOSIS

K. Iwata, Y. Yamamoto, T. Yamashita, H. Uehara, T. Kamoshida and T. Yanai

Department of Microbiology, Kagawa Nutrition College Sakado-Shi, Saitama-ken, Japan

We previously reported that a formol toxoid from candi-toxin, a potent toxin of simple acidic protein nature which we isolated from cells of a highly virulent strain of Candida albicans, showed a delayed type skin reaction in guinea pigs experimentally infected with C. albicans and also in some patients with systemic candidosis (K. Iwata, 1978, Mycopha-thol., 65:141; K. Iwata, 1985, in "Filamentous Fungi," T. Arai (ed) Japan Societies Press, Tokyo). Later we isolated three toxic glycoprotein fractions from cells of another virulent

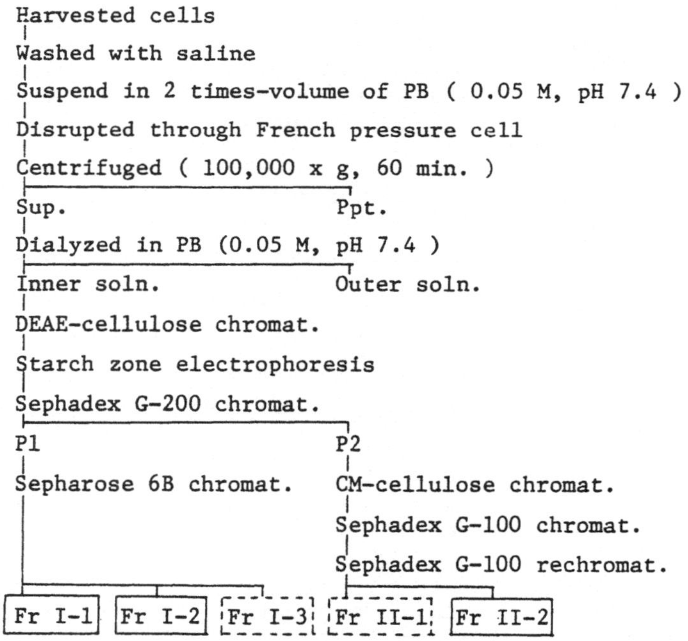

```
Harvested cells
│
Washed with saline
│
Suspend in 2 times-volume of PB ( 0.05 M, pH 7.4 )
│
Disrupted through French pressure cell
│
Centrifuged ( 100,000 x g, 60 min. )
├──────────────────────────────┐
Sup.                           Ppt.
│
Dialyzed in PB (0.05 M, pH 7.4 )
├──────────────────────────────┐
Inner soln.                    Outer soln.
│
DEAE-cellulose chromat.
│
Starch zone electrophoresis
│
Sephadex G-200 chromat.
├──────────────────────────────┐
P1                             P2
│                              │
Sepharose 6B chromat.          CM-cellulose chromat.
│                              │
│                              Sephadex G-100 chromat.
│                              │
│                              Sephadex G-100 rechromat.
│                              │
├──────┬──────┬────────┬───────┤
Fr I-1  Fr I-2  Fr I-3  Fr II-1  Fr II-2
```

Fig. 1. Separation and purification of the toxic glycoprotein fractions.

strain of C. albicans (MTU 12024) grown in a semi-synthetic medium at 37°C for 18h under constant stirring and aerating through the separation and purification procedures similar to those for canditoxin (Fig. 1) (K. Iwata, Y. Yamamoto, 1977, Proc. 4th Int. Conf. on the Mycoses, PAHO, Scientific Publication No. 356:246; K. Iwata, 1985, in "Filamentous Fungi," T. Arai (ed) Japan Societies Press, Tokyo). Of these fractions, Fr II-2 was the most potent in toxicological, biological, immunological and other relevant activities. It exhibited the highest acute toxicity for laboratory animals upon intravenous injection; the LD 50 was 15 ug/g body weight. For these reasons, it was almost exclusively used throughout the present study.

Fr II-2 in the final purification step by Sephadex G-100 chromatography showed a single peak. Its molecular weight was approximately 50,000. The ratio of sugar to protein was 86:14 (w/w). The sugar moiety consisted solely of D-mannose. Its toxicity was so acute and severe that upon intravenous administration of 1 MLD, the majority of the treated animals succumbed within 48h with the symptomatology characterized by anaphylaxis-like reactions and the pathology similar to those caused by canditoxin. It manifested versatile biologic activities such as pyrogenicity, coagulation of Limulus amoebocyte lysate, shortening of plasma coagulation time, a slight antitumor activity against Ehrlich ascites tumor and Sarcoma 180 tumor cells in mice, etc.

Fr II-2 was a potent in immunogen, but could not be changed to toxoid with formol, unlike canditoxin. This communication deals with the results of skin tests with Fr II-2 performed as part of our investigations of experimental and clinical candidosis of the systemic type.

Preliminary skin tests revealed that intradermal administration of 0.1 ug of Fr II-2 in 0.1 ml of buffered saline solution into guinea pigs infected intravenously with 1 x 10^8 cells of a toxic glycoprotein-producing strain of C. albicans (MTU 12024) gave a positive skin reaction of the delayed type, characterized by a series of conspicuous tissue responses, viz. redness, swelling and induration, but gave a negative reaction without manifesting direct toxic effects in normal guinea pigs locally and systemically. Eight of 10 animals infected under the above-mentioned conditions gave a positive skin reaction of the delayed type to that same dose of toxin. Three of 5 animals infected with a canditoxin producing strain and one of 5 animals infected with a C. tropicalis strain showed a similar reaction, whereas all 5 animals infected with a C. krusei strain gave a negative reaction. The positive reaction reached a maximum around 48h after the injection of Fr II-2, maintained that level for an additional 48h and then almost disappeared 72h later.

In human systemic candidosis including broncho-pulmonary candidosis, five out of 6 patients showed a positive reaction of the delayed type to the intradermal administration of 0.2 ug of Fr II-2, whereas the same number of healthy individuals gave a negative reaction.

These results suggest that such toxins produced by C. albicans may be used as a skin antigen for the diagnosis of systemic candidosis to a certain extent in combination with other tests, although their sensitivity and specificity as well as the degree of hypersensitivity due to infection should be studied in more detail with more numerous cases.

CANDIDA MANNAN ANTIGEN DETECTION VERSUS QUANTITATIVE CULTURE IN URINE, SPUTUM, STOOL, AND VAGINAL SWABS

R. Kappe, D. Kubitza, I Scheidecker and J. Müller

Institute of Parasitilogy and Mycology, Center of Hygiene, University of Freiburg, Postfach 820, D-7800 Freiburg, West Germany

A self-prepared latex agglutination test with a sensitivity of 8 ng/ml was used for the detection of Candida mannan antigen. 1 ml of sputum and stool specimens were homogenized with 3 ml and 100 ml trypsin solution (0.25%), respectively. Candida albicans suspensions, sputum and stool homogenates,

Table 1. Results of microscopy and Candida antigen detection in C.albicans suspensions and 1312 clinical specimens with graded numbers of C.albicans colony forming units per ml.

Cfu of Candida albicans per ml	Prepared Yeast-suspension titer	Percent positive (titers 1:4 - 1:512)								Candida albicans estimation of Quantity
		n = 305 urine		n =318 sputum		n = 402 stool		n = 287 vaginal swabs		
	Mi Ag	Mi	Ag	Mi	Ag	Mi	Ag	Mi	Ag	
Negative	neg. neg.	0	0	1	1	53	1	<1	<1	negative
<10^2	neg. neg.	0	0	8	5	-	-	5	5	Broth only
10^2	neg. neg.	17	6	45	5	63	11	29	3	+
<10^3	+ neg.	43	9	44	2	71	21	79	38	++
<10^4	+ 1:4	68	61	84	47	54	2	100	65	+++
<10^5	+ 1:16	95	84	100	100	74	14	100	100	++++
<10^6	++ 1:64	95	89	100	100	81	48			
<10^7	+++ 1:2000	100	100	100	100	100	100			
<10^2 — 10^7 pos.		39	28	49	21	67	16	51	35	Broth pos. ++++

Cfu : Colony forming units on Sabouraud dextrose agar
Mi : Microscopy of a direct Gram stained smear (inspection time : 2 min.)
Ag : Candida mannan antigen detection with a latex agglutination test (8 ng/ml).

urine samples, and extracts of vaginal swabs were treated with pronase, boiled, and centrifuged. The supernatants were tested for Candida antigen.

In clinical specimens of urine, sputum, stool, and vaginal swabs, microscopy using a direct Gram stained smear and Candida mannan antigen detection (sensitivity: 8 ng/ml) are about one thousand times less sensitive than culture. Antigen detection was even less sensitive than microscopy. However, both microscopy and Candida antigen detection are very quick and highly specific (specificity for the presence of yeasts: 100%). One possible advantage of Candida mannan antigen detection in the examined specimens may be a high predictive value for clinical significance of the yeasts in the corresponding tissue.

SYSTEMATIC USE OF A LATEX TEST TO DETECT SOLUBLE CANDIDA

ANTIGENS IN SERODIAGNOSIS OF SYSTEMIC CANDIDIASIS

Y. Le Fichoux, H. Chan and P. Marty

Laboratoire de Parasitologie-Mycologie, Hopital
Pasteur, B.P. 69, 06002 Nice Cedex, France

1198 sera from 560 patients were received for serodiagnosis of candidiasis in our laboratory. A commercially
available test (Gentry L.O., Wilkinson D., Lea A.S., Price,
M.F., Eur. J. Clin. Microbio., 1983, $\underline{2}$:122) was systematically
performed on these sera for the detection of soluble Candida
antigens. We studied, retrospectively, the hospitalization
reports of the patients who had at least one serum with a latex
titer $\geq$ 1/4.
Of the 1198 tested sera, 128 (11%) sera from 68 patients
(12%) were positive. Only 52 patients were studied. Candida
spp. were found in 8 patients (15%) by blood culture and in
another 25 patients (48%) by culture of urine, bronchoalveolar lavage, various punctures or catheters. 16 patients
(30%) presented with hyperthermia without bacteriological
explanation.
The risk factors chosen were: surgery or intensive care,
urinary or blood catheter, broad spectrum antibiotic therapy
and immunodeficiency. 15 patients (29%) presented 4 risk
factors; 21 patients (40%) presented 3 risk factors; 8 patients
(15%) presented 2 risk factors; 6 patients (12%) presented 1
risk factor. Only 2 patients (4%) had no risk factor. Clinical improvement was observed for 11 (55%) of the 20 patients
who received systemic antifungal treatment and for 14 (44%) of
the 32 patients without antifungal treatment. This retrospective study shows that only 11% of the tested sera had a
latex titer $\geq$ 1/4. Systemic candidiasis was not proved for the
52 patients but 96.2% of them presented at least one risk
factor and 69.2% at least 3 risk factors.
Candida spp. was found in biological samples of 63.5% of
the patients. Blood cultures are positive for only 15.4% of
the cases, but the medium used was not optimal. Among the 52
patients, 38.5% received antifungal therapy, but the difference
in clinical improvement between the treated group and the non-
treated group is not significant (55% against 44%).
With the exception of the cases for which a Candida could
be isolated, we think the association of a latex titer $\geq$ 1/4
with three risk factors justifies an early treatment just after
taking mycological samples.

EVALUATION OF IMMUNE COMPLEXES AND DETECTION OF

A M_r48,000 PROTEIN IN CANDIDIASIS

B. T. Maida and H. R. Buckley

Department of Microbiology and Immunology
Temple University School of Medicine
Philadelphia, Pa. 19140, U.S.A.

Serum samples from 12 patients with systemic candidiasis were tested for
the presence of elevated levels of circulating immune complexes (CIC) and
evaluated for specific Candida antigens following isolation methods. Sera
were screened for CIC using the Ortho Serum ELISA test for Clq-IgG. Of
the 104 sera studied, 10 (representing 5 of 12 patients) demonstrated
elevated levels of CIC. These non-specific immune complexes were then
isolated and evaluated for the presence of specific Candida antigens. Our
initial procedure included five steps: (1) enrichment of immune complexes
with 5% PEG; (2) separation of immune complexes from monomeric IgG by gel
filtration; (3) isolation of IgG-containing immune complexes by affinity
chromatography on Protein A-Sepharose; (4) dissociation of antigen-
antibody complexes with subtilisin A; and (5) characterization of antigens
by fused rocket immunoelectrophoresis and immunoblotting. The ultimate
protocol included immunoprecipitation with Staphylococcus aureus Cowan I
cells and characterization of antigens using the immunoblotting technique.
Of the 104 sera obtained from patients with systemic disease, 23 (7 of
12 patients) contained detectable amounts of an immunodominant M_r48,000
protein specific to Candida albicans when tested with both polyclonal
rabbit and monoclonal mouse antibodies. A single plasma evaluated from a
patient with chronic mucocutaneous candidiasis also contained the M_r48,000
protein. Sera obtained from persons with non-invasive disease as well as
a pool of normal human sera and sera obtained from individuals with infections due to fungal pathogens other than C. albicans were negative for
Candida-specific proteins. There was no correlation between CIC levels in
the blood and the ability to detect Candida-specific antigens in serum.
This suggests that CIC do not play a significant role in the difficulties
associated with detecting circulating antigens of C. albicans. However,
we have found immunoprecipitation followed by immunoblotting to be a
reliable method for detecting Candida-specific proteins in the sera of
patients with systemic candidiasis. The ability to detect such proteins
could significantly increase the early diagnosis of invasive disease.

MONOCLONAL ANTIBODIES: ANTI-CANDIDA ALBICANS MANNAN AND ANTI-ASPERGILLUS FUMIGATUS GALACTOMANNAN

C. Munoz[+], J.C. Mazie[+], J.M. Delga*, B. DuPont*,
J.P. Latge and E. Drouhet*

Hybridolab[+] and Unite de Mycologie*, Institut
Pasteur, Paris

Mannan of Candida albicans and galactomannan of Asper-gillus fumigatus are the major antigens of these two fungi which play important roles in the immunology of candidosis and aspergillosis. Two monoclonal antibodies were prepared against these polysaccharidic antigens.

Immunization: 4 mice (Biozzi) were immunized by intraperitoneal injections. All mice received 10^7 whole yeasts of Candida albicans killed by heat (60oC, 1 hour) or spores of Aspergillus fumigatus. Mice were inoculated at day 1, 7, 12 and 18. 47 days later one mouse received a booster under the same conditions and 3 days later the fusion was done.

Fusion: Ag8 x 63 6-5-3 plasmocytoma cell from Balb/c mice was used as myeloma. This line is not immunoglobulin secreting and HAT sensitive. The enzymatic deficiency was used for final selection between hybridoma and plasmocytoma cells. 8 x 10^7 spleen cells from the best immunized mouse were fused with 8 x 10^7 cells from plasmocytoma, according to the Koelher and Milstein technique. For the spleen cell, a viable count with trypan blue was done. This operation was not necessary for hybridoma cells.

Screening: The screening for antibody detection was done by the ELISA technique. Each antigen (mannan or galactomannan) was coated in 96 well plates (1 ug on each well). Supernatants or ascites fluids were added to the coated wells and anti-mouse sera binding with peroxidase was used as the conjugate.

Mannan results: After different studies by dilution, cloning, and inhibition we isolated 5 clones. The ascites fluids could be diluted 1/800,000 and 1/500,000. F92, B33 and B81 gave 50% inhibition with 0.2 ug/ml. C44 needed 0.32 ug/ml and A15, 0.66 ug/ml for obtaining the same results. The results obtained in immunoprocipitation gel technique (Ouchterlony) show a continuous precipitating line with B33 diluted at 1/20. The isotypic characterization led to our conclusion that the B33 monoclonal is an IgM molecule.

The anti-mannan mAb of C. albicans (B33) identified by immunofluorescence and agglutination C. albicans serotypes A and B and C. tropicalis (Figs. 1,2), but not other Candida, Torulopsis, Cryptococcus, or Trichosporon species. The monoclonal can replace a polyclonal antibody for ELISA inhibition test.

Immunocytochemical techniques have demonstrated that the mannan mAb bound exclusively to the outermost and the innermost layers of the wall of Candida (Figs. 3,4).

Galactomannan: 1 clone against galactomannan was isolated (GM2). In ascites fluid monoclonal antibody could be incubated after dilution (1/300,000). This inhibition was obtained with 150 ug of galactomannan antigen. The inhibition was obtained only with 5 ug of glucogalactomannan. The isotypic determination revealed an IgM molecule. We did not observe any cross reactivity between the mannan and galactomannan monoclonal antibodies. GM2 recognized the principal species of Aspergillus (A. fumigatus, A. flavus, A. nidulans, A. terreus, A. niger), but not any other filamentous fungis (Penicillium, Fusarium, Trichophyton, etc.) (Fig. 5). GM2 bound essentially to the intracellular compartment of A. fumigatus (Fig. 6).

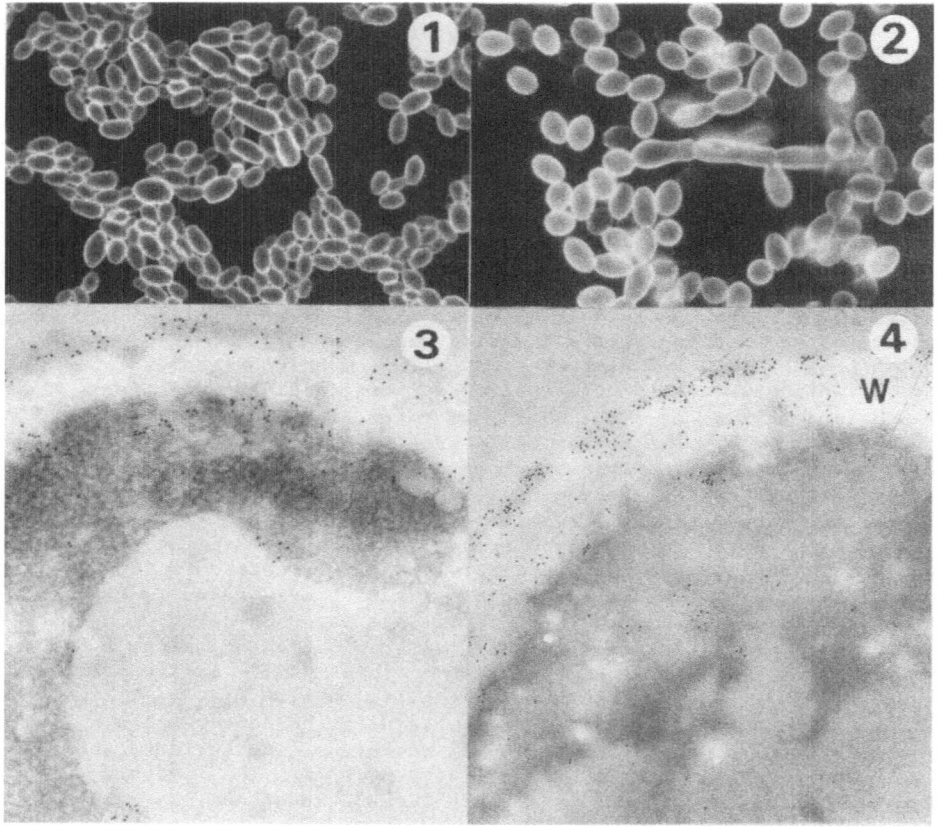

Figs. 1,2. Immunofluorescence of C. albicans yeasts serotype A (Fig. 1) and serotype B (Fig. 2) with anti-mannan mAb B33 (dilution 1/80) (x1000).
Figs. 3,4. Localization of Candida mannan studied with mAb B33 in electron microscopy. Lowicryl thin section of C. albicans serotype A (Fig. 3) and C. tropicalis (Fig. 4) labeled with anti-mannan mAb B33 rabbit IgG anti-mouse IgM-protein A-colloidal gold (dilution 10^{-6}).

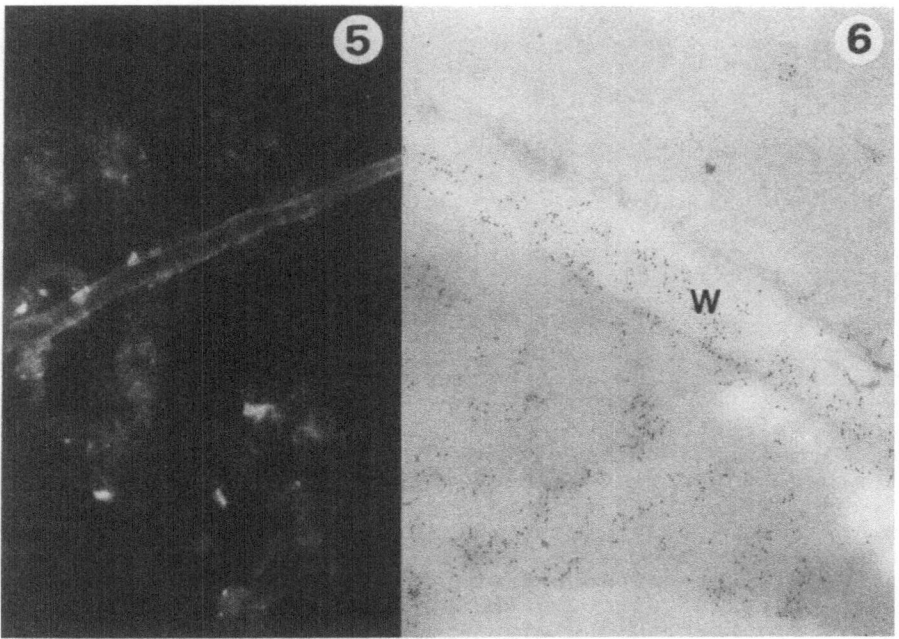

Fig. 5. Immunofluorescence of <u>A. fumigatus</u> with mAb (dilution 1/80).
Fig. 6. Localization of galactomannan antigen of <u>A. fumigatus</u> studied by electron microscopy. Lowicryl thin section of <u>A. fumigatus</u> labeled with mAb GM2, rabbit IgG anti-mouse IgM (dilution 10^{-6}) protein A-colloidal gold.

YEAST KILLER TOXIN MONOCLONAL ANTIBODIES

L. Polonelli*, M. Castagnola** and G. Morace*

Instituto di Microbiologia* and Chimica**, Facolta
di Medicina e Chrirurgia "A. Gemelli," Universita
Cattolica del Sacro Cuore, Roma, Italy

Killer yeasts and toxins have been extensively studied in
our Institute for strain differentiation of pathogenic yeasts
within the species and as epidemiological markers for the control and prevention of yeast nosocomial infections. Subsequently, we observed that the killer phenomenon, which was
previously considered to be restricted to yeasts, also occurs
among unrelated microorganisms. The potential therapeutic
effect of a yeast killer toxin on experimental mycotic infections in laboratory animals prompted us to try purification and
characterization of the yeast killer toxin apparently produced
by Hansenula anomala UCSC 25F. Monoclonal antibodies were
raised against the toxin and used in affinity chromatography.

Monoclonal antibodies of the IgG class were produced and
characterized by the Western blot technique. One of them (KT4)
which showed precipitating properties permitted us to neutral-
ize the killer activity of the toxin used as the immunogen.
Immunodiffusion tests carried out with other extracts of
recognized killer yeasts showed taht all reacted with monoclonal antibody KT4, although the Western blot analysis proved
that they may present a different molecular structure.

The unexpected immunodiffusion reactivity of the extracts
of yeasts isolated from clinical specimens with monoclonal
antibody KT4 was corroborated by the finding of appropriate
sensitive strains. Our results suggest the possibility that
most yeast isolates may display a killer activity under
favorable circumstances.

A QUANTITATIVE IMMUNOFLUORESCENCE TEST: AN EVALUATION OF CANDIDA ANTIGEN IN THE DIAGNOSIS OF INVASIVE CANDIDIASIS

A. Sanchez-Sousa, J.M. Aguiar, C. Torres,
H. Cercenado, A.M. Malo and F. Baquero

Unidad de Micologia, Hospital "Ramon y Cajal,"
Madrid, Spain

There is an increasing interest in the serological diagnosis of deep candidiasis as this type of infection is now more frequent because of the growing number of iv-drug abusers and immunocompromised patients. A modification of the quantitative fluorescence technique (FIAX International Diagnostic Technology, Santa Clara, California) of Estes (G.B. Estes et al., 1980, J. Immunol. Methods, 35:105), using a somatic antigen of C. albicans (Ag S, Institut Pasteur, Paris) was evaluated comparatively with immunodiffusion (ID) in patients with evidence of invasive candidiasis as well as in negative controls. The purpose of this work was to evaluate the possibility of a rapid diagnosis of deep candidiasis.

Two groups of patients were selected: 56 control sera of healthy blood donors, presenting a negative immunodiffusion test to somatic (S) and metabolic (M) antigens of Candida albicans. 12 sera obtained from proven cases of deep candidiasis presenting a positive-immunodiffusion test to the somatic and metabolic antigens. Criteria for invasive infection were: a) positive culture of C. albicans in two or more samples of normally sterile organic fluids and b) microbiological and/or histological evidence of C. albicans in deep tissue samples.

The fluorescence emitted on both sides of the StiQs is measured with a fluorimeter. The FIAX stiQs have a cellulose acetate adsorbitive surface with high hydrophobic capacity mounted on a plastic carrier which can be inserted into tubes containing test solutions and later into the fluorometer. The "target gain" is used to set FSU (fluorescence signal units) at a predetermined number for a reference sample.

Somatic antigen of C. albicans (Institut Pasteur) was diluted at a concentration of 0.08 mg/ml in order to fix each stiQ with a charge of 2 mcg. A simple 1:50 dilution of problem sera in PBS at pH 7.6 was used. Antibodies complexed to Ag were bound to fluorescein labeled anti-human immunoglobulins and the number of bound labeled antibody was read in an automatic fluorometer microprocessor. The calibration was performed by fixing a 160 FSU in such a way that a standard serum (50.640) showed values in the range 23-35 DFSU with norm factors of 0.0003.

The mean result of the 56 control sera processed in duplicate was 25.5 DFSU by FIAX. In 5 cases, figures were between 40 and 90 DFSU. The mean result of the 12 positive sera by FIAX was 135.9 DFSU, more than 5 times the mean of control. In 10 cases the results exceeded 90 and in another 2 were betweem 40 and 90.

The FIAX test has a sensitivity of 80% and specificity of 90% compared to the ID test. With the exception of 2 cases out of 12, the values obtained by the FIAX method clearly differentiated normal sera from those obtained from deep candidiasis patients. Results can be obtained in only 2h (immunodiffusion test needs 48h), and only a small amount of sample is required. This method saves a substantial amount of antigen, provides an objective, quantitative result, is not expensive and is simple to perform.

DIAGNOSIS OF SYSTEMIC CANDIDOSIS BY LATEX PARTICLE AGGLUTINATION

A.M. Sharp, E.G.V. Evans and J.A. Carney*

Department of Microbiology, University of Leeds,
Leeds and *Mercia Diagnostics, Guildford, Surrey, UK

A number of tests which detect Candida antigen in body
fluids have been described for the daignosis of systemic candidosis. Many of these tests have poor sensitivity or involve
complex methodology and none is routinely established. In this
study we compare the diagnostic potential of a new slide latex
agglutination (SLA) test, which detects Candida cell wall antigens with a commercial SLA antigen test (Cand-Tec, Ramco
Laboratories, Houston).

The latex reagent for the new SLA test was sensitised with
rabbit antibody to a partially purified cell wall preparation
from Candida albicans, serotypes A and B, and C. glabrata, and
could detect 15 ng mannan/ml. A control latex, coated with
pre-immune rabbit antibody, was used to detect any non-
specific reactions which were then eliminated by boiling
samples with 0.1M EDTA. The Cand-Tec antigen detection test
was performed as described in the manufacturer's instructions.

Fifty-eight patients were included in the study, 19 had
unequivocal systemic candidosis and 39 were control patients
with similar underlying pathologies (malignancies, post-
surgery and transplant complications, neonates) but no evidence
of Candida infection. The new SLA test detected antigen in 31
(79.4%) out of 39 serum and csf samples from 15 (78.9%) of the
19 patients with candidosis and there were no false positive
results with 69 sera from the 39 control patients. The Cand-
Tec test, which could only be used with serum samples, was
unequivocally positive with only 5 (16.1%) sera from 4 (23.5%)
of the candidosis patients and also gave 2 false positive
results.

The new SLA test appears to be an improvement on Cand-Tec.
It can be used with samples other than serum and a simple
pretreatment allows it to be used with RF positive sera without
loss of test sensitivity. Its specificity is excellent, and
its sensitivity (79.4%) compares favorably with sensitivities
published for other Candida antigen detection tests. Further
prospective testing of this reagent is in progress.

ANTIGENS FOR THE DETECTION OF <u>CANDIDA</u> <u>GUILLIERMONDII</u> VAR. <u>GUILLIERMONDII</u> INFECTION IN RUMINANTS

P. Sutka* and K. Sutka**

*Institute of Animal Breeding and Fodder Control,
Budapest, Keleti Karoly str. 34; ** Drug Research
Institute, Budapest, Szabadsagharcosok str. 47

From the epididymis of bulls with semen disorders, yeast strains were isolated in 1968. They were identified as <u>Candida</u> <u>guilliermondii</u> var. <u>guilliermondii</u>. The following antigens for detection of <u>C. guilliermondii</u> infection in ruminants were prepared according to the new antigen producing technology by the authors (International Hungarian Patent No. 183115): intradermal allergen for early and delayed sensitivity reactions, antigens for agglutination and immunofluorescence tests, soluble antigens for latex agglutination and precipitin tests. Different ruminants (cattle, sheep and buffalo) were examined using these antigens and immunobiological screening methods for the presence of infections with <u>C. guilliermondii</u> var. <u>guilliermondii</u>. It was found that about 45% of the ruminants which had to be slaughtered due to reproductive disorders and udder inflammation were infected by <u>C. guilliermondii</u> var. <u>guilliermondii</u>. After slaughtering, the organs of animals showing positive reaction were taken for pathological, mycological and histopathological examinations. The fungus was observed primarily in the genital organs, the liver, brain, kidney, lung and pleura. It was also found that when the udder was infected, <u>C. guilliermondii</u> was excreted in the milk.

Allergenic
Moulds

ALLERGENS OF ASPERGILLUS FUMIGATUS

Joan L. Longbottom

Senior Lecturer, Department of Allergy and Clinical
Immunology, Cardiothoracic Institute, Brompton Hospital,
Fulham Road, London, SW3 6HP, U.K.

INTRODUCTION

The word 'allergen' was originally introduced to indicate any
substance that gave rise to an immunologically altered response i.e. any
of the four types of specific hypersensitivity reactions as defined by
Gell & Coombs (1975). However, more recently the word has become
synonymous with "atopic allergen" meaning those substances that give rise
to Type I immediate hypersensitivity reactions mediated by IgE antibody,
i.e. "antigens that react with IgE antibodies".

It must be remembered that such 'allergens' may also act as normal
antigens in producing IgG antibody both in man, where a strong association between presence of IgE and IgG antibodies to various allergens has
been found (Platts-Mills 1982) and more especially when injected into
animals. However, even here it has been shown that subcutaneous
injection of timothy pollen extracts into rodents induces a specific IgE
response to the same components as that found in patients presumably
sensitized by inhalation (Nordvall et al 1982).

Our studies of patients with allergic bronchopulmonary aspergillosis
(ABPA), who mount both a specific IgG and IgE response, have also
demonstrated that there is a close relationship between these responses,
with many of the components of Aspergillus fumigatus combining with both
types of antibody - whether to the same or different epitopes remains to
be determined. However, although all these components must be regarded
as allergens, as will be discussed later, some appeared to be more
associated with IgG antibodies, and for convenience have been termed
major 'antigens', whilst others appeared closely associated with IgE
antibodies and were termed major 'allergens'. In this presentation,
although using this terminology, both component types will be included
as "the allergens of A. fumigatus".

Allergen preparations

Most allergen preparations for specific IgE detection both in vivo
and in vitro, are crude extracts containing both allergenic and non-
allergenic substances consisting of varying numbers of high molecular
weight components: proteins, glycoproteins, polysaccharides, nucleic
acids, as well as low molecular weight components: peptides, hexoses,

pigments, salts and with fungi even the possibility of toxins. Standardization of procedures for fungal antigen/allergen production are fraught
with problems which confront all immuno-mycologists in that fungi belong
to a "mutable and treacherous tribe" (Longbottom & Pepys 1978) in which
"everything that can vary will vary" (Salvaggio & Aukrust 1981). Even
within a single fungal strain morphological and thus biochemical changes
may occur, either spontaneously or induced by different cultural
conditions. Furthermore in the actual preparation of fungal extracts,
many variables exist, e.g. the strain used, whether from spores,
mycelium (cytosol) or culture filtrate (metabolic) and as well the
cultural conditions, such as the inoculation procedure, the type of
growth, the growth medium, the temperature and length of culture – all
factors which may influence the ultimate allergenic/antigenic
composition. There no are recognised 'standard' methods. However, the
most important condition when culturing fungi for allergen preparations
is the use of fully dialysable (synthetic) media, in order to avoid the
possibility of allergenic high molecular weight medium contaminants.
For most fungal allergens, especially for commercial preparations,
extracts of dried cultures (consisting of spores and mycelium) have been
used, although long term culture filtrates, not possibly reflecting too
closely the actual airborne material, also contain considerable
allergenic (and of course antigenic) activity.

A. fumigatus extracts

Most studies of A. fumigatus extracts have relied on the detection
of antigenic components, usually as precipitinogens in double diffusion
(DD) and crossed immunoelectrophoresis (XIE) tests, although more
recently as IgG- (and IgE)-combining components in enzyme-linked immunosorbent assays (ELISAs), using either patients' (ABPA or aspergilloma)
sera or rabbit antisera. Attempts to prepare purified antigenic
components, by a variety of extraction procedures of various specific
forms/parts of the organism (spores, mycelium, cell walls, cytoplasm
etc.), and by different fractionation procedures (gel filtration, isoelectric focusing, affinity and hydrophobic interaction chromatography
– and various combinations of these) of the extracts, have mostly
resulted in preparations still comprising several components – i.e.
partially purified components.

Early investigations of allergenicity, based on skin test
reactivity in allergic patients, indicated that allergenic 'potency' of
A. fumigatus extracts paralleled their antigenic content, i.e. number
of antigens present. In particular, the highly antigenic protein
precipitate obtained by ammonium sulphate precipitation of crude culture
filtrate extracts, compared to the predominantly polysaccharide supernatant, proved to be 10-100 times more reactive on a weight/volume basis
in the number and size of positive skin reactions produced. Confirmation
of the peptide nature of the allergenic determinants was obtained by
periodate oxidation which had no effect on the skin reactivity
(Longbottom 1964).

Allergenicity of extracts/fractions have more recently been determined by RAST (radioallergosorbent test), RAST inhibition and ELISA but,
as with skin tests, such tests can only give an indication of overall
potency and not of the allergenic character of the individual components.

ALLERGEN CHARACTERIZATION

Crossed radioimmunoelectrophoresis (XRIE)

It was not until the introduction of crossed radioimmunoelectrophoresis that the actual components reacting with IgE antibodies could be identified within complex mixtures (Weeke & Lowenstein 1973). This test uses the dried crossed immunoelectrophoretic profile (XIE) of the whole extract against a hyperimmune rabbit antiserum as the solid-phase (compare with the allergen-linked paper discs used in RAST). This is then incubated with patient's serum so that specific IgE antibodies bind to allergen-containing immunoprecipitates. After a second incubation with ^{125}I-anti-human IgE the uptake is visualised by autoradiography on X-ray film. Comparison of autoradiograph with XIE stained plate identifies the allergenic components as those antigenic peaks that have also bound the radioactivity and hence specific IgE antibody. This procedure therefore relies on the character of the polyclonal antisera produced, i.e. that there are precipitating antibodies to all the potential allergenic components present in the extract, and furthermore that specific IgE antibodies are still able to bind to the immuno-precipitated antigen.

Studies of XRIE on A. fumigatus allergens, using pooled allergic serum to demonstrate the broadest spectrum of IgE specificities, have shown that of the 32-42 precipitin peaks on XIE (depending on the batch of antigen), 11-14 were identified as allergenic, 4-7 of these showing particularly strong radioactive uptake (Longbottom & Austwick 1986) (Fig. 1). An important finding was that although the antigenic/allergenic (i.e. XIE stained plate/autoradiograph) profiles appeared to vary qualitatively, the differences between the batches were demonstrated by crossed line immunoelectrophoresis to be essentially quantitative.

Evaluation, in the form of an "allergogram", of XRIE tests on individual patient's sera allows the allergenic importance of the various components to be assessed (Lowenstein 1978), and thus major, minor and intermediate allergens have been defined. Components which bind specific IgE from at least 50% of sera tested, with strong uptake by the majority, are 'major' allergens, those to which less than 10% of sera show usually only weak binding are 'minor' allergens, with those falling between these two values as 'intermediate' allergens. In such a study of sera from 26 A. fumigatus allergic patients, 18 of 44 antigens were identified as allergens, 2 major (antigens no. 10 and no. 40), 10 intermediate and 6 as minor allergens (Wallenbeck et al 1984).

Earlier findings on the use of ammonium sulphate precipitation to concentrate both the antigenic (Longbottom & Pepys 1964, Hearn et al 1980) and allergenic (Longbottom 1964, Kim et al 1978) activities in the protein/glycoprotein precipitate have also been confirmed by XIE/XRIE. By contrast, the predominantly polysaccharide supernatant which comprised >30% of the dry weight of the crude antigen contained little of either activity (Longbottom & Austwick 1986).

Further, by incorporation of concanavalin A (Con A) into an intermediate gel of XIE/XRIE (Fig. 2) many of the antigenic components were absorbed or partially absorbed by Con A (Fig. 2A/B) indicating their glycoprotein nature and the presence of α-D manno or α-D gluco-pyranoside terminal residues. [These findings are in accordance with those of other workers who have demonstrated the Con A binding properties of some A. fumigatus antigens (Hearn & Mackenzie 1980, Kauffman et al 1985a, Kurup et al 1983, 1986, Schonheyder & Andersen 1984).] However most of the allergenic components, as seen in Fig. 2 C/D, were not affected, indicating the absence of such end residues and thus their different physicochemical nature.

Gel filtration on Sephacryl S200 of the protein precipitate, monitored by fused rocket and fused radio rocket immunoelectrophoresis

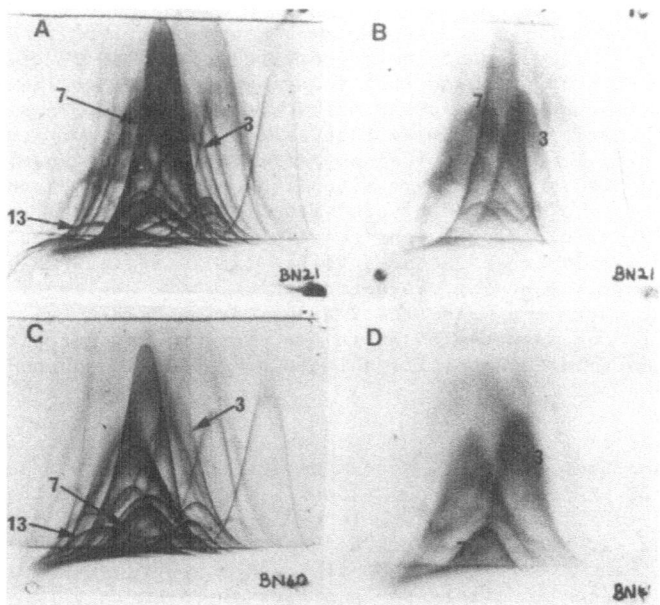

Fig. 1. XRIE of two identically prepared batches of A. fumigatus
culture filtrate antigen, BN21 & BN40, using rabbit serum
incorporated in the gel and an ABPA serum pool as a source of
specific IgE antibodies. **A & C** protein stained plates, **B & D**
corresponding autoradiographs. Peaks corresponding to Ags 3, 7 and
13 are indicated.
Note: variation in amounts of the antigens, particularly of Ag 3,
between the two batches (difference in peak heights) and also the
uptake of radioactivity on both Ag 3, only weakly protein stained,
and Ag 7.

(FRIE/FRRIE), revealed a resolution of its multiple antigenic components
over a wide range of molecular weight (MW) from >200 Kd to 10 Kd, the
majority >43Kd (Fig. 3A), and a paralleled spread of allergenic activity
(Fig. 3B), pointing to the molecular complexity of the allergenic
components. Most components eluted as discrete peaks. However, some
eluted over a large number of fractions, indicative of considerable MW
heterogeneity (Longbottom 1986).

Other studies by RAST have also indicated a spread of allergenic
activity over a wide MW range. Kim et al (1978) showed the greatest
reactivity in the high MW fractions (no MW markers used), whereas
Kauffman et al (1986) found it in the 28-60 Kd fraction.

Immunoblotting

A more recent technique developed to detect allergenic components
in crude extracts is that of immunoblotting. Unlike XRIE, this technique
does not rely on rabbit antibodies to 'fix' the components to the solid
phase, but rather the electrophoretic transfer of proteins/glycoproteins
onto nitrocellulose membranes following electrophoretic separation by
SDS-PAGE (sodium dodecyl sulphate-polyacrylamide gel electrophoresis).
This method is therefore dependent not only on transfer onto the membrane
but also on the migration in SDS-gel of the potential allergens, although

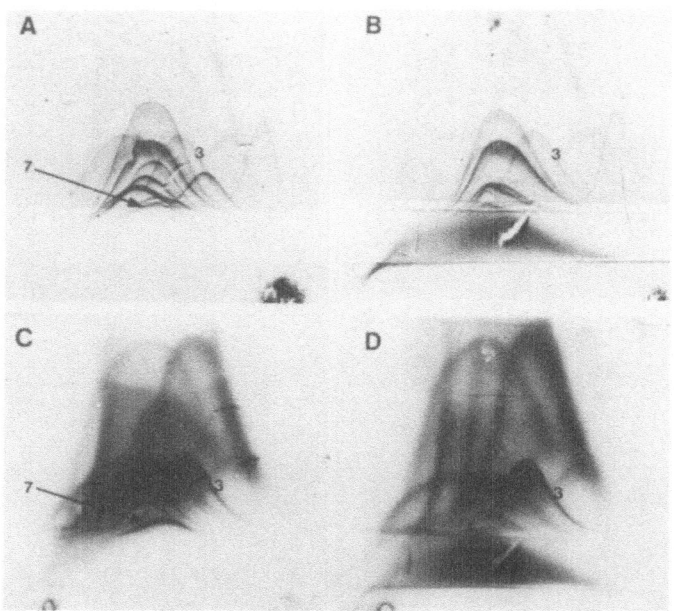

Fig. 2.   XRIE of <u>A. fumigatus</u> antigen, **B & D** with, and **A & C** without
Con A being incorporated into an intermediate gel. **A & B** protein
stained plates, **C & D** corresponding autoradiographs. **Note:** the fewer
numbers of precipitin peaks in **B** compared to **A** - in particular
absorption of Ag 7. By comparison in their autoradiographs **D**
compared with **C**, there are almost the same number of peaks with no
reduction in their heights - in particular Ag 3, exception is Ag 7.
An indication of the different physicochemical nature of most of IgE
binding components.

the native quaternary structure of the molecules may be altered by both
the mercaptoethanol reduction (which is usually carried out) and also
possibly by SDS. Following transfer the membranes are incubated first
with patient's serum and then either radiolabelled (^{125}I) or enzyme-
labelled (alkaline phosphatase or beta galactosidase) anti-IgE and the
uptake identified either by autoradiography or by an appropriate chromogenic substrate. Marked differences in allergenic composition of
batches of <u>A. fumigatus</u> antigens have been demonstrated, and a band of
MW 18 Kd identified as exhibiting strong IgE binding for 3/5 sera tested
(Bengtsson et al 1986).

Identification of specific 'major' components

Our approach to characterization and purification has been
initially to identify the most important or 'major' components, in
respect to the immunological response of both IgG and IgE of ABPA
patients, i.e. components to which the majority of patients had either
IgG (major antigens) or IgE (major allergens) antibodies, and then to
attempt to purify and characterize these specific components. By using
a modification of XRIE in which patient's serum is incorporated in the
second dimension gel, i.e. self-XRIE, it was possible to identify
components to which there were not only IgG (precipitating) antibodies
but also IgE antibodies (Longbottom 1983a). When an ABPA 'serum pool'

227

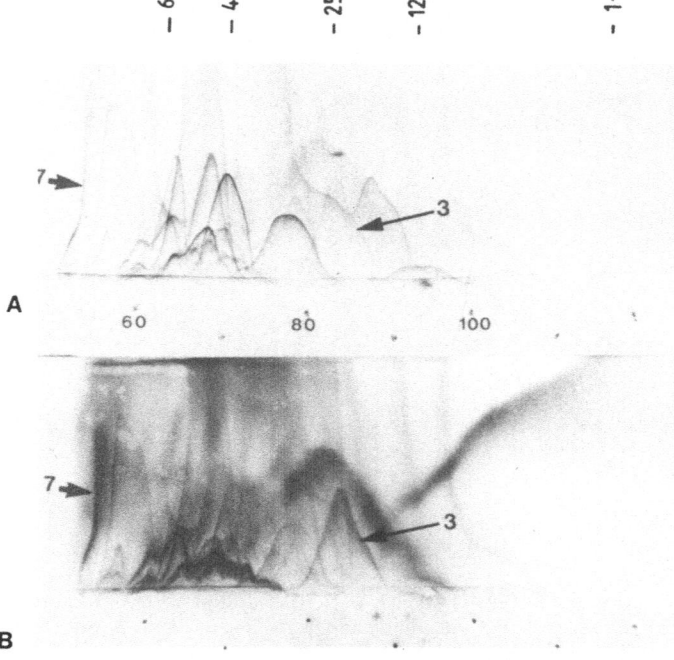

Fig. 3. Fused radio rocket immunoelectrophoresis of Sephacryl S200 fractionation of a protein precipitated fraction of A. fumigatus with molecular weight makers, and peaks corresponding to Ag 7 and Ag 3 indicated. (A, protein stained and B, autoradiograph).

was used, 15 IgG/IgE binding components were identified (Fig. 4 A/B) of which 6 were only visualised in the autoradiograph, i.e. not by protein (Coomassie blue) staining. Although responses varied, by comparing 32 individual serum patterns to that of the numbered reference 'pool', a self-XRIE allergogram was compiled (Fig. 5) (Longbottom 1983b). From this the components were divided into two categories: (1) major antigens (minor allergens) since they produced strong precipitin reactions with the majority of sera tested and only weak radioactive uptake with a few – in particular, these were Ags 7 and 13; and (2) major allergens (minor antigens) being those that appeared poorly precipitating (mostly not visualised by protein staining) but showed strong radioactive uptake, of which three, Ags 2, 3 and 15 were dominant.

'MAJOR' ALLERGEN/ANTIGEN CHARACTERIZATION

Major allergen - Ag 3

Further characterization of A. fumigatus components resulted from the monitoring by self-FRIE/FRRIE of the Sephacryl S200 column fractionation of the highly allergenic protein precipitate (Longbottom 1986). Using the ABPA serum pool in these tests, there was one predominant component on the autoradiograph (not visualised by protein staining) with MW 24 Kd (also seen in FRRIE, Fig. 3). By XIE/XRIE and crossed line immunoelectrophoresis of the fractions containing this component, it was identified as Ag 3, the previously recognised, major allergenic/minor antigenic component, i.e. showing strong IgE binding with at least

228

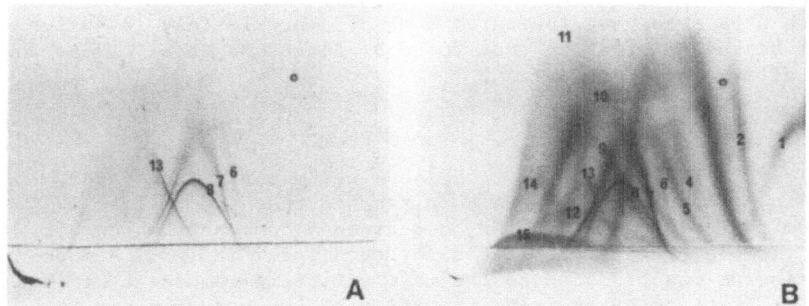

Fig. 4. A & B reference self-XRIE pattern of A. fumigatus using the ABPA serum pool both in the gel (precipitins IgG) and as a source of specific IgE. (A protein stained plate and B corresponding autoradiograph). Note: the increased sensitivity of autoradiography compared to protein staining. Peaks numbered arbitrarily from the anode.

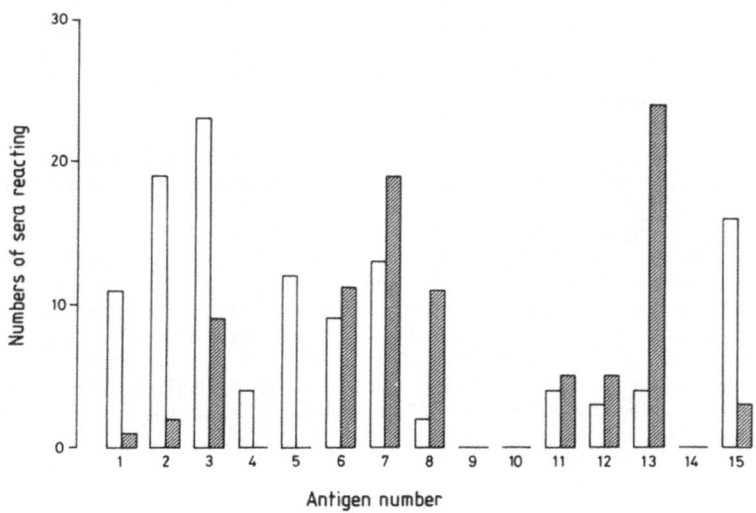

Fig. 5. Self-XRIE allergogram, being the assessment of individual self-XRIE tests on 32 ABPA sera expressed as numbers of sera reacting with visible precipitins (IgG) (hatched bars) and on autoradiography (IgE) (open bars) to the antigenic peaks numbered as in Fig. 4 (Longbottom 1983b).

75% of the sera, their associated precipitin (IgG) reactions usually not visualized (Longbottom 1983b). Ag 3 proved to be heat-labile (completely destroyed 75°C 15 min), and did not bind to Con A. It was also identified in a conventional XRIE (using rabbit antisera) as a dominant fast migrating, strong IgE-binding component (Longbottom 1986, Longbottom & Austwick 1986) which is possibly identical with antigen no. 10 of Wallenbeck et al (1984). More recent studies by SDS-PAGE and immunoblotting with ABPA sera have indicated a MW of approximately 18 Kd for Ag 3 by this method, similar to that found by Bengtsson et al (1986) for their major allergenic component.

Major antigen - Ag 7

The self-FRIE/FRRIE profile of the ABPA serum pool (Longbottom 1986) and also of 6 individual ABPA sera, showed that the initial fraction from the column, MW $\geq$150 Kd, contained a component which not only showed visible precipitates but also bound specific IgE. A monospecific antiserum was produced to the main antigenic component of this high MW column fraction by excising the relevant immunoprecipitin peak from an XIE gel with rabbit antiserum (Fig. 6A), and using it as the immunogen in rabbits (Harvey & Longbottom 1986a). By incorporation of the resultant monospecific antiserum (Fig. 6B) into an intermediate gel of the reference ABPA self-XRIE pattern the component 'absorbed' (Fig. 6 C/D) was identified as Ag 7 - a major antigen/minor allergen, i.e. with precipitins in 62% and IgE binding in 40% ABPA sera (Fig. 5). Confirmation of the MW of Ag 7 as 150-200 Kd was obtained from a FRIE gel filtration profile incorporating the Ag 7 monospecific antisera in an intermediate gel, and SDS-PAGE revealed that it was apparently composed of 36 Kd subunits. Further characterization showed Ag 7 to be a heat stable (100°C 5 min) acidic glycoprotein which binds to Con A suggesting the presence of α-D manno- or α-D-glucopyranoside terminal residues. However, attempts to purify Ag 7 by affinity chromatography proved rather unsuccessful, due to the high binding affinity of Ag 7 to the monospecific antiserum.

This component shares many of the characteristics both physiochemical and immunological of partially purified antigens, CS2 from the cell sap of A. fumigatus (Calvanico et al 1981, Piechura et al 1983) and antigen IIb from culture filtrate (Kurup et al 1983), and also a cell wall-derived glycoprotein antigen used for detection of antigenemia in systemic aspergillosis (Weiner & Coats-Stephen 1979). Further work is required to determine whether these are the same or related molecules. It is also interesting to note that the 40 Kd band regarded as clinically important for patients with invasive aspergillosis by IgG antibody immunoblot analysis (Mathews et al 1985), could also be related to Ag 7, since SDS had a dissociating effect on this molecule resulting in subunits of 36 Kd.

Major antigen - Ag 13

It was of considerable interest having identified two major antigenic components, i.e. Ag 7 and Ag 13, and characterized and partially purified one of them, i.e. Ag 7, to determine whether either of these could be the component previously recognised as the C antigen because of its chymotryptic activity (Tran Van Ky et al 1966, Biguet et al 1967), and which also had reacted with the majority of "aspergillosis" (ABPA and aspergilloma) sera (Dessaint et al 1976). By incorporation of a monospecific antiserum to the C antigen (produced originally by Dr. L.A. Yarzabal et al (1978) and kindly donated to us by Drs. S. de Magaldi and D.W.R. Mackenzie) into the intermediate gel of the reference ABPA self-XIE pattern, the C antigen was revealed as our Ag 13 to which >75% ABPA

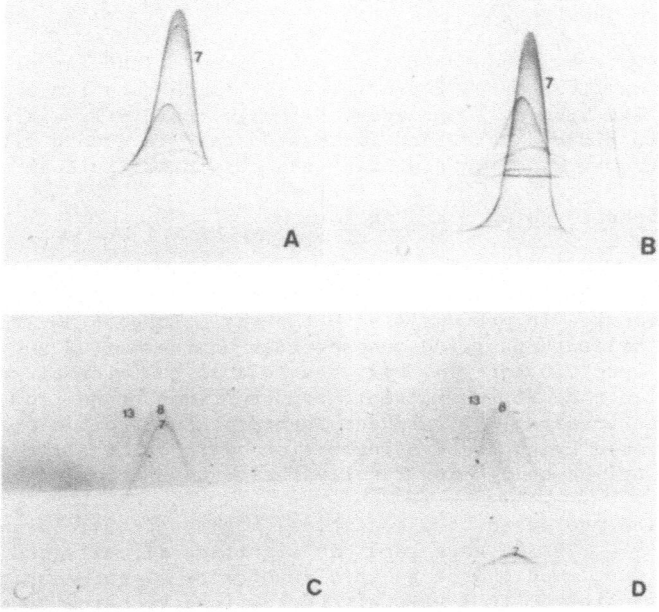

Fig. 6. A & B. XIE of initial high MW Sephacryl S200 column fraction into rabbit antiserum. A control and B with Ag 7 monospecific antiserum in the intermediate gel. **Note:** The single specificity of the monospecific antiserum which was produced using the Ag 7 peak as seen in **A.**

C & D. The reference ABPA self-XIE pattern C control and D with the monospecific Ag 7 antiserum incorporated in an intermediate gel. **Note:** the absorption of peak Ag 7.

sera had precipitins. Unlike Ag 7 the other major antigen, Ag 13 was only weakly allergenic with relatively few ABPA sera (12.5%) showing associated IgE binding (Fig. 5). Further studies have shown that it is relatively heat-labile (partially destroyed 75°C 15 min), binds to Con A and has a MW of 70 Kd by gel filtration (Harvey & Longbottom, manuscript in preparation).

Release of allergens during culture

A study of the release of these three major components during shake-culture of A. fumigatus showed that there was only minimal release into the culture medium after seven days and that all three appeared maximally after 14 days growth, coinciding with an increase in pH and suggesting that they are released due to autolysis (Harvey & Longbottom, manuscript submitted for publication). This is further indicated for Ag 7 and Ag 13 by immunofluorescence tests using the respective monospecific antisera, as these showed that neither antigen was detected on the spore/germling surface (Taylor & Longbottom, to be published). Although all cultures contained significant amounts of Ag 3 and Ag 7, in some cultures amounts of Ag 13 were barely detectable. Other components able to bind specific IgE antibody (one identified as the fast migrating

Ag 1) were however detected even after 24 hours growth, and thereafter a gradual increase in the number of IgE and IgG binding components (and their titres) were observed.

These findings are in contrast to another study (Kauffman et al 1985b) in which the liberation of precipitating components increased continuously but IgE binding components, although rapidly liberated (3 days), appeared to diminish and then increased again in accord with the three phases of growth associated with pH changes in the culture medium.

Specific IgG antibodies to Ag 7 and Ag 13

Using the monospecific antisera produced to these two major components, sandwich ELISAs have been developed to detect Ag 7 and Ag 13 specific IgG antibodies in patient's sera (Harvey & Longbottom 1986b). In these assays, affinity purified monospecific antiserum (as a capture antibody for the specific antigen) was used to coat the microtitre plates, after which the single antigenic component was bound from a partially purified preparation (relevant Sephacryl S200 column fraction) and this was followed by successive incubations with patient's serum, enzyme labelled anti-human IgG and finally the chromogenic substrate.

The Ag 7 ELISA proved to be a sensitive, highly specific assay and the finding that 97% (29/30) ABPA sera had significantly elevated IgG antibody to Ag 7 confirmed the diagnostic importance of this antigen. IgG subclass antibodies to Ag 7 were also determined by using specific monoclonal antisera instead of the polyclonal anti-IgG, and both IgG1 and IgG4 Ag 7 specific antibody levels were found to be significantly raised in the ABPA sera. Comparison to the more conventional indirect ELISAs (in which plates are coated directly with antigenic extracts) using the crude culture filtrate antigen showed that the Ag 7 ELISA was a more highly specific assay for antibody detection in aspergillus related diseases (Harvey et al, in press).

Ag 7 specific antibody (Fig. 7A) was present in all patients with ABPA and aspergilloma with no significant difference between their overall value, and only low titres were found in 3/40 sera in the other disease groups tested. Indirect ELISA with the partially purified CS2 cell sap antigen, which appears to resemble Ag 7, also showed levels of antibody above 'the normal range' in most sera from aspergilloma and ABPA patients, even in some which did not appear to have precipitins to CS2, however, sera from other disease groups were not tested (Piechura et al 1983). Furthermore, Ag 7 is likely to be one of the four Con A binding glycoprotein components in the fraction of a culture filtrate antigen obtained by preparative isoelectric focusing (pH 3.8-4) with which both ABPA and aspergilloma sera were highly reactive by ELISA (Kurup et al 1983).

With the Ag 13 ELISA although there were problems associated with some free antigens (not Ag 13) being present in the affinity purified antiserum, this was overcome by performing simultaneously, tests with and without the antigen (partially purified Ag 13) being added - the difference in absorbence between the tests being attributed to Ag 13 specific antibodies (Harvey & Longbottom in preparation). With sera from aspergillus related and other disease groups, the Ag 13 ELISA (Fig. 7B) displayed sensitivity and specificity comparable to the Ag 7 ELISA in that all ABPA sera had significant levels of Ag 13 IgG antibodies. However, two of five aspergilloma sera gave equivocal results and although the control and atopic groups were negative, of twelve farmer's lung sera, three were weakly positive and two equivocal. These results are very similar to those obtained in an indirect ELISA

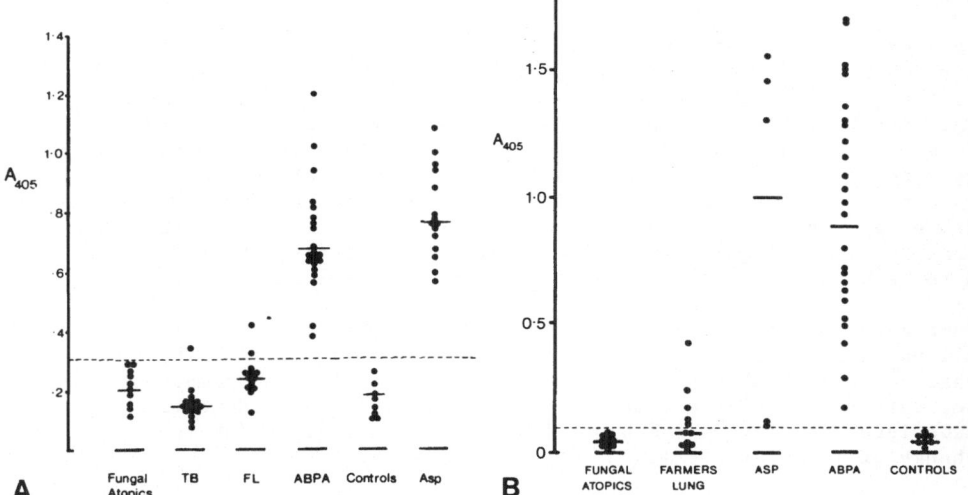

Fig. 7. A. Ag 7 specific IgG measured by sandwich ELISA in sera from
21 ABPA, 15 aspergilloma, 16 tuberculosis, 13 farmer's lung, 10 skin
test positive to fungi and 8 control individuals.
    B. Ag 13 specific IgG measured by sandwich ELISA in sera from
25 ABPA, 5 aspergilloma, 12 farmer's lung, 10 fungal skin test
positive and 8 controls.
    In both A & B the mean values for each group and the mean + 2SD
of the control sera are indicated.

using "affinity-purified" (not apparently tested for its purity)
C-antigen to coat the microtitre plate (de Magaldi & Mackenzie, 1984).
Here 2/14 ABPA and 2/17 aspergilloma sera were non reactive, whilst 8/30
sera from patients with other mycoses (in particular 4/10 with
candidosis) had low levels of C-antigen (Ag 13) specific antibody.

    The partially purified antigen of Kurup et al (1986) which from its
pI value may contain Ag 13, also showed high levels of both IgG and IgE
antibodies in aspergilloma and ABPA sera although absolute values and
tests of other disease groups were not included in their study.
Similarly, for the 70 Kd somatic Fraction VIII (Schonheyder & Anderson
1983) IgG antibodies, by ELISA, were found to be significant in
"aspergillus related diseases".

SUMMARY

    Our studies of the mosaic of components present in extracts of
A.fumigatus identified 15 components which reacted with both IgG and IgE
antibodies from patients with allergic bronchopulmonary aspergillosis.
Depending on their relative reactivity these were divided into major
allergens/minor antigens being those that showed strongest IgE binding,
and minor allergens/major antigens being the most precipitinogenic

components. By applying various quantitative immunoelectrophoretic and separation techniques three of these components, Ag 3 a major 'allergen' and Ag 7 and Ag 13 major 'antigens', have now been further characterized and partially purified.

Ag 3 was shown to be a very heat labile, non-Con A binding component with a MW 24 Kd by gel filtration and 18 Kd by SDS-PAGE. Ag 7 which although a dominant precipitinogen also showed appreciable IgE binding, is a heat stable, Con A binding acidic glycoprotein with MW 150-200 Kd, composed of 36 Kd subunits. Whilst Ag 13, with comparatively low IgE binding was identified as C antigen, and is relatively heat labile, binds to Con A and has a MW of 70 Kd. Sandwich ELISAs developed to measure Ag 7 and Ag 13 specific IgG antibodies in patients' sera have confirmed the immunological significance of these two antigens and proved to be highly sensitive, specific assays for antibody detection.

Thus three major components of importance to the diagnosis of aspergillus related diseases and in the standardization of extracts have been characterized. Also the possible relationship of these components to other partially purified components has been discussed. However, their absolute identity, which hopefully may be realized by future interchange of material, remains to be determined.

REFERENCES

Bengtsson, A., Karlsson, Å., Rolfsen, W., and Einarsson R., 1986, Detection of allergens in mould and mite preparations by a nitrocellulose electroblotting technique, Int. Archs Allergy appl. Immunol., 80:383.

Biguet, J., Tran Van Ky, P., Fruit, J., and Andrieux, S., 1967, Identification d'une activité chymotrypsique au niveau de fractions remarkables de l'extrait antigenique d'_Aspergillus fumigatus_. Répercussions sur le diagnostic immunologique de l'aspergillose, Revue d'Immunologie, 31:317.

Calvanico, N.J., Dupont, B.L., Huang, C.J., Patterson, R., Fink, J.N., and Kurup, V.P., 1981, Antigens of _Aspergillus fumigatus_ I. Purification of a cytoplasmic antigen reactive with sera of patients with Aspergillus related disease, Clin. Exp. Immunol., 45:662.

Coombs, R.R.A., and Gell, P.G.H., 1975, Classification of allergic reactions responsible for clinical hypersensitivity and disease, in: "Clinical Aspects of Immunology", 3rd edition, chapter 25, P.G.H. Gell, R.R.A. Coombs, and P.J. Lachmann, eds., Blackwells, Oxford.

Dessaint, J.P., Bout, D., Fruit, J., and Capron, A., 1976, Serum concentrations of specific IgE antibody against _Aspergillus fumigatus_ and identification of the fungal allergen, Clin. Immunol. Immunopathol., 5:314.

Harvey, C., and Longbottom, J.L., 1986a, Characterization of a major antigenic component of _Aspergillus fumigatus_, Clin. Exp. Immunol. 65:206.

Harvey, C., and Longbottom, J.L., 1986b, Development of a sandwich ELISA to detect IgG and IgG subclass antibodies specific for a major antigen (Ag 7) of _Aspergillus fumigatus_, Clin. Allergy, 16:323.

Harvey, C., Shaw, R.J., and Longbottom J.L., Diagnostic specificity of a sandwich ELISA for Aspergillus-related diseases. J. Allergy Clin. Immunol. (in press).

Hearn, V.M., and Mackenzie, D.W.R., 1980, Mycelial antigens from two strains of _Aspergillus fumigatus_ an analysis by two-dimensional immunoelectrophoresis, Mykosen, 23:549.

Hearn, V.M., Wilson, E.V., Proctor, A.G., and Mackenzie, D.W.R., 1980, Preparation of Aspergillus fumigatus antigens and their analysis by two-dimensional immunoelectrophoresis, J. med. Microbiol., 13:451.

Kauffman, H.F., Van der Heyden, P.J., Van der Laan, S., Van der Heide, S., Beaumont, F., and de Vries, K., 1985a, Antibody determination against Aspergillus fumigatus by means of enzyme-linked immunosorbent assay. II. Physico- and immunochemical properties of the polystyrene-binding components, Int. Archs Allergy appl. Immunol., 78:174.

Kauffman, H.F., Van der Heide, S., Van der Laan, S., Hovenga, H., Beaumont, F., and de Vries, K., 1985b, Standardization of allergenic extracts of Aspergillus fumigatus. Liberation of IgE binding components during cultivation, Int. Archs Allergy appl. Immunol., 76:168.

Kauffman, H.F., Van der Heide, S., Beaumont, F., Blok, H., and de Vries, K., 1986, Class-specific antibody determination against Aspergillus fumigatus by means of enzyme linked immunosorbent assay. III. Comparative study: IgG, IgA, IgM ELISA titres precipitating antibodies and IgE binding after fractionation of the antigen, Int. Archs Allergy appl. Immunol., 80:300.

Kim, S.J., Chaparas S.D., Brown, T.M., and Anderson, M.C., 1978, Characterization of antigens from Aspergillus fumigatus. II. Fractionation and electrophoretic, immunologic and biologic activity, Am. Rev. Respir. Dis., 118:553.

Kurup, V.P., Ting, E.Y., and Fink, J.N., 1983, Immunochemical characterization of Aspergillus fumigatus antigens, Infect. Immun., 41:698.

Kurup, V.P., John, K.V., Rusnick, A., and Fink, J.N., 1986, A partially purified glycoprotein antigen from Aspergillus fumigatus, Int. Archs Allergy appl. Immunol., 79:263.

Longbottom, J.L., 1964, Immunological investigation of Aspergillus fumigatus in relation to disease in man, Ph.D. Thesis, University of London.

Longbottom, J.L., 1983a, Antigens/allergens of Aspergillus fumigatus. Identification of antigenic components reacting with both IgG and IgE antibodies of patients with allergenic bronchopulmonary aspergillosis, Clin. Exp. Immunol., 53:354.

Longbottom, J.L., 1983b, Allergic bronchopulmonary aspergillosis. Reactivity of IgE & IgG antibodies with antigenic components of Aspergillus fumigatus (IgE/IgG antigen complexes), J. Allergy Clin. Immunol., 72:668.

Longbottom, J.L., 1986, Antigens and allergens of Aspergillus fumigatus II. Their further identification and partial characterization of a major allergen (Ag 3), J. Allergy Clin. Immunol., 78:18

Longbottom, J.L., and Austwick P.K.C., 1986, Antigens and allergens of Aspergillus fumigatus I. Characterization by quantitative immunoelectrophoretic methods, J. Allergy Clin. Immunol., 78:9.

Longbottom, J.L., and Pepys J.L. 1964, Pulmonary aspergillosis: Diagnostic and immunological significance of antigens and C-substance in Aspergillus fumigatus, J. Path. Bact., 88:141.

Longbottom, J.L., and Pepys J.L., 1978, Immunological methods in mycology, in: "Handbook of Experimental Immunology" D.M. Weir, ed., 3rd edition, chapter 41, Blackwells, Oxford.

Lowenstein H., 1978, Quantitative immunoelectrophonetic methods as a tool for the analysis and isolation of allergens, Progr. Allergy 25:1.

De Magaldi S.W., and Mackenzie D.W.R., 1984, Specificity of antigens from pathogenic Aspergillus species II. Studies with ELISA and immunofluorescence, Sabouraudia: J. Med. Vet. Mycol., 22:381.

Mathews R., Burnie J.P., Fox A., Tabagchali S., 1985, Immunoblot analysis of serological responses in invasive aspergillosis. J. Clin. Pathol., 38:1300.

Nordvall S.L., Griminer O., Karlsson T., and Björkstén B., 1982, Characterization of the mouse and rat IgE antibody responses to timothy pollen by means of crossed radio immunoelectrophoresis, Allergy 37:259.

Piechura J.E., Huang C.J., Cohen S.H., Kidd J.M., Kurup V.P., and Calvanico N.J., 1982, Antigens of Aspergillus fumigatus II Electrophoretic and clinical studies, Immunology 49:657.

Platts-Mills T.A.E., 1982, Type I or immediate hypersensitivity: Hayfever and asthma, in: "Clinical Aspects of Immunology" P.J. Lachmann and D.K. Peters, eds., 4th Edition, chapter 21, Blackwells, Oxford.

Salvaggio J.L., and Aukrust L., 1981, Mold-induced asthma, J. Allergy Clin. Immunol., 68:327.

Schonheyder H., and Anderson P., 1983, Determination of antibodies to partially purified Aspergillus antigens by an enzyme-linked immunosorbent assay, Int. Archs Allergy appl. Immunol. 70:108

Schonheyder H., and Anderson P., 1984, IgG antibodies to purified Aspergillus fumigatus antigens determined by enzyme linked immunosorbent assay, Int. Archs Allergy appl. Immunol., 74:262

Tran Van Ky P., Biguet J., and Fruit K., 1966, Localisation et frequence des arcs des immunoelectrophoregrammes produits de la serum des malades atteints de mycétomes aspergillaires appliquées contre l'antigène Aspergillus fumigatus, Revue d'Immunologie 30:13.

Wallenbeck I., Aukrust L., and Einarsson R., 1984, Antigenic variability of different strains of Aspergillus fumigatus, Int. Archs Allergy appl. Immunol., 73:166.

Weeke B., and Lowenstein H., 1973, Allergens identified in crossed radio-immunoelectrophonesis, Scand. J. Immunol., 2:149.

Weiner M.H., and Coats-Stephen M., 1979, Immunodiagnosis of systemic aspergillosis. I. Antigenemia detected by radioimmunoassay in experimental infection, J. Lab. Clin. Med., 93:111.

Yarzábal L.A., de Albornez M.B., de Cabral N.A., and Satiago A.R., 1978, Specific double diffusion microtechnique for the diagnosis of aspergillosis and paracocudioidomycosis using monospecific antisera, Sabouraudia 16:55.

ALLERGENS OF ALTERNARIA AND CLADOSPORIUM

Jean-Paul Latge and Sophie Paris

Unite de Mycologie
Institut Pasteur
Paris, France

INTRODUCTION

Cladosporium and Alternaria are among the most widespread airborne fungi. The conidia of these fungi are present indoors and outdoors all year around with a peak during mid and late summer. Eventhough the spore counts vary according to the climatic and geographic conditions, the mean concentration of spores during the peak season is usually higher for Cladosporium than Alternaria (5,000 to 250,000 vs 200 to 5,000 spores per m^3 of air respectively) (Fig. 1). However the volume of one spore of Alternaria is typically 20 to 40 times greater than that of Cladosporium. Consequently, the difference in the realtive abundance of these two genera becomes less significant if the volume contribution is taken into account (Mallea and Charpin, 1980; Lacey, 1981; Larsen, 1981; Al Doory, 1984; Beaumont et al., 1985).

The presence of high concentrations of spores in the air may elicit respiratory allergic symptoms (rhinitis and/or asthma). Lacey (1981) estimates that 20% of the population are atopic and can be sensitized by airborne spores at concentrations up to 10^6 per m^3. Allergy or immediate hypersensitivity results from the sensitization of the immune system of atopic individuals that produce IgE antibodies in presence of specific antigenic molecules (allergens). These IgE antibodies have a high affinity for the receptors of mast cells of the mucosa and basophils of the blood. When a sensitized patient comes into contact with an allergen, the macromolecules will react with the specific IgE bound to the cells, induce degranulation of mast cells and basophils, and cause release of mediators responsible for the allergic host response.

Because of the predominance of Cladosporium and Alternaria spores in the atmosphere, it is understandable that these two fungal genera have been recognized for a long time as the main source of mould allergy. However, their role in inhalant allergy remains unclear. This is primarily due to the poor quality of commercial extracts and lack of standardization of these products which make diagnosis of fungal allergy often unreliable. Most of commercial allergen preparations are extracted from mycelia which are easy and cheap to mass-produce but do not necessarily contain the immunogenic macromolecules to which the host is exposed.

For example, recent immunocytochemical studies on C.cladosporioides

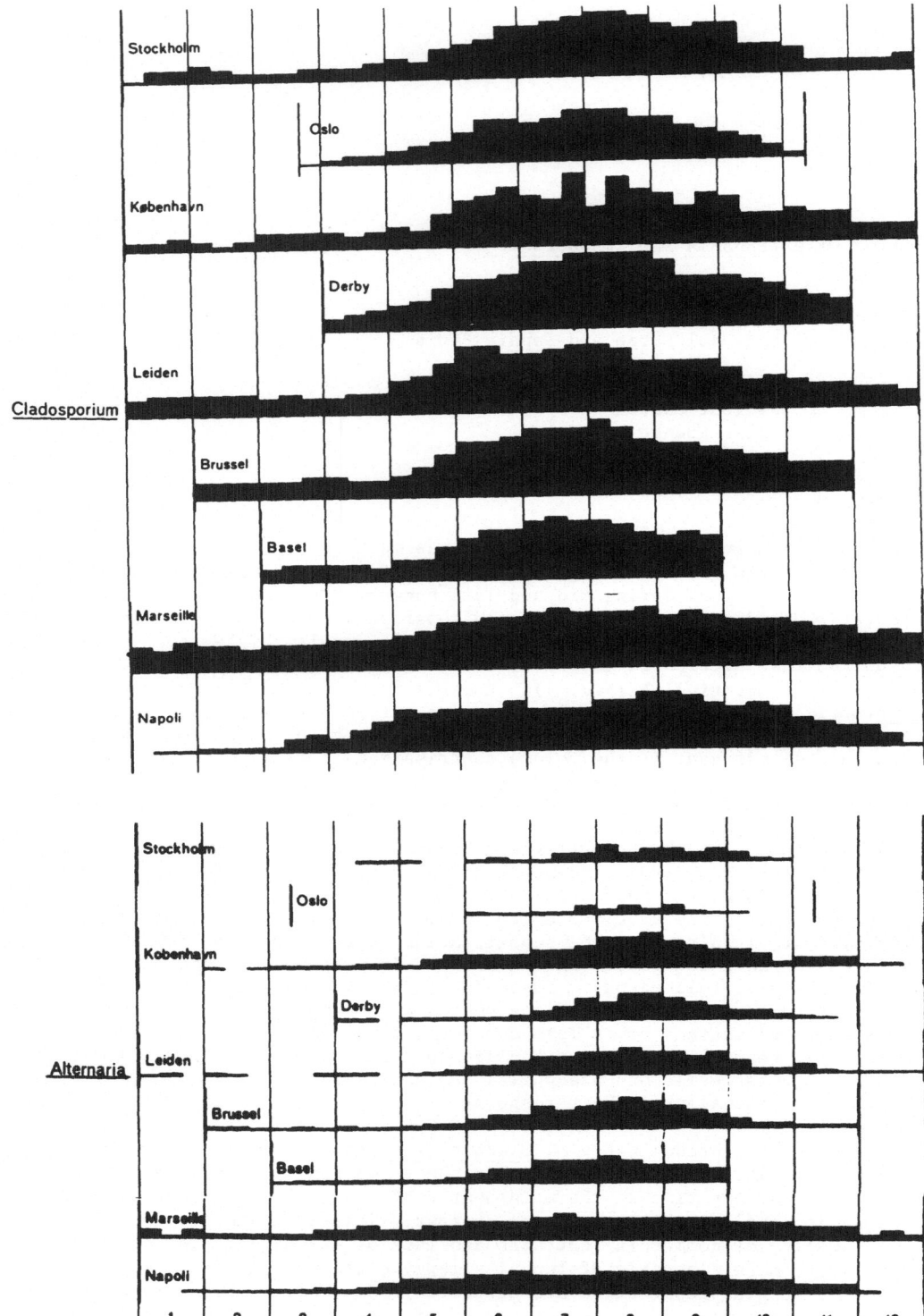

Fig. 1. Relative abundance of airborne conidia of Cladosporium
and Alternaria in Europe (columns in irregularly growing classes as indicated by Spieksma, 1984).

demonstrates that sera of allergic or non-allergic patients are very rich in spore specific IgGs (Latge et al., unpublished). It is also known that fungal allergen concentration is highly dependent on the growth medium and age of the culture (Sward Nordmo et al., 1984a; van der Heide et al., 1985; Vijay et al., 1985). Moreover, fungi have an innate tendency to yield genetic variants during repeated subculturing which may affect allergen production (Schumacher and Jeffery, 1976; Vijay et al., 1985). In spite of these findings, culture media, duration of growth, origin of the strains, number of subcultures, type of propagule, and extraction procedures vary between commercial producers of allergens (Dreborg et al., 1986). If the complexity (over 60 precipitin lines found by CIE) and the absence of characterization of the fungal antigens of Alternaria and Cladosporium is also taken into account, it is understandable that the allergenic quality of commercial allergens is highly variable (Hoffman, 1984).

To better understand the nature of allergic responses elicited by Alternaria and Cladosporium, additional work is needed on compositional analyses of the allergens and improvement of the sensitivity of the methods for their detection. This chapter will present recent advances in these two areas of research.

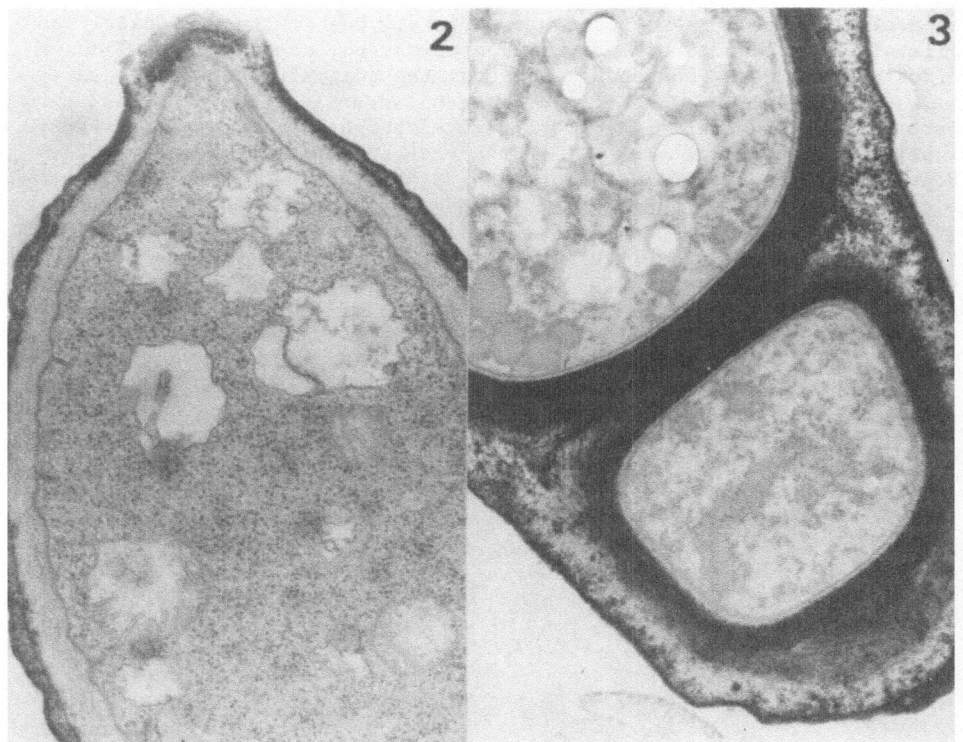

*Figs. 2-3. Conidia of C.cladosporioides (Fig. 2, P.papilla) and Alternaria (Fig. 3) fixed in presence of 1% osmium tetroxide (The inner wall (iw) of A.alternata is electron translucent after fixation in aldehyde alone) (x 25000).

*Results presented in all figures and tables of this chapter were obtained with strain LCP 404 of C.cladosporioides, strain LCP 406 of C.sphaerospermum, strain G1 of C.herbarum from the mycology Unit of Pasteur Institute, strains IP 1563 and MUCL 20297 of A.alternata.

I. MORPHOLOGY AND CHEMISTRY OF CONIDIA OF <u>ALTERNARIA</u> AND <u>CLADOSPORIUM</u>

In order to extract and isolate all the allergenic components from fungal propagules, it is essential to have a good knowledge of their ultra-structure and biochemistry. The conidia of <u>Alternaria</u> and <u>Cladosporium</u> are characterized by a bilayered cell wall. The inner layer is electron trans-lucent, while the outer alyer is very electron dense and contains melanin (Rast et al., 1981) which is responsible for the typical black color of these dematiaceous fungi (Figs. 2 and 3). In the case of <u>C.cladosporioides</u>, the cell wall is rich in B (1,3) glucans (59%) and melanin (15%) and poor in chitin (5%) and proteins (Latge et al., 1987).

The outermost components of the conidial wall may be extremely important in the case of fungal respiratory allergy since they are the first components in contact with the mucus epithelium and may trigger an allergic crisis. Carbon-platinum replicas of the surface of the spores show that the conidia of <u>C.cladosporioides</u> and <u>C.herbarum</u> are covered by a fibrous network of interwoven rodlets (10 x 150 - 350 nm) (Fig. 4). On the other hand, the surface of <u>C.macrocarpum</u> and <u>A.alternata</u> conidia appear amorphous (Fig. 5). However, in <u>C.macrocarpum</u>, a rodlet layer present under the outer amorphous layer can be exposed by freeze etching (Cole and Samson, 1984). Conidia of <u>Alternaria</u> and <u>Cladosporium</u> species are hydrophobic irrespective of the presence of rodlets on the outer spore surface. Due to the easy detection of rodlets by simple shadow replica preparation, the conidia of <u>C.clados-porioides</u> appear to be good models to study the composition of the outer-most layer and its allergenic potency. Techniques have been described to remove the rodlet layer of fungal conidia, although significant differences exist between species with respect to the ease of isolating rodlets (Cole and Pope, 1981, Wu-Yuan and Hashimoto, 1977, Boucias and Latge, 1986, and Latge et al., 1986). In the case of <u>C.cladosporioides</u>, a short ultrasonica-tion (30 min. at 80 Watts) removed the rodlet layer (Fig. 6). The pellet recovered by ultracentrifugation (100.000 x g for 18 h) from 0.15 - 0.3% dry weight conidia, however, does not show a rodlet configuration. The pellet is essentially composed of carbohydrates, melanin, salts, and lipids and is almost devoid of proteins (Latge et al., 1987). These results are

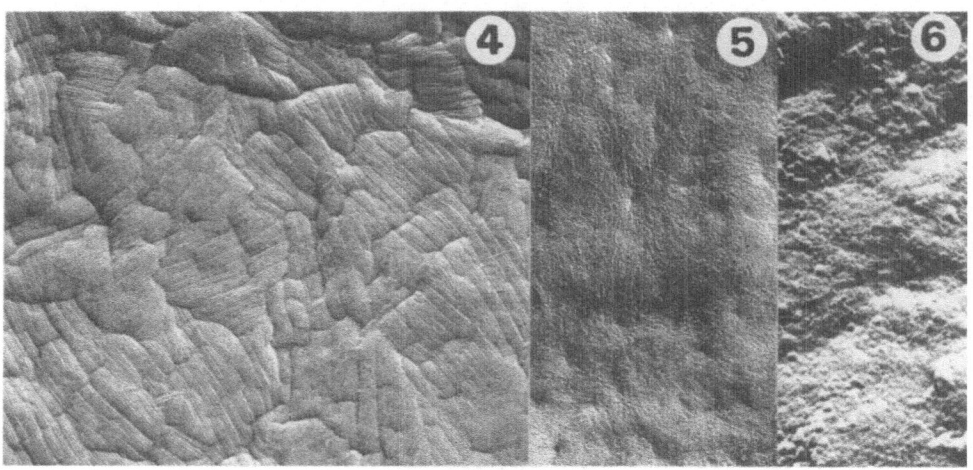

Figs. 4-6. Replicas of air dried conidia of <u>C.cladosporioides</u>
showing the presence of rodlets (Fig. 4) and <u>A.alter-</u>
<u>nata</u> demonstrating an amorphous surface (Fig. 5);
Fig. 6. removal of rodlet fascicles from <u>C.clados-</u>
<u>porioides</u> by ultrasonication (x 60000).

in contrast with reports of the composition of similarly-isolated surface wall components of other conidial fungi (Cole and Pope, 1981, Beever et al., 1979). Conidia of C.cladosporioides also shed their rodlets when incubated with 0.1 N NaOH, 50 mM NaHCO₃ or 50 mM Tris solutions, but not in the presence of 1% NP40 or 1% Triton X100 (Bouziane et al., 1987c). Hydrophobicity of the spores however, is markedly reduced by incubation in all detergent solutions irrespective of their ability to remove rodlet fascicles.

The conidial cytoplasm may not be as important as the outer wall layer in the triggering of early rhinitis and asthma symptoms. Nevertheless, intracellular components may play an important role in the sensitization of individuals since all inhaled spores are eventually destroyed by phagocytic host cells and thereby release their allergens (Roitt et al., 1985). Intracellular constituents account for about 10% of the total dry weight of the spores of C.cladosporioides and are essentially glycoproteins (4:1 w/w) (Bouziane et al., unpublished).

II. PREPARATION OF ALLERGEN EXTRACTS

1. Crude extracts

In spite of the difficulty of mass-producing conidia of Alternaria and Cladosporium, it is essential to prepare fungal extracts from conidia since they are inhaled biological products and contain powerful and specific allergens (Hoffman, 1984). In the case of Alternaria and Cladosporium, two types of extraction procedures have been reported: incubation of cells in solutions and breakage of the cells in various extraction buffers (Fig. 7). Although culture filtrates have recently been proven to be allergenic (van der Heide et al., 1985), no allergenic extracts have been based, until now, on secondary metabolic products.

Release of free sugars and amino acids, as well as polysaccharides (1.5-3%) and proteins (1-2%), occurs when conidia of these two fungi are incubated in various detergents or mineral solutions. The amount and kinetics of products released during the extraction procedure depend on the composition of the solution used and concentration of conidia in the extraction medium (Table 1). Maximum release is attained 3 to 4 hours after the beginning of the incubation (Fig. 8). However, the concentration of proteins released is modified by extracellular serine-proteases (Fig. 8). Proteolysis is very active in 50 mM NaHCO₃ (pH 8.2) which has been the main extraction buffer used for isolation of Cladosporium allergens (Aukrust and Borch, 1979; Sward-Nordmo et al., 1984a, 1985). The proteolytic activity of the exudate is blocked in the presence of protease inhibitors such as PMSF (phenylmethylsulfonylfluoride) or using 5-50 mM Tris buffer (pH 9.0) (Bouziane et al., 1989c). In the case of Alternaria, no proteolytic activity has been detected in the fungal exudates (Paris, unpublished). These results demonstrate that great care has to be taken during extraction procedures to insure reproducibility and maintain the molecular integrity of the fungal extracts. On the other hand, protease may play a role in allergy. Fungal antigens partially hydrolyzed by proteases may become powerful allergens; proteases themselves are also known for their immunoreactivity.

Other allergenic extracts recently isolated from Alternaria and Cladosporium have been obtained by glass bead homogenization of conidia in the presence of either 5-50 mM Tris or 50 mM NaHCO₃ buffer (van der Heide et al., 1985; Bouziane et al., 1987a; Paris et al., unpublished). The 20,000 x g supernatant of this total extract has demonstrated high allergenic potency. This novel extraction method for the recovery of allergens has previously been employed for isolation of fungal antigens (Huppert, 1983).

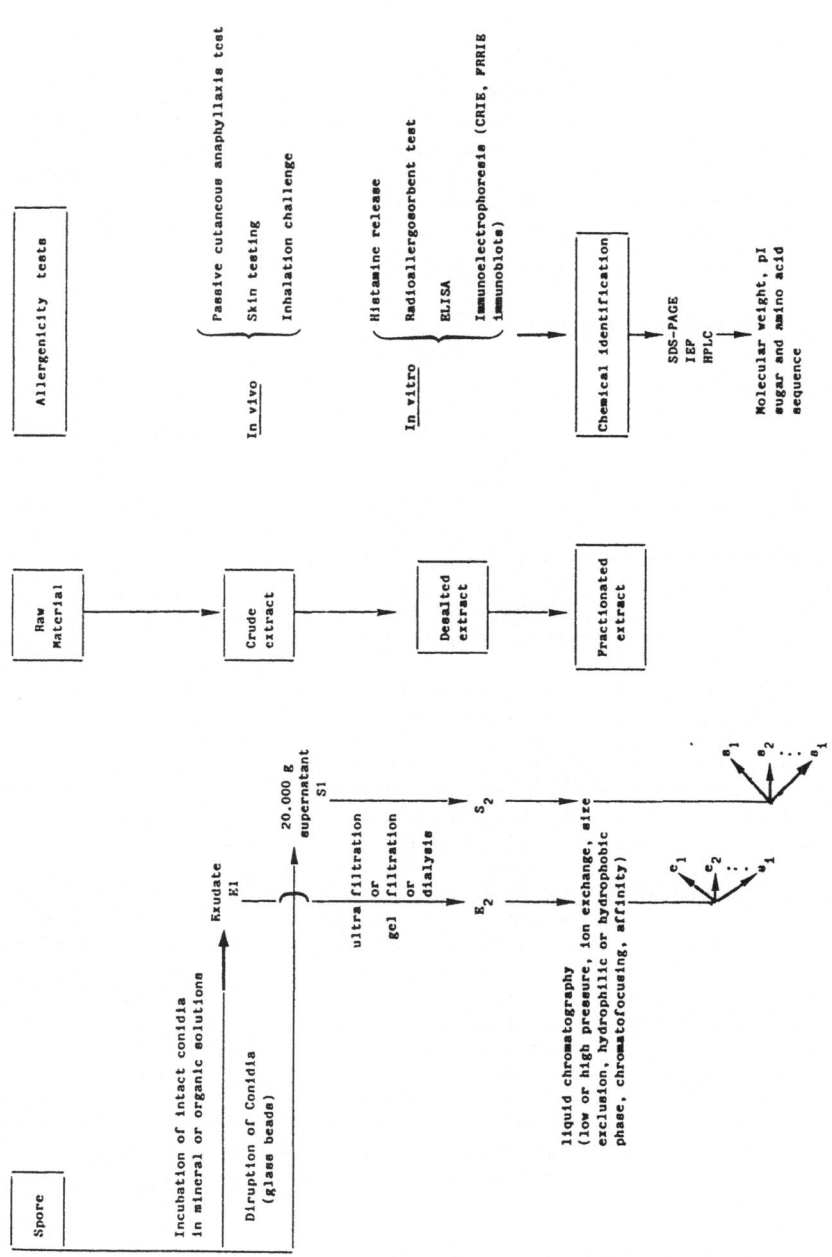

Fig. 7. Flow diagram showing the methods employed for extraction, purification, and identification of dematiaceous allergens.

Table 1. Fungal exudates released by conidia of
C.cladosporioides which were incubated
in various chemical solutions.

| Solution | Concentration of spores (mg dry weight/ml) | Rodlets | Exudate composition | |
			Sugars (a) (%)	Proteins (b) (%)
NP40 1 %	0.12	+(d)	8.5(e)	1.9
NaHCO$_3$ 50mM	0.12	−	8.5	1.9
Coca (c)	0.12	−	8.9	2.1
NaOH 0.IN	0.12	−	12.2	2.2
	0.85	+	5.8	1.0

(a) Sugar content expressed as free glucose determined by the phenol
method ; polysaccharides represented 15-20 % total carbohydrates.
(b) Determined using Bio-Rad method and bovine serum albumin as the standard
(c) Coca : NaHCO$_3$ 0.3 % + NaCl 0.9 % + phenol 0.5 %
(d) Presence (+) or absence (-) of rodlets on the conidial surface determined by C-Pt replicas
(e) Results expressed as percentages of total dry weight of spores

The SDS-PAGE patterns of polypeptide bands for the soluble total extracts
are different from those of the exudates and are dependent on the strain
used and whether conidia or mycelia were extracted, and the extraction
buffer (Fig. 9). Consequently, differences in allergenic responses to
various extracts may be the result of different extraction techniques.

2. Fractionated extracts

Fractionation of crude extracts has been performed until now using the
two step procedure (Fig. 7), described below.

a) Removal of low molecular weight molecules (MW 6,000-10,000) are
performed by dialysis, ultrafiltration or gel filtration chromatography.
Ultrafiltration has been commonly used by commercial allergen producers

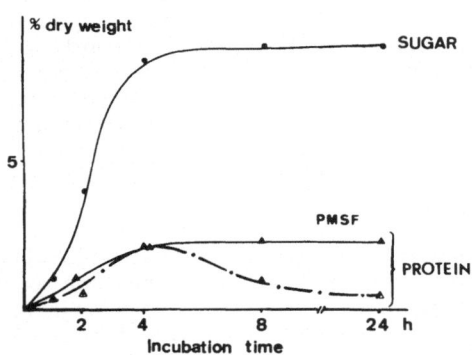

Fig. 8. Kinetics of sugar and proteins released from conidia of
C.cladosporioides during incubation in a coca solution
at 20°C (PMSF = spores incubated in a coca solution containing 1 mM phenylmethylsulfonyl fluoride; sugar estimated
as free glucose using the phenol method; protein determined
using BSA as standard and Bio-Rad method.

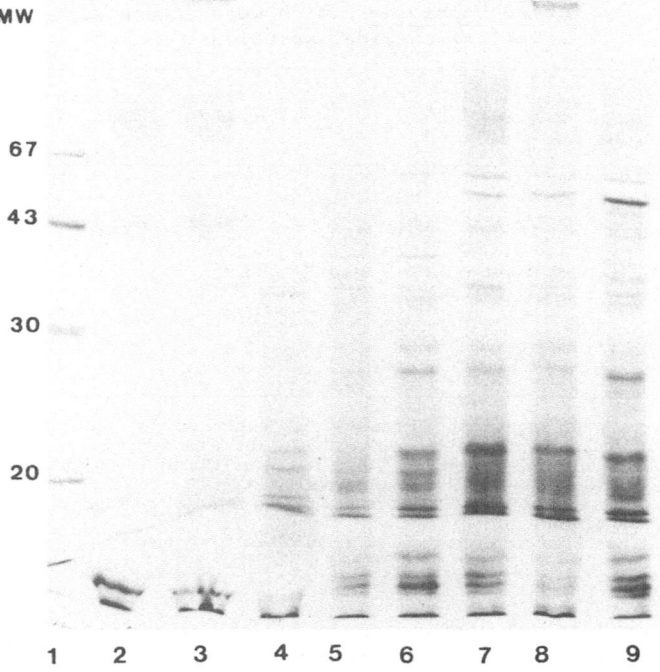

Fig. 9. SDS gel electrophoresis of A.alternata proteins
1. molecular weight standards; 2-4 exudates obtained
after 24 h incubation at 5°C in a coca solution of
spores of IP 1563 (2) MUCL 20297 (3) and mycelium of
IP 1563 (4); 5-9: intracellular proteins recovered by
20.000 g centrifugation after glass bead disruption
of untreated spores of IP 1563 (5) and MUCL 20297
(6) or spores of IP 1563 (7-8) and MUCL 29297 (9)
previously incubated in the coca solution.

while gel filtration has been essentially restricted to research procedures.
Dialysis seems the least efficient technique because of the prolonged de-
salting times with the possibility of chemical modification of the allergenic extract. However, dialysis avoided the non-specific blockage or
retention of certain components of the extract when agarose, dextran, or
polyacrylamide gels or ultrafiltration cartridges were employed (Bouziane
et al., unpublished). Ammonium sulfate precipitation has also been used
as a preliminary step in Alternaria and Cladosporium purification procedures (Lynch et al., 1975, Yunginger et al., 1980, Bouziane et al., unpublished).

    b) Separation of high molecular weight molecules (10,000-100,000
daltons) is usually performed by size exclusion and ion exchange chromatography. Usually DEAE anionic exchange resins are most efficient to separate
Alternaria and Cladosporium allergens since these fungal allergens demonstrate acid pI (4-6.5) (Vijay et al., 1979, 1981; Yunginger et al., 1980;
Sward-Nordmo et al., 1984b; Bouziane et al., unpublished). Affinity
chromatography (lectin - Sepharose) has also been used to separate allergens of Alternaria (Meyers et al., 1985). Separation of allergens has
generally been performed by low pressure chromatography technique. However, more recently HPLC size exclusion columns have been used to separate

Cladosporium and Alternaria allergens (Vijay et al., 1985, Landmark and Aukrust, 1985, Sward-Nordmo et al., 1985). This latter procedure has the advantage of using very dilute allergenic solutions which minimizes hydrophobic protein binding occuring during the fractionation process when the protein concentration of the extract exceeds 0.5-1.0 mg/ml. HPLC separations are also very rapid avoiding any possible degradation of the allergens by enzymes released during extraction. The composition of fractions recovered by the chromatographic procedures are examined by SDS-PAGE. It is essential to know if the allergenicity of apparent high MW fractions is actually due to the adsorption of low molecular weight allergens to larger molecules.

Preparative electrofusing has also been used to separate allergens of Cladosporium and Alternaria (Aukrust and Borch, 1979, Yunginger et al., 1980).

III. ALLERGEN DETECTION AND QUANTIFICATION

Methods used to measure the allergenic potency of an extract are different from those used for fungal antigens. This is due to the characteristics of IgE and the specific function of this class of immunoblobulin in the induction of an allergy crisis. In the serum, IgE are present at very low concentration (0.001% of total serum immunoglobulins) and are non-precipitating immunoglubulin. IgE antibodies are either circulating and directed toward basophils or fixed on mucosa mast cells. In both cases, the target cell degranulates and releases specific mediators after the binding of one molecule of allergen per two molecules of IgE antibodies (Roitt et al., 1985).

Allergenic potency of an extract can be studied using in vivo or in vitro tests, in either patients or animal models.

1. Animal Models

Rabbit, guinea pigs, and mice have been sensitized to Alternaria or Cladosporium by intranasal, intramuscular, or intraperitoneal challenge using extracts mixed with aluminium hydroxide (Vijay et al., 1979, 1981, and 1984; Bouziane et al., 1987 b,c). The production of IgE is monitored by passive cutaneous anaphylaxis tests (PCA) developed by Ovary (1985). This assay consists of two steps: (1) different dilutions of the serum of sensitized animals are injected intradermally into non-sensitive animals and specific IgE bind to the mast cells at the site of injection; (2) 2 or 48 hours later (for rats or guinea pigs, respectively), the allergen extract mixed with Evans blue as an indicator of degranulation, is injected intravenously into the animal. The presence of blue spots under the skin at the site of injection indicates the degranulation of sensitized mast cells. The lowest positive dilution of the serum recorded as the PCA titer is indicative of the allergenic potency of the extract.

Total cytoplasmic or mycelial extracts of Alternaria and Cladosporium have induced the synthesis of specific IgE in mice and guinea pigs with respective PCA titers of 270 and 90 (Table 2). No cross reactivity was observed between Alternaria and Cladosporium, or between conidia and mycelium of Cladosporium indicating the presence of specific allergens in each cell type. On the other hand, common allergens were detected in the conidia and mycelium of Alternaria and in different commercial batches of this fungus (Vijay et al., 1984; Bouziane et al., 1987 b,c). Conidial exudates are also allergenic. However, their allergenic potency depended on incubation medium and duration of incubation. Material recovered from C.cladosporioides conidia incubated in 0.15 M NaCl, 0.1 N NaOH or 50 mM

Table 2. Passive cutaneous anaphylaxis test obtained
in guinea pigs and rats with A.alternata and
C.cladosporioides extracts.

Animal (a)	sensitized to		Antigens(b) used for PCA Challenge in guinea pigs							
			Sp Aa	Myc Aa	Sp Cc	Myc Cc	Sp Ch	Myc Ch	Sp Cs	Myc Cs
Guinea pig	A.alternata	spore	+(c)	+	-	-	NT(d)	NT	NT	NT
	C.cladosporioides	spore	-	-	+	-	+	-	+	-
		mycelium	-	-	-	+	-	+	-	+
			Antigens used for PCA Challenge in rats							
			Sp Cc	Myc Cc	Ex NP40 1%	Ex NaHCO$_3$ 50mM	Ex NaOH 100mM	Ex Tris 50mM	Ex NaCl 15mM	US
Mice	C.cladosporioides	spore	-	-	-	-	+	+	+	+

(a) Sensitization performed as described by Bouziane et al., (1987 b, c).
(b) Abbreviations used: Aa: A.alternata, Cc: C.cladosporioides, Ch: C.herbarum Cs: C.sphaerospermum, Sp, Myc: Intracellular fractions of spore (sp) and mycelium (myc) recovered as 20,000 g supernatant after cell disruption by glass beads. Ex: exudates recovered by incubation of spores during 4 h at 25°C in various chemical solutions. US: outer layer of spore recovered by ultrasonication and 100,000 g ultracentrifugation.
(c) Positive (+) or negative (-) PCA tests after injection of antigens to the animals (100 ug protein per animal).
(d) NT: not tested.

Tris gave positive PCA tests, whereas NP 40 and NaHCO$_3$ extracts were negative (Table 2). A 24 h incubation period was better than 4 h suggesting either a more tightly binding or an internal position of the allergens. However, the surface components removed by ultrasonication were also PCA

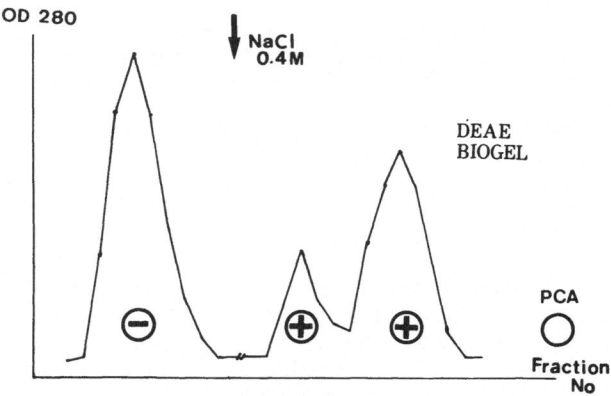

Fig. 10. Fractionation of C.cladosporioides extract (20.000 g supernatant of glass bead disrupted conidia homogenate) on a DEAE anionic exchange column eluted with 5 mM Tris pH 8.0 and 0.4 M NaCl in the same buffer. Positive (+) or negative (-) PCA tests obtained with the three peaks recovered from the column.

positive (Table 2). The reaginic activity of the different fractions obtained by chromatography can also be tested by PCA. For example, after anionic exchange chromatography of a C.cladosporioides extract, the polar fraction recovered by 0.4 M NaCl was positive whereas the neutral fraction is PCA negative (Fig. 10). After gel filtration chromatography of an Alternaria extract, Vijay et al., (1981), recovered only one fraction which was PCA positive with a molecular weight of 30,000-40,000 daltons.

Tests using animal models also permit in situ observation of the effects of allergens on the nasal mucosa of sensitized animals. Experiments using allergens derived from Alternaria or Cladosporium show that both fungi induce degranulation of nasal mast cells of sensitized guinea pigs (Fig. 11) without blockage of ciliary beating or any alteration of the cilia epithelium (Bouziane et al., 1987b). The role of fungal allergens during degranulation seems different from that of pollen, which alter the ciliary epithelium and arrest beating after 5-10 min exposure (Etievant et al., 1980).

2. Diagnostic Procedures in Human

Skin testing (prick or intradermal tests) is the usual method employed in allergy diagnosis. However, due to the lack of standardization of the Alternaria and Cladosporium extracts, skin test results vary depending on the batch (Aas et al., 1980; Gumovski et al., unpublished). For example, the incidence of positive skin tests to C.herbarum extracts varied from 12

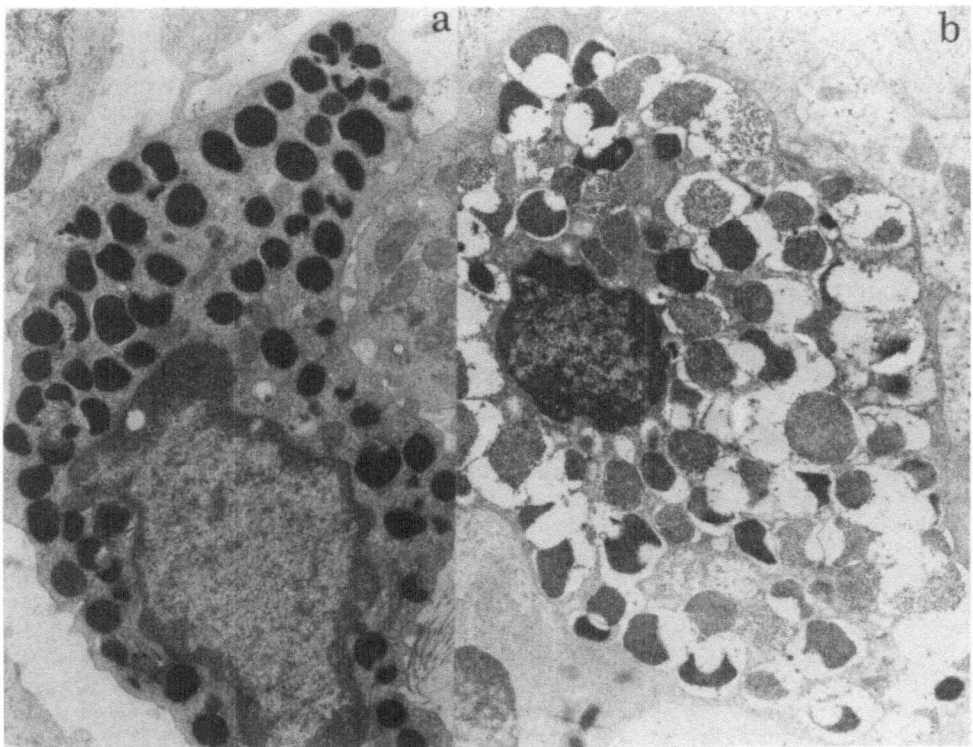

Fig. 11. Normal (a) and degranulated (b) mast cell of the
nasal mucosa of untreated and sensitized to C.clados-
porioides guinea pigs challenged with intracellular
extract of C.cladosporioides obtained by cell disruption.

to 60% depending on the preparative procedure (Aas et al., 1980). In any case, a dose-response curve must be determined for comparison of allergens. The minimal concentration of extract giving a positive reaction (equivalent to a control with 1 mg/ml histamine in prick test for example) mut also be determined.

When contradictory data are obtained from clinical history and skin testing (ST), inhalation challenge (IH) may be used to establish cause and effect relationship between the allergen and the symptomatology. Although few studies have involved this type of comparative analysis (Aas et al., 1980; Kozak and Hoffman, 1984; Licorish et al., 1985; Malling et al., 1986), the results obtained are very promising. A combination of ST and IH permit distinction between a real immunological reaction and an irritative phenomenon. IHs are currently performed in our laboratory using soluble antigens of C.cladosporioides or conidia killed by irradiation (Gumovski and Latge, unpublished).

In vivo tests using animals and humans are usually well correlated in Alternaria. However, differences are observed for Cladosporium. For example, negative PCA tests were obtained when an antiserum against conidia of C.cladosporioides was injected into a rat followed by challenge with a mycelial extract of the same fungus (Table 2). In contrast, most patients showing positive reaction to Cladosporium, react both to mycelial and conidial extracts (Bouziane et al., 1987a). Hyperimmunization seems to exacerbate the difference in immunological responses to the allergens.

In vivo tests also can be compared to in vitro tests: histamine release, radioallergosorbent test. The test for histamine release is based on the quantitation of histamine liberated by basophils of patient blood in presence of the fungal allergens (Lebel, 1983). This test has been rarely employed with Alternaria and Cladosporium. In the Cladosporium system, this assay gives 18% false negative and 18% false positive (Malling et al., 1986). All Alternaria sensitive patients screened by Fadel et al. (1987) gave positive histamine release tests. The quantity of the conidial extract required to release 50% of the patients histamine was 2 times lower than the amount of mycelial extract required for equivalent response and 5 times lower than a commercial extract (Fig. 12).

The most common in vitro test used in immuno-allergy is the radio-allergosorbent test (RAST). Allergens are coupled to cyanogen bromide (CNBr) activited cellulose matrix. Specific IgE of the patients serum that bind to allergen are quantitatively determined with an ^{123}I - antihuman IgE antiserum (Yunginger et al., 1976b; Yunginger, 1978). A modification of this test is the RAST inhibition, in which the patient serum is incubated with the allergens before incubation with the CNBr-allergen disc (Yunginger et al., 1976a). This test is very useful to compare the potency of various allergic fractions and their eventual antigenic communities. These tests have demonstrated that conidial extracts of Cladosporium and Alternaria are active at much lower doses than mycelial extracts and contain specific allergens (Fig. 13) (Hoffman et al., 1981; Hoffman, 1984; Fadel et al., 1986; Bouziane et al., 1987a). Usually a good correlation is observed between RAST and in vivo tests (Aas et al., 1980; Malling et al., 1985; Fadel et al., 1987). However some discrepancies have been found between the RAST and in vivo tests in the case of Cladosporium (Duc et al., 1986, Malling et al., 1986). Several explanations are possible. RAST discs are usually prepared with total crude extracts and it has been shown that use of more refined allergenic extracts have yielded higher RAST titers probably because of a more efficient binding on the disc. Only very small quantities of crude C.cladosporioides extracts (defatted or not) bind to glutaraldehyde activated Magnogel (Latge, unpublished). In addition, positive in vivo tests may be due essentially to polysaccharide-containing antigens which do not

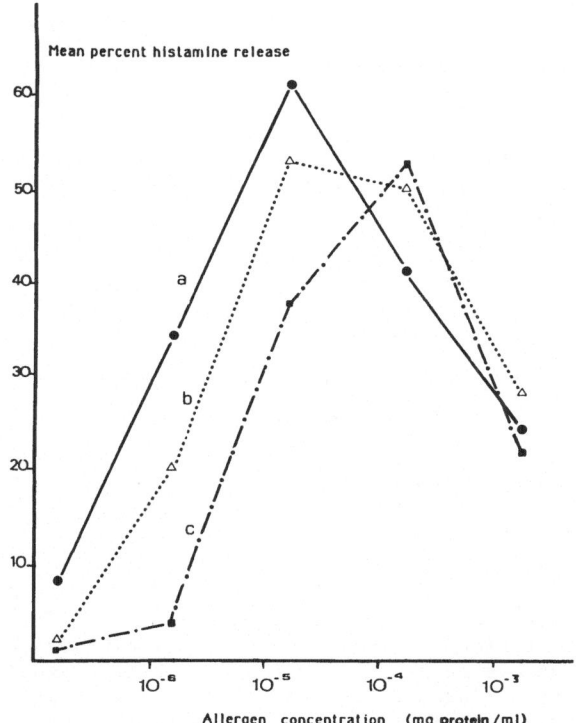

Fig. 12. Histamine release from basophils of a patient
allergic to A.alternata induced by different
concentrations of sporal (a) mycelial (b) extracts
(coca exudate). The curve (c) is obtained with a
commercial extract.

bind to the CNBr activated discs. Other methods such as benzoquinone or
epichlorohydrine activating techniques, or ELISA may be successfully
employed for testing IgE against fungal allergens of predominantly polysaccharide nature. Another possibility is that patients sensitve to
Alternaria and Cladosporium are polysensitized (Malling et al., 1985 and
1986) and this condition could decrease the efficiency of the RAST. Mono-
sensitized patients, like children, usually show a better correlation between RAST and skin tests with allergenic extracts (Paupe, personal communication). The extracts tested in vivo may contain molecules that will provoke
a non-specific skin or bronchial reaction in the absence of any IgE inter-
vention (Guinnepain, Gumovski, personal communications). Finally, patient
sera may have high titers of specific IgG that can compete actively with
IgE in the RAST tests (Lynch et al., 1975, Gumovski, personal communication).

IV. CHARACTERIZATION AND LOCALIZATION OF THE ALLERGENS

1. Localization

Investigations of the location of allergens in fungal cells have not
yet been reported. Cellular localization would be essential in order to
select the most efficient extraction procedures. The difficulty in local-
ising fungal allergens is due to their low concentration and specificity
for IgE. To visulaize allergens, IgE antibodies have been concentrated
from a pool of sera from patients positive to Cladosporium on a monoclonal
antihuman IgE antibody affinity column (Bourgeois et al., 1984).

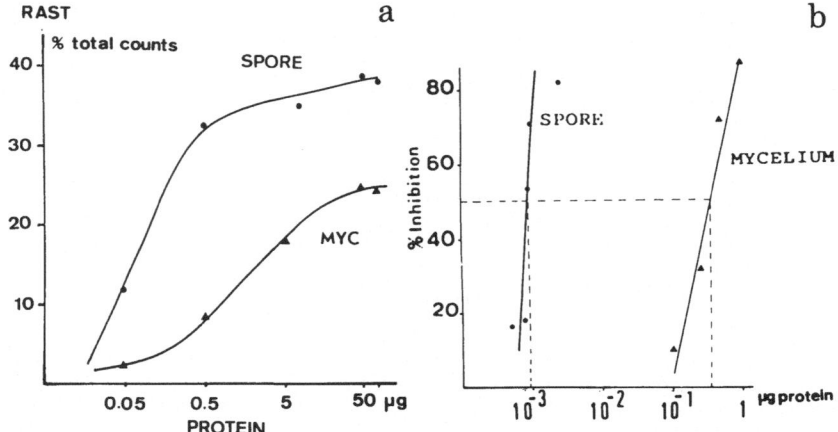

Fig. 13. RAST (a) and RAST-inhibition (b) obtained with
C.cladosporioides; a: RAST values obtained when
CNBr activated discs are incubated with different
concentrations of spore and mycelium extract; b:
RAST values obtained in a RAST-inhibition experiment
where the liquid phase was the spore or mycelium
extract and the solid phase the spore extract. All
extracts were 20.000 g supernatant of disrupted conidia. Note a 500 fold difference in allergen potency
as measured by 50% inhibition point between spore and
mycelium.

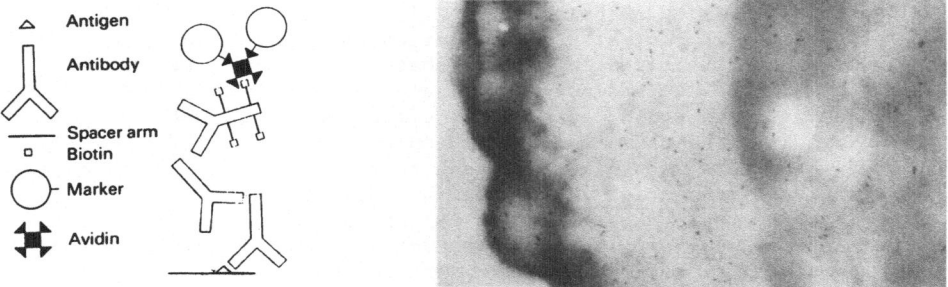

Fig. 14. Localization of allergens in a conidium of C.clados-
porioides on Lowicryl ultra thin section. Allergen
triangle is localized with sequential incubation with
patient serum, an antihuman IgE rabbit IgG, an anti-
rabbit IgG conjugated to gold or an antirabbit IgG
biotinylated detected with a streptavidin-gold complex (x 60000).

Indirect triple immunolabelling has been employed using gold conjugated immunoglobulins and streptavidin on thin sections of spores embedded in Lowicryl (Fig. 14) (Hodges et al., 1984; Coggie et al., 1986; Varndell and Polak, 1986). These immunocytochemical techniques demonstrated that allergens are present in the cell wall and cytoplasm of the conidia at very low concentrations. Sera of patients also contain high levels of specific IgG directed toward <u>Cladosporium</u>. These IgG bind essentially to the spore wall (Fig. 15).

The presence of high amounts of allergens in the soluble fraction obtained after glass bead homogenization of the conidia seems in disagreement with the presence of allergens in the conidial wall. However, immunocytochemical techniques (gold conjugated antigens applied to thin sections of spore previously incubated in a rabbit antiserum directed toward the whole spore) revealed wall components in the soluble fraction (Larsson, 1984) (Fig. 16). This result is indicative of the solubility of wall release after inhalation of the spore. Similar parietal localization of the allergens of <u>Candida albicans</u> has been found using biochemical and immunotransfer methods after enzymatic lysis of the yeast cell wall (Grange et al., 1986).

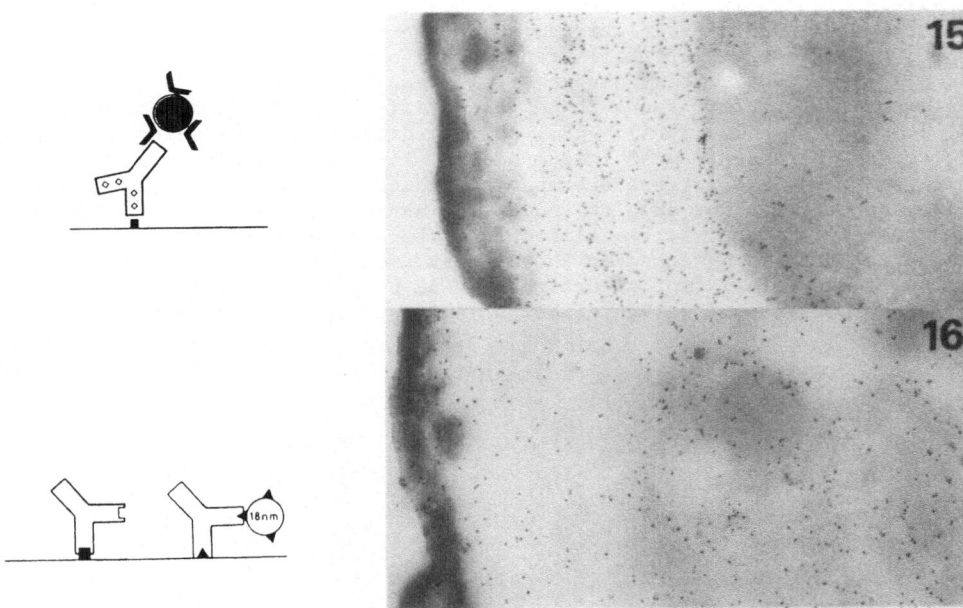

Fig. 15. Localization of IgG-specific antigens in a spore of C.cladosporioides (x 60000). Antigen square is localized with specific patient IgG which can then be detected using protein A-gold particles.

Fig. 16. Localization of the components of the fraction recovered as the 20,000 xg supernatant of an homogenate of disrupted spores (x 60000). Polyvalent anti-C.cladosporioides rabbit Ig bind to all antigens (triangle, square, ...) of the sections. Triangle antigens present in the supernatant and previously conjugated to gold bind to specific IgG already attached to triangle antigen of the thin section.

2. Composition

The content of each allergen in a potential allergenic extract can be estimated by several immunoelectrophoresis techniques described elsewhere: fused radio rocket immunoelectrophoresis (FRRIE), crossed radio immuno-electrophoresis (CRIE), nitrocellulose immunoblot (NCIB)

FRRIE is performed as rocket immunoelectrophoresis and the sample wells are filled with aliquots from a column fractionation. After electrophoresis, the plate is incubated with patient serum, and the specific IgE bound to the allergen is visualized by a second incubation with an 125 I-anti-IgE. This technique has been used by Landmark and Aukrust (1985) to identify the aller-genicity of the fractions recovered after HPLC separation of C.herbarum extracts.

CRIE is probably the most widely used technique amongst immuno aller-gologists to identify individual allergens in a crude extract (Lowenstein, 1978). In CRIE, the proteins separated according to classical two dimen-sional crossed immunoelectrophoresis (Axelsen, 1983) are incubated with patient serum, then the allergen specific IgE bound to the precipitates are visualized by an 125 I-anti-IgE. This technique provides information on the frequency of patients having specific IgE directed toward a particular fungal allergen and relative amount of IgE specific to individual allergen. Major intermediate and minor allergens have been defined in Alternaria and Cladosporium on the basis of their frequency of occurence in CRIE using individual patient sera. Only one major allergen has been detected for C.herbarum and A.alternata using this procedure (Aukrust 1979, Nyholm and Lowenstein, 1982, Nyholm, 1984). CRIE has the disadvantage of being an indirect method dependent on an intermediate anti-fungus rabbit IgG. It is known that not all allergens are able to induce IgG antibodies in rabbits. The allergens that are not able to induce IgG antibodies will not precipitate and consequently will not be visualized in CRIE. For example, the allergen of Alternaria which is detected in the highest amount in CRIE by a pool of positive patients sera is not visualized in CIE by protein staining (Fig. 17). Moreover, the presence of a high number of antigens is not necessarily an

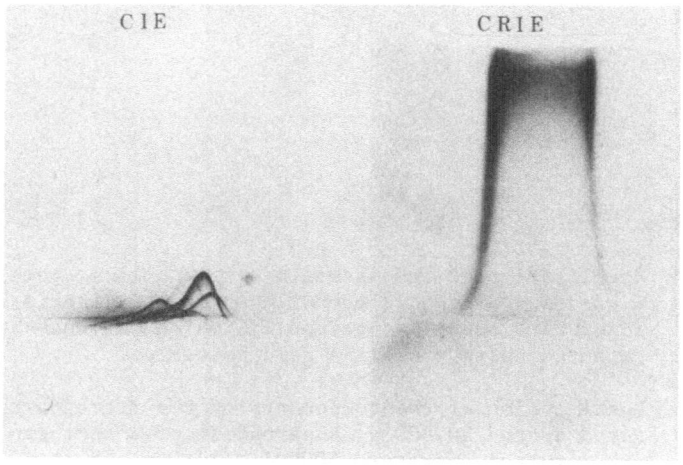

Fig. 17. A.alternata extract analyzed by CIE and CRIE.
CIE: Crossed immunoelectrophoresis against hyperimmune rabbit antisera. CRIE: X-ray film from the
same CIE plate and a pool of sera of patients
allergic to Alternaria.

indication of allergenic potency. For example, the soluble 20,000 x g
conidial supernatant of Alternaria obtained by cell homogenization in
NaHCO$_3$ buffer yielded the highest number of antigens, whereas a coca exudate
characterized by a small number of antigens contains relatively more allergens (Paris, unpublished).

Nitrocellulose (NC) immunoblots also have been used recently to study
individual allergens after their separation by SDS-PAGE (Fig. 18). After
incubation of nitrocellulose in sera, specific IgE antibodies bound to NC
are visualized by 125 I or enzyme labelled anti-IgE. This method has the
advantage of avoiding the intermediate step of the anti-fungus rabbit antiserum (Bengtsson et al., 1982). No attempts have been made to identify
fungal allergens using IEF and passive transfer.

Unlike other allergenic agents such as pollen or mites, the physico-
chemical characteristics and molecular weight of Alternaria and Cladosporium
allergens are poorly defined (Dreborg et al., 1986).

One major allergen, Alt 1, has been studied in A.alternata. This
glycoproteinaceous fraction (7% N) has a molecular weight of about 25,000
to 50,000 daltons. It is composed of at least 12 isoelectric variants in
the range of 4 - 4.3 (Yunginger et al., 1980, Miles et al., 1983). Other
allergenic fractions called Ag i, G2D3 and G2D4 appear to be identical to
Alt I. (Vijay et al., 1981; Nylhom et al., 1983). A basic peptide pI
9.5-9.8 of 6,000 daltons has also been isolated from A.alternata (Budd
et al., 1983a, b).

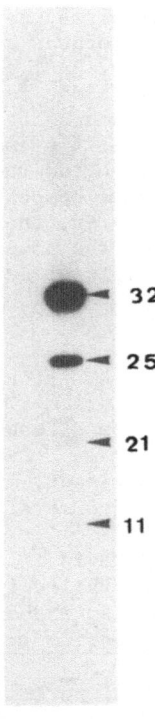

Fig. 18. Nitrocellulose immunoblot of A.alternata
         extract separated by SDS gel electro-
         phoresis showing the presence of two
         major allergens.

In C.herbarum more than 20 allergens were detected out of the 60 antigens visualized by CIE (Lowenstein et al., 1977 and Aukrust 1979). Molecular
weights of these allergens ranged from 10,000 to 300,000 daltons and their
pI are essentially acid (between 3 and 5) (Aukrust and Borch, 1979 and 1980).
Two of the purified allergens are low molecular weight proteins: the major
allergen (Ag 32) is a 10-13,000 dalton protein with five molecular variants
differing in pI from 3.4 to 4.4 and Ag 54, a 20-25,000 dalton protein with
pI 5. It is interesting to note that Ag 54, which is the only purified
allergen, is a glycoprotein containing 80% sugar (mannose:galactose: glucose,
1:0.6:1.3) and only 20% protein (Sward-Nordmo, 1984 and 1985).   However,
this allergen is only found in 1/4 of the positive patients (Aukrust, 1979).
Allergens of C.cladosporioides also occur in a wide range of molecular
weight (10,000 to over 500,000 daltons), have an acid pI (comprised between
4 and 6.5) and are very glycosylated (80% sugar for several fractions
(Bouziane et al., unpublished).

V. PERSPECTIVE FOR FUTURE RESEARCH

     Purification and isolation of allergens should be pursued. Their
characterization will permit standardization of the fungal extracts which
has not yet been realized. Such standardization is essential for clinical
practice, and particularly for specific diagnosis and immunotherapy. The
localization of several allergens in the fungal wall and characterization
of their sugar moiety will open a new field in the immunoallergy. Allergens
have so far been only associated with proteins and it seems that polysaccharides can be very important in triggering allergy crises. Technology in
sugar chemistry (separation and identification) is completely different
from protein chemistry. The search for polysaccharidic allergens will
require the adaptation of these specific methods as well as existing techniques used in immunoallergy. For example, RAST should be performed with
cellulose discs activated with ligands specific for sugars instead of CNBr
which is only specific for proteins. In zoopathogenic fungi, IgG is mostly
directed toward polysaccharidic antigens. It seems logical that other
immunoglobulins like IgE would also be directed toward the more conservative
fungal structures, the polysaccharidic wall components. The understanding
of the cellular reactions involved in allergy, the attachment of the allergens to the IgE on their target cells and the competition with other immunoglobulins will also require the availability of pure allergenic molecules.

REFERENCES

1. K. Aas, J. Leegaard, L. Aukrust, and O.Grimmer, 1980, Immediate type
     hypersensitivity to common molds, comparison of different diagnostic
     materials, Allergy, 35:443-451.
2. M.K. Agarwal, R.T. Jones, and J.W. Yunginger, 1982, Immunochemical and
     physicochemical characterization of commercial Alternaria extracts:
     J.Allergy Clin.Immunol., 70:432-436.
3. Y. Al-Doory, 1984, Airborne fungi in: Mould Allergy Y. Al-Doory and
     J.F. Domson, eds., Lea and Febiger, Philadelphia, pp. 27-40.
4. L. Aukrust, 1979, Crossed radioimmunoelectrophoretic studies of distinct
     allergens in two extracts of Cladosporium herbarum, Int.Archs.Allergy
     appl.Immun., 58:375-390.
5. L. Aukrust and S.M. Borch, 1979, Partial purification and characterization
     of two Cladosporium herbarum allergens, Int.Archs.Allergy Appl.Immun.,
     60:68-79.
6. L. Aukrust and S.M. Borch, 1980, Purification of allergens in Cladosporium
     herbarum, Allergy, 35:206-207.
7. N.H. Axelsen, 1983, Handbook of immunoprecipitation in gel techniques,
     Scand.J.Immunol., 17, suppl. 10.

8. F. Beaumont, H.F. Kauffman, J.G.R. de Monchy, H.J. Bluiter, and K. de Vries, 1985, Volumetric aerobiological survey of conidial fungi in the North-East Netherlands. Comparison of aerobiological data and skin tests with mould extracts in an asthmatic population. Allergy, 40:181-186.

9. R.E. Beever, R.J. Redgwell, and G.P. Dempsey, 1979, Purification and chemical characterization of the rodlet layer of Neurospora crassa conidia, J. Bacteriol., 140:1063-1070.

10. A. Bengtsson, A. Karlsson, W. Rolfsen, and R. Einarsson, 1986, Detection of allergens in mould and mite preparations by a nitrocellulose electroblotting technique, Int. Archs Allergy appl. Immun., 80:383-390.

11. D.G. Boucias and J.P. Latge, 1986, Adhesion of entomopathogenic fungi on their host cuticle, in: Fundamental and applied aspects of invertebrate pathology, R.A. Samson, J.M. Vlak, D. Peters, eds., Foundation ICIP, Wageningen, Holland, 432-434.

12. J.P. Bourgeois, M. Pere, J.C. Mazie, and J.F. Delagneau, 1984, Monoclonal antibodies to human IgE: utilization for total IgE quantification and estimation of allergenspecific IgE antibodies, Develop. Biol, Standard, 57:371-379, F. Karger, Basel.

13. H. Bouziane, C. Fitting, B. David, and J.P. Latge. 1987a, Allergenic potency of spore and mycelium extract of Cladosporium, Allergy, (submitted for publication).

14. H. Bouziane, J.P. Latge, S. Paris, M.C. Prevost, and L.G. Chevance, 1987b, Degranulation of nasal mast cells of guinea pigs induced by Cladosporium and Alternaria allergens, Laboratory Investigation, (submitted for publication).

15. H. Bouziane, J.P. Latge, S. Mecheri, and M.C. Prevost, 1987c, Passive release of allergenic components from Cladosporium conidia, Int. Arch. Allergy Immunol., (submitted for publication).

16. T.W. Budd, C.Y. Kuo, T.J. Yoo, W.R. McKenna, and J. Cazin, 1983a, Antigens of Alternaria. Isolation and partial characterization of a basic peptide allergen, J. Allergy Clin. Immunol., 71;277-282.

17. T.W. Budd, C.Y. Kuo, J. Cazin, T.J. Yoo, 1983b, Allergens of Alternaria: Further characterization of a basic allergen fraction, Int. Archs. Allergy appl. Immun., 71:83-87.

18. G. Coggi, P. Dell'orto, and G. Viale, 1986, Avidin-biotin methods, in: Immunocytochemistry modern methods and applications, J.M. Polak and S. Van Noorden, eds., Wright, Bristol, 54-70.

19. G.T. Cole and L.M. Pope, 1981, Surface wall components of Aspergillus niger conidia, in: The fungal spore morphogenic controls. G. Turian and H.R. Hohl, eds., Academic Press, London, 195-215.

20. G.T. Cole and R.A. Samson, 1984, The conidia, in: Mould allergy, Y. Al-Doory and J.F Domson, eds., Lea and Febiger, Philadelphia, 66-103.

21. S. Dreborg, R. Einarsson, and J.L. Longbottom, 1986, The chemistry and standardization of allergens, in: Handbook of experimental immunology, Volume 1: Immunochemistry, D.M. Weir and L.A. Herzenberg, eds., Blackwell Scientific publications, 10.1-10.28.

22. J. Duc, M. Kolly, and A. Pecoud, 1986, Frequence des allergenes respiratoires impliques dans la rhinite et l'asthme bronchique de l'adulte. Etude prospective. Schweiz. Med. Wschr., 116:1205-1210.

23. M. Etievant, L.G. Chevance, M. Lesourd, and H. Ohayon, 1980, Complement activation and cyto-immunological alterations of the respiratory mucosa, Ann. Immunol, (Inst. Pasteur). 131D:13-42.

24. R. Fadel, S. Paris, C. Fitting, R. Rassemont, and B. David, 1986, A comparison of extracts from Alternaria spores and mycelium, J. Allergy Clin. Immunol., 77, 242.

25. F. Grange, B. de La Parra, P.I. Gumovski, and J.P. Girard, 1986, Distinct cell-wall and cytoplasmic proteinic antigenic fractions from Candida albicans inducing IgE antibodies in hypersensitive patients, in: Fungal antigens, E. Drouhet, G.T. Cole, L. de Repentigny J.P. Latge, and B. Dupont, Plenum Press, New York (this issue).

26. G.M. Hodges, M.A. Smolira, and D.C. Livingston, 1984, Scanning electron microscope immunocytochemistry in practice, in: Immunolabelling for electron microscopy, J.M. Polak and I.M. Varndell, eds., Elsevier Science, Amsterdam, 198-233.

27. D.R. Hoffman, P.P. Kozak, S.S. Gillman, L.H. Cummins, and J. Gallup, 1981, Isolation of spore specific allergens from *Alternaria*. *Ann. Allerg.*, 46:310-316.

28. D.R. Hoffman, 1984, Mould allergens, in: Mould allergy, Y. Al-Doory and J.F. Domson, eds., Lea and Febiger, Philadelphia, 104-116.

29. M. Huppert, 1983, Antigens used for measuring immunological reactivity, in: Fungi pathogenic for humans and animals, B. Pathogenicity and Detection. D.H. Howard, ed., Dekker, New York, 279-302.

30. P.P. Kozak and D.R. Hoffman, 1984, Critical review of diagnostic procedure for mould allergy, in: Mould allergy, Y. Al-Doory and J.F. Domson, eds., Lea and Febiger, Philadelphia, 157-186.

31. J. Lacey, 1981, The aerobiology of conidial fungi, in: Biology of conidial fungi, Volume 1, G.T. Cole and B. Kendrick, eds., Academic Press, New York, 373-416.

32. E. Landmark and L. Aukrust, 1985, High-performance liquid chromatography of *Cladosporium herbarum*. Identification of allergens with immunological techniques, *Int. Archs Allergy appl. Immun.*, 78:71-76.

33. L.S. Larsen, 1981, A three-year-survey of microfungi, in the air of Copenhagen, 1977-79, *Allergy*, 36:15-22.

34. L.I. Larsson, 1984, Labelled antigen detection methods, in: Immunolabelling for electron microscopy, J.M. Polak and I.M. Varndell, eds., Elsevier Science, Amsterdam, 123-128.

35. J.P. Latge, G.T. Cole, M. Harisberger, and M.C. Prevost, 1986, Ultrastructure and chemical composition of the ballistospore wall of *Conidiobolus obscurus*, *Exp. Mycol.*, 10:99-113.

36. J.P. Latge, H. Bouziane, and M. Diaquin, 1987, The conidial wall of *Cladosporium cladosporioides*, *J. Med. Vet. Myc.*, (submitted).

37. B. Lebel, 1983, A high-sampling-rate automated continuous-flow fluorometric technique for the anlaysis of nanogram levels of histamine in biological samples, *Anal. Biochem.*, 133:16-29.

38. K. Licorish, H.S. Novey, P. Kozak, R.D. Fairshter, and A.F. Wilson, 1985, Role of *Alternaria* and *Penicillium* spores in the pathogenesis of asthma, *J. Allergy Clin. Immunol.*, 76:819-825.

39. H. Lowenstein, 1978, Quantitative immunoelectrophoretic methods as a tool for the analysis and isolation to allergens, *Prog. Allergy*, 25:1-62.

40. H. Lowenstein, L. Aukrust, and S. Gravensen, 1977, *Cladosporium herbarum* extract characterized by means of quantitative immunoelectrophoretic methods with special attention to immediate type allergy. *Int. Archs. Allergy appl. Immun.* 55:1-12.

41. N.R. Lynch, P. Dunand, R.W. Newcomb, H. Chai, and J. Bingley, 1975, Influence of IgG antibody and glycopeptide allergens on the correlation between the radioallergosorbent test (RAST) and skin testing or bronchial challenge with *Alternaria*, *Clin. exp. Immunol.* 22:35-46.

42. M. Mallea and J. Charpin, 1980, Moisissures, in: Allergologie, J. Charpin, ed., Flammarion Medecine, Paris, 230-241.

43. H.J. Malling, B. Agrell, S. Croner, S. Dreborg, T. Fouchard, M. Kjellman, A. Koivikko, A. Roth, and B. Weeke, 1985, Diagnosis and immunotherapy of mould allergy. Screening for mould allergy. *Allergy*, 40:108-114.

44. H.J. Malling, S. Dreborg, and B. Weeke, 1986, Diagnosis and immunotherapy of mould allergy. Diagnosis of *Cladosporium* allergy by means of symptom score, bronchial provocation test, skin prick test, RAST, CRIE, and histamine release, *Allergy*, 41:57-67.

45. S. Meyers, H. Burge, E. Smith, M. Muilenberg, and W. Solomon, 1985, Lectin-specific antigens in *Alternaria*, *J. Allergy Clin. Immunol.* 75:118.

46. R.M. Miles, J.L. Parker, R.T. Jones, M.J.E. Dahlberg, and J.W. Yunginger, 1983, Studies on Alternaria allergens. Biological activity of a purified Alternaria fraction (Alt-I). J. Allergy Clin. Immunol., 71:36-39.

47. L. Nyholm, H. Lowenstein, and J.W. Yunginger, 1983, Immunochemical partial identity between two independently identified and isolated major allergens from Alternaria alternata (Alt-1 and Ag 1). J. Allergy Clin. Immunol. 71:461-467.

48. L. Nyholm, 1984, Allergens of Alternaria, in: Atlas of moulds in Europe causing respiratory allergy, K. Wilken-Jensen and S. Gravesen, eds., Foundation for allergy research in Europe, ASK, Copenhagen, 14-15.

49. L. Nyholm and H. Lowenstein, 1982, Identification and characterization of allergens from Alternaria alternata, Allergy, 37, suppl. 1:40.

50. Z. Ovary, 1958, Immediate reactions in the skin of experimental animals provoked by antibody-antigen interactions, Prog. Allergy, Volume 5, pp. 459, Karger, Basel.

51. G. Peltre, J. Lapeyre, and B. David, 1982, Heterogeneity of grass pollen allergens (Dactylis glomerata) recognized by IgE antibodies in human patients sera by a new nitrocellulose immunoprint technique, Immunology letters, 5:127-131.

52. D.M. Rast, H. Stussi, H. Hegnauer, and L.E. Nyhlen, 1981, Melanins, in: The fungal spore, Morphogenetic controls, G. Turian and H.R. Hohl, eds., Academic Press, London and New York, 507-531.

53. I.M. Roitt, J. Brostoff, and D.K. Male, 1985, Immunology, Gower Medical Publication, London.

54. M.J. Schumacher and S.E. Jeffery, 1976, Variability of Alternaria alternata: biochemical and immunological characteristics of culture filtrates from seven isolates, J. Allergy Clin. Immunol., 58:263-277.

55. F.Th.M. Spieksma, 1984, Presentation of spore counts according to type, in: Atlas of moulds in Europe causing respiratory allergy, K. Wilken-Jensen and S. Gravesen, eds., Foundation for allergy research in Europe. ASK, Copenhagen, 83-87.

56. M. Sward-Nordmo, T.L. Almeland, and L. Aukrust, 1984a, Variability in different strains of Cladosporium herbarum with special attention to carbohydrates and contents of two important allergens (Ag-32 and Ag-54). Allergy, 39:387-394.

57. M. Sward-Nordmo, B. Paulsen Smestad, J.K. Wold, and O. Grimmer, 1984b, Characterization of the carbohydrate moiety in a partly purified allergen preparation from the mould Cladosporium herbarum and its possible importance for allergenic activity as tested by RAST-inhibition, Int. Archs. Allergy appl. Immun. 75:149-156.

58. M. Sward-Nordo, J.K. Wold, B. Paulsen Smestad, and L. Aukrust, 1985, Purification and partial characterization of the allergen Ag-54 from Cladosporium herbarum, Int. Archs. Allergy appl. Immun., 78:249-255.

59. S. van der Heide, H.F. Kauffman, and K. de Vries, 1985, Cultivation of fungi in synthetic and semi-synthetic liquid medium. Immunochemical properties of the antigenic and allergenic extracts, Allergy, 40:592-598.

60. I.M. Varndell and J.M. Polak, 1986, Electron microscopical immunocytochemistry, in: Immunocytochemistry modern methods and applications J.M. Polak and S. Van Noorden, eds., Wright, Bristol, 146-166.

61. H.M. Vijay, H. Huang, N.M. Young, and I.L. Bernstein, 1979, Studies on Alternaria allergens, Isolation of allergens from Alternaria tenuis and Alternaria solani, Int. Archs. Allergy appl. Immun., 60:229-239.

62. H.M. Vijay, H. Huang, G. Lavergne, N.M. Young, and I.L. Bernstein, 1981, Studies on Alternaria allergens. Presence of two related antigens with contrasting allergenic properties in Alternaria tenuis extracts, Int. Archs. Allergy appl. Immun., 65:410-416.

63. H.M. Vijay, H. Huang, N.M. Young, and I.L. Bernstein, 1982, Studies on *Alternaria* allergens. Effect of *Alternaria tenuis* on the humoral response to three T cell-dependent antigens in rats, *Int. Archs. Allergy appl. Immun.* 69:34-39.

64. H.M. Vijay, H. Huang, N.M. Young, and I.L. Bernstein, 1984, Studies on *Alternaria* allergens. Comparative biochemical and immunological studies of commercial *Alternaria tenuis* batches, *Int. Archs. Allergy appl. Immun.*, 74:256-261.

65. H.M. Vijay, N.M. Young, G.E.D. Jackson, G.P. White, and I.L. Bernstein, 1985, Studies on *Alternaria* allergens. Comparative biochemical and immunological studies of three isolates of *Alternaria tenuis* cultured on synthetic media, *Int. Archs. Allergy appl. Immun.*, 78:37-42.

66. C.D. Wu-Yuan and T. Hashimoto, 1977, Architecture and chemistry of conidial walls of *Trichophyton metagrophytes*, *J. Bacteriol.*, 129:1584-1592.

67. J.W. Yunginger, R.T. Jones, and G.J. Gleich, 1976a, Studies on *Alternaria* allergens. Measurement of the relative potency of commercial *Alternaria* extracts by the direct RAST and by RAST inhibition, *J. Allergy Clin. Immunol.*, 58:405-413.

68. J.W. Yunginger, G.D. Roberts, and G.J. Gleich, 1976b, Studies on *Alternaria* allergens. Establishment of the radioallergosorbent test for measurement of *Alternaria* allergens, *J. Allergy Clin. Immunol.*, 57:293-301.

69. J.W. Yunginger, 1978, Use of the RAST test, *J. Allergy Clin. Immunol.*, 61:213-216.

70. J.W. Yunginger, R.T. Jones, M.E. Nesheim, and M. Geller, 1980, Studies on *Alternaria* allergens. Isolation of a major allergenic fraction (Alt-I), *J. Allergy Clin. Immunol.*, 66:138-147.

DISTINCT CELL-WALL AND CYTOPLASMIC PROTEINIC ANTIGENIC FRACTIONS FROM CANDIDA ALBICANS INDUCING IgE ANTIBODIES IN HYPERSENSITIVE PATIENTS

F. Grange, B. de la Parra, P.I. Gumowski and
J.P. Girard

Department of Immunology and Allergology, Hopital
Cantonal Universitaire, CH-1211 Geneve 4,
Switzerland

In an attempt to differentiate between various clinical
aspects in patients with Candida albicans (Ca), hypersensitivity, intradermal and serological responses to different
antigenic determinants of Ca were studied.
Four extracts were used in this study. One was a commercial extract made from partially-purified polysaccharides
(Poly-S). Three extracts were prepared by us from the same Ca
culture (HCU 046) that had been grown in a synthetic and
anergic medium:

Extracts used for the in vivo tests included:
- EXO-fraction, that was passively extracted at 4°C in
a Coca solution. After extraction, the remaining
intact cells were broken using glass beads to prepare
another extract containing cytoplasmic proteins and
polysaccharides (ENDO-fraction).
Extracts used for in vitro tests only included:
- An aliquot of cells digested with lyticase immobilized in a solid support, and a cell wall
fraction (CWP).

Electrophoretic analysis (SDS-Page) of these fractions
showed that most of the proteins present in the EXO and ENDO
fractions migrated between 70 and 17 kilodaltons (kd), while
most of the proteins of CWP migrated between 110 and 17 kd.
The EXO and ENDO fractions showed some common bands, but the
presence of distinct protein bands was also observed. Only a
few bands were common between the EXO and CWP fractions.
The extracts were tested in 120 patients whose hypersensitivity towards Ca had previously been demonstrated using a
"crude" extract (strong immediate-type intradermal test,
positive provocative test and/or > class 2 IgE-RAST).
Immediate type 1 skin reactivity towards the EXO fraction
was found in 78 patients, 32 of whom had no other reactions to

the ENDO or Poly-S fractions. This suggested that protein components extractable from the cell surface were also responsible for type 1 hypersensitivity reactions. Immuno-prints of the EXO, ENDO and CWP fractions after incubation with the sera of 21 patients allowed the detection of IgE and IgG antibodies to several proteins of the three fractions. Some proteins apparently induced dual responses (IgE and IgG). IgE antibodies against proteins only shared by the EXO and CWP fractions were detected, as were IgE antibodies against pro-teins contained only in the CWP fractions.

Classic, immediate-type reactions have been attributed to IgE reacting to cytoplasmic protein fractions and/or to mannan polysaccharides. These experiments show _in vivo_ and _in vitro_ evidence that protein components of the cell surface and cell wall are also involved in IgE antibody responses.

SERIAL SPECIFIC IgG MEASUREMENTS TO COMMON MOULDS

B. Guerin, A.M. Chaissac and B. Gouyon

Laboratoire d'Immunologie, 160 quai de Polangis
94340 Joinville-le-Pont, France

The IgG immunodot technique provides a rapid evaluation of the immunogenic potency of environmental substances.

The profiles of allergic and non-allergic populations in France, where the climate is temperate, and on the island of Mauritius, where the climate is sub-tropical, indicate reactivity to the following common moulds and yeasts: <u>Candida albicans</u>, <u>Alternaria</u>, <u>Hormodendrum</u>, <u>Penicillium</u> and <u>Aspergillus</u>.

For common species such as <u>Alternaria</u>, there are significant differences in immunological response according to the strain of the fungus, even with the same method of cultivation. The differences are independent of the population studied. One of the strains studied seemed to be less immunogenic for the production of IgG though it detected IgE titres in clinically allergic subjects. The level of response was at least as high as that elicited by other strains.

All subjects studied had high IgG titres specific to moulds, but these high titres did not seem to interfere with the measurement of specific IgE. This was confirmed when measured by an exhaustive technique which neutralizes possible competition of IgG and IgE responses.

THE ANTIGENIC AND ALLERGENIC PROPERTIES OF BOTRYTIS CINEREA

H.F. Kauffman, S. van der Heide, F. Beaumont and
K. de Vries

Department of Allergology, Clinic for Internal
Medicine, State University Hospital, Groningen,
Netherlands

During aerobiological studies of fungal prevalence in the
Netherlands (F. Beaumont et al., 1984, Ann. Allergy, 53:487) it
was found that next to the commonly occurring genera such as
Cladosporium, Penicillium, Alternaria and Aspergillus, the
conidia of Botrytis belonged to the most prevailing airborne
moulds. It was also found that Botrytis showed a high sensitization rate (6.1%) in a group of patients with bronchial
obstructive complaints (F. Beaumont et al., 1985, Allergy,
40:181). Reports of Botrytis as a causative agent of human
allergies are limited, but severe, early, and late obstructive
reactions after inhalation of Botrytis extracts have been
reported (Morrow-Brown, meeting of the "Foundation for Allergy
Research in Europe").

Studies on the antigenic and allergenic properties of
Botrytis extracts have not been published. We therefore
studied the IgE- and IgG-binding properties of culture filtrates and mycelial extracts of Botrytis cinerea using sera of
patients which showed high IgE-binding titers in a direct
enzyme allergosorbent test (EAST).

During cultivation of B. cinerea, marked changes in pH of
the culture broth were used as markers for the stage of growth
of the microorganism, in a manner similar to that described
for A. fumigatus (H.F. Kauffman, K. de Vries, 1980, Int. Archs.
Allergy Appl. Immunol., 62:252). It was found that IgE- and
IgG-binding components, as measured by EAST and ELISA, were
rapidly released into the culture medium and showed similar
binding capacities for early (Phase I) and late (Phase III)
periods of growth. In contrast, the number of pre-cipitating
antigenic components reached a maximum during later phases of
growth.

Results of studies of a group of sera from patients with
high levels of IgE antibodies against A. fumigatus antigen,
showed that about half of the sera also reacted with Botrytis
extracts, whereas the other half were negative.

Cryptococcus

CORRELATES OF SEROTYPE IN CRYPTOCOCCUS NEOFORMANS

John E. Bennett, and K. J Kwon-Chung

National Institute of Allergy and Infectious Diseases
National Institutes of Health, Bethesda, MD 20892

Serotypic differentiation of C. neoformans isolates was first reported by E. E. Evans in 1949, based upon agglutination reactions with absorbed rabbit serum (1). Evans studied only 12 isolates, 8 of which were designated as type A, 3 as type B and one as type C. Significantly, the author had collected his type B and C isolates from patients in California. It was later shown that only about one of 16 isolates from elsewhere in the USA are types B or C (2). A new serotype was reported in 1966 based upon fluorescent antibody studies with a human serum against one isolate of the new serotype (3). In 1968, existence of a fourth serotype was confirmed in five isolates, using agglutination by absorbed rabbit serum (4). Since this time, no new serotypes have been described.

Serotype of C. neoformans is determined by the capsular glucuronoxylomannan of the fungus. The epitopes determining serotype are still unclear for types B and C but the glucuronyl and O-acetyl groups appear important in recognition of types A and D (5). Preparation of typing serum remains an inexact science, although selection of appropriate monoclonal antibodies may assist this effort. The easiest typing serum to prepare is for type C because very little cross absorption is required. Distinction between type A and D is more problematic in our hands. Almost as many isolates type with both A and D antiserum as type with D alone (2). The so-called AD isolates usually revert to one type or the other with multiple cloning procedures. Only 2% of the isolates are untypable (2). Such isolates come from highly mucoid colonies and have large capsules around the cell. It may be that excess antigen interferes with the agglutination reaction.

Genetic control of serotype has been found to be linked to a surprisingly large number of other phenotypic markers in C. neoformans (Table 1). Each one of these will be considered in detail.

C. neoformans is a heterothallic fungus, containing two mating types, (α) and (a). The α type predominates in isolates from patients and natural sites by 30:1 and 40:1, respectively (6). When two appropriately selected strains are mixed on the appropriate agar medium, cells of the opposite mating type conjugate to form diploid

Table 1. Correlates of Serotype
in Cryptococcus neoformans

Sexual State: morphology and intervarietal crosses
Physiology
     metabolism of creatinine and glycine
     uptake of ℓ-malic acid
     growth inhibition by ℓ-canavanine
     enzyme electrophoretic mobility
     growth rate at 37°C
     mouse virulence
Ecology
     saprophytic state
     geographic distribution

hyphae. Meiosis occurs in a basidium, producing haploid basidiospores. Isolates of C. neoformans have been divided into two varieties, depending upon the appearance of the sexual state and upon ease of obtaining fertile crosses (7). Basidiospores of the variety gattii are shaped like bacilli, whereas those of the variety neoformans are rounded (7). To an extent, the yeast-like cells mirror this change in that elongated cells are often seen in brains of mice inoculated with the variety gattii (7). To date, all isolates of the variety gattii have been of serotype B or C, whereas all isolates of the variety neoformans have been serotype A or D. Fertile crosses between the varieties have been obtained (7), confirming the varietal status, but such crosses are usually unsuccessful or the progeny are infertile. DNA homology also supports the existence of two varieties (8).

Physiologic differences between isolates of C. neoformans tend to parallel serotype or, more broadly, variety. Isolates of serotypes B, C or D utilize creatinine very well as a source of nitrogen (9). An inducible enzyme, creatinine deiminase, converts creatinine to ammonia and methylhydantoin (10). Most serotype A isolates form very little of this enzyme probably because very little creatinine enters the cell (unpublished data). Serotype B and C isolates produce substantial amounts of ammonia that is utilized initially but, as the carbon source in the medium is consumed, ammonium accumulates and alkalinizes the culture medium. This alkalinization does not occur when serotype D isolates use creatinine as a nitrogen source because any accumulation of ammonium represses the synthesis of creatinine (10). Ammonia production also occurs on Guizotia abyssinica agar, which ammonia turns greenish (9). Brom thymol blue is a more appropriate pH indicator, turning from yellow to blue between pH 6 and 7.6. Phenol red has also been used, but requires a higher pH to turn from yellow to red.

The first culture medium described to distinguish serotypes A and D from B and C contained creatinine, glucose and brom thymol blue (11). This medium requires close attention to pH, glucose concentration and inoculum size. Even then, 10% false positive reactions were obtained with serotypes A and D (12). Further, 13% of serotypes B and C were falsely negative. The next medium used glycine as a carbon source, cycloheximide and phenol red (13). Glycine is the major contributor to the specificity of this medium. Serotype B and C isolates have a more rapid uptake of glycine, converting glycine to ammonia, carbon dioxide and serine (14). Although serotype B and C isolates were reported to be more resistant to cycloheximide (13), our analysis of 15 isolates of each serotype found an insignificant difference between serotypes, making the contribution of cycloheximide to the medium unreliable. This may be why the medium is not completely satisfactory, producing 11% false negative results (12). The best medium in our hands has been canavanine-glycine-bromthymol blue (CGB) agar (12). Glycine in this medium serves both as carbon and nitrogen source. Serotype B and C isolates have a more rapid uptake of glycine, converting the amino acid to ammonia, carbon dioxide and serine, initially by the inducible enzyme, glycine decarboxylase. Only a minority of serotype A isolates and serotype D isolates (11%) can utilize glycine. Canavanine at 30 µg/ml inhibits the growth of all serotype D isolates and a third of serotype A isolates. The canavanine-resistant isolates of serotype A, however, are unable to utilize glycine on CGB agar (14).

Another physiologic difference between serotypes is the ability to utilize certain unionized dicarboxylic acids. Utilization of ℓ-malic acid as carbon source is poor in serotype A and D isolates compared to serotype B and C, largely because of differences in the permease (9). Among the other differences between serotypes, isolates of serotype B and C grow somewhat slower at 37°C and tend to be somewhat less pathogenic for mice, though these differences are not at all absolute (7). Recently, Safrin and coworkers reported differences in electrophoretic mobility of glucose phosphate isomerase and phosphoglucomutase between isolates (15). Mobility tended to parallel serotype in the 20 isolates they studied, though their seven type B isolates contained two different populations with entirely different mobility. This work warrants confirmation.

A much more absolute difference between serotypes has been the ability to isolate the fungus in a saprophytic state in nature. Only serotypes A, D or AD have been isolated, not B or C. This has been true even in regions where types B and C are relatively common (2).

The last difference to be detailed here is in geographic distribution (16). Before proceeding, it must be pointed out that data on geographic origin of isolates has to be based upon patients with a limited travel history. Although isolates from animals would be preferable from this point of view, too few are available to allow detailed analysis. What information is available suggests that serotype does not differ between animals and humans.

Serotype C has been found in only 1% of isolates in all areas except Southern California, where it comprised 12.7% of isolates (Table 2). This figure may be changing, in that no type C isolate was obtained in a recent survey of 20 isolates from this area. Serotype D is somewhat more prevalent, comprising 4 - 8% of isolates in most regions. We have encountered two areas of higher prevalence, Italy and Denmark, based upon only a few isolates (Table 3). Environmental isolates from these countries also were largely type D. Others have reported a high prevalence of type D in Switzerland (17) and Germany (18). Strangely, our isolates from France, Holland, Belgium and Great Britain have not had an excess prevalence of type D (16).

In all regions we have studied, type A is the most common (Table 4). For this reason, most of our knowledge of clinical manifestations and treatment is based upon experience with this serotype. We do not know if differences based upon serotype exist.

Table 2. Prevalence of Type C

Southern California	11/86	12.8%
USA, except S. Calif.	5/269	1.9%
South America	1/76	1.3%
Canada	1/80	1.2%
Europe	1/117	0.9%
New Zealand, Australia	0/50	0%

Table 3. Prevalence of Type D

Denmark (12), Italy (7)	16/19	84.2%
Switzerland*	17/23	73.9%
Germany (Mishra)	5/17	29.4%
Europe, elsewhere	3/98	3.1%
New Zealand, Australia	2/50	4.0%
USA	21/355	5.9%
Canada	5/78	6.4%
South America	6/76	7.9%

*Includes 18 of Scholer

Table 4. Prevalence of Type A

USA,	262/355	73.8%
Excluding S. Calif.	215/269	79.9%
Canada	62/78	79.5%
New Zealand, Australia	35/50	70.0%
Europe	78/117	66.7%
South America	44/76	57.9%
S.E. Asia	9/18	50.0%
Japan (ref. 19)	58/62	93.5%

Evidence has become quite convincing that serotype B isolates are more often encountered from infections in tropical or subtropical climates (20). Correlation appears best with a warm winter temperature, rather than rainfall. Countries with an excess prevalence of type B include Southeast Asia, Australia, Venezuela, Brazil and Southern California (Table 5). In temperate climates, the prevalence of type B is most often around 3 - 6%. A modest number of isolates from other countries supports this conclusion. These data do not appear in Table 5 because the number of isolates from any one country was insufficient to draw conclusions (16).

Table 5. Prevalence of Type B

S.E. Asia	9/18	50.0%
S.E. Asia (Scholer)	9/23	39.1%
Australia	9/24	37.5%
Venezuela	5/16	31.2%
Brazil	14/48	29.2%
S. California	22/86	25.6%
Argentina	1/10	10.0%
Canada	5/78	6.4%
USA, except S. Calif.	13/269	4.8%
New Zealand	1/26	3.8%
Europe	4/117	3.4%

Table 6. Cryptococcosis in AIDS

Serotype A, D, OR AD
    USA   13/13
    Zaire   15/15
    Burundi (ref. 24)   11/11
Var. Neoformans (Biochem.)
    USA (ref. 22)   14/14
    USA (ref. 23)   10/10

The type B isolates from the USA outside of Southern California have shown little tendency to cluster geographically, with a possible exception of a suggestive area comprising Southern Alabama, Southern Georgia and Northern Florida. Rather, type B has been isolated in small numbers from all areas of the USA. The excess prevalence of type B in warmer climates has suggested that the natural habitat may be on plants that are more common in this environment. Such an ecologic niche deserves study.

A preliminary study, reported only in abstract, suggested that previously normal patients who were infected with serotype B and C required longer therapy. There was no significant difference in death rates. Nor were immunosuppressed patients studied. It seems likely that the degree of immunosuppression would override any minor differences in effect of serotype. Our small, retrospective study prompted us to examine susceptibility of 60 isolates to amphotericin B and flucytosine. Fifteen from each serotype were examined. No differences were found, confirming a study in which isolates were grouped only biochemically (21).

The acquired immune deficiency disorder is fast becoming the leading predisposing factor to cryptococcosis. Recent reports have noted that only isolates biochemically consistent with serotype A or D were being encountered (22-24). Our experience with 13 isolates from the USA and 15 from Zaire has confirmed the uniform occurrence of type A or D (Table 6). The report from Southern California noted that 2 of 10 patients without AIDS were biochemically B or C whereas none of the 10 isolates from AIDS were B or C. The marked decrease in serotype B and C isolates from Zaire may therefore be due to the rising incidence of AIDS in that country (25). None of the data is convincing but it remains possible that immunodeficient patients are less likely to acquire infections with serotypes B or C. No reason this might occur has been proffered.

Summary

Antigenic composition of the capsular glucuronoxylomannan in C. neoformans tends to parallel a wide variety of other characteristics, including varietal status, physiology and ecology. Most of the differences place serotypes A and D in one group with B and C in another. A major unexplained difference that does not fall across these distinctions is that infections due to serotype B are more prevalent in tropical and subtropical climates. Based upon smaller numbers of isolates, serotype D appears to be more prevalent in infections acquired in certain European countries. These differences may prove useful clues to the natural sources from which infection is acquired.

REFERENCES

1. Evans EE. An immunologic comparison of twelve strains of Cryptococcus neoformans. Proc Soc Exp Biol Med 71:644-6, 1949.

2. Bennett JE, Kwon-Chung KJ, Howard DH. Epidemiologic differences among serotypes of Cryptococcus neoformans. Am J Epidemiol 105:582-6, 1977.

3. Vogel RA. The indirect fluorescent antibody test for the detection of antibody in human cryptococcal disease. J Infect Dis 116:573-80, 1966.

4. Wilson DE, Bennett JE, Bailey JW. Serologic grouping of Cryptococcus neoformans. Proc Soc Exp Biol Med 127:820-3, 1968.

5. Bhattacharjee AK, Bennett JE, Glaudemans CPJ. Capsular polysaccharides of Cryptococcus neoformans. Rev Infect Dis 6:619-24, 1984.

6. Kwon-Chung KJ, Bennett JE. Distribution of α and a mating types of Cryptococcus neoformans among natural and clinical isolates. Am J Epidemiol 108:337-40, 1978.

7. Kwon-Chung KJ, Bennett JE, Rhodes JC. Taxonomic studies on Filobasidiella species and their anamorphs. Antonie van Leeuwenhoek 48:25-38, 1982.

8. Aulakh HS, Straus SE, Kwon-Chung KJ. Genetic relatedness of Filobasidiella neoformans (Cryptococcus neoformans) and Filobasidiella bacillispora (Cryptococcus bacillisporus) as determined by DNA base composition and sequence homology studies. Int J Syst Bacteriol 31:97-103, 1981.

9. Bennett JE, Kwon-Chung KJ, Theodore TS. Biochemical differences between serotypes of Cryptococcus neoformans. Sabouraudia 16:167-74, 1978.

10. Polacheck I, Kwon-Chung KJ. Creatinine metabolism in Cryptococcus neoformans and Cryptococcus bacillisporus. J Bacteriol 142:15-20, 1980.

11. Kwon-Chung KJ, Bennett JE, Theodore TS. Cryptococcus bacillisporus sp. nov.: serotype B-C of Cryptococcus neoformans. Int J Syst Bacteriol 28:616-20, 1978.

12. Kwon-Chung KJ, Polacheck I, Bennett JE. Improved diagnostic medium for separation of Cryptococcus neoformans var. neoformans (serotypes A and D) and Cryptococcus neoformans var. gattii (serotypes B and C). J Clin Microbiol 15:535-7, 1982.

13. Salkin IF, Hurd NJ. New medium for differentiation of Cryptococcus neoformans serotype pairs. J Clin Microbiol 15:169-71, 1982.

14. Min KH, Kwon-Chung KJ. Biochemical bases for the distinction of the two Cryptococcus neoformans varieties with CGB medium. Zentralbl Bakteriol Mikrobiol Hyg, 261:471-80, 1986.

15. Safrin RE, Lancaster LA, Davis CE, Braude AI. Differentiation of Cryptococcus neoformans serotypes by isoenzyme electrophoresis. Am J Clin Pathol 86:204-8, 1986.

16. Kwon-Chung KJ, Bennett JE. Epidemiologic differences between the two varieties of Cryptococcus neoformans. Am J Epidemiol 120:123-30, 1984.

17. HJ Scholer personal communication.

18. Mishra SK, Staib F, Folkens U, Fromtling RA. Serotypes of Cryptococcus neoformans strains isolated in Germany. J Clin Microbiol 14:106-7, 1981.

19. Ikeda R, Shinoda T, Fukazawa Y, Kaufman L. Antigenic characterization of Cryptococcus neoformans serotypes and its application to serotyping of clinical isolates. J Clin Microbiol 16:22-9, 1982.

20. Kwon-Chung KJ, Bennett JE. High prevalence of Cryptococcus neoformans var. gattii in tropical and subtropical regions. Zentralbl Bakteriol Mikrobiol Hyg 257:213-8, 1984.

21. Fromtling RA, Abruzzo GK, Bulmer GS. Cryptococcus neoformans: comparisons of in vitro antifungal susceptibilities of serotypes AD and BC. Mycopathologia 94:27-30, 1986.

22. Rinaldi MG, Drutz DJ, Howell A, Sande MA, Wofsy CB, Hadley WK. Serotypes of Cryptococcus neoformans in patients with AIDS. J Infect Dis 153:642, 1986.

23. Shimizu RY, Howard DH, Clancy MN. The distribution of Cryptococcus neoformans var. neoformans and C. neoformans var. gattii among patients at UCLA Medical Center (#F-39, p. 404). Abstracts of the Annual Meeting of the American Society for Microbiology, Washington, DC, 1986.

24. Fournon M, Petat E, de Closets F, Himana TH, Barin F, du Sorbier CM. A propos de 11 souches de Cryptococcus neoformans isolees du L.C.R. de patients africains. Bull Soc Fr Mycol Med 15:85-8, 1986.

25. Swinne D. Study of Cryptococcus neoformans varieties. Mykosen 27:137-41, 1984.

CHARACTERIZATION OF CROSS-REACTIVITY BETWEEN CRYPTOCOCCAL POLYSACCHARIDES

BY USE OF ENZYME-LINKED IMMUNOSORBENT ASSAYS AND MONOCLONAL ANTIBODIES

Thomas R. Kozel and Thomas F. Eckert

Department of Microbiology and the Cell and Molecular
Biology Program, School of Medicine, University of Nevada
Reno, NV

INTRODUCTION

Cryptococcus neoformans is surrounded by a polysaccharide capsule. Four serotypes of the polysaccharide, serotypes A, B, C, and D, were identified on the basis of agglutination reactions (Wilson, Bennett and Bailey, 1968). Recent studies identified a possible fifth serotype, serotype A-D (Ikeda et al., 1982). Typically, there is extensive cross-reactivity between the various serotypes.

Serotyping of cryptococcal isolates is done with polyclonal antibody raised in rabbits by immunization with whole cryptococci. Assays used for serotyping include agglutination (Ikeda et al., 1982; Wilson, Bennett, and Bailey, 1968) and immunofluorescence (Kaplan et al., 1981). In all cases, whole cryptococci are used as antigens. Presumably, the major high molecular weight polysaccharide, termed the glucuronoxylomannan (GXM; Cherniak, Reiss, and Turner; 1982) accounts for most if not all, of the serotype specificity. However, existing techniques using whole cells cannot exclude a role for the galactoxylomannan (GalXM) or the mannoprotein in type specificity.

We conducted the present study to assess the cross-reactivity between cryptococcal serotypes when purified GXM was used in an ELISA assay. We used polyclonal antibodies obtained from animals immunized with the purified polysaccharide. In addition, we sought to produce a monoclonal antibody which takes advantage of the cross-reactivity between cryptococcal polysaccharides and is reactive with all four serotypes.

The results of our studies i) demonstrate the use of ELISA for study of the immune response to cryptococcal polysaccharide (GXM), ii) confirm the extensive cross-reactivity between cryptococcal polysaccharides, and iii) demonstrate the value of a monoclonal antibody with reactivity for all four serotypes as a possible reagent for detection of cryptococcal antigen in a latex-coupled monoclonal antibody assay.

MATERIALS AND METHODS

Cultures and isolation of cryptococcal polysaccharides

C. neoformans isolates of serotype A (ATCC 24064), serotype B (ATCC

24065), serotype C (ATCC 24066), and serotype D (ATCC 24067) were obtained from the American Type Culture Collection (Rockville, Md.). Additional strains of serotype A (strains 288, 371, 271, 289, and 104), serotype B (strains 444, 184, 435, 182, and 3939), serotype C (strains 836, 1134, 191, 401, and 917), and serotype D (strains 3501, 430, 3502, 529, and 52) were generously provided by Dr. K.J. Kwon-Chung, National Institutes of Health, Bethesda, Md. All yeast cells were grown in a synthetic medium (Cherniak et al., 1980) and were killed with formalin.

Cryptococcal GXM was isolated from culture supernatant fluids of the prototype ATCC strains 24064, 24065, 24066, and 24067. The polysaccharide was purified by sequential precipitation with ethanol and cetyltrimethylammonium bromide (CTAB) as described (Cherniak, Reiss, and Turner, 1982). Galactose was not detected by gas-liquid chromatography analysis of the polysaccharides (Kozel and Gotschlich, 1982), indicating little or no contamination by GalXM.

ELISA assay for antibody to cryptococcal polysaccharides

An ELISA assay was developed to detect rabbit or mouse IgG antibody specific for cryptococcal polysaccharide. The ELISA was a modification of the assay described by Leinonen and Frasch (1982) for detection of antibody to meningococcal polysaccharides. Polystyrene plates were precoated with 1.0 ug poly-L-lysine/well (150,000 MW; Sigma Chemical Co.) in 200 ul 0.05 M sodium phosphate buffer, pH 7.0 (coating buffer) for four h at 20°C. The wells were washed with coating buffer and incubated overnight at 20°C with 0.8 ug cryptococcal polysaccharide/well in 200 ul coating buffer. The wells were washed with coating buffer and blocked by incubation for 2 h at 37°C with 1% gelatin in coating buffer. The wells were washed three times before use in a washing buffer that consisted of 0.05% Tween 20 and 0.5% gelatin in phosphate buffered saline (PBS).

Antibody assays were done by diluting rabbit sera or hybridoma supernatant fluids with washing buffer. Dilutions of antibody (200 ul) were incubated with the polysaccharide-coated wells for 2 h at 37°C. The ELISA plates were washed three times with washing buffer, and an optimal dilution of a peroxidase-labeled second antibody was added for an additional 2 h at 37°C. Peroxidase-labeled, affinity-purified goat antibody to rabbit heavy chains was obtained from Biorad Laboratories. Peroxidase-labeled, affinity-purified goat antibody to mouse IgG was obtained from Bio-Rad Laboratories or Cappel Laboratories. After incubation with the second antibody, the wells were washed three times with washing buffer. The substrate (O-phenylenediamine; Voller and Bidwell, 1986) was added and incubated for 30 min at 20°C. The reaction was stopped by addition of 50 ul of 4 N H_2SO_4, and optical densities were read at 492 nm with a Bio-Tek EIA Reader. An optical density of 0.2 was chosen as the endpoint of a titration.

Polyclonal antisera

Rabbits were immunized with a methylated bovine serum albumin conjugate of purified cryptococcal GXM. Bovine serum albumin was methylated as described (Kozel and Cazin, 1974). A saline solution containing 500 ug cryptococcal polysaccharide was mixed with 500 ug aqueous methylated bovine serum albumin. The volume was brought to 2 ml with sterile distilled water, and the antigen preparation was emulsified with an equal volume of Freund's complete adjuvant. Rabbits were given three weekly injections of an antigen consisting of 250 ug polysaccharide and 250 ug protein. Each weekly immunization was divided between subscapular and intramuscular sites. Rabbits were bled and serum was collected 10-14 days after the final injection.

Monoclonal antibodies

Balb/C mice were immunized with serotype A polysaccharide coupled to sheep erythrocytes. The procedure for coupling polysaccharide to sheep erythrocytes has been described (Kozel and Cazin, 1971). The mice were primed by intravenous injection of 0.2 ml of a 0.1% suspension of sheep erythrocytes. The mice received subsequent intravenous injections of 0.2 ml of a 1% solution of erythrocytes coupled to cryptococcal polysaccharide 13, 76, 114, 121, and 122 days after the priming injection. Splenocytes were collected for fusion 2 days after the last immunization.

Splenocytes from immune mice were fused with SP2.0-Ag14 mouse myeloma cells at a 1:1 ratio of lymphocytes to myeloma cells. The fusion was done with 50% polyethylene glycol (PEG 1450, Eastman Kodak Co.) by the procedures described by Lane, Crissman, and Lachman (1984). Peritoneal cells from Balb/C mice were used as feeder cells. Selection of hybrids by use of hypoxanthine, aminopterin and thymidine (HAT) medium was as described (Lerner, 1981). Supernatant fluids from hybrid colonies were tested by ELISA for specific antibody 11 days after the fusion. Hybridoma colonies secreting anti-cryptococcal antibody were cloned three times by limiting dilution.

Ascites tumors were produced in pristane-primed Balb/C mice. IgG antibody was isolated from ascites fluid by caprylic acid precipitation (Steinbuch and Audran, 1969). The isotype of the monoclonal antibody was determined by incubation of 200 ul of a 1:1000 dilution of purified antibody with polysaccharide-coated polystyrene plates. The plates were washed and incubated with rabbit antibody specific for murine IgG1, IgG2a, IgG2b, IgG3, IgA, or IgM (Hyclone Laboratories, Inc.) for 90 min at 37°C. The plates were washed with washing buffer and incubated with peroxidase-labeled, affinity purified antibody specific for rabbit immunoglobulins. The plates were incubated with substrate in the manner described for the ELISA assay.

Immunochemical assays for analysis of monoclonal antibodies

Monoclonal antibodies were screened and analyzed by ELISA as described above. Double diffusion in agar was used to assess the precipitating activity of the monoclonal antibody. Purified cryptococcal GXM from the prototype strains at a concentration of 2 mg/ml were used as the soluble antigen. Immunodiffusion was done in plates prepared from 1% agarose in PBS. Reactivity patterns were assessed after 16 h.

Tube agglutination assays were done using Formalin-killed yeast cells. Yeast cells (400 ul of 10^7 cells/ml in PBS) were incubated with 400 ul of dilutions of IgG isolated from ascites fluid. The antibody concentration was adjusted to 320 ug/ml before dilution. A slide agglutination assay was used to assess the reactivity of monoclonal antibody with different isolates of the yeast. Monoclonal antibody (100 ul, 2 mg antibody/ml) was mixed with 100 ul yeast cells (10^8/ml) in PBS. The slides were rotated for 5 min., and the results were recorded.

Latex agglutination assay for cryptococcal polysaccharide

Coupling of MAb 471 to latex beads was done by Dr. R.V. Krishna of Difco Laboratories. The assay for antigen was done by mixing 50 ul of a dilution of polysaccharide with 20 ul of latex-coupled monoclonal antibody reagent. The slide was placed on a clinical rotator for 4 min at 125 rpm. The slides were read as follows: 4+, large clumps of aggregated beads with a clear background; 3+, large clumps with a slightly cloudy background; 2+, small but clearly visible clumps with a slightly milky background; 1+, grainy clumps with a slightly milky background. A 1+ result was considered a positive result. The specimens were prepared and read in a double blinded manner.

Purified cryptococcal polysaccharides of serotypes A, B, C, and D were used as the test antigens. Dilutions of the antigens were prepared in phosphate buffered saline.

RESULTS

Cross-reactivity shown by ELISA of polyclonal antibodies

Previous studies of the immunological relationship between serotypes of C. neoformans examined the reactivity between polyclonal antiserum raised against cryptococcal yeasts and whole cryptococci in immunofluorescence or agglutination assays. We sought to determine whether a similar pattern of cross reactivity occurred between antisera raised against purified polysaccharides and polysaccharide-coated plates in an ELISA assay. Polystyrene plates were coated with polysaccharide of serotypes A, B, C, or D. The titers of antisera raised against polysaccharides of each serotype were determined. The results in Table 1 demonstrate the extensive cross-reactivity that occurs between the four serotypes. Table 2 is an analysis of the serological relationship between the four serotypes.

Production and characterization of a monoclonal antibody reactive with serotypes A, B, C, and D

Analysis of the reactivity between polyclonal antisera and cryptococcal polysaccharide confirmed the extensive cross-reactivity between cryptococcal polysaccharides. This data suggested that one or more determinants are common to all four serotypes. As a consequence, we attempted to produce a hybridoma that secreted antibody specific for this putative common determinant. A total of 2000 fusion wells were screened by ELISA for IgG antibody specific for cryptococcal polysaccharide. One of the positive clones, designated MAb 471, was reactive with polysaccharides of all four serotypes. The antibody secreted by this cell line was of the IgG1 isotype. This cell line was cloned by repeated limiting dilution, and antibodies were obtained from ascites fluid.

Table 1. Titers of antibody and cross-reactivity between serotypes of cryptococcal polysaccharides in ELISA

Polysaccharide Serotype	Reciprocal ELISA titer of antisera			
	Anti-A	Anti-B	Anti-C	Anti-D
A	212,000	146,000	3,100	330,000
B	92,000	90,000	7,000	516,000
C	115,000	223,000	46,267	106,659
D	84,000	30,000	16,000	400,000

Table 2.   Serological relationship between cryptococcal polysaccharides based on reciprocal ELISA titers.

Serotype	A	B	C	D
A	1.00			
B	.84	1.00		
C	.19	.61	1.00	
D	.57	.65	.30	1.00

$$\% \text{ relatedness} = \sqrt{\gamma 1 \times \gamma 2} \times 100$$

$$\text{where } \gamma 1 = \frac{\text{heterologous titer strain 2}}{\text{homologous titer strain 1}}$$

$$\text{where } \gamma 2 = \frac{\text{heterologous titer strain 1}}{\text{homologous titer strain 2}}$$

MAb 471 was further characterized by reactivity with cryptococcal polysaccharides in Ouchterlony diffusion and by agglutination of whole cryptococci. The immunodiffusion results (Fig. 1) showed that MAb 471 could precipitate polysaccharides of all four serotypes. Table 3 shows results from tube agglutination of whole cryptococci by MAb 471. These agglutination assays confirmed the broad specificity of MAb 471.

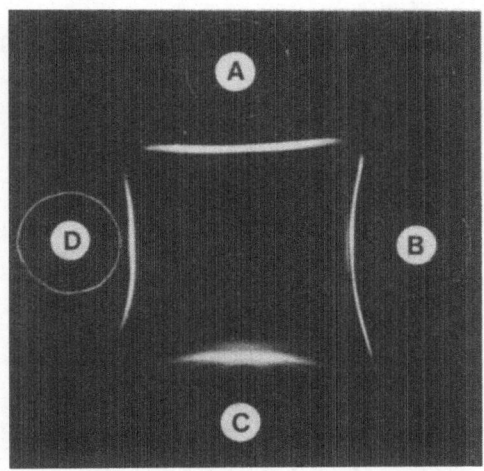

Figure 1.   Analysis of the reactivity of MAb 471 by double diffusion in agar. Cryptococcal polysaccharides of serotypes A, B, C, and D were placed in the outer wells, and MAb 471 was placed in the center well.

Table 3. Tube agglutination of cryptococci by MAb 471

Yeast serotype	Reciprocal agglutination titer
A	2016
B	800
C	8063
D	800

As a further test of the broad specificity of MAb 471, cells of twenty three isolates of known serotypes were examined for their reactivity in a slide agglutination assay with the monoclonal antibody. The results showed that each of six isolates of serotype A, each of six isolates of serotype B, each of five isolates of serotype C, and each of six isolates of serotype D were reactive with the monoclonal antibody.

Latex agglutination assay for cryptococcal polysaccharide based on a broadly reactive monoclonal antibody

The broad reactivity of MAb 471 suggested its utility as an antibody for coating latex beads for detection of cryptococcal polysaccharide. Serial two-fold dilutions of polysaccharides of each serotype were prepared, and the reactivities of the polysaccharides with antibody-coated latex beads were determined. The lowest concentration of polysaccharide producing a 1+ result was considered the detection limit of the assay. The results in Table 4 showed that latex beads coated with MAb 471 were reactive with polysaccharides of all four serotypes. The lowest concentrations of polysaccharide detected by the latex reagent ranged from 10 ng/ml for serotype A polysaccharide to 100 ng/ml for serotype C.

DISCUSSION

Previous studies of the serological relationship between the four serotypes of C. neoformans used whole yeast cells as an antigen. An objective of our study was to determine if we could demonstrate a similar relationship by use of purified cryptococcal GXM and antibody raised

Table 4. Sensitivity of latex agglutination assay for cryptococcal polysaccharides.

Polysaccharide	Detection limit
Serotype A	10 ng/ml
Serotype B	20 ng/ml
Serotype C	100 ng/ml
Serotype D	20 ng/ml

against cryptococcal polysaccharide. Cross-ELISA titers (Table 1) demonstrated a pattern of reactivity similar to the pattern observed when whole cell agglutination was used by Ikeda et al. (1982). Thus, it is likely that the distribution of epitopes on the purified polysaccharide and on the cryptococcal capsule are quite similar if not identical. Further, these results suggest that the galactoxylomannan and the mannoprotein are not major factors in determining serotype specificity.

The serological relatedness between serotypes of C. neoformans was analyzed in Table 2. A similar method of analysis was used to assess antigenic relatedness of T-mycoplasmas (Lin, Kendrick, and Kass, 1972; Lin and Kass, 1980) and Influenza viruses (Archeite and Horsfall, 1950). The results of this analysis largely parallel the known structural similarities between the serotypes of C. neoformans. That is, there is a close relationship between serotypes A and D, and between serotypes B and C. Further, serotypes A and D are only weakly cross-reactive with serotype C. The results suggest a closer relationship between Serotypes A and D and Serotype B than might be inferred from structural studies (Bhattacharjee, Kwon-Chung, and Glaudemans, 1981; Reiss, 1986). We do not know whether this relationship is real or whether it represents an artifact due to the manner in which cryptococcal polysaccharide is bound to ELISA plates.

It is evident from the cross-reactivity observed between crypto-coccal serotypes (Tables 1 and 2) that one or more epitopes are likely shared among all four serotypes. Ikeda et al. (1982) obtained this same result from agglutination and reciprocal adsorption experiments. These data suggested the possibility that a monoclonal antibody could be produced that is reactive with all four serotypes. As a consequence, we immunized mice with serotype A polysaccharide and screened colonies for antibodies that were reactive with polysaccharides of all four serotypes. One clone, designated MAb 471 was reactive with prototype strains of all four serotypes in agglutination assays, ELISA assays, and in Ouchterlony diffusion. In addition, MAb 471 produced a positive slide agglutination with each of six isolates of serotype A, each of six isolates of serotype B, each of five isolates of serotype C, and each of six isolates of serotype D. It should be noted, however, that MAb 471 did not display identical agglutination titers for prototype strains of each serotype (Table 3). The variation in titers may reflect differing affinities for the four serotypes. Alternatively, the various titers might be due to differing amounts of capsule on the prototype strains.

The latex agglutination assay for polysaccharide is one potential use for a monoclonal antibody reactive with cryptococcal polysaccharides. Ideally, such an antibody would be reactive with polysaccharides of all four serotypes. MAb 471 coupled to latex beads was able to detect nanogram amounts of cryptococcal polysaccharide (Table 4). It is possible that variations in the coupling or assay procedures may effect a further increase in sensitivity of the assay. Polysaccharides of all four serotypes were detected. The latex assay exhibited a similar sensitivity for polysaccharides A, B, and D, and a slightly reduced sensitivity for serotype C. These results confirm the potential value of monoclonal antibodies as reagents in the construction of latex agglutin-ation assays for cryptococcal polysaccharides. It is worth noting that a latex-coupled monoclonal assay has been developed for detection of group B streptococcal antigen in body fluids (Rench, Metzger, and Baker, 1984; Ruch and Smith, 1982). If a monoclonal antibody can be identified with an appropriate specificity, monoclonal antibodies offer distinct advant-ages over the hyperimmune rabbit antisera used in most commercially available assay kits. The advantages of monoclonal antibodies include elimination of problems due to lot variability in specificity and titer.

ACKNOWLEDGMENT

We thank K.J. Kwon-Chung for providing isolates of C. neoformans. We are also indebted to Judith Domer who suggested the use of sheep erythrocytes as carriers for immunization of mice. This study was supported in part by Public Health Service grant AI-14209 from the National Institutes of Health.

LITERATURE CITED

Bhattacharjee, A. K., Kwon-Chung, K. J., and Glaudemans, C. P. J., 1981, Capsular polysaccharides from a parent strain and from a possible mutant strain of Cryptococcus neoformans serotype A, Carb. Res., 95:237.

Cherniak, R., Reiss, E., Slodki, M. E., Plattner, R. D., and Blumer, S. O., 1980, Structure and antigenic activity of the capsular polysaccharide of Cryptococcus neoformans serotype A, Mol. Immunol., 17:1025.

Cherniak, R., Reiss, E., and Turner, S. H., 1982, A galactoxylomannan of Cryptococcus neoformans serotype A, Carb. Res., 103:239.

Ikeda, R., Shinoda, T., Fukazawa, Y., and Kaufman, L., 1982, Antigenic characterization of Cryptococcus neoformans serotypes and its application to serotyping of clinical isolates, J. Clin. Microbiol., 16:22.

Kaplan, W., Bragg, S. L., Crane, S., and Ahearn, D. G., 1981, Serotyping of Cryptococcus neoformans by immunofluorescence, J. Clin. Microbiol., 14:313.

Kozel. T. R., and Cazin, J., Jr., 1971, Non-encapsulated variant of Cryptococcus neoformans. I. Virulence studies and characterization of soluble polysaccharide, Infect. Immun., 3:287.

Kozel, T. R., and Cazin, J., Jr., 1974, Induction of humoral antibody response by soluble polysaccharide of Cryptococcus neoformans, Mycopathol. Mycol. Appl., 54:21.

Kozel, T. R., and Gotschlich, E. C., 1982. The capsule of Cryptococcus neoformans passively inhibits phagocytosis of the yeast by macrophages, J. Immunol., 129:1675.

Lane, R. D., Crissman, R. S., and Lachman, M. F., 1984, Comparison of polyethylene glycols as fusogens for producing lymphocyte-myeloma hybrids, J. Immunol. Methods, 72:71.

Lerner, E. A., 1981, How to make a hybridoma, Yale J. Biol. Med., 54:387.

Reiss, E., 1986, Molecular immunology of mycotic and actinomycotic infections, Elseiver, New York.

Rench, M. A., Metzger, T. G., and Baker, C. J., 1984, Detection of group B streptococcal antigen in body fluids by a latex-coupled monoclonal antibody assay, J. Clin. Microbiol., 20:852.

Ruch, F. E., and Smith, L., 1982, Monoclonal antibody to streptococcal group B carbohydrate: applications in latex agglutination and immunoprecipitation assays, J. Clin. Microbiol., 16:145.

Steinbuch, M., and Audran, R., 1969, The isolation of IgG from mammalian sera with the aid of caprylic acid, <u>Arch. Biochem. Biophys</u>., 134:279.

Voller, A., and Bidwell, D., 1986, Enzyme-linked immunosorbent assay, <u>in</u>: Manual of Clinical Laboratory Immunology, 3rd Ed., N.R. Rose, H. Friedman, and J.L. Fahey. eds. American Society for Microbiology.

Wilson, D. E., Bennett, J. E., and Bailey, J. W., 1968, Serologic grouping of <u>Cryptococcus</u> <u>neoformans</u>, <u>Proc. Soc. Exp. Biol. Med</u>., 127:820.

CYTOPLASMIC ANTIGENS OF CRYPTOCOCCUS NEOFORMANS

Guadalupe Reyes[1][*], Jean-Pierre Dandeu[2],
Edouard Drouhet[1] and Jean-Paul Latgé[1]

Institut Pasteur, Unité de Mycologie[1] et Unité
d'Allergoimmunologie[2], Paris

Introduction

Sanfelice (1895) first described <u>Cryptococcus neoformans</u>
(as <u>Saccharomyces neoformans</u>) from fruit juice, as a saprophytic, encapsulated, yeast-like fungus producing lymphatic abdominal masses mimicking tumors after intrabdominal inoculation
in guinea-pigs. Since this discovery, human and animal natural
infections were observed occasionally in the immunocompromized
host. After a "sleeping" period (Ajello, 1970), the mycosis
produced by this yeast became "the awakening giant" (Kaufman
and Blumer, 1978) and now it is considered as the most dramatic
agent among the opportunistic fungi occuring in immunodepressed
patients and particularly in AIDS subjects.

<u>Cryptococcus neoformans</u> is a yeast with a multilamellate
cell wall surrounded by a thick polysaccharidic capsule (Fig.
1). Drouhet et al., (1950) showed for the first time that the
capsular polysaccharide consists principally of a glucuronoxylomannan and is a virulence factor capable of inhibiting
leukocyte migration (Drouhet and Segretain, 1951). Numerous
studies of capsular antigens followed and led to the discovery
of 4 serotypes, A, B, C and D (Evans and Kessel, 1951, Bennett
and Hasenclever, 1965, Wilson et al., 1968, Bhattacharjee and
Bennett, 1984). These serotypes correspond to the two tele-
omorphs (<u>Filobasidiella neoformans</u> var.<u>neoformans,</u> <u>Filobasi-</u>
<u>diella neoformans</u> var.<u>bacillispora</u>) of <u>C.neoformans</u> (Kwon-
Chung, 1975, 1976). These basidiomycetous fungi reveal with
biochemical (Bennett et al., 1978), epidemiological (Kwon-Chung
and Hill, 1981, Kwon-Chung and Bennett, 1984) and pathological
differences. The quantity of capsular material produced in the
host may be so prolific that it can be found as free antigen in
the body fluids, and easily detected by a latex particle agglutination test (Bloomfield et al., 1963). These soluble antigens
are responsible for an altered cellular immunity with weak
production of antibodies, leading to immunologic paralysis
or tolerance (Bennett and Hasenclever, 1965, Fromtling and
Shadomy, 1982, Murphy and Cozad, 1972, Robinson et al., 1982).

---

[*] Present address : Department of Chemistry, Georgia State
University, Atlanta, Georgia 30303, USA

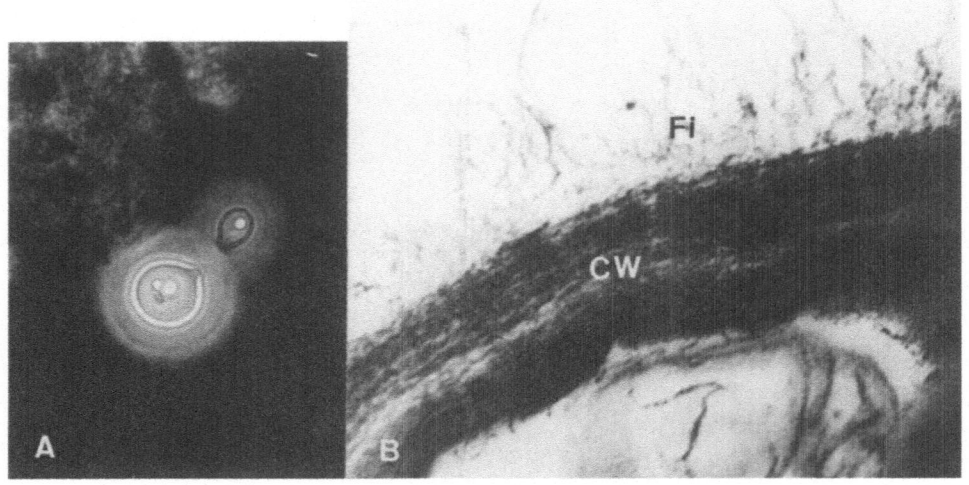

Fig. 1 <u>Cryptococcus neoformans</u>. A : Budding, encapsulated yeast in CSF of a immunodepressed patient with cryptococcosis. Indian ink preparation. Obj. x 100. B : Ultrastructure of cytoplasm (CYT) surrounded by the multilamellate cell (CW) wall and capsular fimbriae (Fi) x 108000.

In spite of numerous biochemical and immunological studies of the capsule as well as extracellular and cell wall antigens (Bhattacharjee et al., 1984, Cherniak et al., 1982, Evans and Kessel, 1951, Huppert, 1983, Reiss et al., 1985), few investigators have been concerned with cellular soluble antigens (Atkinson and Bennett, 1968, Graybill et al., 1982, Hay and Reiss, 1978, Jones et al., 1981). Most of these studies dealt only with the immunological properties of a few crude or refined antigens.

Atkinson and Bennett (1968) extracted, from undisrupted yeasts incubated in urea, an antigen capable of eliciting delayed hypersensitivity reactions in patients with cryptococcosis and in apparently healthy laboratory workers with known contact with <u>C.neoformans</u>. Comparatively few reactions occurred in normal subjects without exposure to the pathogen. This antigen contained over 90 % protein.

Four subcellular fractions of <u>C.neoformans</u> prepared by differential centrifugation of disrupted whole yeasts were examined by Hay and Reiss (1978) for their ability to elicit delayed-type hypersensitivity in sensitized animals. The chemical composition of three subcellular fractions from the "postmitochondrial supernatant" had relatively high concentrations of carbohydrate (carbohydrate : protein, 2 : 1), whereas the extracellular antigen from the culture filtrate was highly concentrated in carbohydrate. This is probably due to the capsular polysaccharide released into the medium. Studies by gel permeation chromatography on Sephadex G-100 of 100,000 xg supernatant fractions showed that only one fraction contained demonstrable activity in gel immunodiffusion against a rabbit antiserum to whole C.neoformans cells injected by intravenous (i.v.) route. This fraction which was capable of eliciting an early 4 h mouse footpad swelling in immunized animals, produced 2 or 3 lines of precipitation and one of these was identical to one or 2 lines produced with cryptococcal polysaccharide. These antigens are also active in the in vitro macrophage migration inhibition test.

Jones et al. (1981) showed that differential centrifuga-tion of a homogenate from a mechanically disrupted, acapsular isolate of C.neoformans resulted in a 105.000 xg supernatant (105 K) and a microsomal fraction (MS), containing membrane and ribosomes both of which were capable of eliciting specific delayed cutaneous hypersensitivity and lymphocyte transforma-tion in infected guinea pigs. Polyacrylamide gel electrophore-sis revealed two major proteins in the MS and seven proteins in the 105 K fractions. In vitro lymphocyte transformation eli-cited by 1 to 10 μg/ml of antigens was maximal after 4 days of incubation. The reacting cell populations were peripheral blood leukocytes and peritoneal exudate cells. Spleen cells of infec-ted animals were unresponsive to in vitro antigenic stimula-tion.

In further studies by Graybill et al. (1982), C.neoformans was mechanically disrupted cells and culture supernatant ex-tracts were fractionated by ion exchange and gel filtration column chromatography. Four glycoprotein cytoplasmic fractions contained between 1.88-11.0 mg protein per ml and 6-17.5 mg of carbohydrate per ml depending on the fraction isolated. They corresponded to the fractions 16-24 obtained by Hay and Reiss (1981) using Sephadex G100 gel filtration. While these authors did not obtain satisfactory delayed type hypersensitivity, Graybill et al. (1982) found both in vitro and in vivo antigen specific reactivity to be concentrated in fraction I. The authors were uncertain whether the capsular polysaccharide from the culture filtrate contributed to delayed type hypersensiti-vity reactivity or any other component of cell mediated immu-nity.

The molecular nature of immunologically active prepara-tions other than the capsular polysaccharide has not been studied extensively according to Huppert (1983). All the sub-cellular fractions prepared by Hay and Reiss (1978) contained both carbohydrate and protein. The 100,000 xg supernatant pro-duced two or three precipitation bands in ID tests and one of these was identical to one of the two formed by cryptococcal polysaccharide. It seems logical, therefore, that non-encapsu-lated cells may be a better source of subcellular fractions. Such preparations from earlier studies have been shown to contain multiple antigens and their molecular characterization is uncertain. Isolation and purification of immunologicaly active cytoplasmic antigens, has been largely ignored in con-trast to the numerous studies on extracellular antigens. For that reason, we have performed an immunochemical analysis of an aqueous cellular extract from disrupted yeast cells, utilizing crossed immunoelectrophoresis, crossed-line immunoelectrophoresis, gel filtration and affinity chromatography. From these data, fundamental and practical implications are considered.

Preparation of cytoplasmic antigens and immune sera

Table 1 summarizes the preparation conditions of the antigens studied. These include yeast cytoplasmic extract (YCE) from broken cells, yeast protoplast extract (YPE) from cells treated by cell wall lytic enzymes according to Bastide et al. (1979) and Rhodes and Kwon-Chung (1985), culture filtrate antigen (CF Ag), capsular polysaccharide (CPS) and circulating soluble antigens from biological fluids (CSF, serum, urines) from patients with cryptococcosis associated with AIDS. The best method for the extraction of antigens from YCE was disruption in distilled water at pH 5.6 when compared to results obtained with Tris-HCl buffer pH 8.6, glycine-HCl buffer (pH 2.5), phosphate buffer (pH 7.4) or borate buffer (pH 10.6). Antigens of the water-soluble extract were best resolved when examined by iso-electrofocusing in polyacrylamide gels between pH 3 and 10, followed by immunoblotting using rabbit antisera against YCE and whole yeast cell (WYC) as well as by the latex particle agglutination test specific immunogenic polysaccharide (Bloomfield et al., 1963). This latter test can detect 1 to 25 nanograms of cryptococcal polysaccharide depanding on the purity of the preparations.

TABLE 1 - ANTIGENS STUDIED

Somatic, intracellular, cytoplasmic Antigens	
Yeast cytoplasmic extract (YCE)	**Yeast protoplast extract (YPE)**
• Y cells, stable clone, serotype A	• Y cells cultured on YNB (Difco) + glucose
• 72h cultures at 28°C, grown in Sabouraud glucose broth	• enzyme treatment (Novozym 234, Novoindustry, Denmark) protoplasts obtained by 4h digestion
• cell disruption in Braun MSK homogenizer	• washing, lysis, lyophilization
• Supernatant from centrifuged broken cells (35,000xg) in d.water pH 5.6	
• dialysis, lyophilization	

Metabolic, extracellular Antigens	
Culture fitrate antigen CFAg	**Circulating soluble Antigens**
• Y cells cultured as for somatic antigens	C S F from patients with
• supernatant from centrifuged whole cells	serum Ags* cryptococcosis
• dialysis, lyophilization	urine associated with AIDS
Capsular polysaccharide CPS	
• supernatant of same Y cells culture filtrate	* high titers of polysaccharide determined by latex
• 95 % ethanol precipitation (2.5 vol.),deproteinization, using classical purification (Drouhet et al., 1950)	particle agglutination test

In Table 2, we have summarized the data on the preparation of rabbit antisera against 1) the whole yeast cells using the intravenous route, and 2) the yeast cytoplasmic extract by multiple injection technique (subcutaneous, intramuscular, intraperitoneal, intradermal and intravenous) with Freund adjuvant leading to the production of numerous precipitating antibodies. Immunoglobulins from hyperimmunsera were prepared by precipitation with 18 % sodium sulfate solution.

TABLE 2 - SERA STUDIED

Immunized animals : New Zealand white rabbits (2.5 kg weight)

Serum Anti-whole yeast cells (WYC)	Serum Anti-yeast cytoplasmic extract (YCE)
. I.V. inoculum : 1ml 10^8 cells/ml in PBS	. 50 mg YCE in 5 ml PBS + 5 ml complete Freund adjuvant in multiple local deposit : subcutaneously intraperitoneally
. formalin (2.5 %) killed cells	
. 1 ml/day x 5 days, 5 days rest	. after 3 weeks, 25 mg YCE intradermal + intraperitoneal injection
. immunization repeated each 10-15 days until titer $\geq$ 1:640	. after 2 weeks, 10 mg YCE in 1 ml PBS intravenously once a week for 3 weeks

---

Localisation of cellular antigens studied by immunofluorescence and immunocytochemistry

The two hyperimmune experimental sera prepared were utilized for immunological localization of antigen in the Cryptococcus neoformans cells cultivated in vitro and in tissues.

a) Immunofluorescence

The indirect immunofluorescence technique was used for examination of histological sections of brain and lungs of infected mice with C.neoformans. Sections were successively incubated with rabbit antibodies and goat anti-rabbit IgG fluorescein isothiocyanate (FITC) conjugate. The anti-whole cell antiserum strongly reacted with the outlayer of C.neoformans polysaccharidic capsule. On the other hand the anti-YCE antiserum did not label the capsule, but stained the cytoplasm of the cells (Fig. 2 A and B).

These results demonstrated that the antibodies from the anti-WYC serum are directed against the capsular antigens while the antibodies of the anti-YEC were mainly directed against the cytoplasmic antigens.

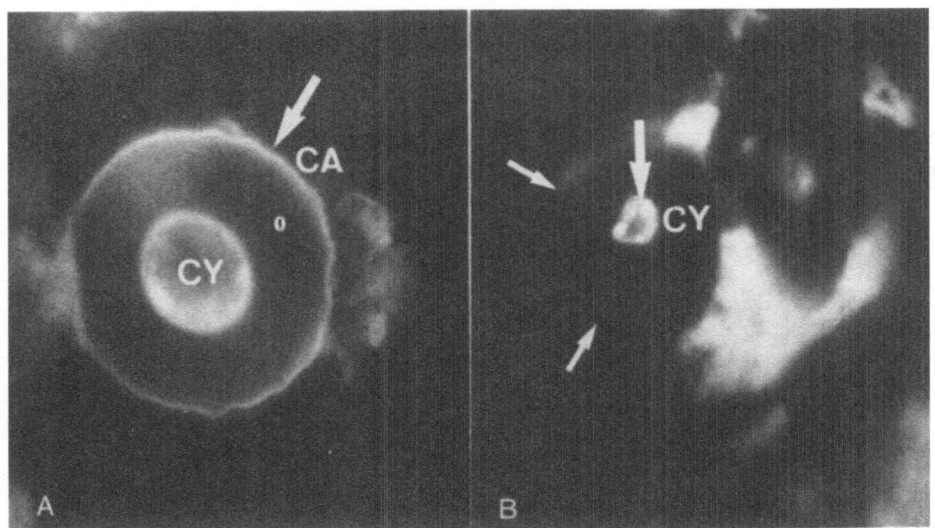

Fig. 2   Brain sections of Balb/c mice inoculated intravenously with 2 X 10^6 C.neoformans cells 15 days before sacrifice.   A : Section treated by a rabbit antiserum against whole yeast cells (WYC).   Dilution 1/25.   B : Same section treated with antiserum against yeast cytoplasmic extract (YCE).   Dilution 1/100.   Both sections incubated with a 1/100 dilution of a goat antirabbit IgG/FITC conjugate (Biosys, Paris). Contrast staining in Evans blue (1/10,000 in PBS) for both figures.   X 100.   In A, the outline of the capsule (CA) shows a strong green brilliant immunofluorescence while the cytoplasm (CY) is orange red. In B the capsule is not stained but the cytoplasm shows an inten-sively green immunofluorescence. The tissue cells are red-orange.

b) Immunocytological study of C.neoformans

Electron immunocytochemical studies of these sections re-vealed that anti WYC labeled mainly the polysaccharidic capsule (Fig. 3). However, some intracellular antigens were also recognized by this antiserum. Anti-YCE essentially bound to the intracellular compartments of the cells but some sparse labe-ling was also observed on the capsule.

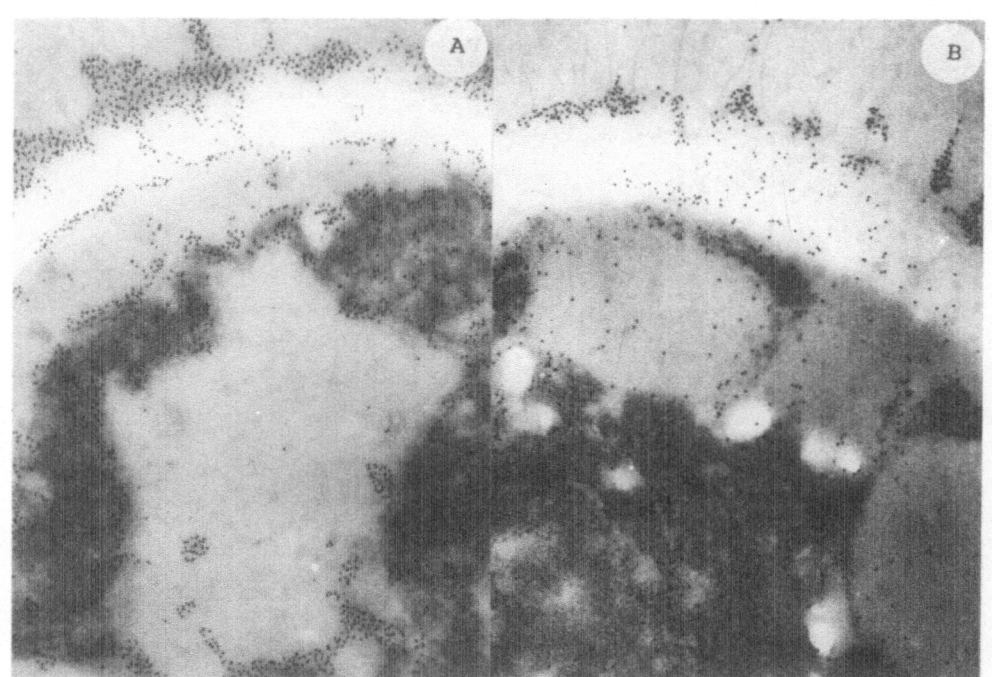

Fig.3.A,B Thin sections of yeasts fixed in 2.5 % formaldehyde + 0.1 % glutaraldehyde embedded in Lowicryl K4M incubated in anti-WYC (Fig. 3A) and anti-YCE (Fig. 3B) antisera. Antigen-antibody binding sites were visualized using a protein A/5 nm colloidal gold complex as previously described (Latgé et al., 1986).

Immunochemical analysis of the yeast cytoplasmic extract

Fourty-two proteins were detected in YCE after isoelectro-focusing in polyacrylamide gels between pH 3 and 10, (Fig. 4). The majority of bands (31 bands) had an isoelectric point between pH 4 and 7.

Twenty-one precipitin lines (Fig. 5) were revealed in the yeast cytoplasmic extract of C.neoformans analysed by cross immunoelectrophoresis using the anti-YCE antiserum. One precipitin line was observed when the same extract was analyzed with anti-whole yeast cells IgG (Fig. 6a). Crossed-line immunoelec-trophoresis performed with the capsular polysaccharide in the intermediate gel, showed that the precipitin line detected by anti-whole yeast cell IgG was a constituent of the capsular polysaccharide, since a precipitin line of identity with inter-mediate gel was formed (Fig. 6b). This antigen was not revealed by the rabbit antisera directed against the yeast cytoplasmic extract.

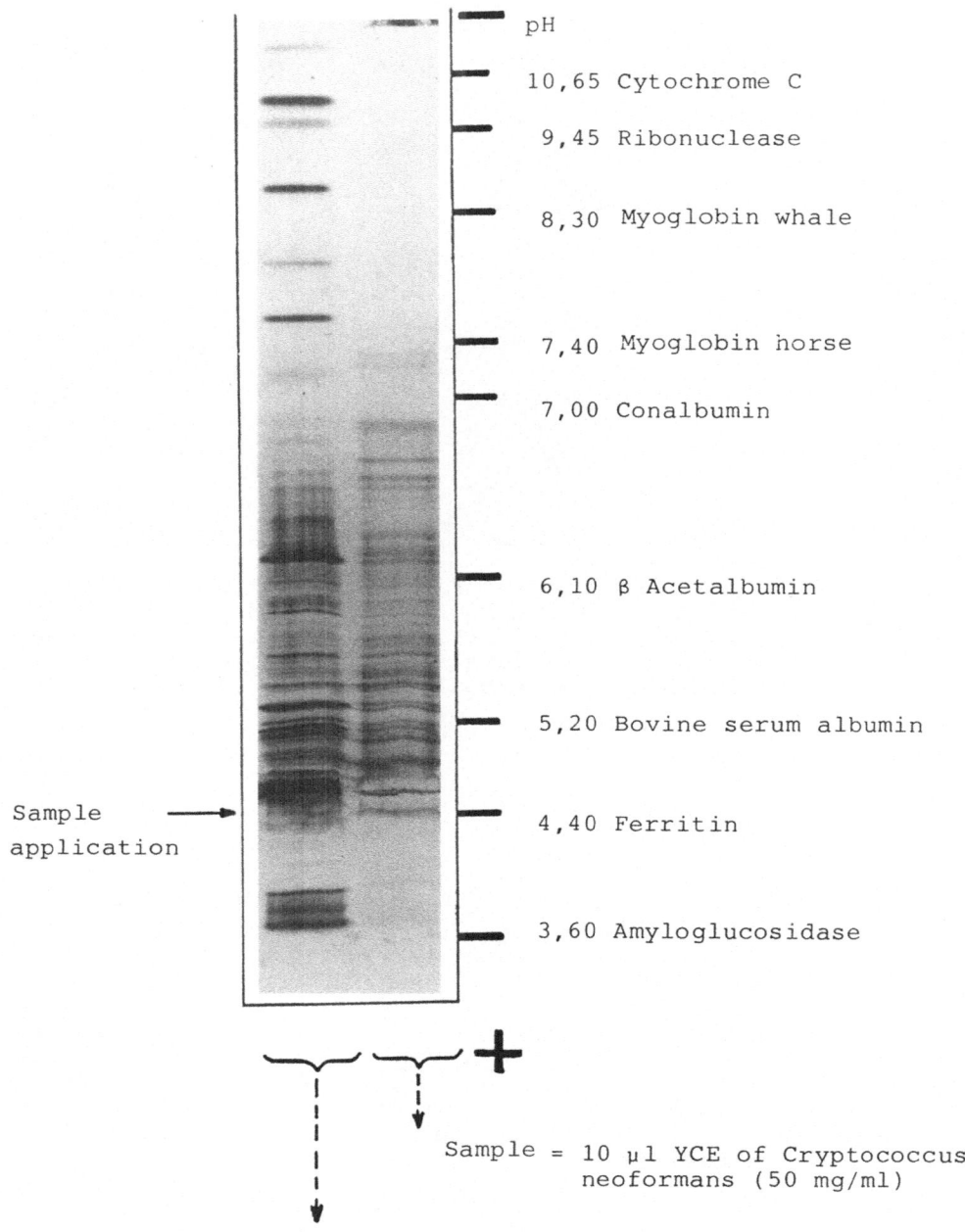

pH

10,65 Cytochrome C

9,45 Ribonuclease

8,30 Myoglobin whale

7,40 Myoglobin horse

7,00 Conalbumin

6,10 β Acetalbumin

5,20 Bovine serum albumin

Sample application →

4,40 Ferritin

3,60 Amyloglucosidase

Sample = 10 µl YCE of Cryptococcus neoformans (50 mg/ml)

Protein test mixture of YCE in PAG. Blue Commassie staining

Fig. 4 Isoelectrofocalisation of the yeast cytoplasmic extract (YCE) of polyacrylamide gel Coomassie blue stain

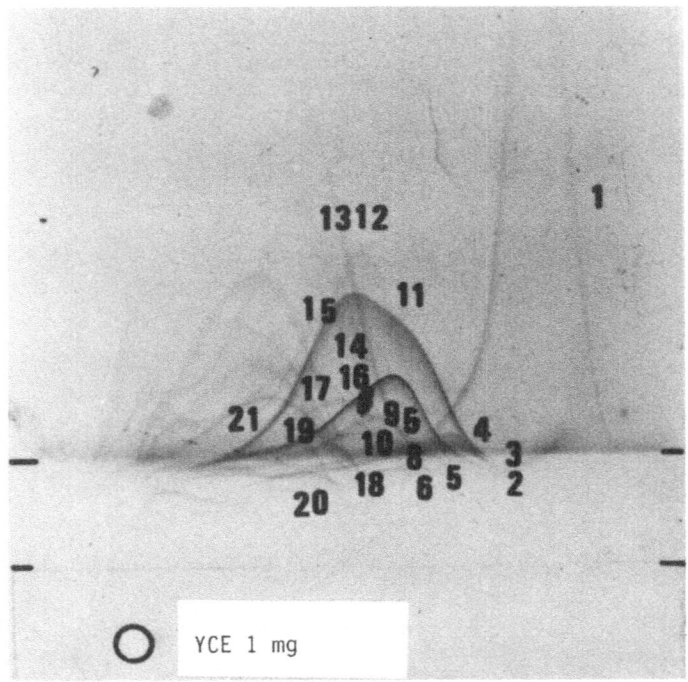

Fig. 5 Crossed immunoelectrophoresis (CIE) gel of yeast cellular extract (YCE ; 1 mg) antigen against rabbit sera anti-YCE. Upper gel contains 20 ul of serum/cm² in Veronal buffer (pH 8.6).

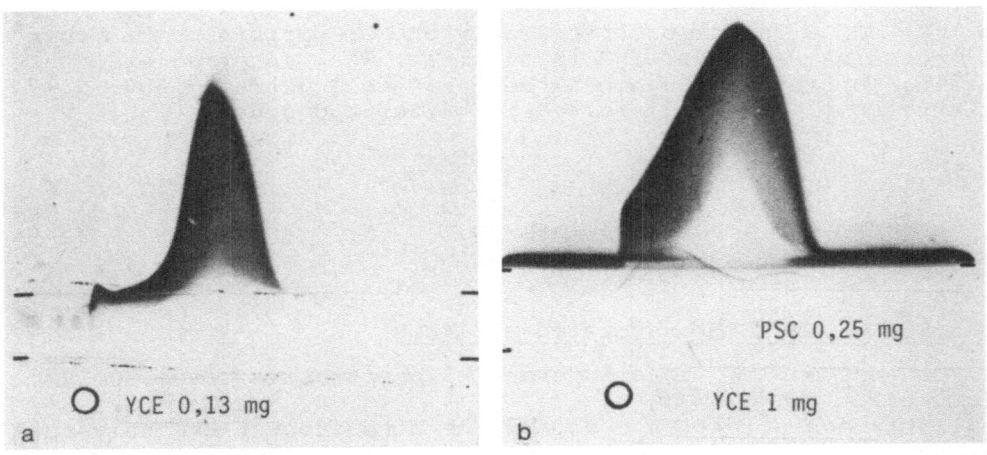

Fig. 6 (a) CIE gel of YCE against anti-whole yeast cell (WYC) IgG. Upper gel contains IgG anti-WYC (2 ul/cm²). Antigen YCE (0.13 mg) dissolved in buffer (pH 8.6) (b) Crossed-line immuno-electrophoresis (CLIE) gel demonstrating one precipitin peak which cross reacts with the capsular polysaccharide antigen. Upper gel contains anti-WYC IgG (2 ul/cm²). The intermediate gel contains the capsular polysaccharide (0.25 mg). The antigen, yeast cellular extract (1 mg), is dissolved in buffer pH (8.6).

The yeast cytoplasmic extract contains only one antigen that is similar or has epitopes in common with the capsular polysaccharide (Fig. 6a). This antigen was not precipitated by YCE antibodies present in the intermediate gel, while all the other antigens were totally precipitated (Fig. 7).

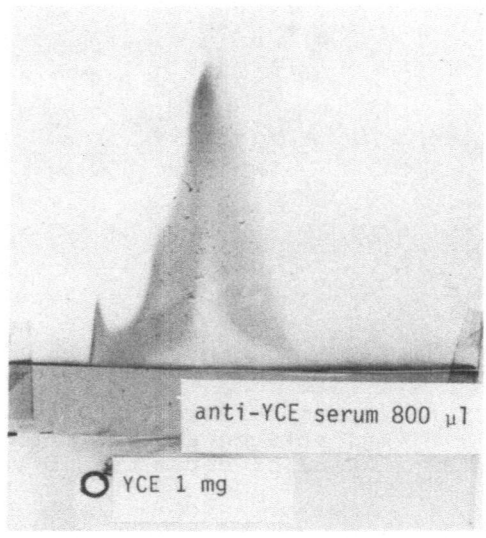

Fig. 7 - CLIE gel demonstrates only one precipitin peak. Upper gel contains anti-WYC IgG (2 ul/cm²). The intermediate gel contains antibodies against the yeast cell extract (800 ul). YCE antigen (1 mg) is dissolved in buffer (pH 8.6).

Isolation of the CPS associated antigens

YCE was applied to a Superose 12 preparative grade column (Fig. 8).
The latex agglutination test for the detection of the polysaccharide antigen of C.neoformans was used to determine which fraction contained antigens that reacted with WYC antibodies, i.e. anti-capsular polysaccharide antibodies. One fraction Fs, which appeared polydispersed by gel filtration, was very active.
The chromatography of fraction Fs was performed on Superose 12 to purify this antigen (Fig. 9). Two fractions, Fs1 (2×10^6d, 10^4d) and Fs2 (6×10^4d), were positive by the latex agglutination test (titers $1/10^3$ and $1/10^4$ respectively) while the remaining fractions Fs3-Fs8, showed no activity (Fig. 9).

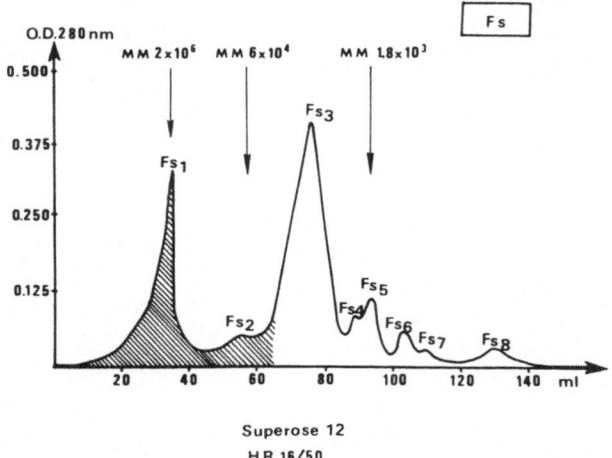

Fig. 8   Elution profile of YCE (100 mg of lyophilized YCE) was dissolved in 0.02 M Tris-HCl pH 8.6 buffer containing 0.5 M NaCl. A 2000 μl sample was applied to a Superose 12 Preparative grade column HR 16/50 and monitored at 280 nm. Elution was carried out at a flow rate of 1 ml/min., chart speed 0.25 cm/min with the same buffer.

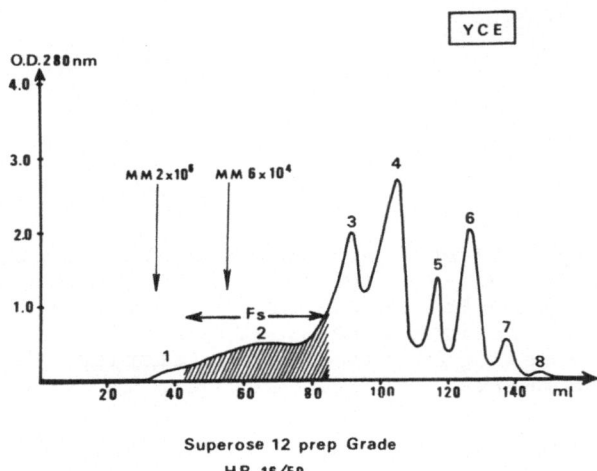

Fig. 9   Elution profile of 500 μl of Fs fractions of YCE from Superose 12 (Pharmacia) column. Sample fractions were monitored at 280 nm. Molecular weight standards were eluted separately and used to estimate molecular weight of fractions (Fs1-Fs8).

Fractions Fs1 (Fig. 10a) and Fs2 (Fig. 10b) reacted with the anti-WYC IgG, but not with the anti-YCE using the CIE technique.

Fs2 was further purified by passage through an immunosorbant column prepared by coupling anti-YCE serum raised in rabbits to CNBr activated Sepharose 4B. The result is that Fs2 appeared to contain a pure antigen specifically reacting with IgG raised against WYC (Fig. 10b).

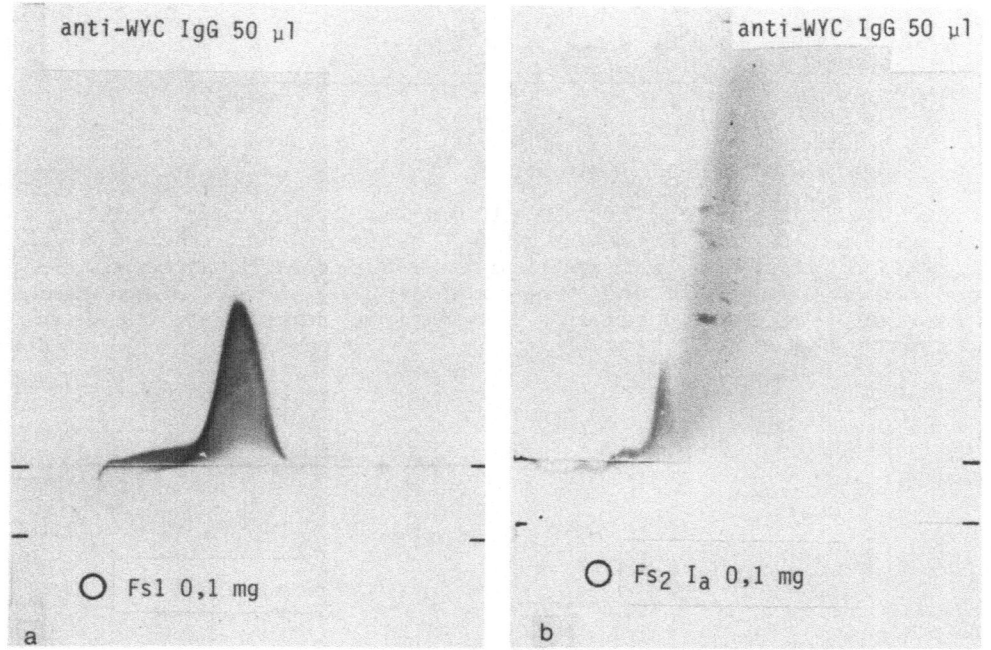

Fig. 10 (a) : CIE gel demonstrating one precipitin peak. Upper gel contains anti-WYC IgG (1 µl/cm²). Anodal well contains antigen fraction Fs1 in buffer pH 8.6 (0.1 µg). (b) : CIE gel revealing one precipitin peak. Upper gel anti-WYC IgG (1 µl/cm²). Anodal well contains antigen fraction Fs2 purified with Sepharose 4B CNBr.

Rather than being a cytoplasmic component of C.neoformans, the antigen present in Fs2 could be derived by partial hydrolysis of the components of Fs1, a high molecular weight of the yeast cell wall. We attempted to isolate a similar fraction from the yeast protoplast extract (YPE).

Cytoplasmic antigens from C.neoformans protoplasts

The following experiments were performed using protoplasts. An aqueous extract was prepared following the method defined for the whole yeast cells. The extract was fractionnated by exclusion chromatography on Superose 12, (Fig. 11).

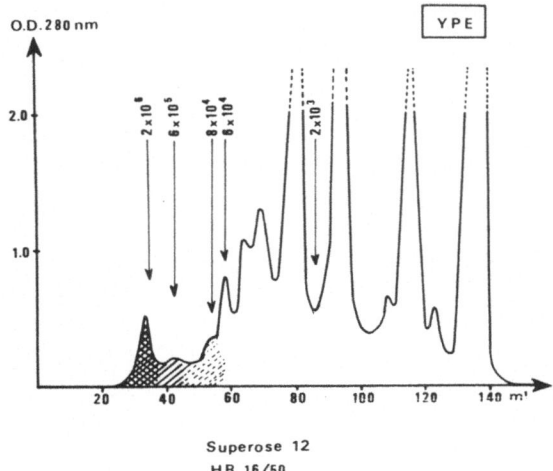

Superose 12
HR 16/50

Fig. 11 Chromatographic elution profile of yeast protoplast extract (YPE). 500 μl of YPE containing 25 mg were deposited on the Superose 12 column HR 16/50 monitored at 280 nm, equilibrated with 0.02 M Tris-HCl buffer (pH 8.6) containing 0.5 M NaCl. Elution was performed at a flow rate of 1 ml/min, and chart speed of 0.5 cm/min.

Only the first two fractions PFs1 and PFs2 reacted with the specific anti-WYC or anti-CPS antibodies. Respective titers in the agglutination test were $1/10^5$ and $1/10^3$. Since PFs2 has an apparent molecular weight of 6.10^4d, it can be assumed to be similar to the Fs2 fraction isolated from the YCE. Nevertheless it appears as a minor constituent of the protoplast extract (PE). Fig. 12, shows the pattern obtained in CIE when YPE reacted with rabbit serum anti-YCE. YPE contained several common antigens but it also contained an antigen sharing common epitopes with CPS.

C.neoformans circulating antigen

Application of the classical latex particle agglutination test indicated that sera from patients with cryptococcosis contain an antigen that shares common epitopes with the capsular polysaccharide. IgG antibodies raised in rabbits against the whole yeast cells (i.e. anti-capsular polysaccharide of the cell wall) showed that this circulating antigen shared common epitopes with the Fs2 fraction (Fig. 13a). Anti-whole yeast cells IgG reveals only one antigen in the serum KAS (Fig. 13b) of an AIDS patient with cryptococcosis. This antigen represents a reaction of total identity with the capsular polysaccharide antigen as well as with the Fs2 fraction (Fig. 13c). The same results were obtained with sera, CSF and urine of two other patients with cryptococcosis associated with AIDS.

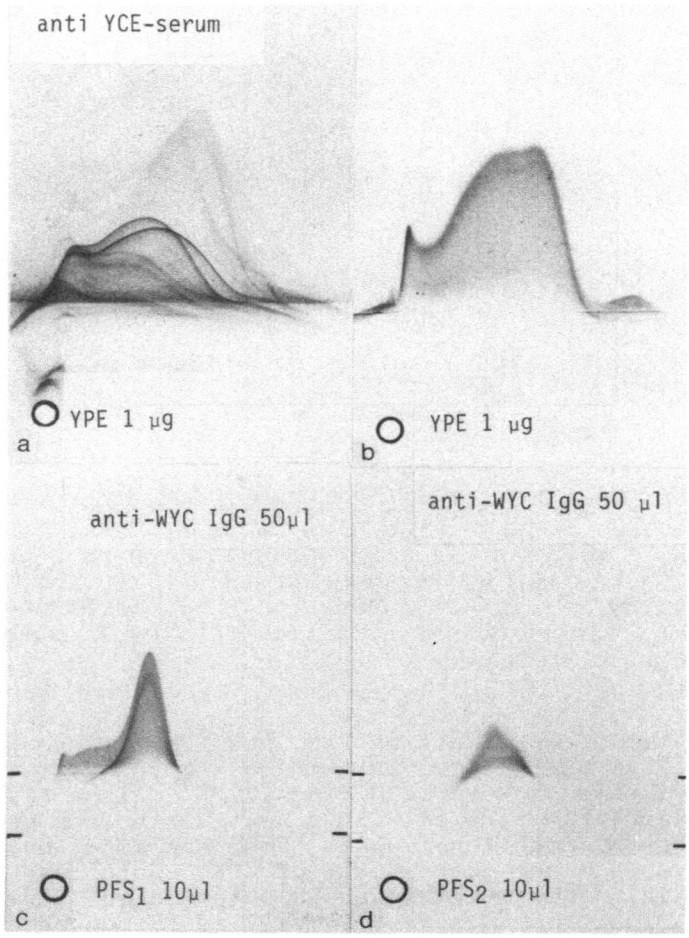

Fig. 12 (a) CIE gel of YPE against anti-YCE serum. Upper gel anti-yeast cell extract serum above anodal well which contains the antigen (YPE) at the concentration of 1 ug. (b) CIE gel revealing one precipitin peak. Upper gel contains anti-WYC IgG (2 ul/cm²), above anodal well which contains the YPE (1 ug). (c) CIE gel reveals one precipitin peak. Upper gel serum anti-whole yeast cells, above the anodal wells which contain the fractions PFs1 (10 μl) and (d) PFs2 (10 μl).

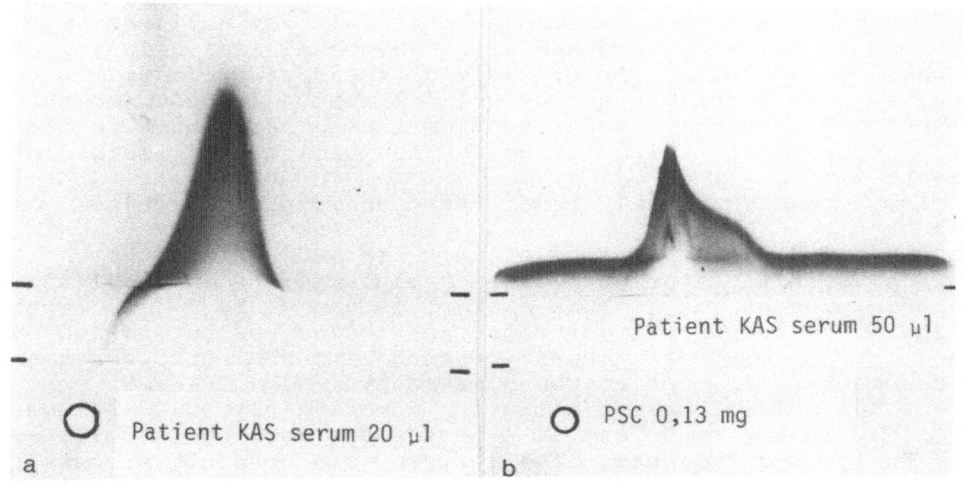

Patient KAS serum 50 µl

Patient KAS serum 20 µl

PSC 0,13 mg

a

b

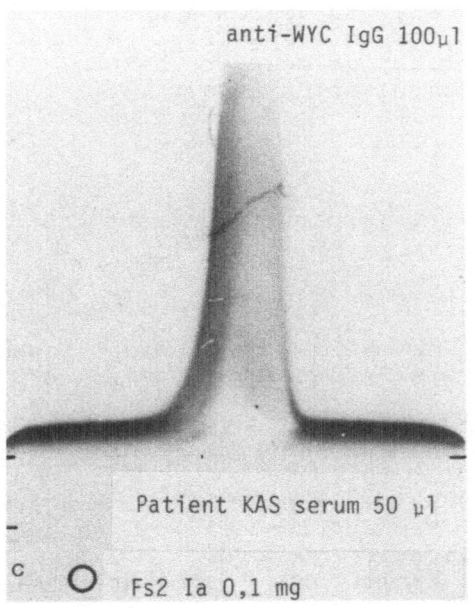

anti-WYC IgG 100µl

Patient KAS serum 50 µl

c   Fs2 Ia 0,1 mg

Fig. 13   (a) CIE gel of a cryptococcosis patient (KAS) serum
against anti-WYC serum.   KAS serum (10 µl) in the anodal well.
Upper gel contains anti-WYC IgG (2 µl/cm²).   (b) CLIE gel demonstrating identity between CPS and circulating Ag from serum
of a patient with cryptococcosis. Upper gel contains anti-WYC
IgG (2 µl/cm²). Intermediate gel contains pateint KAS serum (50
µl).   Anodal well contains CPS (0.13 mg).   (c) CLIE gel demonstrating identity between fraction F2Ia (purified on immunosorbant column) and circulating Ag from the same patient serum.
Upper gel contains anti-WYC IgG (2µl/cm²). Anodal well contains
Fs2 Ia (0.1 mg).   Intermediate gel contains pateint KAS serum
(50 µl).

Conclusions

Isolation, purification and characterization of two different preparations of C.neoformans intracellular antigens were performed on extracts obtained either by mechanical grinding of the whole yeast cells, or by lysis of their protoplasts. Both extracts were fractionnated by gel filtration in HPLC and then examined. The antigens obtained from these fractions were compared to antigens circulating in the biological fluids of patients with cryptococcosis.

Immunochemical analysis of these antigenic fractions as shown by the latex agglutination test was performed using rabbit antibodies with strong affinity raised against the whole yeast cells, anti-capsular polysaccharide IgG, or intracellular antigens present in a crude extract of the whole cells (anti YCE-IgG). The specificity of these antibodies was determined by immunological methods such as crossed immunoelectrophoresis, fluorescent and immunoelectron microscopy.

A low molecular mass antigen, aproximately 5×10^4 d was identified in the cytoplasm of C.neoformans from whole cells as well as from protoplasts. The antigen is a possible precursor of the capsular polysaccharide. Presence of a similar antigen was identified in CFS, the serum and urine of AIDS patients with cryptococcosis. Such a specific and purified antigen and its corresponding antibodies could be helpful for diagnosis of cryptococcosis and immunization trials.

Acknowledgement

This sudy was supported in part by a grant INSERM-CNAMTS 1984-1987, CSSN° 3.

References

Ajello, L. 1970. The medical mycological iceberg, p 3-12. In Proc : Xth Intern.Symposium Mycoses, PAHO Scient.Publ., Washington D.C.

Atkinson, A.J., and J.E. Bennett. 1968. Experience with a new skin test antigen prepared from Cryptococcus neoformans. Amer.Rev.Resp.Dis., 97:637.

Axelsen, N.H. 1983. Handbook of immunoprecipitation in gel techniques. Scand.J.Immunol., 17 (suppl. 10) Blackwell Scientific Publication, Oxford.

Bastide, M., E.H. Hadibi, D. Scheiber, M Miegeville, C. Vermeil, et J.M. Bastide. 1979. Modalités de libération des protoplastes de Saccharomyces cerevisiae : étude en microscopie électronique à balayage. Ann.Microbiol. (Inst.Pasteur). 130A:419.

Bennett, J.E., and H.F. Hasenclever. 1965. Cryptococcus neoformans polysaccharide : studies of serologic properties and role in infection. J.Immunol. 94:916.

Bennett, J.E., K.J. Kwon-Chung, and T.S. Theodore. 1978. Biochemical differences between serotypes of Cryptococcus neoformans. Sabouraudia 16:167

Bhattacharjee, A.K., and J.E. Bennett. 1984. Capsular polysaccharides of Cryptococcus neoformans. Rev.Infect.Dis. 5:619.

Bloomfield, N., M.A. Gordon, and D.F. Elmendorf. 1963. Detection of Cryptococcus neoformans antigen in body fluids by latex particle agglutination. Proc. Soc. Exp. Biol. Med. 114:64.

Cherniak, R., E. Reiss, and S.H. Turner. 1982. A galactoxylomannan antigen of Cryptococcus neoformans serotype A. Carbohyd.Res. 103:239.

Dandeu, J.P., M. Colin, J. Rabillon, and B. David. 1985. Immunochemical study of bee hemolymph : I. Comparative study of two gel filtration experiments on superose 12 and superose 12 prep grade using FPLC system, p 36. In : Abst. 5th Intern.Symposium HPLC proteins, peptides, polynucleotides, Toronto, Ontario, 4-6 november.

Dandeu, J.P., J. Le Mao, M. Lux, J. Rabillon, and B. David. 1982. Antigens and allergens in Dermatophagoides farinae mite. II. Purification of AG11, a major allergen in Dermatophagoides farinae. Immunology. 46:679.

Drouhet, E., G. Segretain, et J.P. Aubert. 1950. Polyoside capsulaire d'un champignon pathogène Torulopsis neoformans. Relation avec la virulence. Ann.Inst.Pasteur (Paris). 79 A:891.

Drouhet, E., et G. Segretain. 1951. Inhibition de la migration leucocytaire in vitro par un polyoside capsulaire de Torulopsis neoformans. Ann.Inst.Pasteur (Paris), 81 B:674.

Evans, E.E., and J.F. Kessel. 1951. The antigenic composition of Cryptococcus neoformans. II. Serologic studies with the capsular polysaccharide. J.Immunol. 67:109.

Fromtling, R.A. and Shadomy H.J. - Immunity in cryptococcosis: an overview. Mycopathologia. 77:183.

Graybill, J.R., D.C. Straus, T.J. Nealon, M. Hague, and R.E. Paque. 1978. Immunogenic fractions of Cryptococcus neoformans. Mycopathologia. 1982, 78:31.

Hay, R.J., and E. Reiss. 1978. Delayed type hypersensivity responses in infected mice elicited by cytoplasmic fractions of Cryptococcus neoformans. Infect. Immunity. 22:72.

Huppert, M. 1983. Antigens used for measuring immunological reactivity, p. 219-302. In:DH. Howard (ed), Fungi pathogenic for humans and animals, Part B, Pathogenicity and detection : I. Marcel Dekker, Inc, New York.

Ikeda, R., T. Shinoda, Y. Fukazawa, and L. Kaufman. 1982. Antigenic characterization of Cryptococcus neoformans serotypes and its application to serotyping of clinical isolates. J.Clin.Microbiol. 16:22.

Jones, A.E., E. Reiss, and T.J. Spira. 1981. A microsomal fraction of Cryptococcus neoformans induces lymphocytes blastogenesis in infected guinea pigs. Mycopathologia. 75:129.

Kaufman, L., and S. Blumer. 1978. Cryptococcosis : the awakening giant, p 176-182. In : Proc.IVth Intern.Conf.Mycoses. PAHO Scient.Publ. 356, Washington D.C.

Kozel, T.R., and E.C. Gotschlich. 1982. The capsule of Cryptococcus neoformans passively inhibits phagocytosis of the yeast by macrophages. J.Immunol. 129:1675.

Kwon-Chung, K.J. 1975. A new genus Filobasidiella, the perfect state of Cryptococcus neoformans. Mycologia. 67: 1197.

Kwon-Chung, K.J. 1976. A new species of Filobasidiella, the perfect state of Cryptococcus neoformans B and C sero-types. Mycologia. 68:942.

Kwon-Chung, K.J., and J.E. Bennett. 1984. Epidemiologic diffe-rences between the two varieties of Cryptococcus neo-formans. Amer. J. Epidemiol. 120:123.

Kwon-Chung, K.J., and W.B. Hill. 1981. Sexuality and patho-genicity of Filobasidiella neoformans (Cryptococcus neo-formans), p 243-250. In : R. Vanbreuseghem and C. De Vroey (eds), Sexuality and pathogenicity of fungi. Masson, Paris.

Latgé, J.P., T.C., Garry, M. Horisberger and M.R., Prévost. 1986. Ultrastructure and chemical composition of the ballistospore wall of Conidiobolus obscurus. Experimental Mycology, 10:99.

Le Mao, J., J.P. Dandeu, J. Rabillon, M. Lux M., and B. David. 1981. Antigens and allergens in Dermatophagoides farinae mite. I. Immunochemical and physicochemical study of two allergenic fractions from a partially-purified Dermato-phagoides farinae mite extract. Immunology. 44:239.

Murphy, J.W., and L. Cozad. 1972. Immunological unresponsi-veness induced by cryptococcal capsular polysaccharide as-sayed by hemolytic plaque technique. Infect.Immunol. 5:896.

Reiss, E., M. Huppert, and R. Cherniak. 1985. Characterization of protein and mannan polysaccharide antigens of yeasts, moulds and actinomycetes, vol. I, pp. 172-207. in: M.R. McGinnis (ed.), Current topics in medical mycology. Springer Verlag, Berlin.

Rhodes, J.C., and K.J. Kwon-Chung. 1985. Production and regene-ration of protoplasts from Cryptococcus. Sabouraudia J.Med.Vet.Mycol. 23:77.

Robinson, B.E., N.K. Hall, G.S. Bulmer, and R. Blackstock R. 1982. Suppression of responses to cryptococcal antigen in murine cryptococcosis. Mycopathologia. 80:157.

Stockman, L., and G.D. Roberts. 1982. Specificity of the latex test for cryptococcal antigen : a rapid, simple method for eliminating interference factors. J.Clin. Microbiol., 1982, 16, 965.

CHARACTERIZATION, SPECIFICITY AND PROTECTIVE EFFECT OF A MURINE MONOCLONAL ANTIBODY REACTIVE WITH CRYPTOCOCCUS NEOFORMANS (CN) CAPSULAR POLYSACCHARIDE (CNPS)

F. Dromer, J. Salamero, J. Charreire, A. Contrepois, C. Carbon and P. Yeni

Laboratoire de Infections Experimentales, Faculte Xavier Bichat and INSERM U 238, Hopital Cochin Paris, France

Polyclonal anti-CN antibodies are used for the serotyping of CN clinical isolates and the diagnosis of cryptococcosis. They contribute to the characterization of CN capsular components. Many in vitro studies suggest that anti-CN antibodies, essentially IgG, could, if present, participate in yeast killing in vivo (T.R. Kozel, J.L. Follette, 1981, Infect. Immun., 31:978). However, previous attempts to protect mice against experimental cryptococcosis by passively transferred immune sera led to rather negative results. We report the production and characterization of a monoclonal mouse anti-CNPS antibody and evaluate its protective effect against experimental murine cryptococcosis.

Production of Anti-CNPS Monoclonal Antibodies

CNPS was purified from a CN serotype A strain. It was used for the detection of anti-CNPS antibodies by ELISA and for active immunization of C3H/HeJ mice (1 ug i.v.). Fusion between myeloma P3x63 and immune spleen cells (1:10) was performed with polyethyleneglycol. Positive hybridomas were cloned by limiting dilution and expanded in ascites.

Immunological Characteristics of Anti-CNPS Monoclonal Antibody

Monoclonal antibodies were purified after affinity chromatography using a protein A-Sepharose column and analysed by isoelectrofocusing on polyacrylamide gel. Isotypes were determined by ELISA with rabbit anti-mouse IgG subclasses conjugated with horseradish peroxidase. Antibody activity was assessed by five methods. Agglutination and indirect immunofluorescence assay with CN cells of the four serotypes, precipitation and direct ELISA binding with CNPS of the four serotypes were performed using ten-fold dilutions of monoclonal antibodies. Antibody activity was also studied by competitive binding in ELISA using soluble antigens (CNPS and glucuronoxylomannan [GXM] purified from the four serotypes, galactoxylomannan [GalXM], mannoprotein, CNPS constituent monosac-

charides, and a mannan purified from Saccharomyces cerevisiae)
or whole cells (CN of the four serotypes, C. laurentii, Candida
albicans, Trichosporon beigelii, Histoplasma capsulatum and H.
duboisii). These antigens, at increasing concentrations, were
incubated with an antibody solution (20 ng/ml; v/v), in which
residual anti-CNPS activity was measured by ELISA.

In vivo Activity of anti-CNPS Monoclonal Antibody (E_1)

C5 deficient mice (DBA/2) which are highly susceptible to
crypto-coccosis were used (J.C. Rhodes et al., 1980, Infect.
Immun., 29:494). The protective effect of E_1 was studied by
the survival rate after i.v. infection with a virulent CN
serotype A strain (3-5x10^6 cells). The mice received i.p. 24
h before infection various doses of E_1 (100-0.1 ug), 100 ug of
an unrelated monoclonal antibody (anti-Tg), or saline. The
protective effect was also measured five days after infection
with a smaller inoculum (3x10^4 and 3x10^3) in protected (100 ug
of E_1) or unprotected (100 ug of anti-Tg) mice. CN tissue
counts (CFUs) were measured in the brain and spleen. Anti-
CNPS antibody levels were determined by ELISA and antigen
levels by the latex agglutination test in appropirately diluted
sera collected five days after infection.

Results

An IgG$_1$ monoclonal anti-CNPS antibody (E_1) was produced
which agglutinated CN serotype A cells and was precipitated
with CNPS serotype A. It reacted with the glucuronoxylomannan
component of CNPS but not with the constituent monosaccharides
nor with the mannose
(1-->3)-linked oligosaccharide structures present on CNPS. It
appeared to be specific for CN serotype A by agglutination of
whole cells and for soluble CNPS serotype A by gel immunoprecipitation. However, indirect immunofluorescence, direct
ELISA and competitive binding experiments showed low levels of
cross-reactivity with serotype B and D, but not with serotype
C. Among the other yeasts tested, a cross-reaction was only
detected with T. beigelii. The four CN serotypes could be
distinguished by an indirect immunofluorescence assay using E_1,
according to intensity and the pattern of fluorescence.
    In vivo studies showed that E_1 was protective. Heavily
infected (3x10^6 cells) untreated mice died in 2.9 ± 0.5 days.
Survival time was 17.9 ± 1.6 days for mice treated with 100 ug
of E_1 and 3.0 ± 0.7 days for mice treated with 100 ug of anti-
Tg. Protection was dose-dependent and required at least 10 ug
of E_1 (mean serum antibody concentration = 4.5 ± 0.9 ug/ml).
After infection with a smaller inoculum (5x10^4 and 5x10^3),
treated mice had negative soluble antigen in serum and fewer
CFUs in their spleen and brains than controls.
    These results suggest that monoclonal antibodies will be
useful for fundamental studies of the glucuronoxylomannan
structure as well as for clinical applications, such as C.
neoformans serotyping and possibly serological diagnosis of
cryptococcosis. Furthermore, protective studies confirm the
potential role of specific antibodies during cryptococcosis.
Therefore, attempts to enhance humoral immunity deserve further
evaluation.
    We thank Dr. J.E. Bennett (Bethesda, MD), Dr. R. Cherniak
(Atlanta, GA) and Dr. E. Drouhet (Paris) for their generous
gifts.

KINETIC STUDY OF HUMORAL AND CELLULAR IMMUNE RESPONSES TO CRYPTOCOCCUS

NEOFORMANS MEASURED BY ELISA AND IMMUNOCYTE MIGRATION INHIBITION

S. Gauthier-Rahman*, S. Wahab** and E. Drouhet**

* Institut d'Immunologie, ER 281 CNRS, Hopital
Broussais, Paris
** Institut Pasteur,Unite de Mycologie, Paris

The isolation of purified metabolic (polysaccharidic) and somatic (glycoproteinic) antigens (Ag) of Cryptococcus neo-formans (Cn), paves the way to a more precise study of the immune response to this fungus (Reyes, Dandeu, Drouhet, Abstracts, IX Inter. Congress ISHAM, Atlanta, May 1985, pp. 3-4). It is thought that resistance to Cn is related to the development of cell-mediated immunity (CMI), the role of antibody being only indirect. The inhibition of migration (MI) of immunocytes in the presence of specific Ag, has been considered an in vitro correlate of delayed type hypersensitivity, and more recently as a parameter of CMI. The capsular polysaccharide of Cn produced inhibition of the migration of leukocytes (E. Drouhet and G. Segretain, 1951, Ann. Inst. Pasteur, 78A:891). The development of a rapid and sensitive photoelectric method of reading cell migrations (Gauthier-Rahman et al., J. Immunol. Methods, 1982, 53:77) has enabled us to use this otherwise laborious technique to explore CMI in response to Cn in presence of a wide range (10 logs) of in vitro concentrations of purified polysaccharidic cryptococcal Ag.

While IgG antibody measured by ELISA in sera of individual mice was found to be present 25 days after a single i.v. injection of different doses of live Cn. The mean titre was proportional to the size of the dose of Cn injected (Fig. 1). CMI response showed a variable pattern. The proliferative response to mitogens, PHA and LPS, and to metabolic Ag, showed an initial phase of increasing stimulation after injection of a lethal dose of live Cn followed by decrease shortly before the death of the animals.

Intravenous injection of live Cn led to the development of MI, which was studied in both spleens and lymph nodes of indi-vidual mice and in both strains of mice (Swiss and Balb/c). This MI could be observed as early as 24 hours. The cells migrated from 1 ul agarose microdroplets and appeared in two zones of antigen concentration (metabolic or somatic) in vitro, one low, between 10^{-5} to 10^{-2} ug/ml or less, and one high, 10 ug to 1000 ug/ml. While the kinetics of MI in spleen and lymph node cells was related to the immunizing dose of Cn, the general pattern showed early development of MI followed by a gradual decrease, suggesting the existence of a suppressive phenomenon inhibiting expression of this type of CMI (Fig. 2).

After five days culture of the migrating spleen and lymph node cells, Cn colonies developed, no doubt from Cn already present inside macrophages of the cell suspensions. The number of migrations showing colonies was related to the immunizing dose of Cn (higher number of colonies the larger the dose). Furthermore, it was observed that in certain wells of the migration plate, containing concentrations of metabolic or somatic antigen which liberate MIF optimally, i.e. in the high or low zone of Ag concentrations, fewer or no colonies were observed, suggesting the destruction of Cn by macrophages activated by MIF (Fig. 3).

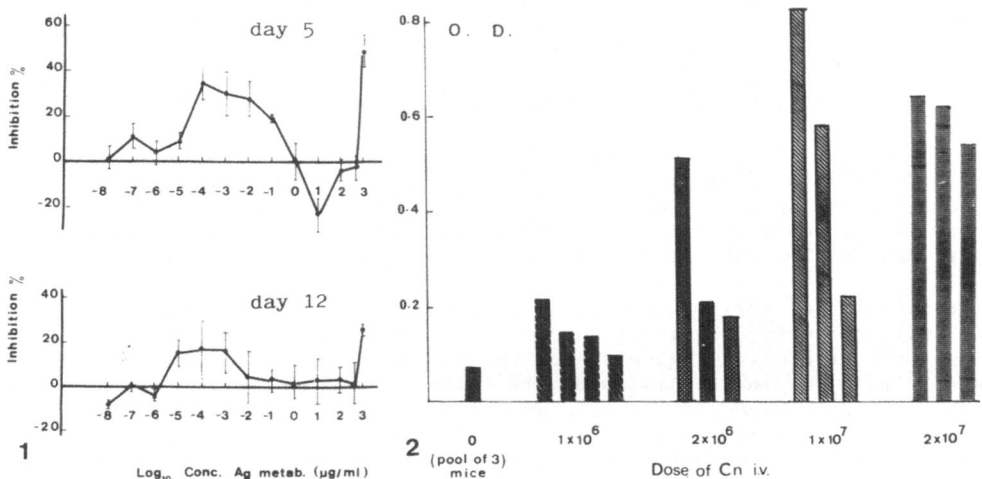

Fig. 1. Detection by ELISA of anti-Cn antibody in sera of individual mice 25 days after i.v. injection of different doses of live Cn.

Fig. 2. The inhibition of migration in presence of Cn antigen of spleen cells of male Balb/c mice immunized by the i.v. route with s x 10^6 live Cn.

Parallel studies of MI in spleen and lymph node cells of the same mice showed that MI in lymph nodes appeared and disappeared earlier than in the spleen. The disappearance of MI activity in lymph nodes was accompanied by a dramatic increase (up to tenfold) in the number of cryptococcal colonies as compared to the spleen where MI was still present (Fig. 4). The enhanced colony count in lymph nodes was accompanied by obvious dissemination of Cn infection in the mouse.

These observations suggest a role for MIF in the early control of multiplication of Cn in vivo, the suppression of MI activity being accompanied by enhanced multiplication and dissemination of Cn.

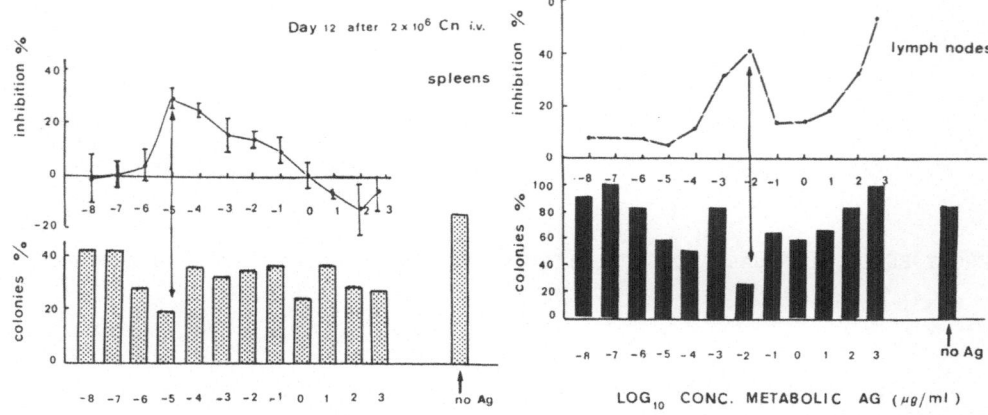

Fig. 3. The presence of MI appears to inhibit the growth of Cn colonies
in migrating spleen and lymph node cells of infected mice.

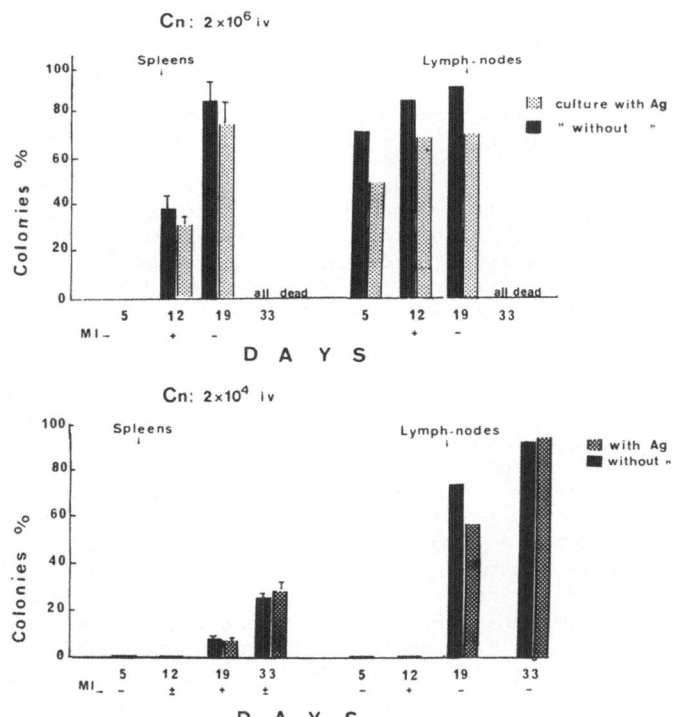

Fig. 4. Macroscopic colony counts of Cn in migrations of spleen and lymph
node cells from agarose microdroplets observed on different days
after i.v. injection of live Cn in Balb/c mice.

SEROTYPES OF CLINICAL AND SAPROPHYTIC ISOLATES OF CRYPTOCOCCUS NEOFORMANS

E. G. V. Evans, D. Hall, D. Swinne* and C. de Vroey*

Department of Microbiology, University of Leeds, Leeds, U.K.

and * Institute for Tropical Medicine, Antwerp, Belgium

Infections due to Cryptococcus neoformans occur worldwide but they are now seen most often in patients with AIDS. There are 2 varieties and several serotypes of Cr. neoformans. Cr. neoformans var. neoformans comprises serotypes A, D and AD and Cr. neoformans var. gattii serotypes B and C. Relatively little work has been done on the distribution of the serotypes of Cr. neoformans. Generally, what is known is that (i) the majority of infections are due to serotype A and that so far all isolates from AIDS patients have been serotype A (ii) serotype D is more common in Europe than elsewhere (iii) serotypes B and C are rare except in certain tropical or subtropical countries and (iv) only serotype A and occasionally serotypes D and AD are found as saprophytes in the environment. In this study we establish the serotypes of clinical and saprophytic isolates of Cr. neoformans from Europe, Central Africa and Asia. The isolates were serotyped using a slide agglutination test where whole yeast cells were reacted with serotype specific rabbit antisera. Isolates were also tested on canavanine medium as a confirmatory test.

A total of 148 Cr. neoformans isolates were examined: 115 were serotype A, 21 serotype B, 6 serotype AD and 6 serotype D. None was serotype C. Among the 115 serotype A isolates, 86 were recovered from patients (73 AIDS) and 29 from the environment (16 Africa, 8 Belgium, 5 Malaysia). The 21 B serotypes were all clinical isolates: 13 from Malaysia, 2 from China (= var. shanghaiensis) and 6 from Zaire, 5 of which were isolated before 1970 and only 1 recently from an AIDS patient living in Lubumbashi. Three of the 6 AD isolates were recovered from the environment in Belgium and the other 3 from AIDS patients (2 Rwanda, 1 Zaire). All the D serotypes were clinical isolates: one from an AIDS patient in Zaire and 5 from non-AIDS patients in Europe. It is noteworthy that 3 of these European patients had primary cutaneous cryptococcosis.

The results further confirm the predominance of serotype A as the cause of cryptococcosis, particularly in AIDS patients. However, the isolation of Cr. neoformans serotype B from a patient with AIDS in Zaire is believed to be the first report of Cr. neoformans var. gattii as the cause of cryptococcosis in this group of patients. As in previous studies, no serotype B isolates were recovered from the environment to account for the B serotype infections.

ABOUT SOLUBLE CRYPTOCOCCAL ANTIGEN (SCA): INTEREST IN SCREENING OF PATIENTS WITH AIDS

P. Roux*, J.L. Touboul**, C. Mayaud**, D. Basset*,
J.L. Poirot*, M. Denis** and F. Lancastre*

Laboratoire de Parasitologie-Mycologie*, Service de
Pneumologie**, Hopital Tenon, Paris, France

Cryptococcosis is frequent in AIDS patients. Diagnosis is sometimes difficult. Results of our prospective study suggest the diagnostic potential of SCA in these patients.

Between February 1985 and February 1986, 129 AIDS patients were tested. Most of them were hospitalized for respiratory symptoms: pneumonitis or fever of unknown origin. The test was performed using a cryptococcal antigen latex agglutination method (Whittaker. MA. Bioproducts). We looked for SCA in 144 sera, 27 cerebrospinal fluids, 44 bronchoalveolar lavages and 44 urines.

SCA was present in 7 patients (5,3%):
- in 3 patients with meningitis, cryptococcosis had been suspected upon their admittance, and Cryptococcus neoformans was cultured from various body sites.
- in 3 patients with pneumonitis (n=2) or with an unexplained fever (n=1), Pneumocystis carinii had been diagnosed and was being treated. The discovery of SCA was the first suggestion of associated cryptococcosis. Cryptococcus neoformans was present in samples obtained thereafter (blood, cerebrospinal fluid, urine, bronchoalveolar lavage), attesting to the existence of disseminated cryptococcosis.
- in the last patient, who had Pneumocystis carinii pneumonia, cultures were negative for Cryptococcus.

These results demonstrate that screening for SCA in AIDS patients seems to be a simple and reliable test, especially when clinical presentation may be misleading.

Dermatophytes

ANTIGENS OF DERMATOPHYTES AND THEIR CHARACTERIZATION

USING MONOCLONAL ANTIBODIES

Luciano Polonelli and Biulia Morace

Istituto di Microbiologia, Facolta' di Medicina e Chirurgia
"A.Gemelli", Universita' Cattolica del Sacro Cuore
L.go F. Vito, 1-00168 Roma, Italy

INTRODUCTION

One of the most interesting roles for monoclonal antibodies in
Medical Mycology is as probes to separate and identify antigenic fractions
from complex mixtures of fungal macromolecules.

Fungi are organisms with complex antigenic compositions and it is
this complexity which has frequently hampered the appropriate sensitivity
and specificity. The advent of mydridoma technology could provide an
invaluable tool for the analysis and isolation of specific fungal antigens.
Monoclonal antibodies are produced, at present against a wide range of
fungi. This study will attempt to illustrate some aspects of the potential
usefulness with particular reference to a representative model of dermatophyte: Microsporum canis.

MATERIALS AND METHODS

For this study, there was a preliminary choice to be made between
using mycelium homogenate or soluble antigen as immunogen. A decision was
also necessary to be taken between spending time purifying the antigen and
then obtaining the relevant hybridomas easily, or looking for the one
hybridoma that makes the required antibody out of several hundreds.

Because this study was intended to be directed to the immunoidentifca-
tion of dermatophytes, we chose to use a soluble exoantigen. This type
of antigen had constantly proved to be highly effective in the immunoidentification of mycelial cultures of pathogenic fungi (3). It is of
primary importance, moreover, that such antigen might be prepared very
easily and rapidly from the dermatophyte cultures to be identified. The
50 isolates belonging to 25 species and varieties of 3 genera used in this
study were from our collection, Universita' Cattolica del Sacro Cuore
(UCSC), or were kindly furnished by the Division of Mycotic Diseases,
Centers for Disease Control (CDC), Atlanta, Georgia (Table 1).

M.canis CDC B2094 was the strain used for production of the reference
antigen. It was derived from a Merthiolate-treated solution (1L5,000) and
concentrated 50x by lyophilization, obtained from many Sabouraud dextrose
agar slant cultures of different age. Chemical and serological analysis

Table 1. List of dermatophyte isolates investigated
with M.canis monoclonal antibodies

Epidermophyton floccosum: (CDC B3807) (UCSC 1, 8)

Microsporum audouinii: (CDC B3800)

M.canis: (CDC B2094) (UCSC 0, 1, 2, 4, 6, 8, 10, 12, 14)

M.cookei: (CDC B3803)

M.distortum: (CDC B2174)

M.equinum: (CDC B2699)

M.ferrugineum: (CDC 83-056097)

M.fulvum: (UCSC 0)

M.gallinae: (CDC B3801) (UCSC 0)

M.gypseum: (CDC B3816) (UCSC 0, 1, 2, 4)

M.nanum: (CDC B3815)

M.persicolor: (CDC B1923)

M.racemosum: (UCSC 0)

M.ripariae: (UCSC 0)

Trichophyton megninii: (UCSC 0)

T.mentagrophytes var. erinacei: (CDC B1865)

T.mentagrophytes var. interdigitale: (UCSC 3, 5, 6, 7)

T.mentagrophytes var. mentagrophytes: (CDC B6209)

T.mentagrophytes var. quinckeanum: (CDC B2331)

T.rubrum: (CDC 3806) (UCSC 5, 9, 10, 11, 13)

T.soudanense: (CDC 83-048430)

T.tonsurans: (CDC B3810)

T.tonsurans var. sulphureum: (UCSC 0)

T.violaceum: (CDC B3610) (UCSC 2, 4)

T.violaceum var. glabrum: (UCSC 0)

----

had previously shown that this antigen was comparable to the one obtained
via an acetone treated filtrate (2). All the antigens of the dermatophyte
isolates investigated in this study were analogously produced.

BALB/c mice were immunized intraperitoneally with the reference antigen according to a standard protocol (4).

The NS1 cell line was the myeloma cell line used in this study. It
was derived from a BALB/c myeloma line. The NS1 cell line and hybrids
derived from the fusion of NS1 cells with BALB/c mouse spleen cells were
grown in RMPI 1640 medium supplemented with 20% fetal bovine serum.

Three days after the booster injections, the hyperimmune mice were
sacrificed and their spleens were removed. The spleen cells were fused
to an equal number of NS1 cells by polyethylene glycol 1000. The cell
culture was subjected to a 1 month selection regime by adding hypoxanthine,
diaminopterin and thymidine to the culture medium. For screening the
hybrids, the hybrid fluids were tested by enzyme linked immunoassay (ELISA).

In addition to specific recognition of the antigen, other antibody properties considered included cross reactivity, avidity and other thermodynamic or kinetic properties, cytotoxicity or agglutination and precipitation properties. All these considerations are of great importance for the selection of hybrid lines and the experimenter should use screening methods which will test the activity of prospective clones not only for one property, namely recognition of the antigen, but also for other desired ones.

Hybridoma cultures secreting the desired antibody were propagated and clones by limiting dilution in the presence of mouse thymocytes. Clones continuing to produce the antibody were expanded. The monoclonality of the antibody produced by the various clones was ascertained by testing them by immunodiffusion against rabbit anti-mouse immunoglobulin antisera. For producing ascites fluid from which concentrated immunoglobulins could be obtained, an intraperitoneal injection of 10^7 hybridoma cells was administered to intra peritoneally pristane treated syngeneic BALB/c mice. Purification of immune ascites has been performed by ammonium sulphate precipitation of the immunoglobulins followed by centrifugation and dialysis. The clones were frozen at −70°C overniht and then stored in liquid nitrogen (4).

Two hundred clones were obtained from the fusions of NS1 myeloma cells with spleen cells from mice immunized with the reference antigen of M.canis.

RESULTS

Monoclonal antibodies of the immunoglobulin G class were produced. When the antigen was electrophoretically separated in denaturing gels and then immobilized on nitrocellulose strips (Western blot technique), we detected a diversity of monoclonal antibodies to fungal glycoproteins and the study of the fungal polypeptides reactive with the different monoclonal antibodies permitted us to investigate several properties of the antigens. Eight of the 44 antibody-rich ascitic fluids produced from each of the different expanded clones reacted in immunodiffusion which permitted detection of an unexpected distribution of antigenic determinants in the dermatophyte isolates considered in this study. The production of ascitic fluids has been necessary in immunodiffusion for obtaining the optimal concentration of monoclonal antibody reacting with a standard concentration of the reference antigen. Ascitic fludi produced from type 1 monoclonal antibody reacted in immunodiffusion either with all of the heterologous or homologous dermatophyte antigens. The presence of an antigen with a determinant shared by all of the dermatophyte isolates tested is obviously suggestive of a very common function. Type 2 monoclonal antibody did not react with the antigens obtained from three different isolates of M.canis, thus demonstrating the existence of different serotypes within the species (Fig. 1). The antigens of the same three M.canis isolates that were negative with the type 2 monoclonal antibody regularly reacted with their homologous rabbit polyvalent antiserum. This result suggested that possibility that different serotypes may also exist within other species. Effectively, type 3 monoclonal antibody produced against M.canis permitted the detection of different serotypes within the species Trychophyton rubrum (Table 2), (2).

Type specific monoclonal antibody should more readily differentiate between the serotypes and it is conceivable that some isolates lack the antigenic determinant necessary for reactivity with the monoclonal antibody.

These preliminary data show that monoclonal antibodies are powerful reagents for serotyping dermatophyte isolates and provide clear demonstration of intratypic antigenic variation. The effectiveness of monoclonal antibodies as serotyping reagents would be considerably enhanced by the use of a panel of type-specific antibodies. We expect that subgroups are

Table 2. Immunodiffusion serotyping of dermatophytic
isolated by monoclonal antibodies against M.canis

Isolates	Monoclonal antibody reaction		
	Type 1	Type 2	Type 3
Microsporun canis			
CDC B2094	+	+	+
UCSC 0	+	+	+
UCSC 1	+	+	+
UCSC 2	+	−	+
UCSC 4	+	−	+
UCSC 6	+	+	+
UCSC 8	+	−	+
UCSC 10	+	+	+
UCSC 12	+	+	+
UCSC 14	+	+	+
Trichophyton rubrum			
CDC B3806	+	+	−
UCSC 5	+	+	+
UCSC 9	+	+	+
UCSC 10	+	+	−
UCSC 11	+	+	+
UCSC 13	+	+	−

dependent upon the number of monoclonal antibodies used for anlaysis and that additional groupings are likely to emerge as more hybridomas are used.

Monoclonal antibodies should be readily adapted as serotyping reagents and may prove suitable for epidemiological studies of dermatophyte isolates.

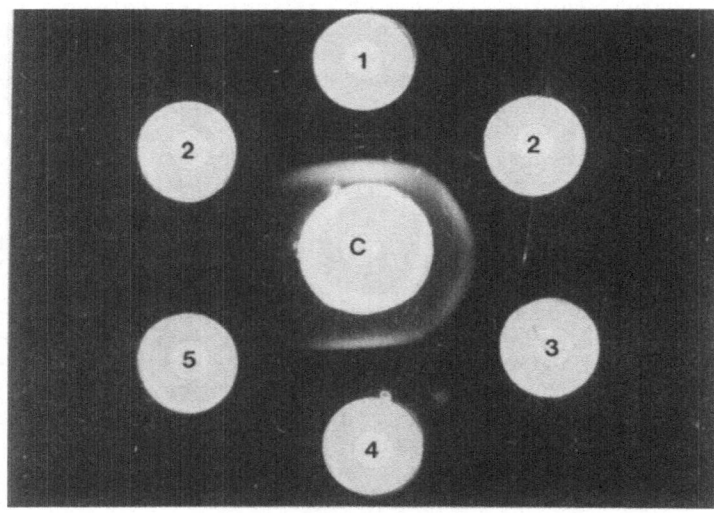

Fig. 1. Occurrence of serotypes within M.canis isolates by
monoclonal antibodies in immunodiffusion test. Well
contents: c, type 2 monoclonal antibody ascites fluid;
1 and 4, M.canis CDC B2094 antigen; 2, M.canis UCSC 6
antigen; 3, M.canis UCSC 12 antigen; 5, M.canis UCSC 2
antigen; 6, M.canis UCSC 8.

Subdivision of species by serotyping is of value for epidemiological purposes, particularly when other systems are not available.

Our results demonstrate the feasibility of using monoclonal antibodies for the immunoidentification and antigenic characterization of dermatophytes by the Western blot technique (1, 6).

In this paper we report on the screening of monoclonal antibodies for reactivity with dermatophyte antigens electrophoretically separated on polyacrylamide denaturing gels and then electrically transferred to nitrocellulose strips. The potential of the technique is discussed with 4 different monoclonal antibodies produced against M.canis (UCSC 3, 5, 7, and 13). Several aspects of the results merit further discussion.

The potential of the Western blot technique may be completely exhausted by using a gel filtration calibration kit of substances of known molecular weight. In our study the markers were mixed together with the soluble antigens before the test. Each strip of nitrocellulose could be differentially stained either by Coomassie blue staining for proteins or immunoperoxidase for evidence of antigen-monoclonal antibody binding.

Consequently, the Western blot technique permitted the easy determination of the molecular weight of each single antigenic determinant reacting with each monoclonal antibody. the specificity of a monoclonal antibody derives from its purity, but does not preclude cross reactivity due to recognition of similar antigen or identical determinant(s) located on different antigenic mixtures. This is particularly relevant in complex situation, like expression of cell surface antigens where a common antigenic determinant could be expressed in different molecules: for instance a common segment of evolutionarily related proteins, a carbohydrate moiety etc. On the other hand, lack of reactivity with a monoclonal antibody does not prove absence of the antigen, since such a reaction could be affected by changes in the environment of the antigenic determinant.

This property was particularly clear for relatively nonspecific monoclonal antibodies. M.canis monoclonal antibody 3 was recognizing all of the M.canis isolates tested by reacting with specific antigenic determinants of molecular weight 74 K and 95 K daltons. The detection of other reactive antigenic determinants in the different isolates within the species clearly shows the existence of numerous serotypes. No heterologous species, showing cross reactivities, presented the simultaneous presence of the 74 and 95 K daltons antigenic determinants.

Analogous was the Western blot behaviour of M.canis monoclonal antibody 5; it was characterizing single M.canis isolates either by species-specific 75 and 95 K daltons or strain specific different molecular weight antigenic determinants.

In M.canis monoclonal antibody 7, the only species-specific antigenic determinant was detected at a molecular weight of 98 K daltons. It is interesting to note that all 5 isolates of M.gypseum tested showed a common domain at a molecular weight of 25 K daltons.

M.canis monoclonal antibody 13 proved to be the most interesting and specific one. It reacted only with a single antigenic determinant in each M.canis isolate tested at a molecular weight of 16 K daltons (Fig. 2). This was common also to the M.distortum isolate tested, thus confirming the identity of the two species as demonstrated by mating experiments. Cross reactions, although at different molecular weight, were observed only in the M.ferrugineum, M.gallinae, and T.soudanense isolates tested, which are considered morphologically related (Table 3), (5).

Table 3. Antigenic characterization of dermatophyte isolates with
M.canis monoclonal antibody UCSC 13 by Western blot technique

----

Isolates*	Molecular size (kDa)		

Microsporum canis			
CDC B2094	16		
UCSC 0	16		
UCSC 1	16		
UCSC 2	16		
UCSC 4	16		
UCSC 6	16		
UCSC 8	16		
UCSC 10	16		
UCSC 12	16		
UCSC 14	16		
M.distortum CDC B2174	16		
M.ferrugineum CDC 83-056097		11	
M.gallinae CDC B3801			23
Trichophyton soudanense CDC 83-048430	12		

----

*The other dermatophytes listed in Table 1 were negative with this mono-
clonal antibody.

Our results show that when the antigens were electrophoretically
separated in denaturing gels and then immobilized on nitrocellulose strips
we detected a greater diversity of monoclonal antibodies to dermatophyte
glycoproteins than when we used the technique of immunoprecipitation of
soluble non-denaturated dermatophyte antigens or ELISA. The primary advantage of the Western blot technique is in the detection of non-precipitating
antibody and of antibody to poorly soluble antigens not available for reactions in other preparations. Studies of the dermatophyte protein reactive
with a panel of monoclonal antibodies indicated that the technique can be
used to investigate several properties of the separated antigens.

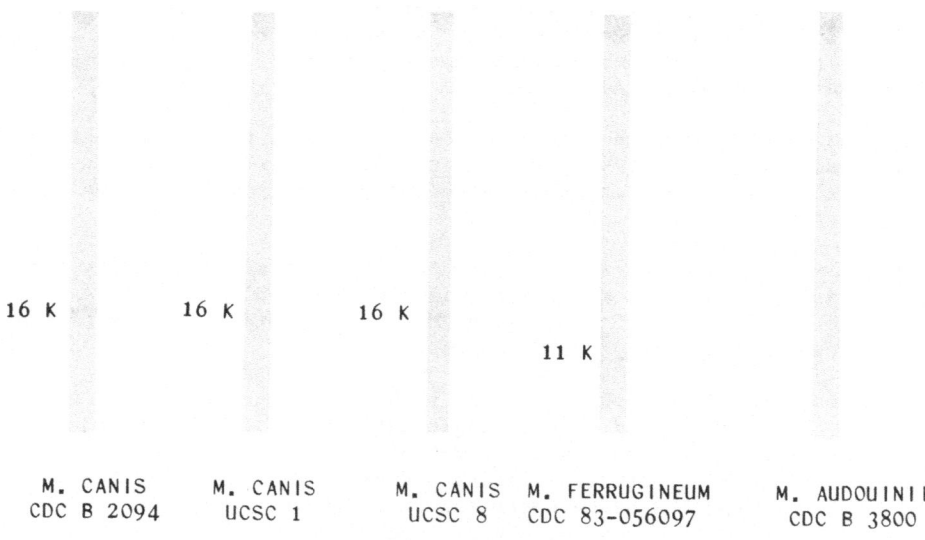

16 K   16 K   16 K

                                        11 K

M. CANIS   M. CANIS   M. CANIS   M. FERRUGINEUM   M. AUDOUINII
CDC B 2094   UCSC 1   UCSC 8   CDC 83-056097   CDC B 3800

Fig. 2.   Western blot analysis of dermatophyte antigens by
microsporum canis monoclonal antibody UCSC 13

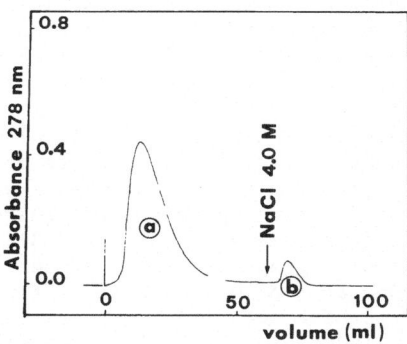

Fig. 3. Affinity chromatography of M.canis CDC B2094 total
antigen against monoclonal antibody USCS 7 coupled
to Affigel 10 resin. Peak a represents the unadsorbed
antigenic fraction, and peak b represents the fraction
eluted by sodium chloride (4.0 M) in elution buffer
arrow. (Ref. 5).

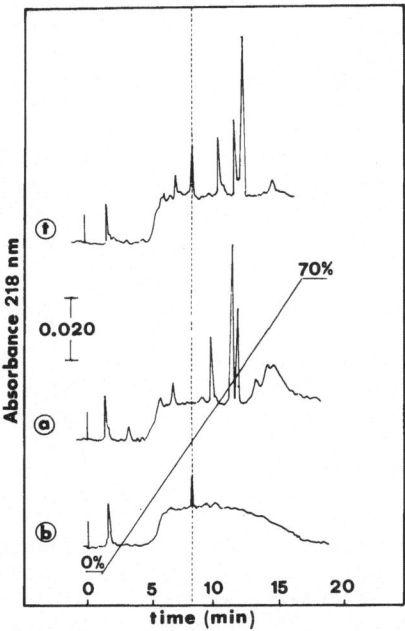

Fig. 4. Reverse-phase HPLC. T, M.canis CDC B2094 total antigen;
a, unadsorbed antigenic fraction; b, the fraction eluted
by sodium chloride (4.0 M) in elution buffer after affinity
chromatography against monoclonal antibody UCSC 7 on Affigel
10 resin. The dashed line shows the positin of the reactive
antigenic determinant in the chromatographic pattern. The
acetonitrile gradient applied in the chromatography is also
indicated. (Ref. 5).

Complex antigenic mixtures, on the other hand, may contain dominant immunogenic components. These may effectively decrease the variability on the response to a given antigen; in which case partial purification of the antigen in question may become essential. One general way to achieve such partial purification is by using monoclonal antibodies as immunoadsorbants to remove such impurities.

In our study we have found that single monoclonal antibodies were directed against single peaks of the antigen mixture of M.canis CDC B2094 used for immunization. After affinity chromatography with immunoadsorbed M.canis monoclonal antibody 7, it was possible to separate peak 1 (unadsorbed antigen) and peak 2 (adsorbed antigen) (Fig. 3). Reverse-phase high performance liquid chromatography (RPHPLC) of peak 1 (unadsorbed antigen) showed the disappearance of a single peak of the original pattern (Fig. 4). The immunospot carried out with the total antigen, peak 1 (unadsorbed antigen) and peak 2 (adsorbed antigen) tested with M.canis monoclonal antibody 7 confirmed the loss of reactivity of peak 1 (unadsorbed antigen).

DISCUSSION

Our study clearly demonstrates the potential of monoclonal antibodies in the immunoidentification f dermatophytes. Their effectiveness may be completely exhausted by using them with a highly sensitive and analytical procedure as the Western blot technique. The primary results could be dependent on the luck of the experimenter in finding relatively specific monoclonal antibodies devoid of too many cross reactions. The procedure of affinity chromatography, however, may permit to superate unfavourable circumstances by immunoadsorbing the nonspecific monoclonal antibodies to obtain a purified antigen depleted of the common antigenic determinants. The production of the new monoclonal antibodies of second generation could permit to achieve the desired specificity or to follow up on the purification of the antigen.

REFERENCES

1. W.N. Burnette, 1980, "Western blotting": Electrophoretic Transfer of Proteins from Sodium Dodecyl Sulphate-polyacrylamide Gels to Unmodified Nitrocellulose and Radiographic Detection with Antibody and Radio-iodinated Protein, A. Anal. Biochem., 112:195-203.
2. G. Morace, G. Amalfitano, and L. Polonelli, 1986, Serotyping of Fungal Isolates by Precipitating Monoclonal Antibodies, Mycopathologia, 94:53-57.
3. L. Polonelli and G. Morace, 1984, Rapid Immunoidentification of Pathogenic Fungi, in: "New Horizons in Microbiology", A. Sanna and G. Morace (eds), pp. 203-209, Elsevier Sciences Publishers, Amsterdam.
4. L. Polonelli and G. Morace, 1985, Serological Anlaysis of Dermatophyte Isolates with Monoclonal Antibodies Produced Aganist Microsporum canis. J. Clin. Microbiol., 21:138-139.
5. L. Polonelli, M. Castagnola, and G. Morace, 1986, Identification and Serotyping of Microsporum canis Isolates by Monoclonal Antibodies. J. Clin. Microbiol, 23:609-615.
6. H. Towbin, T. Staehelin, and J. Gordon, 1979, Electrophoretic Transfer of Protein from Polyacrylamide Gels to Nitrocellulose Sheets Procedure and some Applications. Proc. Natl. Acad. Sci. U.S.A., 76:4350-4354.

ANTIGENS INVOLVED IN THE PATHOLOGY OF

FONSECAEA PEDROSOI : IMMUNOCHEMICAL STUDY

Oumaîma Ibrahim-Granet and Claude de Bièvre

Institut Pasteur, Unité de Mycologie
25,rue du Dr. Roux
75015 Paris, France

INTRODUCTION

The dimorphic fungus Fonsecaea pedrosoi is the principal agent of chromomycosis, which is a human verrucous dermatitis.

Chromomycosis may be caused by any of seven dematiaceous fungi having different saprophytic filamentous phases and an identical parasitic phase characterized by unicellular sclerotial bodies.

Thus far, the fungal antigens have been principally useful for studying taxonomic relationships, among the agents of chromomycosis and taxonomically related fungi (1-8,). Two types of antigens have been used by investigators :
 - Crude antigens made either of fungal cells ground in saline or obtained from broth filtrate precipitated with acetone and dissolved in saline.
 - Purified carbohydrate antigens extracted from killed cells after precipitation with ethanol. It has been demonstrated that the galactomannan and, more precisely, the acid-labile galactofuranosyl residue is an important antigenic determinant. Serological precedures used were complement fixation, tube precipitation, immunodiffusion and immunoelectrophoresis. The involvement of specific protein antigens in the pathology of chromomycosis has not been investigated until now.

Such study was undertaken in our laboratory for F.pedrosoi, the principal agent of chromomycosis. When grown on usual media at 23°C, this fungus produces a filamentous phase characterized by the presence of hyphal cells and three types of conidiogenesis. At 37°C, a modified filamentous phase consisting of normal hyphal cells and some budding blastoconidia appears. After several buddings, the blastoconidia and hyphae may give Monilia-like cells.

This modified phase can also be observed in human beings 15 days after infection with F.pedrosoi. The final parasitic phase in vivo consists of dark-brown unicellular bodies. In our work, we were interested in studying the protein patterns of extracts from cultures incubated at 23°C containing the filamentous phase, at 37°C which contains the modified phase and from cultures grown at 23°C and shifted to 37°C. By this study we attempted to determine whether transformation in vivo of the filamentous to the modified phase might be a response to thermal stress.

In addition, modifications of the protein pattern produced by thermal stress were studied serologically to determine which proteins are most immunogenic during the course of
infection. Finally the potential role of antibodies as growth
inhibitors of F.pedrosoi was considered.

THERMAL STRESS RESPONSE

Two strains of F.pedrosoi were used for this study.

Strain 1335

This strain, isolated from a human case of chromomycosis
in Martinique, was grown in a defined medium (9) at 23°C for 8
days (active phase). Cultures were then labelled with L-^{35}S
methionine and incubated at 37°C for 1.5h, 2h, 6h, 18h or 24h.
A control culture was labelled and maintained at 23°C for 24h.
Protein extracts were prepared after cell disruption as
previously described (10) and studied by one dimensional SDS-
PAGE, according to the methods of Laemmli and Bonner (11, 12).
Results are shown in Fig. 1. Extracts from cells at 23°C and
the heat shocked cultures at 37°C exhibited different protein
patterns. Polypeptides of Mr 95,000,   75,000, 65,000, 63,000,
42,000 and 35,000 were enhanced after incubation at 37°C for
24h. Three polypeptides of Mr 70,000, 46,000 and 23,000 were
induced at 37°C. In contrast, six polypeptides disappeared in
the cells shifted to 37°C. These heat stroked proteins had Mrs
of 72,000, 64,000, 50,000, 40,000, 41,000 and 24,000.

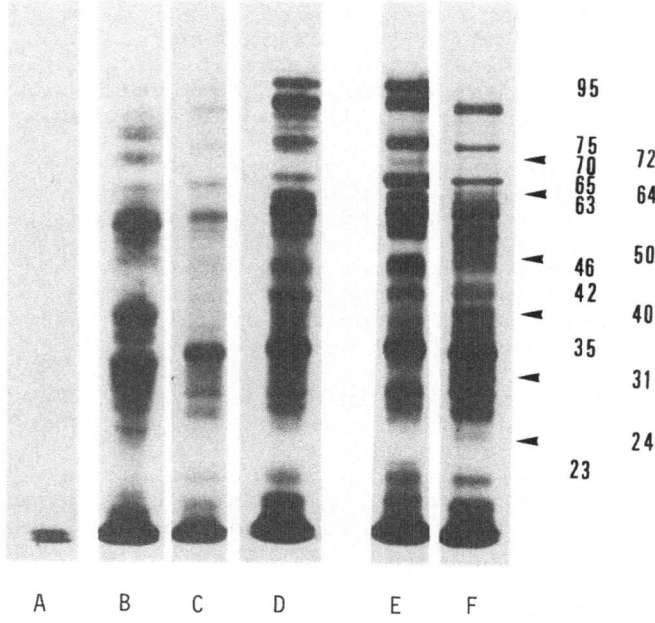

Fig. 1   Fluorogram of 10 % one dimensional SDS-PAGE of ^{35}S
methionine labelled proteins synthesized in F.pedrosoi cells
grown at 23°C during 8 days and shifted to 37°C. Lanes A, B,
C, D and E represent cells transferred to 37°C for 1.5h, 2h,
6h, 18h and 24h. Lane F represents labelled cells maintained
at 23°C. Arrows in E indicate the heat shock proteins and
arrows in F the heat stroke proteins.

Strain 1314

This strain, isolated from a human case in Gabon, was grown in Sabouraud liquid medium. Cultures were incubated at 23°C or at 37°C for 12 days (stationary phase) ; one additional culture was grown for 7 days at 23°C and then transferred to 37°C and grown for 5 days at this temperature. Proteins were extracted as mentioned above (10) and studied by one and two dimensional SDS-PAGE according to Laemmli and O'Farrell (11, 13).

The protein patterns obtained by one dimensional SDS-PAGE stained with Coomassie blue are shown in Fig. 2.

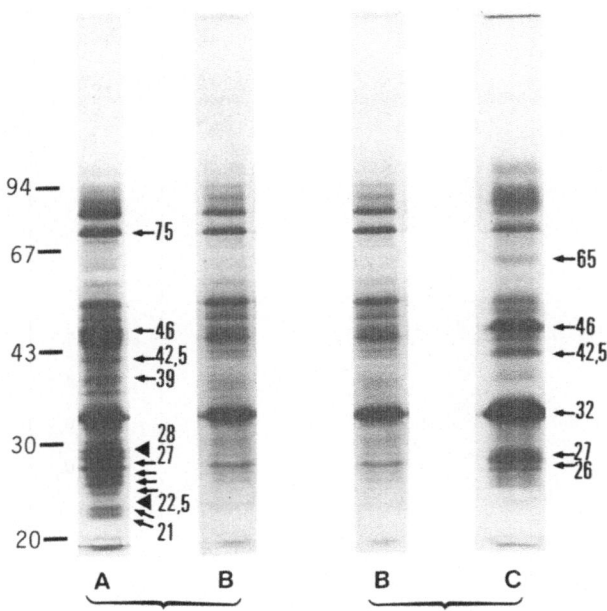

Fig. 2   Coomassie blue - stained 10 % one dimensional SDS-PAGE of 23°C and 37°C cell extracts of F.pedrosoi. Lanes were loaded with 40 ug of protein extract and show (from left to right) the polypeptide patterns of 37°C, 12 days continuous culture extract (A) ; 23°C, 12 days continuous culture (B) and the extract from 23°C, 7 days cells shifted to 37°C for 5 days (C). ← : proteins enhanced at 37°C. ◀ : proteins induced at 37°C.

Comparison of lanes A (37° continuous culture) and B (23°C, continuous culture) shows enhancement at 37°C of proteins with Mrs of 75,000, 46,000, 42,500, 39,000, seven other polypeptides with Mrs ranging between 27,000 and 21,000 were also enhanced. Moreover, two polypeptides with Mrs of 28,000 comfirmed the previous results (Fig. 3). Seven proteins appear to be enhanced at 37°C and seven are present only at 37°C. They are the proteins specific to cultures grown at 37°C for 12 days. The relative molecular mass, isoelectric point, and the spots integration* of the different spots obtained at 37°C and 23°C are shown in Table 1.

Table 1. Results of the two-dimensional electrophoresis
applied to strain 1314 grown at 23°C and 37°C for 12 days

Relative Molecular Mass	isoelectric point	Spots Integration at	
		37°C	23°C
75000	6.4	50000	39600
71000	6	12529	3500
69000	6.15	11697	-
63000	6.25	15000	-
60500	6.25	8000	-
56000	6.3	10909	-
55000	5.15	14341	4153
52000	5.9	22844	9307
51000	6.3	39695	27074
46000	4.65	3000	800
41000	6	5000	1000
39000	4.75	3000	-
32000	5.25	6500	-
28000	4.5	15000	-

The results demonstrate that hyperthermia promotes the
appearence, enhancement or decrease of several polypeptides 18
h after transfer of active phase F.pedrosoi culture to 37°C.
Proteins induced or enhanced at 37°C may be considered heat
shock proteins, because they appeared 18 h after the transfer
to 37°C. Temperature also influenced the protein pattern during the stationary phase of the growth cycle. Comparison of
protein patterns of extracts from cultures grown for twelve
days at 23°C or 37°C revealed that proteins were enhanced or
induced at 37°C. The same phenomenon was observed when cells
grown at 23°C were transferred to 37°C for the last 5 days.

---

* Spots integration is numerical data quantified by a digital
image processor connected to a photoscan densitometer.

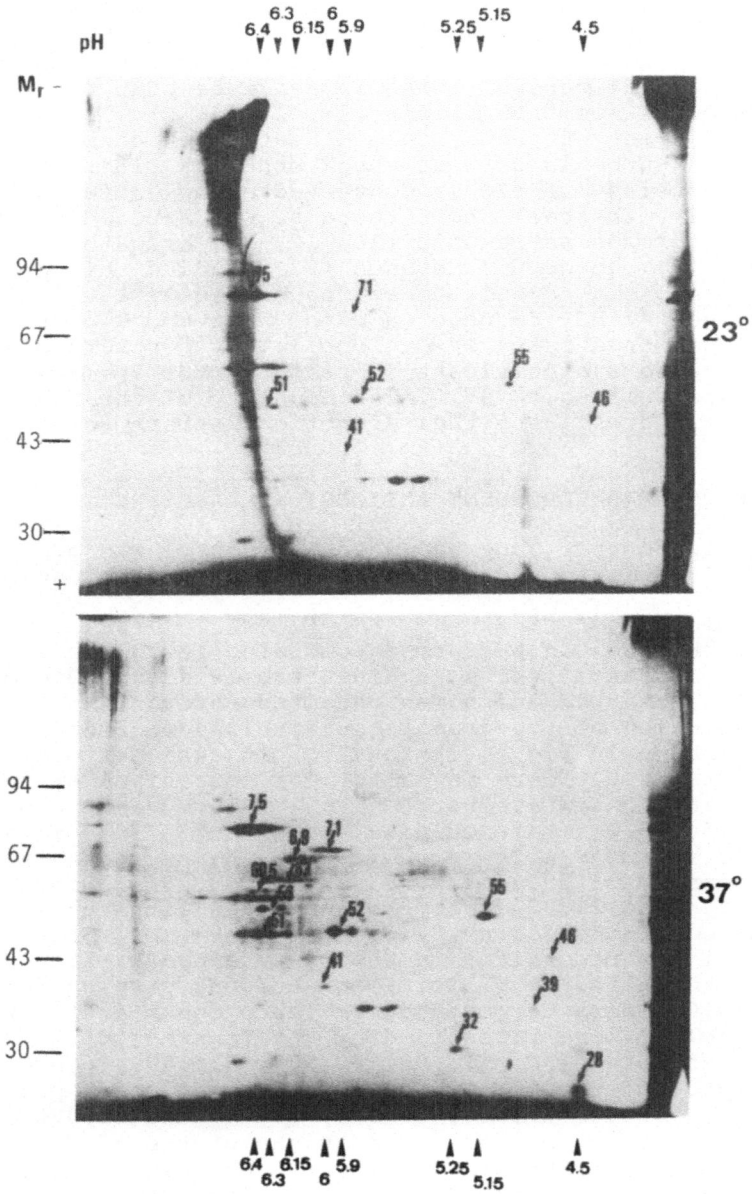

Fig. 3　　Two dimensional gel electrophoresis of the 12 days, 23°C or 37°C cell extracts of <u>F.pedrosoi</u>. Proteins (20 μg) was loaded on isoelectric focusing gels using ampholines of pH range 3.5 to 10 and 4 to 6.5. The second dimension was carried out on 10 % SDS-PAGE and stained with silver nitrate. Positions of molecular mass standards are indicated in the left margin and in the right margin the protein extracts separated by one dimensional electrophoresis.

The thermal stress response in F.pedrosoi could facilitate understanding of the dimorphism of this fungus. The induction of new polypeptides and the enhancement of others at 37°C may be related to the thermotolerance of the fungus and its survival in vivo. Several studies of the heat shock response indicate that heat shock proteins are synthesized in response to thermal stress. There is evidence that the response protects cells from the stress (14, 15, 16, 17) by stimulating a specific set of genes, called heat shock genes, responsible for the synthesis of heat shock proteins (18). These heat shock proteins are highly conserved among different organisms. Recently, the heat shock response was studied in fungi which are either non pathogenic (16, 19, 20) or pathogenic to man (21, 22, 23). Sacco and Maresca (23) studied this phenomenon in Histoplasma capsulatum at the molecular level. The authors cloned the HSP 70 gene (a major heat shock protein conserved among organisms) in Drosophila. By Northern blots, they demonstrated a significant increase of HSP 70 mRNA once the cells were shifted to 37°C. The identity of the heat shock proteins in F.pedrosoi and their functions remain to be established.

PATTERNS OF THERMAL DEPENDENT ANTIGENS

The antigens that appeared when F.pedrosoi was grown at 37°C and 23°C were studied by immunoblotting and immunoprecipitation.
Immunoblotting
This technique was performed according to Towbin (24) using rabbit antisera raised against the 12 days, 23°C or 37°C protein extracts, and human infected sera.
The detection of F.pedrosoi (strain 1314), 12 days 37°C antigens is shown in Fig. 4. A total of 16 antigens of Mr varying from 94,000 to 18,000 were revealed. The comparison of lanes A and B shows that only five antigens of Mr 55,000, 51,000, 46,000, 43,000 and 35,000 were detected by the anti-23°C rabbit serum. All the 37°C antigens were recognized by the patient sera. From these antigens Mr 75,000, 55,000, 51,000, 46,000, 39,000, 32,000, 28,000, 25,000 and 21,000 were, in agreement with the 1 and 2-D gels, enhanced or induced at 37°C.
The antigens recognized in the culture grown at 23°C for 12 days are shown in Fig.5. Thirteen antigens varying from 94,000 to 18,000 were detected by the autologous anti serum from which only 7 were recognized by the "anti-37°C" serum. These antigens were of Mr 75,000, 55,000, 51,000, 46,000, 43,000, 39,000 and 25,000. Of the thirteen antigens, polypeptides with Mr 51,000, 39,000, 35,000, 32,000 and 29,000 were recognized by both human sera. Comparison of the 37°C and 23°C antigens showed that the Mr 28,000, 27,000, 25,000, 21,000, 19,000 and 18,000 were specific to the 37°C extract. Two of them (25,000 and 18,000) were only weakly detected in the cultures grown at 23°C.

Fig. 4   Analysis of the 12 day 37°C (strain 1314) antigens by immunoblotting using the anti-37°C total rabbit serum diluted 1/10000 (A) ; the anti-23°C total rabbit serum diluted 1/10000 (B) and human patient sera (C) and (D) diluted 1/6.

Fig. 5   Analysis of the 12 day 23°C (strain 1314) antigens by immunoblotting using the anti-23°C total rabbit serum diluted 1/10000 (A) ; the anti-37°C total rabbit serum diluted 1/10000 (B) and human patient sera (C) and (D) diluted 1/6.

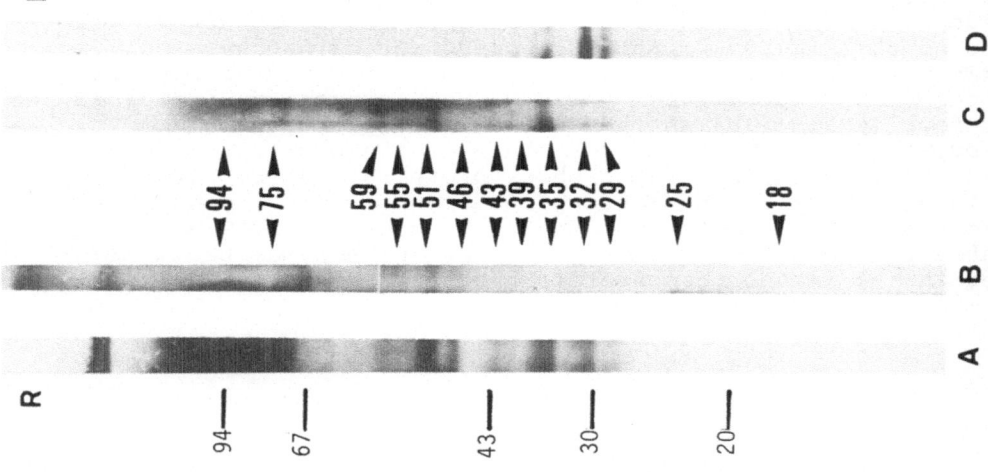

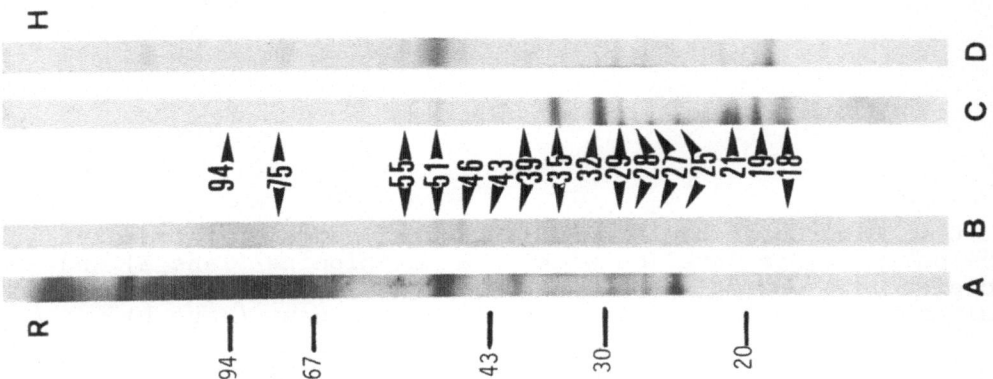

Immunoprecipitation

This technique was performed according to Kessler (25).
Fig.6 shows the immunoprecipitation of _F.pedrosoi_ strain 1335
soluble antigens labelled with ^{35}S methionine using normal
rabbit serum, the anti-37°C rabbit serum and A human, infected
serum. Thirteen antigens varying between 115,000 and 32,000
were recognized by the anti-37°C rabbit serum. All these
antigens were also immunoprecipitated by the human serum. The
antigens of Mr 84,000 ; 55,000, 35,000 and 32,000 were only
weakly visible. The Mr 11,5000, 80,000 and 51,000 were precipitated by the normal rabbit serum. This experiment indicated
that four strain 1335 antigens of Mr 75,000, 70,000 and
35,000, as demonstrated in the study of the heat shock response in this strain, were heat shock proteins. The results
obtained both by immunoblotting and immunoprecipitation thus
indicated that proteins enhanced or induced at 37°C are
related to the pathogenicity of _F.pedrosoi_ and may be implicated in the immune response against this fungus.

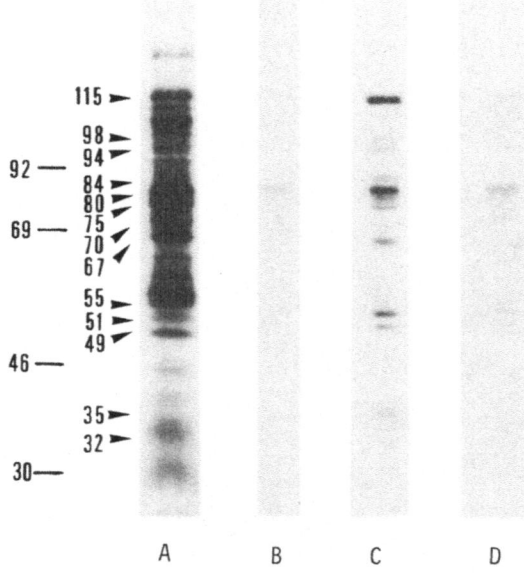

Fig. 6   Fluorogram of 10 % SDS-PAGE of _F.pedrosoi_ (strain
1335) polypeptides labelled with ^{35}S methionine (lane A) and
precipitated by normal rabbit serum (lane B), by anti-37°C
rabbit serum (lane C) and by human infected serum (lane D).

Protein iodination

Eight day cells of _F.pedrosoi_ strain 1335 were iodinated
with ^{125}I using the chloramine T technique of Hunter and
Greenwood (26). Solubilized iodinated antigens were studied by
immunoprecipitation. As shown in Fig.7, five polypeptides of
Mr 80,000, 75,000, 63,000, 55,000 and 35,000 were iodinated.
Two of these polypeptides (Mr 75,000 and 55,000) were precipitated by the anti-37°C rabbit serum and by the human infected
serum.

Protein synthesis

Protein synthesis was measured by incubating separately eight (strain 1335) fungal balls approximately 2 mm in diameter in 2 ml of the above mentioned defined medium (9). ImmunoglobulinG (IgG), purified from the anti-37°C rabbit serum, was added to each flask at the concentration of 1 mg/ml. Cultures were labelled with L-^{35}S menthionine and incubated at 37°C. Control cultures were incubated with IgG purified from normal rabbit serum. Cultures were stopped after 24h, 48h, 8 and 10 days. Protein synthesis was measured for both cell extracts and culture filtrates. Results are shown in Fig. 8. All the curves of remaining methionine in the culture medium, protein synthesis and immunoprecipitation indicate a greater protein synthesis in cells growing with normal IgG and an inhibition of about 50 % by the anti-F.pedrosoi IgG. This observation was confirmed by comparing for each age the cpm values from culture grown in he presence of normal or anti-F.pedrosoi IgG. serum. These results show that some F.pedrosoi soluble proteins could be iodinated, suggesting that they may be present in the fungal cell wall. Two of these were antigenic and so may be involved in stimulating protective immunity against chromomycosis.

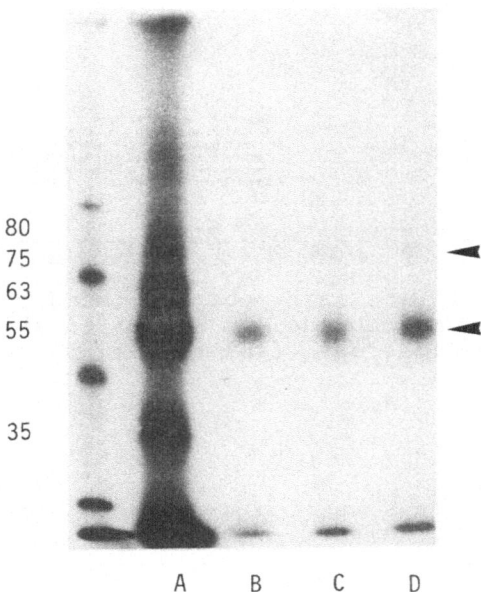

Fig. 7   10 % SDS-PAGE of F.pedrosoi (strain 1335) polypeptides labelled with ^{125}I using the chloramine T technique (lane A) and precipitated by normal rabbit serum (lane B), by anti-37°C F.pedrosoi rabbit serum (lane C) and by the human infected serum (lane D).

ROLE OF SPECIFIC IMMUNOGLOBULINES ON GROWTH OF F.PEDROSOI

The influence of specific IgG on fungal growth was studied by measuring protein and DNA synthesis in cultures grown in the presence of specific IgG purified from the anti-37°C F.pedrosoi rabbit serum.

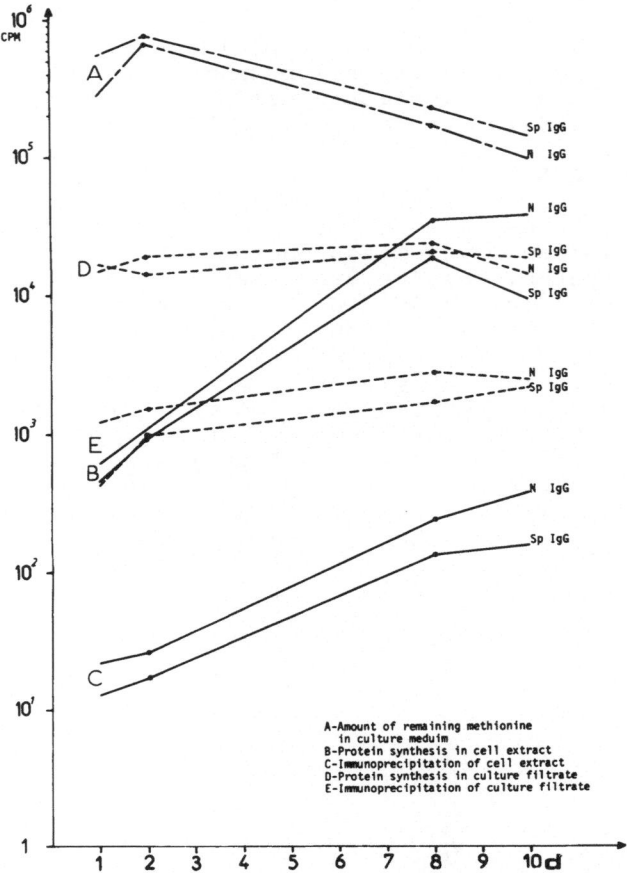

Fig. 8   Growth kinetics of F.pedrosoi (strain 1335) balls labelled with ^{35}S methionine and incubated at 37°C with normal (N) or anti-F.pedrosoi (sp) specific immunoglobuling.

Electrophoresis of proteins for all cell extracts and culture filtrates are shown in Figs. 9 and 10, respectively.

The protein patterns of cell extracts showed that polypeptides of Mr 172,500, 136,000 and 20,000 were specific to cells grown in the presence of anti-F.pedrosoi IgG. Polypeptides of Mr 72,000, 58,000, 43,000 and 40,000 were specific to cells grown in the presence of normal IgG. For culture filtrates (Fig. 10), a polypeptide of 63,000 was specific to cells grown in presence of anti- F.pedrosoi IgG and a polypeptide of about 68,500 was specific to cells grown in the presence of normal IgG.

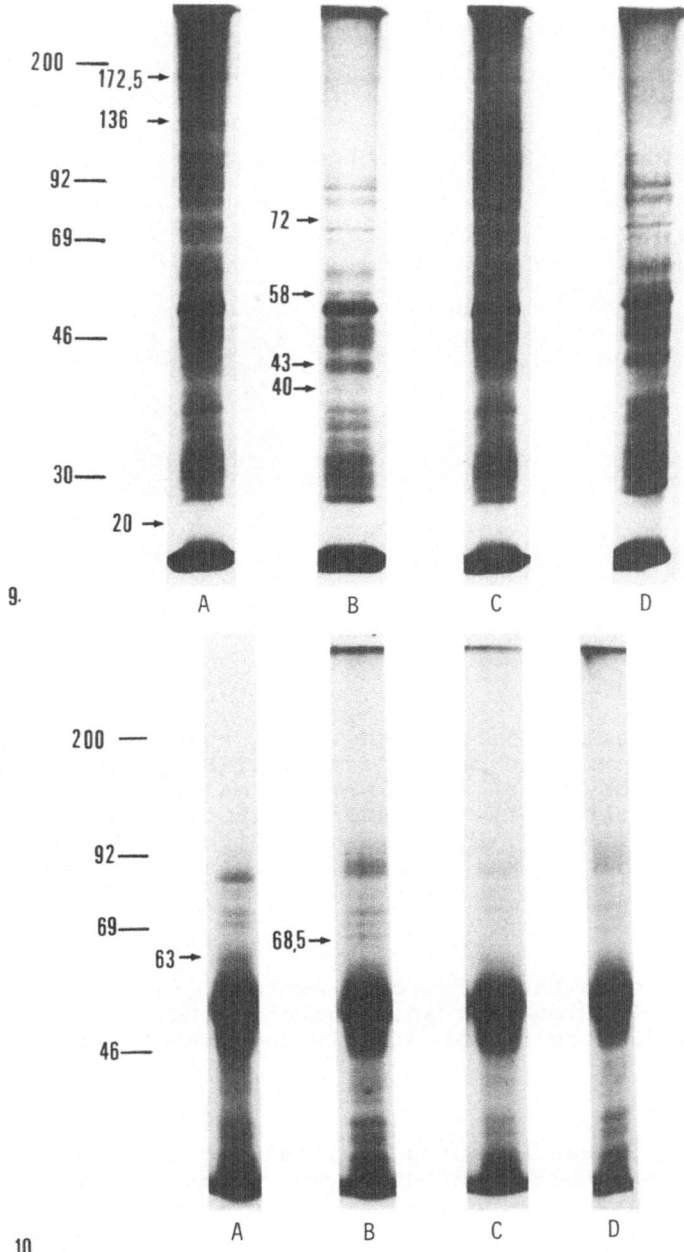

Fig. 9   Fluorogram of 10 % SDS-PAGE of <u>F.pedrosoi</u> polypeptides (cell extract) labelled with ^{35}S methionine and incubated with anti-<u>F.pedrosoi</u> IgG for 8 or 10 days (lanes A and C) and with normal IgG for 8 or 10 days (lanes B and D).

Fig. 10   Fluorogram of 10 % SDS-PAGE of <u>F.pedrosoi</u> polypeptides (culture filtrate labelled with ^{35}S methionine and incubated with anti-<u>F.pedrosoi</u> IgG for 8 or 10 days (lanes A and C) and with normal IgG for 8 or 10 days (lanes B and D).

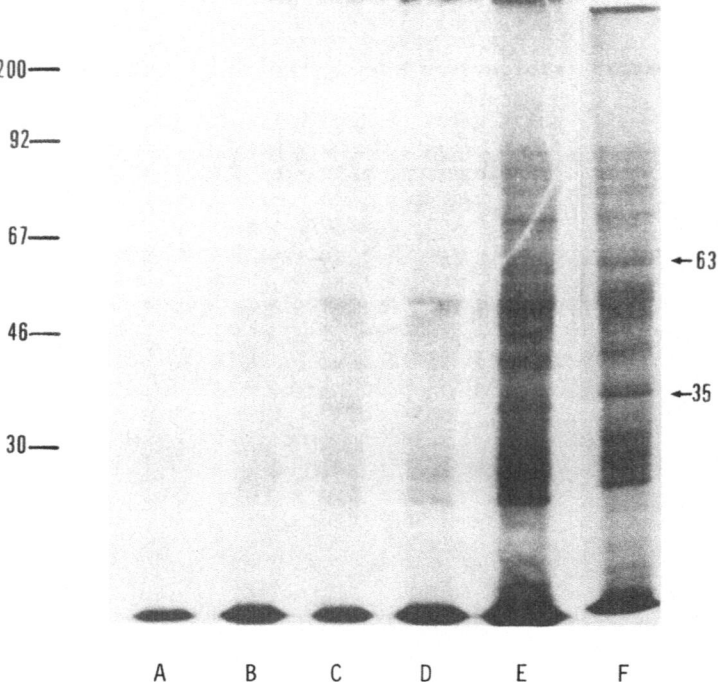

Fig. 11. Immunoprecipitation of the F. pedrosoi polypeptides (cell extract) growing in presence of anti-F. pedrosoi IgG for 8 or 10 days (lanes A and C) and with normal IgG (lanes B and D). Lanes E and F represent the immunoprecipitation of the cell extract growing with anti-F. pedrosoi IgG to which 5 µl of the total 37°C anti-F. pedrosoi rabbit serum was added. SDS-PAGE was conducted at 10% polyacrylamide.

By immunoprecipitation of cell extract (Fig. 11) and culture filtrates (Fig. 12), antigens of 63,000, 35,000, 27,000 and 17,000 were preferentially detected by the anti-F.pedrosoi IgG.

DNA synthesis

DNA$_3$synthesis was measured by labelling the fungal balls with L ^{3}H hypoxanthine. Cultures were prepared as described above, and growth was stopped at 8, 9 and 12 days. Results are shown in Fig. 13. DNA synthesis was maximum at 9 days and diminished at 12 days. Comparison between the cpm values of cells grown with normal IgG and cells grown with anti-F.pedrosoi IgG indicated that the inhibition caused by the latter was about 60%.

This work demonstrated the possibility of inhibiting the growth of F.pedrosoi by specific immunoglobulins. Protein and DNA synthesis can be inhibited by as much as 50 and 60 %. We observed in immunoprecipitation experiments that antigens of 63,000, 35,000, 27,000 and 17,000 were preferentially reco-gnized by the anti-F.pedrosoi IgG and, as we mentioned above, the 63,000 and 35000 candidate proteins could be iodinated. These four antigens may be important in stimulating protective immunity against chromomycosis.

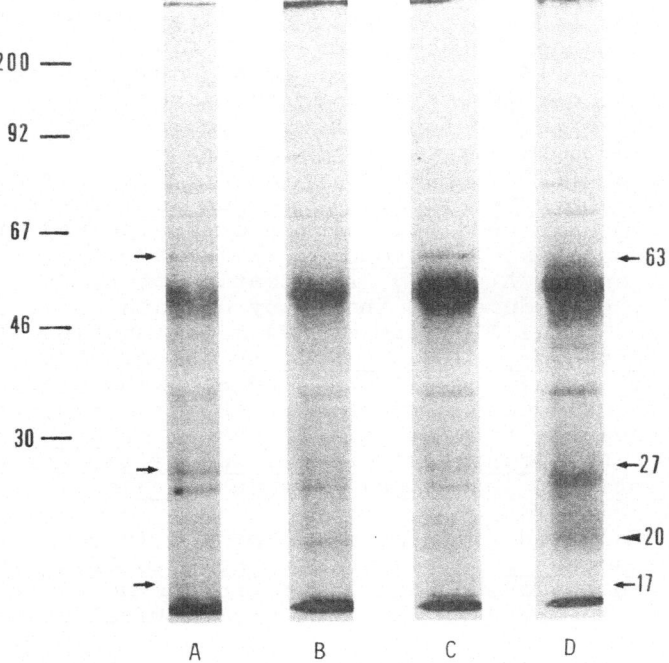

Fig. 12 Immunoprecipitation of the F.pedrosoi polypeptides
(culture filtrate) growing in the presence of anti-F.pedrosoi
IgG for 8 or 10 days (lanes A and C) and with normal IgG
(lanes B and D). SDS-PAGE was conducted at 10 % polyacrylamide.

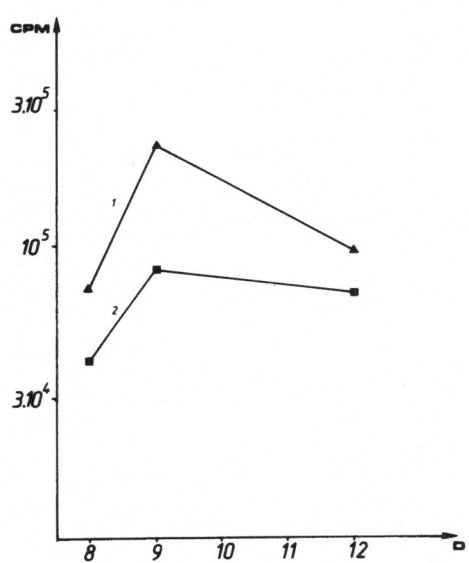

Fig. 13 Growth kinetics of F.pedrosoi (strain 1335) balls
labelled with ^{3}H hypoxanthine incubated at 37°C with normal
(curve 1) or anti-F.pedrosoi immunoglobulinG (curve 2).

REFERENCES

1. H. R. Buckley and I.G. Murray, Precipitating antibodies in chromomycosis, _Sabouraudia_ 5:78 (1966).
2. N.F. Conant and D.S. Martin, The morphologic and serological relationships of the various fungi causing dermatitis verrucosa (chromoblastomycosis), _Am.J.of Trop. Med._ 17:553 (1937).
3. B.H. Cooper and J.D. Schneidau, A serological comparison of _Phialophora verrucosa, Fonsecaea pedrosoi_ and _Cladosporium carrionii_ using immunodiffusion and immunoelectrophoresis, _Sabouraudia_ 8:217 (1970).
4. M.A. Gordon and Y. Al Doory, Application of fluorescent antibody procedures to the study of pathogenic dematiaceous fungi II. Serological relationships of the genus _Fonsecaea,_ _J.Bacteriol._ 80:551 (1965).
5. T. Iwatsu, M. Miyaji, H. Taguchi and S. Okamoto, Evaluation of skin test for chromoblastomycosis using antigens prepared from culture filtrates of _Fonsecaea pedrosoi, Phialophora verrucosa, Wangiella dermatitidis_ and _Exophiala jeanselmei, Mycopathologia_ 77:59 (1982).
6. N. Kurita, Cell-mediated immune responses in mice infected with _Fonsecaea pedrosoi, Mycopathologia_ 68:1 (1979).
7. S. Suzuki and N. Takeda, Serologic cross reactivity of the D- galacto-D mannans isolated from several pathogenic fungi against anti-_Hormodendrum pedrosoi_ serum, _Carbohydr.Res._ 40:193 (1975).
8. W. Torinuki, K. Okohchi, H.Kakematsu and H.Tagami, Activation of the alternative complement pathway by _Fonsecaea pedrosoi, J.Invest.Dermatology_ 83:308 (1984).
9. E. Drouhet and F. Mariat, Etude des facteurs déterminant le développement de la phase levure de _Sporotrichum schenckii, Annales I.Pasteur_ 83:506 (1952).
10. O. Ibrahim-Granet, C. de Bièvre, E. Romain and S. Letoffe, Comparative electrophoresis, isoelectric focusing and numerical taxonomy of some isolates of _Fonsecaea pedrosoi_ and allied fungi, _Sabouraudia J.Med.Vet.Mycol._ 23:253 (1985).
11. U.K. Laemmli, Cleavage of structural proteins during the assembly of the head of bacteriophage T4, _Nature_ 227:680 (1970).
12. W.M. Bonner and R.A. Laskey, A film detection method for tritium labelled proteins and nucleic acids in polyacrylamide gels, _Eur.J.Biochem._ 46:83 (1974).
13. P.H. O'Farrell, High resolution two dimensional electrophoresis of proteins, _J.Biol.Chem._ 250:4007 (1975).
14. G.C. Li and Z. Werb, Correlation between synthesis of heat shock proteins and development of thermotolerance in chinese hamster fibroblasts, _Proc.Natl.Acad.Sci.USA_ 79:3218 (1982).
15. L. Mc Alister and D.B. Finkelstein, Heat shock proteins and thermal resistance in yeast, _Biochem.Biophys.Res. Commun._ 93:819 (1980).
16. H.B. Lejohn and C.E. Braithwaite, Heat and nutritional shock induced proteins of the fungus Achlya are different and under independant transcriptional control. _Can.J.Biochem. Cell.Biol._ 62:837 (1984).
17. T. Yamamori and T. Yura, Genetic control of heat-shock

protein synthesis and its bearing on growth and thermal resistance in Escherichia coli K-12, Proc.Natl. Acad.Sci. USA 79:860 (1982).

18. M.J. Schlesinger, M. Ashburner and A. Tissieres (Eds), Heat shock : from bacteria to man, Cold Spring Harbor (1982).

19. M. Kapoor and A.W.L. Chow, A two dimensional immunoelectrophoretic analysis of the heat shock response exhibited by Neurospora crassa cells, Can.J.Biochem.Cell. Biol. 62:691 (1984).

20. M. Kapoor, A study of the effect of heat and metal ions on protein synthesis in Neurospora crassa cells. Int.J.Biochem. 18:15 (1986)

21. N. Dabrowa and D.H. Howard, Heat shock and heat stroke proteins observed during germination of the blastoconidia of Candida albicans. Infect.Immun. 44:537 (1984).

22. C. Hilton, D. Marki, E. Rikkerink and R. Poulter, Heat shock induces chromosome loss in the yeast Candida albicans. Mol.Gen.Genet. 200:162 (1985).

23. M. Sacco, M. Caruso and B. Maresca, Cloning of Histoplasma capsulatum gene coding for major heat shock protein Hsp 70 and its expression during differentiation. p.75. In abstract meeting on heat shock, cold spring Harbor, New York (1985).

24. H. Towbin, T. Stacehelin and J. Gordon, Electrophoretic transfer of proteins from polyacrylamide gels to nitrocellulose sheets : Procedure and some applications. Proc.Natl.Acad.Sci.USA. 76:4350 (1979).

25. S.W. Kessler, Rapid isolation of antigens from cells with staphylococcal protein A antibody absorbent : parameters of the interaction of antibody antigen complexes with protein. A.J.Immunol. 115:1617.

26. W.M. Hunter and F.C. Greenwood, Preparation of iodine 131-labeled human growth hormone of high specific activity. Nature 194:495.

IMMUNOLOGICAL ASPECTS OF A KERATINASE ISOLATED FROM

TRICHOPHYTON MENTAGROPHYTES

J. Abbink and M. Plempel

Bayer AG, 5090 Leverkusen, FRG

It is well established that dermatophytes and certain
Candida species which are capable of forming mycelia can be
regarded as biphasic fungi, since there is a marked difference,
both morphologically and biochemically, in the in vitro
saprophytic growth form and the in vivo parasitic growth form.
Trichophyton species also exhibit characteristic changes in
their morphology when cultured in vitro under parasitic conditions on nutrient media containing keratin material as a sole
nitrogen source. Under these conditions Trichophyton is able
to secrete specific proteolytic enzymes capable of degrading
keratin (J. Abbink et al., 1987, Proc. Int. Symp. on Bifonazole, New Isenburg, in press). Apparently, these keratinases
are enzymes inducible by the addition of keratin sources as
substitute substrates instead of peptone. In fact, the addition of peptone inhibits the enzyme induction, as does the
absence of keratin. These biochemical results indicate that
the secretion of keratinases plays a pivotal role in the
course of infection. In addition, morphological studies with
the electron microscope suggest a possible occurence of enzymatic digestion in keratinized tissue by dermatophytes and
Candida (D. Grappel, F. Blank, 1972, Dermatologica, 145:245).
The keratinases of several dermatophyte species differ with
respect to biochemical properties such as isoelectric point,
substrate specificity and activity. It is of great significance that azoles like bifonazole exert a pronounced effect on
the secretion of such keratinases at the nannogram level, which
is far lower than the MIC value obtained in normal in vitro
test conditions. For instance, in T. mentagrophytes hair
culture medium, bifonazole has an inhibitory concentration
(IC) 50% value of 0.7 ng/ml and an IC95% of 20 ng/ml with
respect to the bifonazole free medium. Other azoles and
naftifin differ in the intensity of this inhibition (e.g.,
ketoconazole and naftifin have an IC50% of about 10 ng/ml and
an IC95% of 100 ng/ml). Protein biosynthesis inhibitors, like
griseofulvin, do not affect the secretion and activity of these
extracellular keratinases. Therefore, we suggest that the
effect of bifonazole on the secretion of the enzyme is caused
by a membrane structural change due to an impaired ergosterol
biosynthesis. It is interesting to note that the extracellular
keratinases obtained from T. menta-grophytes cultures grown

under superficial, semiparasitic conditions have antigenic properties.

In an Ouchterlony gel diffusion test we could clearly demonstrate an immunulogical reaction with serum obtained from T. mentagrophytes infected guinea pigs three weeks post-infection. Control sera did not show any immunological reaction at all. The IgG fraction of serum from infected guinea pigs was able to inhibit the expression of keratinolytic activity. Guinea pig skin test studies demonstrated the ability of the keratinases secreted by T. mentagrophytes to provoke a phlogistic effect in vivo for a short period of time (probably due to the instability of the enzyme under such conditions), a reaction compatible with a T. mentagrophytes infection. It remains to be determined whether the biochemical and immunological differences between the keratinases of several dermatophyte species are related to different clinical aspects of infections.

DIVERSITY OF ANTIGENIC EXTRACTS FROM THE DERMATOPHYTE

TRICHOPHYTON RUBRUM

P. de Haan and J. Wikler

Department of Dermatology, Academic Hospital, Free
University, 1081 HV Amsterdam, The Netherlands

Patients with Trichophyton rubrum infections show, after
intracutaneous testing with dermatophyte antigens (trichophytin), immunological reactions varying from immediate to
delayed type reactions. In vitro tests demonstrated the
occurence of specific antibodies as well as specific cell-
mediated reactivity using antigens extracted with ethylene
glycol from acetone dried mycelium. However, commercial trichophytins are prepared mainly from heat-treated, concentrated
culture filtrates obtained from 1 to 15 species. It is well
known that the composition of antigens in culture filtrates
produced by the same strain may differ in chemical composition
depending on growth conditions (e.g. substrate and oxygen ten-
sion). The purpose of this investigation was to determine the
possible antigenic diversity of polysaccharides and glycopeptides in culture filtrates of T. rubrum and to compare these
preparations with that of an ethylene glycol extract of acetone dried matieral of T. rubrum.

Seven culture filtrates (I-VII) obtained after inoculation
of glucose peptone medium containing penicillin-streptomycin
were concentrated (20 times) in an Amicon concentration cell
(YM-10 filter). After three washes, the glycoproteins were
isolated from the filtrates by alcohol precipitation (Cruickshank et al., 1960 J. Invest. Dermatol., 35:219). The precipitate was spun off, redissolved in water and lyophilized. A
reference composition of glycoproteins (I-VII) was prepared by
mixing equal amounts of the seven isolated glycoproteins.
Ethylene glycol extraction of acetone dried material was done
according to Cruickshank et al. (1960) and coded as VIII.
Rabbits were immunized with reference antigens I-VII (Cult) or
with VIII (Rubrac). Immunodiffusion in agarose gels was used
for the detection of the number of precipitin lines. ELISA
and ELISA-inhibition were done according to de Haan et al.,
(1983, Clin. Allergy, 13:563; 1985, Int. Archives Allergy Clin.
Immunol., 76:42). The extracts were analyzed in 10% separating
polyacrylamide gels (PAGE). The samples applied consisted of
20 ul extract solution (20 mg/ml) and 10 ul sample buffer
(glycerol/SDS/Tris-H_3PO_4, pH 6.8). The antigens were electro-
blotted onto nitrocellulose paper. The nitrocellulose paper
was cut in 9 strips, each strip containing all blotted glycoproteins or polysaccharides of a sample. Each strip was cut

into 2 identical halves and incubated with a 1:10 dilution of either Cult or Rubrac antiserum in PBS/Tween 20, followed by an incubation with a peroxidase conjugate (concentration 2.5 mg/ml).

In the precipitin text, Rubrac gave only one precipitin line using the ethylene glycol extract as antigen, while 3 lines were found with Cult. Two precipitin lines were found with Rubrac + the mixture of culture filtrates and 5 lines when Cult + the mixture of culture filtrates were used. In the ELISA, using different antigens (culture filtrates and ethylene glycol extract) the titres of the antisera differed only slightly. In the ELISA-inhibition tests, for example, using filtrate as antigen and Rubrac as antiserum only slight differences were found in the quantity of antigen (culture filtrate or ethylene glycol extract) needed for 50% inhibition.

Protein staining of PAGE revealed many bands. The main bands were shown at 20, 50, 68 and 90 kD. In extract VIII only a 50 kD band was observed. Immunoblotting showed only a limited number of bands with Cult. The reference preparation showed bands with 12, 28, 49, 51, 68 and 100 kD. Sometimes antigenic material was detected in the 38-45 kD region. In extract VIII only two bands were observed. Using the Rubrac antiserum, the 28 and 100 kD bands could not be demonstrated in both the reference preparation and the individual extracts. On the contrary, Rubrac demonstrates only the 100 kD band in extract VIII.

The present study clearly demonstrates the diversity of extracts of individual T. rubrum isolates. The results indicate that immunoblotting is very useful to evaluate differences between the preparations. It appears that not only the two types of extracts (from mycelia by ethylene glycol and from culture filtrates), but also the culture filtrate extracts differ in their composition. Moreover, these two types of extracts give rise, after immunization of rabbits, to antisera which detect a number of different components in the preparations. The results obtained with ELISA and ELISA-inhibition suggest that there is a definite diversity in antigenic material. However, these methods cannot discriminate between the individual components. Immunodiffusion and immunoblotting provided more detailed information on the number of components present. Immunoblotting bands which are not detected after protein stain are also observed. From the pathogenetic point of view, one must bear in mind that the diversity reported here reflects differences elicited after in vitro culturing of the fungus. It remains unknown if the host reaction to the invading organism is influenced by these antigenic diversities.

MONOCLONAL ANTIBODIES AGAINST <u>TRICHOPHYTON</u> <u>RUBRUM</u> AND THEIR CROSS-ACTIVITY WITH RELATED AND NON-RELATED ANTIGENS

P. de Haan, E.M.H. van der Raay-Helmer and J. Wikler

Department of Dermatology, Academic Hospital,
Free University, 1081 HV Amsterdam, The Netherlands

The antigenic composition of culture filtrates of <u>Tricho-phyton</u> <u>rubrum</u> is highly variable (P. de Haan & J. Wikler, this book). These results indicated that dermatophyte extract must be standardized. Since polyclonal antisera against antigens of dermatophytes are limited in quantity, we decided to prepare monoclonal antibodies against an extract of <u>T. rubrum</u> with the purpose of examining the antigenic composition of the culture filtrates of <u>T. rubrum</u>. The monoclonal antibodies were also used for the determination of cross-reactivity with related and non-related fungal extracts.

Six monoclonal antibodies were prepared against culture filtrates of <u>T. rubrum</u>. Using an immunoblotting assay, differences in specificity were detected. The 7 culture filtrates I-VII described previously (de Haan & Wikler, this book) were tested. Three out of the 7 culture filtrates gave reactions with all monoclonal antibodies. The 4 other culture filtrates reacted only to 3 or 4 monoclonals.

Cross-reactivity with culture filtrates of <u>T. mentagro-phytes</u> were determined with an immunospot assay. One monoclonal antibody (64) showed cross-reactivity with all <u>T. mentagrophytes</u> culture filtrate antigens. However, staining intensities were less than in the case of <u>T. rubrum</u>. The monoclonal antibody (64) also cross-reacted with <u>Alternaria alternata</u>, <u>Aspergillus</u> <u>fumigatus</u> and <u>A. niger</u>. All fungal extracts were Con A positive. However, the cross-reactivity was not inhibited by α-methylmannoside, a Con A inhibitor.

ENZYME-LINKED IMMUNOSORBENT ASSAY (ELISA) FOR DETECTION

OF ANTIBODIES AGAINST PITYROSPORUM ORBICULARE

J. Faergemann* and S. Johansson**

*Department of Dermatology, University of Gothenburg
Sahlgren's Hospital, S-413 Goteborg, Sweden;
**Institute for Clinical Bacteriology, University
of Uppsala, S-751 05 Uppsala, Sweden

Serum IgG antibodies against Pityrosporum orbiculare were detected by enzyme-linked immunosorbent assay (ELISA). Preliminary experiments with different microtiter plates, various techniques for coating, incubation and washing indicated that the following procedures were most suitable. Titertec P.V.C. Microplates, high activated (HAC) (Flow, England) were used. Wells were coated with whole P. orbiculare cells suspended in a 0.05 M carbonate buffer (10^7 cells ml^{-1}), pH 9.6 for 3h at 37°C. After each incubation step the wells were washed 3 times with phosphate buffered saline (PBS) con-taining 0.05% Tween 20 (PBS-Tw), pH 7.4. The wells were filled with bovine serum albumin (BSA) 20 g L^{-1} in PBS and left overnight at 4°C. The plates were incubated for 2.5h at 37°C with serum diluted in PBS-Tw. Alkaline phosphatase-conjugated rabbit anti-human IgG (Dako-Pads, Denmark) diluted in PBS-Tw was added and left overnight at 4°C. Finally 4-nitrophenyl phos-phate (Sigma, USA) 1 mg ml^{-1} in 1 M diethanolamine buffer, pH 9.8 (Merck, West Germany) was allowed to react. The absorbance was read after 50 min with a Titertec-Multiscan MCC photometer (Flow, England). Ten sera from healthy adult volunteers were used. The experiment was done with 4 different samples from each of the sera and each sample was run in triplicate. The ELISA-method was compared with the indirect immunofluorescent technique.

The sensitivity of the ELISA method was significantly higher than that of the indirect immunofluorescent technique. The ELISA technique for detection of antibodies against P. orbiculare is sensitive and valuable in screening many sera. The development of a method with a high sensitivity and capacity is necessary for complete characterization of the antigen.

STANDARDIZATION OF DERMATOPHYTE ANTIGENS (TRICHOPHYTINS) FOR SKIN TEST PURPOSES

T. Kaaman

Karolinska Institute, Department of Dermatology
Sodersjukhuset, S-100 64 Stockholm, Sweden

The standardization of dermatophyte antigens has in the past been a neglected topic. Various preparations with non-defined antigenic properties have been used for delayed skin test in humans. The results with respect to clinical significance of the skin test have often been conflicting and unreliable. Many different procedures can be used in order to produce an ideal antigen. The final product should be a specific antigen -- characterized, standardized, and adapted for clinical use in humans.

One approach is described here starting from a viable inflammatory dermatophyte isolate which was further cultured under standardized conditions (T. Kaaman, 1985, Current Topics in Medical Mycology, Ed. M.R. McGinnis, Springer-Verlag, New York, 1:117). The resulting dermatophyte mass was then purified according to the ethylene glycol method ending up with a purified trichophytin antigen (a defined glycopeptide). The chemical composition glycopeptide was further investigated including its protein and carbohydrate content. For clinical purposes the cell mediated immune (CMI) response gives the most relevant information about the immune status of the host. Consequently the standardization of a dermatophyte antigen for clinical use should be adapted to these circum-stances, i.e. the specificity and sensitivity of the antigen preparation should be measured with respect to CMI.

It is important to stress the fact that each batch of the purified dermatophyte antigen preparation must be monitored with respect to immunological activity. It is well known that despite a carefully controlled antigen production line, biological activity of the antigen shows considerable variation.

The skin test can be used for this purpose as it reflects all components of the CMI response. The ideal standardization would be testing a group of sensitized or infected humans with the available antigen preparations in question. However, this approach is no longer allowed.

Experimental systems with sensitized guinea pigs have been used and appear to produce reliable results. However, animal experiments are both expensive and time consuming and an exploration of alternative methods have taken place. As CMI is the factor of interst in this context, the search has focused on different lymphocyte function tests. We have previously shown that sensitized guinea pigs develop a CMI response which is expressed as positive skin test and antigen induced lymphocyte stimulation (LST) in vitro (T. Kaaman & J. Wasserman, 1981, Acta Derm. Venereol., 61:213). Furthermore, the results confirm the generally accepted notion of common, major antigenic determinants among dermatophyte species.

This investigation included a cross-reactivity experiment with groups of sensitized guinea pigs exposed to different dermatophyte antigen preparations. Antigenic activity was measured by both skin test and LST. Trichophyton mentagrophytes antigen appeared as a potent and reliable preparation and was applied in subsequent experiments with human lymphocyte donors (J. Torssander et al., 1987, Acta Derm. Venereol, in press). Different batches of purified T. mentagrophytes antigen preparation were exposed to human lymphocytes. The different preparations showed significant variations in their effect on cord lymphocyte reactivity as well as on lymphocytes from patients with dermatophytosis. LST is thus another way to obtain information about immunological specificity and sensitivity and may be useful as an alternative means for standardization of dermatophyte antigen preparations.

STUDIES ON THE ANTIGENS OF <u>PITYROSPORUM</u> <u>(MALASSEZIA)</u> SPECIES

G. Midgley

Department of Medical Mycology, Institute of
Dermatology, London, WC2H 7BJ, UK

Isolates of <u>Pityrosporum</u> <u>(Malassezia)</u> species from human
and animal skin have been maintained as stable forms and
classified according to their lipid dependence and morphology
of the cells. They were divided into three species; <u>P. pachy-
dermatis</u>, <u>P. orbiculare</u> and <u>P. ovale</u>, the last species being
subdivided further into three distinct forms (<u>P. ovale</u> Form 1,
Form 2 and Form 3). The lipid dependent species were isolated
and maintained on a bile salt/oleic acid medium (N.J. van Abbe,
1964, <u>J. Soc. Cosmetic Chemists</u>, 15:609).

Soluble cytoplasmic extracts were prepared from several
isolates of each of the five <u>Pityosporum</u> variants after
disruption in a Dynomill and these were used to raise hyperimmune antisera in rabbits. The interrelationships of isolates was studied by various immunoprecipitation techniques
and these showed that, although cross-reactions occurred
between all the cultures, those which exhibited a similar
morphology were found to have a greater number of common
antigens. Further investigations with adsorbed antisera
revealed that the <u>Pityrosporum</u> variants each contained some
specific antigens.

Proteins of the extracts were compared after treatment
with sodium dodecyl sulphate and separated by polyacrylamide
gel electrophoresis. Grouping the traces of each extract
confirmed the divisions made by the morphological classification although <u>P. ovale</u> Form 1, of which only one extract was
available, was included within the cluster of Form 3 isolates.

Initial studies with concentrated human sera showed that
the number containing antibodies to <u>Pityrosporum</u> was significantly higher among 40 patients with seborrhoeic dermatitis
(73%) than in 40 healthy control subjects (23%), $p<0.0005$.
However, the relevance of this finding in the relationship
between <u>Pityrosporum</u> species and seborrhoeic dermatitis has
still to be determined by further investigations. Although all
the <u>Pityrosporum</u> extracts were equally effective in inducing
antibody formation in rabbits, their response with human sera
was not identical and antibodies were detected most frequently
with <u>P. orbiculare</u> antigen followed by <u>P. ovale</u> Form 3, the two
variants isolated most frequently from human skin. This fact
will be of value when selecting cultures for the preparation of
antigens in further immunological studies with <u>Pityrosporum</u>.

Aspergillus

ANTIGENIC DIFFERENCES BETWEEN EXOANTIGENS, MYCELIAL AND CONIDIAL ANTIGENS IN ASPERGILLUS

Emil J. Bardana, Shirley Craig, Kathe Strangfeld
and Antony Montanaro

Oregon Health Sciences University
Portland, Oregon
USA

INTRODUCTION

Antibodies to the genus Aspergillus have been detected or measured by a variety of fundamentally different immunologic techniques. A classification of antibody detection systems into primary, secondary and tertiary tests has been previously proposed[1] and discussed in relation to Aspergillus antibody [2,3]. Quantitative primary binding tests have permitted sensitive measurement of the total humoral as well as class-specific immune response to a spectrum of Aspergillus antigens, and have been utilized with increasing frequency in the past decade. The availability of a variety of sensitive immunoassays has focused increasing attention on the isolation and characterization of immunoreactive Aspergillus antigens.[4-11] The purpose of many recent studies has been the isolation of antigens which when employed in an antibody detection system are both sensitive and specific in their detection of clinically important variants of aspergillosis. In most instances of human aspergillosis, particularly in aspergilloma, mycelial antigens are felt to play the predominant role in pathogenesis. In the hypersensitivity-induced aspergilloses, inhalation of conidial antigens are felt to be directly related to exacerbations of disease. Exoantigens probably participate in all clinical variants. The purpose of this study was to compare the utility of measuring specific antibody against mycelial, conidial and culture filtrate antigens from A. fumigatus in the serodiagnosis of several aspergillosis variants.

Human test sera were obtained from the patient population and selected employees of the Oregon Health Sciences University. Sera from 177 control and aspergillosis patients were divided into nine major disease categories as shown in Table 1. Criteria used for the diagnosis of various forms of clinical aspergilloses have been previously described.[12,13] The hypogammaglobulinemia category included 8 patients with common variable immunodeficiency and 3 patients with x-linked infantile agammaglobulinemia (Bruton type). All were receiving intravenous serum immune globulin on a regular basis as part of an ongoing study. Individuals with bronchial asthma included 13 individuals with an atopic background with mixed allergic and non-allergic causation and the remainder were individuals with adult-onset, non-allergic asthma. The non-Aspergillus mycoses group included 10 individuals with Farmer's lung disease, 4 patients with severe oral and esophageal candidiasis, 3 with nocardiosis, 2 with severe Trichophyton dermatoses, and one each with geotrichosis, histoplasmosis and coccidiomycosis. The chronic pulmonary infection category included 11 individuals with tuberculous or non-tuberculous mycobacterial disease, 6 with cystic fibrosis, 3 with bronchiectasis, and the remainder a variety of acute and subacute pneumonidities.

Relatively crude and purified antigens were isolated from a stock culture collection of A. fumigatus (Fresenius). Culture conditions and extraction for the various Aspergillus antigens were varied to yield maximum immunological reactivity. Mycelial antigen was produced by culturing in continuously agitated, submerged cultures in Sabouraud broth at 25^0 C for 3 weeks. The mycelial mat was heat killed, filtered, washed, sonified, centrifuged and the supernatant (MS) concentrated 10 times. A portion of the MS antigen was used in the immunodiffusion (ID) testing. The remainder was further purified by ultracentrifugation, radiolabeled with ^{125}I, precipitated with 10% trichloroacetic acid (TCA) and the supernatant (TCA-SS) used in the ammonium sulfate test (Farr assay). Relatively pure conidial antigen was obtained by culturing at 25^0 C for 10 days on Sabouraud agar. Spores were harvested by surface scraping and then killed with merthiolate. The conidia were then sonified, centrifuged and the supernatant (CS) concentrated 10 times. An aliquot was dialyzed for ID studies and the remainder was labeled with ^{125}I, TCA precipitated and the supernatant (TCA-CS) used in the Farr assay. Exoantigen (EA) was produced by culturing for 2 months at

25° C on asparagine broth and inactivating with merthiolate. Separation
was accomplished by filtration with subsequent concentration and
dialysis for use in ID studies. The nitrogen (N) content of the
antigens was determined by a modification of the Micro-Kjeldahl method
using a Technicon Autoanalyzer.

Anti-_Aspergillus_ antibodies to mycelial and conidial antigens were
measured by the ammonium sulfate test in all 177 test sera. Details of
this procedure and control experiments to establish specificity of the
reaction have been described previously.[14,15] Results of this
radioassay are expressed in percent radiolabeled antigen (TCA-SS or TCA-
CS) at a concentration of 0.02 ug N/ml bound by antibody in a 1:5
dilution of patient serum.

Undiluted sera from 25 patients with aspergilloma and 25 patients
with allergic bronchopulmonary aspergillosis (ABPA) were tested for
precipitating antibody to mycelial (MS), conidial (CS) and culture
filtrate (EA) antigens. All antigens were employed at a concentration
of 2 mg N/ml. The number of sera reacting from each disease category as
well as the number of precipitin bands in each reaction were used as
indicators for comparison.

Prior to the estimation of significance between sample means, the
equality of the variances of the sample was tested by variance ratio
(Snedecor's F test). Where the variances were not significantly
different, Student's t test was applied using small sample formulas.
Where variances were significantly different, appropriate formulas for
the calculation of an approximation to Student's t were used.

RESULTS

Circulating antibody to both mycelial and conidial antigen
components were noted in all sera studied using the ammonium sulfate
test. These results are summarized in Table 1. The universal presence
of _Aspergillus_ antibody is consistent with previous studies using a
similar assay system. There was no significant difference in binding to
either the radiolabeled mycelial or conidial antigen in any disease
category studied. However, binding results to either labeled antigen
identified nearly all patients with either aspergilloma or ABPA from
other non-aspergillosis disease categories. When considering all 55
patients with either aspergilloma, ABPA or both, binding to ^{125}I-TCA-SS
ranged between 15.1 and 73.4% with a mean of 47.7%. Using ^{125}I-TCA-CS,
binding in these 55 patients ranged between 17 and 66.2% with a mean of

45.1%. Only one patient with aspergilloma failed to bind above two standard deviations from the normal control population with either labeled antigen.

Table 1. Summary of binding of 177 test sera to both mycelial and conidial antigen using the Ammonium Sulfate Test (Farr Assay)

Percent ^{125}I antigen bound to 1:5 dilution of test sera

DISEASE CATEGORY	No.	TCA-SS (mean ± SD)	Range	TCA-CS (mean± SD)	Range
1. Normal individuals	25	14.2 ± 3.7	4.5-23.6	14.5 ± 3.7	9.2-25.3
2. Hypogammaglobulinemia	11	6.6 ± 1.8*	3.0- 9.9	7.6 ± 2.1*	3.5-10.3
3. Bronchial asthma	25	17.0 ± 5.0	8.4-33.9	17.2 ± 4.5	10.9-26.8
4. Non-Aspergillus mycoses	22	19.5 ± 7.2	9.6-34.5	23.7 ± 9.4	10.6-42.5
5. Chronic lung infection	25	18.7 ± 5.1	11.9-27.9	20.4 ± 7.1	11.0-39.8
6. Aspergilloma	25	50.4± 14.0**	15.1-73.4	48.7 ±12.7**	17.0-66.2
7. ABPA	25	45.3± 11.7**	29.3-66.5	41.9 ±10.0**	21.8-60.4
8. Aspergilloma and ABPA	5	46.7± 11.3**	34.4-59.2	43.5 ±13.2**	26.7-61.3
9. Invasive aspergillosis	14	31.6± 16.9+	9.2-60.0	29.3 ±15.2++	10.1-56.6

SD: Standard deviation; No. = Number of individuals in category

* Significance of difference from all other categories: p< 0.001
** Significance of difference from categories 1, 3, 4 & 5: p< 0.001
+ Significance of difference from other categories 1, 3, 4 & 5: p< 0.001
++ Significance of difference from other categories 1, 3, 4 & 5: p< 0.01

Binding of sera from patients with invasive aspergillosis to either antigen varied from 9.2 to 60% and tended to be higher in patients with limited invasive disease or preexisting saprophytic disease. Binding was normal or diminished in patients with advanced invasive or disseminated aspergillosis. Sera from patients with hypogammaglobulinemia bound labeled antigen significantly less than that found in all other disease categories, ranging between 3 and 10.3%. The reduced antibody that was measured was felt to reflect Aspergillus antibody in the replacement serum immune globulin each of these patients was regularly receiving. This has been confirmed by

direct measurement on several lots of gammaglobulin manufactured by
several pharmaceutical firms.

Sera from 16 of 72 individuals (22%) with chronic pulmonary disease
(Category 3, 4 and 5) bound ^{125}I-TCA-SS above 2 standard deviations from
the normal binding. Binding in these 72 sera ranged from 8.4 to 34.5%
with a mean of 17.4%. Twenty of 72 sera from the same group (27%)
demonstrated elevated binding to the ^{125}I-TCA-CS component with a range
of 10.6 to 42.5% and a mean of 18.9%. Increased binding was most
notable in the non-Aspergillus mycoses and chronic lung infection groups
(Category 4 and 5).

Precipitating antibody was detected in 23 of 25 sera from patients
with aspergilloma and in 22 of 25 sera from patients with ABPA with the
antigen preparations employed. The results are summarized in Table 2.
The mycelial and culture filtrate antigens were nearly equally effective
in detecting precipitating antibody in either disease category. The
mycelial antigen was slightly more reactive than the exoantigen, and
both were significantly more reactive than sporal antigen in detecting
precipitating antibody (p 0.001).

Table 2. Summary of immunodiffusion results in 50 test sera with
mycelial, conidial and culture filtrate antigens

Mean number of precipitin bands and number of undiluted sera
reacting with antigen (2 mg N/ml)

Disease Category	No.	MS		CS		EA	
		Bands	Positive Sera	Bands	Positive Sera	Bands	Positive Sera
Aspergilloma	25	2.86	22	1.48	20	2.52	21
ABPA	25	1.94	21	1.40	16	1.76	22

DISCUSSION

The universality of circulating anti-Aspergillus antibodies
reflects the ubiquity of Aspergillus spores in the atmosphere, soil and
other organic material.[3,12] Prior studies have demonstrated antibody
within most classes of immunoglobulin against Aspergillus antigens. The
premise of this study was based on apparent pathogenetic differences
suggesting that mycelial and conidial antigens may play selective roles

in the induction of saprophytic and hypersensitivity-induced aspergilloses. This concept was initially proposed and demonstrated in animals by Scholer.[16] However, it has not been widely tested in human disease. Scholer felt that the overwhelming predominance of mycelial elements in aspergilloma would make mycelial components ideal for detection of specific antibody in this condition. The same could be hypothesized for early phases of invasive aspergillosis, although Trull et al. reported culture filtrate antigen to be more reactive in such cases.[17] On the other hand, conidial immunogens were thought to play a more significant role in the development of both IgG and IgE-specific antibody in ABPA.[16] This was felt to account for the reduced levels of precipitating antibody in sera from ABPA patients reported by conventional methods.

In order to test the clinical applicability of this premise, mycelial and conidial antigens were prepared and employed in a primary binding test. The mycelial and conidial components represented sonicated supernates that were further purified by precipitating with TCA and radiolabeled for use in the ammonium sulfate test. The major radiolabeled mycelial ligand has been shown to be an immunologically non-precipitating anionic mucopolysaccharide.[14] Significant differences in antigen binding were not observed between sera from patients with aspergilloma and ABPA with the antigen preparations used. However, both antigen components appeared to discriminate patients with aspergilloma and ABPA reasonably well. Patients with invasive aspergillosis were noted to have increased antigen binding only when there was limited invasion or where the condition was preceded by saprophytic disease. This is consistent with the observations of other investigators.[18] Recent studies have suggested that serial determination of _Aspergillus_ antibody using a primary binding system in vulnerable patients may prove helpful in diagnosis.[17] In acutely invasive or widely disseminated aspergillosis, antibody production is inhibited by the secondary humoral and cellular immune deficiency characteristic in these severely ill individuals. As well, expanding _Aspergillus_ components may also be active in suppressing cellular immunity.[19] In such patients, assays for circulating _Aspergillus_ antigen have proven to be more rewarding.[20]

Increased antigen binding to either labeled mycelial or conidial antigen was also demonstrated in some sera from each of the control disease populations with the exception of hypogammaglobulinemia. In some of the patients with non-_Aspergillus_ mycotic infections and chronic

lung infection, increased antigen binding could be secondary to the
presence of shared antigenicity between aspergilli and other fungi,
mycobacteria and a variety of taxonomically unrelated
bacteria.[2,14,15,21,22] The latter might also account for a portion of
the increased binding noted in other disease categories. However, the
presence of saprophytic Aspergillus colonization with resultant antibody
formation cannot be excluded. This has been noted frequently in
patients with tuberculosis, cystic fibrosis and bronchiectasis.[12,23]

Similar results were noted by immunodiffusion with concentrated
mycelial and conidial supernatant antigens and a concentrated culture
filtrate antigen. The mycelial antigen had the highest degree of
reactivity in detecting Aspergillus antibody in both the aspergilloma
and ABPA sera. However, culture filtrate antigen was nearly as
effective as mycelial antigen in detecting precipitating antibodies.
Conidial antigen was the least reactive antigen in detecting
precipitating antibody in either aspergilloma or ABPA sera and failed to
facilitate identification of ABPA patients.

In summary, despite obvious differences in pathogenetic mechanisms
between variants of aspergillosis, enhanced serodiagnosis could not be
demonstrated by employing relatively purified mycelial and conidial
antigen components in a primary binding test, or less purified mycelial,
conidial or culture filtrate antigens in a secondary capacity test.
However, both the mycelial and conidial antigens were useful in the
diagnosis of aspergilloma, ABPA and, to a lesser extent, invasive
aspergillosis. In our experience, the ammonium sulfate test with either
the mycelial or conidial labeled antigen and measurement of total serum
IgE is usually sufficient with clinical and radiologic parameters to
establish a secure diagnosis of either aspergilloma or ABPA. Where
invasive aspergillosis is suspected, a primary binding test for antibody
should be coupled with a sensitive assay for circulating antigen and
used in conjunction with the most intense clinical surveillance.[12]

REFERENCES

1. R.S. Farr, P. Minden, The measurement of antibodies, Ann N.Y. Acad
    Sci 154:107 (1968).
2. E.J. Bardana, Measurement of humoral antibodies to aspergilli, Ann
    N.Y. Acad Sci 221:64 (1974).
3. E.J. Bardana, The clinical spectrum of aspergillosis, I.

Epidemiology, pathogenicity, infection in animals and immunology of Aspergillus, Crit Rev Clin Lab Sci 13:21 (1981).

4. E.J. Bardana, Culture and antigen variants of Aspergillus, J. Allergy Clin Immunol 61:225 (1978).

5. S.J. Kim, S.D. Chaparas, H.R. Buckley, Characterization of antigens from Aspergillus fumigatus. IV. Evaluation of commercial and experimental preparation and fractions in the detection of antibody in aspergillosis. Am Rev Resp Dis 120:1305 (1979).

6. N.J. Calvanico, B.L. DuPont, C.J. Huang, R. Patterson, J.N. Fink, V.P. Kurup, Antigens of Aspergillus fumigatus. I. Purification of a cytoplasmic antigen reactive with sera of patients with Aspergillus-related disease, Clin Exp Immunol 45:662 (1981).

7. E.V. Wilson, S.W. DeMagaldi, V.M. Hearn, Preparative isoelectric focusing of immunologically reactive components of Aspergillus fumigatus mycelium, J Gen Microbiol 130:919 (1984).

8. J.E. Piechura, V.P.Kurup,J.N. Fink, N.J. Calvanico, Antigens of Aspergillus fumigatus. III. Comparative immunochemical analyses of clinically relevant aspergilli and related fungal taxa, Clin Exp Immunol 59:716 (1985).

9. H.F. Kauffman, S. vander Heide, S. van der Laan, H. Hovenga, F. Beaumont, K. deVries, Standardization of allergenic extracts of Aspergillus fumigatus, Int Arch Allergy Appl Immunol 76:168 (1985).

10. J.L. Longbottom, P.K.C. Austwick, Antigens and allergens of Aspergillus fumigatus. I. Characterization by quantitative immunoelectrophoretic techniques, J Allergy Clin Immunol 78:9 (1986).

11. C. Harvey, J.L. Longbottom, Development of a sandwich ELISA to detect IgG and IgG sub-class antibodies specific for a major antigen (Ag7) of Aspergillus fumigatus, Clin Allergy 16:323 (1986).

12. E.J. Bardana, The clinical spectrum of aspergillosis. II. Classification and description of saprophytic, allergic, and invasive variants of human disease, Crit Rev Clin Lab Sci 13:85 (1980).

13. E.J. Bardana, Pulmonary aspergillosis, in: Aspergillosis Y. Al-Doory, G.E. Wagner, Eds., Charles C. Thomas, Springfield (1985).

14. E.J. Bardana, J.K. McClatchy, R.S. Farr, P. Minden, The primary interaction of antibody to components of aspergilli. I. Immunologic and chemical characteristics of a nonprecipitating antigen, J Allergy Clin Immunol 50:208 (1972).

15. E.J. Bardana, J.K. McClatchy, R.S. Farr, P. Minden, The primary interaction of antibody to components of aspergilli. II. Antibodies in sera from normal persons and from patients with aspergillosis. J Allergy Clin Immunol 50:222 (1972).

16. H.J. Scholer, Specific mycelial and conidial antigens from Aspergillus fumigatus, in: Aspergillosis and Farmer's Lung in Man and Animals, R. deHaller, F. Suter Eds., Hans Huber, Bern (1974).

17. A.K. Trull, J. Parker, R.E. Warren, IgG enzyme linked immunosorbent assay for diagnosis of invasive aspergillosis: retrospective study over 15 years of transplant recipients, J Clin Pathol 38:1045 (1985).

18. R.C. Young, J.E. Bennett, Invasive aspergillosis: absence of detectable antibody response, Am Rev Respir Dis 104:710 (1971).

19. S.D. Chaparas, P.A. Morgan, P. Holobaugh, S.J. Kim, Inhibition of cellular immunity by products of Aspergillus fumigatus, J Med Vet Mycol 24:67 (1986).

20. J.R. Sabetta, P. Miniter, V.T. Andriole, The diagnosis of invasive aspergillosis by an enzyme-linked immunosorbent assay for circulating antigen, J Infect Dis 152:946 (1985).

21. P. Tran Van Ky, J. Biguet, T. Vaucelle, Etude d'une fraction antigenique d' Aspergillus fumigatus support d'une activite catalasique, Rev Immunol (Paris) 32:37 (1968).

22. P. Minden, J.K. McClatchy, R.S. Farr, Shared antigens between heterologous bacterial species. Infect Immun 6:574 (1972).

23. A. Kahanpaa, Bronchopulmonary occurrence of fungi in adults, Acta Pathol Microbiol Scand, Sect B (Suppl) 227:9, (1972).

CHARACTERIZATION OF PURIFIED WALL ANTIGENS

OBTAINED FROM ASPERGILLUS SPECIES

Veronica M. Hearn, Elaine V. Wilson and D.W.R. Mackenzie

Mycological Reference Laboratory, CPHL

61 Colindale Avenue London NW9 5HT

INTRODUCTION

Aspergillus belongs to the group of fungi characterized by chitin-glucan walls. Evidence for a layered arrangement of its hyphal cell wall has been known for some time.[2,3] With fluorescein-conjugated lectins as probes, Stoddart and Herbertson[4] reported strong hyphal staining of A.fumigatus incubated with lectins specific for N-acetyl-D-galactosamine and β-D-galactose and moderate staining with one specific for α-D-glucose and α-D-mannose residues. In an earlier report from this laboratory,[5] A.fumigatus surface structure and function were disrupted when exposed to hydrolases with specificities for $(1\rightarrow3)$-β-and $(1\rightarrow4)$-β-D-glucans, $(1\rightarrow4)$α-D-glucans and α-D-mannosyl residues. While fungal polysaccharides located in the outer layer of the cell wall are usually covalently bound to form peptido-polysaccharide complexes, such as the galactomannan of A. fumigatus,[6] little is known about the infrastructure of these macromolecules in the aspergilli.

The application of sodium dodecyl sulphate-polyacrylamide gel electrophoresis (SDS-PAGE) as a means of resolving complex mixtures, coupled with the electrophoretic transfer of these separated substances to nitro cellulose (NC) paper and their localisation with specific ligands,[7] has proved an important tool in structural and immunological analyses. This system has been applied in our laboratory to study the cell wall structure of A.fumigatus. We have removed components of the outer layers from intact mycelium by detergent extraction procedures. By similar means, we have obtained components from an isolated wall fraction and separated them by selective precipitation and affinity chromatography before subjecting them to analysis.

By SDS-PAGE these detergent-solubilized mixtures were resolved and their individual constituents identified by a double-staining silver-Coomassie Brilliant Blue technique. These components were subsequently transferred to NC sheets and their protein/glycoprotein moieties investigated using peroxidase bound either to immunoglobulins or to different lectins. While lectins have lower specificity than immunoglobulins, by combining lectins of different specificities, various aspects of glycoprotein structure can be elucidated. Further information has been obtained by studying the effect of polysaccharolytic enzymes on antigen-antibody binding. In this way an attempt has been made to identify constituent molecules of the outer and inner layers of the cell wall of A.fumigatus.

METHODS

Water-soluble antigens from whole mycelium. A water-soluble (WS) extract was prepared from 3-day-old mycelium of A.fumigatus (strain No. NCPF 2109) by physical rupture of the organism, separation of the soluble constituents from the insoluble residue by centrifugation and concentration of the supernatant against polyethylene glycol 6000.[8]

Concanavalin A (ConA)-affinity purified antigens. A WS preparation was separated by affinity chromatography on ConA-Sepharose. Two peaks were obtained, one of which eluted with the void volume and one which could be eluted only after application of methyl α-D-mannoside.[8] This bound fraction (BF) was carbohydrate enriched.

Wall antigens. The insoluble residue, obtained during the preparation of the WS antigen, was washed repeatedly in hot water (50°-60°C) until the washings were clear. The pellet was extracted with 0.5% (v/v) Triton X-100 (BDH, scintillation grade) in 0.05M NH_4HCO_3 and 30°C for 2x2 hr. periods with moderate stirring. The combined supernates constituted the TE antigen.

Fractionation of TE (wall) antigens. Triton X-100 was removed from the TE extract by a combination of the Bio-Bead SM-2 method of Holloway[10] and by ultra-filtration, following disruption of the detergent micelles with an ethanol-ethylene glycol mixture.[11] The retained material, after dialysis and concentration, served as a source of WS/TE antigens.[12]

A TE preparation was subjected to affinity chromatography on ConA-Sepharose after a pre-equilibration step in 0.1M Tris-HCl buffer, pH 7.2 containing 0.001M $CaCl_2$, 0.001M $MnCl_2$ and 0.2M NaCl. Unbound material and detergent were eluted in this buffer; bound water-soluble material (TBF) was eluted from the column when 0.2M α-methyl mannoside was added to the same buffer.[13]

Surface antigens. Neutral glucose peptone (100 ml) was inoculated with $2x10^6$ spores of A.fumigatus which were then grown for 40 hr. in an orbital shaker incubator (Gallenkamp) at 30°C and 100 r.p.m. The mycelium was harvested as compact spheres (approximate diameter of 1.5-2.0 mm) which were separated from the medium by filtration through a nylon membrane followed by copious washings of distilled water and a final wash in 0.02M Tris-HCl, pH 7.4. For detergent extraction 1 vol. of spheres was suspended in 2 vols. of 0.02M Tris-HCl, pH 7.4, containing either 2% (v/v) Triton X-100 or 2% (w/v) sodium dodecyl sulphate (SDS from BDH, specially pure). Suspensions were incubated for 2 hr. in a shaker water-bath at 50°C. The spheres were pelleted in a microfuge and the supernatants were siphoned off, they constituted the TSE (triton sphere extract) and the SSE (SDS sphere extract).

Antigens from Other Fungal Species. Candida albicans cytoplasmic and mannan antigens (strain No. NCPF 3153) were the routine Mycological Reference Laboratory diagnostic reagents. Coccidioides immitis antigen was a commercial sample from Immuno-Mycologics, U.S.A. Paracoccidioides brasieliensis was a culture filtrate antigen kindly provided by Professor S. DeMagaldi (Caracas, Venezuela).

<u>Production of Rabbit Antisera</u>. Antibodies to <u>A.fumigatus</u> were obtained by the hyperimmunisation of New Zealand White rabbits. Antisera to total WS material and crude wall were obtained as previously described.[5,14] Control serum was obtained by bleeding the rabbits prior to inoculation with antigen.

<u>Human sera</u>. Sera which were strongly positive in an ELISA system to detect antibodies to <u>A.fumigatus</u> were selected and pooled. A control was prepared by combining sera which gave negative readings for <u>A.fumigatus</u> by ELISA.[8]

<u>SDS-Polyacrylamide Gel Electrophoresis (SDS-PAGE)</u>. Samples were subjected to electrophoresis at room temperature on vertical gradient 5-15% SDS-PAGE (1.5mm thickness), using the discontinuous buffer system described by Laemmli.[15] The gels were run at a constant current of 30 mA until the dye front had reached the bottom of the gel.

<u>Preparation of Samples for SDS-PAGE</u>. Sample concentration and the removal of excessive amounts of detergent from TE, TSE and SSE was achieved by treatment of 1 vol. of each extract with 2 vols. of ethanol overnight at 4°C. The resultant precipitates were pelleted in a microfuge and the supernatants were discarded. The pellets were than boiled for 10 min. in "dissolving buffer" (containing 2% SDS, 10% glycerol, 5% mercaptoethanol and a trace of bromophenol blue in 0.064M Tris-HC1 buffer, pH 6.8). Water-soluble antigens (WS, BF, WS/TE and TBF) were boiled for 3 min. in "dissolving buffer" before application to the gel surface.

<u>Detection of Proteins and Glycoproteins on Gels</u>. Gel slabs were fixed in a methanol/acetic acid mixture followed by two washes in ethanol/acetic acid to remove residual SDS. The silver-staining procedure was based on that of Merril et al,[16] using the Bio-Rad kit.

Each silver-stained gel was counter-stained with 0.1% (w/v) Coomassie Brilliant Blue R-250 in 25% (v/v) methanol and 7.5% (v/v) acetic acid for 1 hr. and destained in the same methanol/acetic acid mixture.[17] Separate gel slabs were stained for carbohydrate-containing components by the periodic acid-Schiff (PAS) method.[18]

<u>Electrophoretic Transfer of the Resolved Proteins to Nitro Cellulose (Western Blotting)</u>. The electrophoretically-separated proteins and glyco-proteins together with protein standards were transferred to nitro cellulose (NC) sheets in 0.025M Tris-0.192M glycine - 20% (v/v) methanol at 70V for 2 hr.[19]

The high-molecular weight protein standards (Gibco, U.K.) were visual-ized on blots stained with 0.1% (w/v) Amido Black in 45% methanol (v/v) and 7% acetic acid (v/v) according to Hancock and Tsang.[20] For antigen detection, the NC papers were coated for 1 hr. at 40°C on a rocking platform in either (i) phosphate-buffered saline (PBS) containing 0.15% (v/v), Tween-20 (PBS-T) prior to exposure of the blotted proteins/glycoproteins to antibody or (ii) 0.05M Tris-HC1, pH 7.4-0.2M NaC1 (TBS) containing 3% (w/v) BSA (TBS-A) prior to exposure to ConA-peroxidase and peanut (PNA) agglutinin conjugated to peroxidase or (iii) TBS containing 2% (w/v) polyvinylpyrolidone (PVP) prior to exposure to wheat germ agglutinin (WGA) conjugated to peroxidase.[21]

<u>Detection of Antigen on NC Blots</u>. Partially-purified 1gG fractions, prepared from rabbit and human serum samples by the caprylic acid procedure[22] were used at a dilution of 1:800, in PBS-T (blocking buffer) and incubated

with NC blots at $4°C$ overnight with continuous agitation. The sheets were washed in blocking buffer before exposure of the blots to either swine antiserum to rabbit IgG conjugated to peroxidase or rabbit antiserum to human IgG conjugated to peroxidase, respectively. The peroxidase conjugates were obtained from Dako (Denmark) and each was used at a 1:1000 dilution in blocking buffer. Following several washes the antigen/antibody complexes were localized by staining for peroxidase activity in 0.05M-phosphate buffer, pH 7.4, containing 0.4 mg/ml of 3,3'-diaminobenzidine (DAB) and 0.005% (v/v) of hydrogen peroxide.

Detection of Glycoproteins on NC Blots. Lectins coupled to peroxidase were diluted appropriately to get a suitable colour development. BSA-saturated sheets were overlaid with diluted ConA-peroxidase (2 µg protein/ml TBS-A, containing 84 milliunits purpurogallin, from Sigma) and rocked continuously for 30 min. at room temperature. PVP-saturated sheets were overlaid with diluted WGA-peroxidase (2.5 µg protein/ml TBS-PVP, containing 300 milliunits purpurogallin; Sigma) and rocked continuously for 2 hr. at room temperature. BSA-treated sheets were similarly incubated with diluted PNA-peroxidase (1.25 µg protein/ml TBS-A containing 60 milliunits purpurogallin; Sigma). Following exhaustive washing steps the bound lectin conjugates were reacted with DAB substrate.

Effect of Polysaccharolytic Enzymes on Antigen. Lytic enzyme L_1 from Cytophaga (BDH) was incubated at $37°C$ for 1 hr. with antigens which had been blotted onto NC membranes. The crude enzyme mixture (containing several endo-$(1\longrightarrow3)$-β-and endo-$(1\longrightarrow4)$-β-(glucanases) was used at a concentration of 1% (w/v) in 0.05M phosphate buffer, pH 6.4. The treated antigens were then incubated with antibodies specific for A.fumigatus and the complexes detected as outlined above.

RESULTS

Detergent-extraction of a crude wall preparation of A.fumigatus gave a TE fraction which constituted 0.2% of the wet weight of the mycelial mat (based on protein and carbohydrate content). Further purification of this extract by removal of the detergent molecules yielded a water-soluble fraction (WS/TE) which constituted 0.02% of the starting material. The yield of a TBF peak, following affinity chromatography, resulted in a recovery of 0.01% of the original mycelium. These figures may be compared to an average recovery of 2.5% of the WS preparation from an equal amount of starting material.

Recoveries of detergent-solubilized fractions from the walls of intact mycelium were not quantified except to the extent that SDS-extraction gave higher yields than did Triton X-100. Initial extractions were performed at $30°C$ but, owing to the low yields obtained with Triton, the temperature of the extraction procedure was increased to $50°C$. Microscopic examination by phase-contrast showed no rupture of the hyphal wall under these conditions.

SDS-PAGE Analysis. Components obtained from mycelial wall by the different extraction and separation procedures were analysed on gradient gels (Fig.1a,b). The TE fraction was difficult to assess, since it tended to smear when stained with the silver reagent: this occurred at all concentrations tested. However, its component molecules were found mainly in the range 43-12 kilodaltons (kd). It contained two discrete bands which were sensitive only to Coomassie Blue staining, one of apparent mol.wt. = 26.9 kd, the other with an apparent mol.wt. = 22.4 kd; a band with the same mobility as the former was seen in both the BF and TBF preparations.

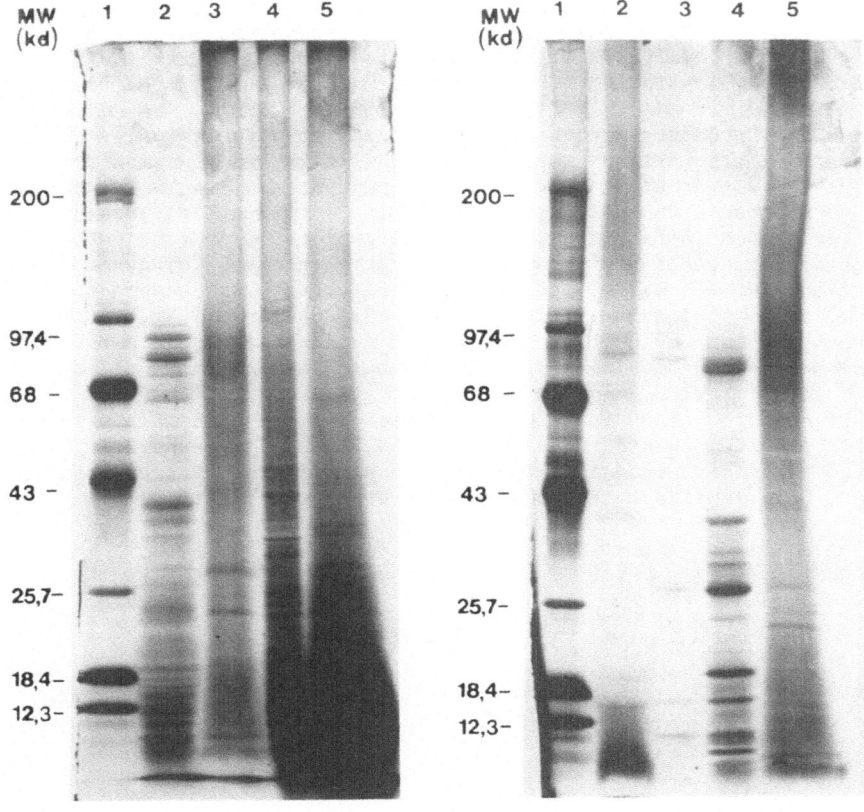

A   B

Fig. 1. SDS-PAGE on 5-15% gradient gel of antigens from A.fumigatus
        mycelium.   The gels were subjected to the combined silver-
        Coomassie Blue-staining procedure.   Lanes(A):1, mol.wt.
        protein standards; 2,WS; 3,TE; 4,TSE and 5,SSE.
        Lanes(B):1,mol.wt. protein standards; 2,WS/TE; 3,TBF;
        4,BF and 5,TE.

        Both WS/TE and TBF were essentially Coomassie Blue-staining fractions,
each containing a relatively small number of major bands.   The most
prominent components in WS/TE had apparent mol.wts. of 91.2, 72.4, 58.8
and 39.8kd while those of the TBF fraction had apparent mol.wts. of 87.0,
38.9, 26.9 and 15.8kd.

        The TSE fraction yielded a very large range of molecular species,
varying in size from >100 to <12kd.   The SSE procedure, on the other hand,
favoured the release of smaller mol.wt. entities (43-12kd).

        The WS and BF samples, included for comparative purposes, proved to
be very complex mixtures with mol.wts. ranging from approximately 100 to
<12kd.   Both stained predominently with Coomassie Blue, the BF bands in
particular, showed intense staining.   The BF and TBF preparations showed
two readily identifiable bands with apparently identical mol.wts. of 87.0
and 15·8kd.

Gels treated with the PAS reagent showed virtually no staining with any mycelial extract. This indicated that few of these fractions had sufficient carbohydrate moieties to allow them to be detected with this relatively insensitive stain.

Western Blots Developed with Rabbit Antibodies to A.fumigatus. Blots of A.fumigatus mycelial antigens were incubated with an 1gG fraction from rabbits inoculated with the WS antigen. All preparations, including the wall extracts, reacted with these antibodies (Fig.2). Similarly, when an IgG fraction raised in rabbits to a crude wall preparation was tested in this system, cross-reactivity again occurred, in this case with the water-soluble fractions (WS and BF). The degree of reactivity was somewhat less than with homologous antibodies. In general, the smaller mol.wt. entities in each extract were either poorly or non-immunogenic. Control rabbit immunoglobulin was unreactive with Aspergillus antigens.

When antigens from other fungal species were monitored in this system, some P.brasiliensis components (mol.wts. 60-40kd) showed trace reactivity towards an IgG fraction raised in rabbits to an A.fumigatus WS preparation. C.albicans and Cocc.immitis antigens showed no cross-reactivity.

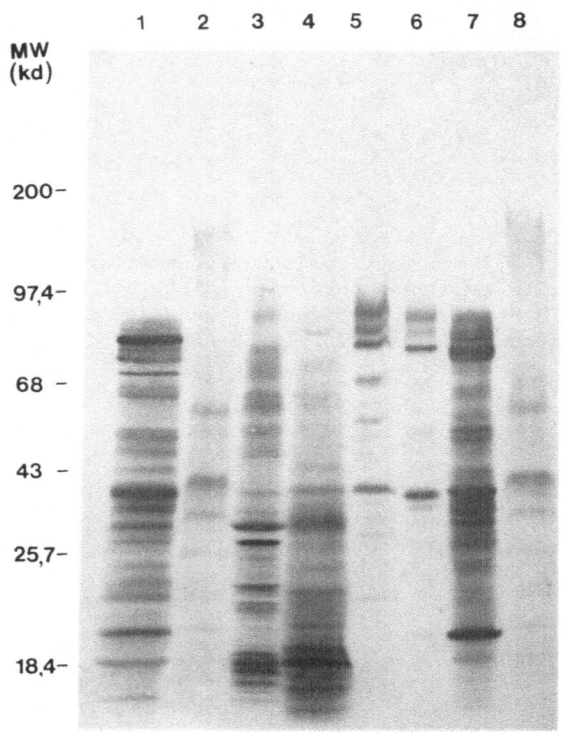

Fig. 2. A.fumigatus mycelial antigens separated by SDS-PAGE and transferred to NC sheets and stained with rabbit anti-WS 1gG. Lanes:1,WS; 2,TE; 3,TSE; 4,SSE; 5,WS/TE; 6,TBF; 7,BF and 8,TE. Mol.wts. of standard proteins (in kilodaltons) are listed to the left of the picture.

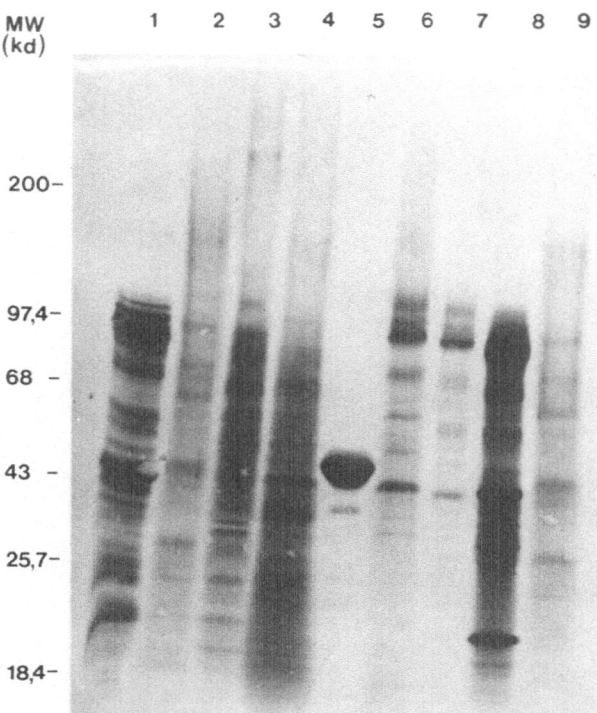

Fig. 3. A.fumigatus mycelial antigens
separated by SDS-PAGE and
transferred to NC sheets and
stained with a pooled IgG fraction
from patients showing Aspergillus
antibodies on ELISA.   Lanes: 1,WS;
2,TE; 3,TSE; 4,SSE; 5,WS/TE; 6,TBF;
7,BF and 8,TE.

Western Blots Developed with Human Antibodies to A.fumigatus.   When
antigen blots were exposed to a pooled sample of human antibodies to
A.fumigatus, reactions were again obtained with all extracts (Fig.3).
However, the number of reactive components was somewhat less than that
seen with the rabbit system.   The WS preparation gave three major band
groupings, one with mol.wts. between 100-70kd, a second with mol.wts. between
45-35kd and a final group with mol.wts. ranging between 23-20kd.   A somewhat
similar, but not identical distribution of components was seen with the BF
fraction.   The TE extract gave a dominant band of mol.wt. = 150kd which
was not visible on protein staining; a similar but possibly non-identical
band was seen in the rabbit system.

WS/TE and TBF showed a triplet of large mol.wt. entities (97, 87 and
79.4kd) which reacted strongly with human antibodies.   A small number of
less-reactive, smaller mol.wt. components (in the 43-30kd region) were also
detected.   The TSE and SSE extracts in contrast, demonstrated a very
heterogenous array of molecules when reacted with pooled human antibodies.
The SSE looked less reactive than the TSE fractions.   Control human sera
showed a low level of antibody reactivity towards some of the larger mol.wt.
components of the WS preparation (those in the 100-70 and 23-20kd areas).

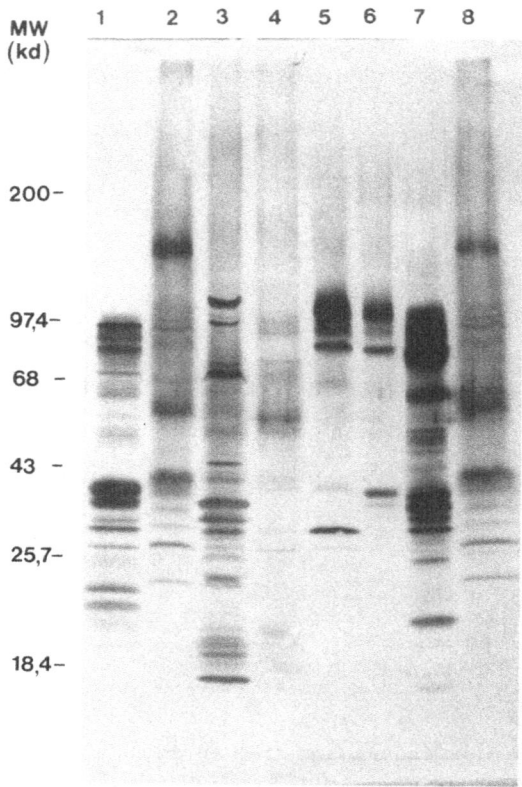

Fig. 4. A.fumigatus mycelial antigens separated
by SDS-PAGE and transferred to NC sheets
for detection of N-glycoproteins with a
ConA-peroxidase reagent. Lanes: 1,WS;
2,TE; 3,TSE; 4,SSE; 5,OVA; 6,WS/TE;
7,TBF; 8,BF and 9,TE. OVA (ovalbumin,
mol.wt. 43,000) was included as a positive
control. Mol.wts. of standard proteins
(in kilodaltons) are listed to the left
of the picture.

Western Blots Developed with Lectins. Blots of gels loaded with
antigenic extracts exhibited a wide range of molecular species when stained
with ConA conjugated to peroxidase. Moreover, it appeared that most, if
not all, of those bands which stained positively for protein also contained
sugar residues which bound to ConA. Only the small mol.wt. moieties in
these extracts failed to respond to the lectin (Fig.4).

When stained with WGA virtually all WS and BF moieties reacted. Many
bands in the TSE and SSE (especially the medium to small mol.wt. components)
were positively stained as were some of the major components of the WS/TE
and TBF fractions. Only a few, small mol.wt. components (25-18kd) of the
TSE and SEE extracts bound to PNA. None of the other antigenic preparations
bound to this lectin under the conditions tested.

Effect of Hydrolytic Enzymes on Antigen-Antibody Binding.
Cytophaga lytic enzyme was incubated with blotted antigens for 1 hr. at 37°C
before exposure to human IgG specific for A.fumigatus.   This enzyme
virtually abolished all antigen-antibody binding under these conditions.

DISCUSSION

Detergent extraction of the hyphal wall material yielded a complex
array of molecules varying in unit site from >100,000-10,000 mol.wt.
Most of these moieties appear to be glycoproteins, staining positively
both for protein and sugars, especially with ConA which has a requirement
for at least two non-substituted or 2-0-substituted α-mannosyl residues.
Many of the same molecules also bound WGA, which reacts with sialic acid
or N-acetyl glucosamine residues.   Some components of TSE and SSE reacted
with PNA which binds preferentially to galactose residues linked
β-1,3 to N-acetylgalactosamine.   Interestingly, some of the wall fractions
reacted almost exclusively with Coomassie Blue while others were more
sensitive to the silver stain.

Previous work had provided some evidence for the complicated nature
of these extracts.   Crossed immunoelectrophoresis of TE and WS/TE analysed
with rabbit and human antisera revealed up to 20 precipitating systems
with the former and as many as 14 with the latter.   Many of these antigen/
antibody precipitates could be abolished by incorporating ConA in the
system.[14]   In addition TE, TBF and WS/TE were all sensitive to pronase,
as judged by the disappearance of precipitin lines in double-diffusion
tests.[12,13]   The WS/TE fraction was monitored in an ELISA system, and
proved to be comparable to the original WS antigen in its ability to detect
Aspergillus antibodies in serum from patients suffering from a variety
of Aspergillus-related diseases.[23]

All extracts proved highly immunogenic and showed a wide spectrum of
molecules which reacted with IgG fractions raised in rabbits to specific
A.fumigatus antigens.   Reactions of both wall and what are predominantly
cell-sap antigens were comparable with homologous or heterologous antiserum.
This degree of cross-reactivity was not entirely unexpected.   Earlier
experiments had shown that when cytoplasmic antigens derived from
protoplasts and wall antigens (WS/TE) were tested in an ELISA system with
their respective antisera, a high level of cross-reactivity occurred in
the heterologous systems.[23]

All fractions incubated with a pool of human sera positive for
Aspergillus, were again highly reactive.   Only the small molecular weight
species in most fractions failed to respond; these molecules also lacked
ConA-binding sites.   Exposure of antigens to Cytophaga lytic enzyme
substantially abolished antigen-antibody binding, which implicates (1→3)-
and (1→4)-linked β-glucans in this reaction, cf Hearn.[5]   Some antigenic
components were recognised by immunoglobulins from control sera.   This
phenomenon has previously been noted by Schønhyder and Andersen,[24] who
prepared 11 antigenic fractions of A.fumigatus and found that healthy
controls had antibodies to 7 of the fractions.   Recently, the same authors
have reported antibodies in healthy subjects to a 470kd antigen.[25]   In
a study of patient response to A.fumigatus antigens using an SDS-PAGE/
immunoblot system, Matthews et al.,[26] reported two relatively large
molecular weight components which responded to antibodies present in the
control subjects.   They described a 40,000 mol.wt. component which was
not recognised by control sera and gave a positive response with 13 of
16 patient's sera.   It was suggested that this constituted a clinically
significant antigen.

The predominant antigens of the WS/TE and TBF fractions have mol.wts. in the region of 91-58,8kd. The two preparations appear to contain a similar but not an identical range of molecules. The 40,000 dalton antigen described by Matthews & Burnie[26] may occur, at least in WS/TE (mol.wt. = 39,800). It may well be present in sphere extracts, given the large variety of molecules released during this process. Triton X-100 appeared to release components with a greater range of molecular sizes and these also seemed somewhat more immunogenic than their SDS-counterparts. This may be a function of the milder extraction procedure represented by the use of an non-ionic detergent. Both extraction procedures emphasise the complex nature of the outer wall of A.fumigatus.

Results obtained from the analysis of Western blots indicate that cell sap and hyphal wall share many antigenic determinants. There is some evidence for the presence in these extracts of antigens which are expressed preferentially, or perhaps exclusively, in the wall layers. The TSE extracts contain two Coomassie-blue staining bands with mol.wts. >100kd, one of which appeared to react strongly with Aspergillus-specific antibodies. These bands were not detectable in the WS and BF fractions. The TE extract contained a component (approx. mol.wt. of 150kd) which also reacted to Aspergillus antibodies. This component bound ConA but was not apparently present in sufficient quantity to be visualized by either silver or Coomassie Blue staining. Molecules as large as the antigen with catalase activity, described by Schønhyder, were not seen although he has found antigens with high antibody reactivity in the 160-50kd range.[24] It may be that under milder extraction conditions a greater number of large-molecular moieties could be released from the cell surface. Taken in conjunction with the extraction procedure used here, they could be expected to yield many components for further study of the wall architecture of Aspergillus.

REFERENCES

1. S. Bartnicki-Garcia, Cell wall chemistry, morphogenesis and taxonomy Ann. Rev. Microbiol. 22:87 (1968).
2. K. Horikoshi and K. Arima, X-ray diffraction patterns of the cell wall of Aspergillus oryzae, Biochim. Biophys. Acta. 57:392 (1962).
3. A. T. Bull, Chemical composition of wild-type and mutant Aspergillus nidulans cell walls. The nature of polysaccharide, and melanin constituents, J. gen. Microbiol. 63:75 (1970).
4. R. W. Stoddart and B. M. Herbertson, The use of fluorescein-labelled lectins in the detection and identification of fungi pathogenic for man: a preliminary study, J. med. Microbiol. 11,315 (1978).
5. V. M. Hearn, Surface antigens of intact Aspergillus fumigatus mycelium: their localization using radiolabelled protein A as marker, J. gen. Microbiol. 130:907 (1984).
6. I. Azuma, H. Kimura, F. Hirao, E. Tsubura, Y. Yamamura and A. Misaki, Biochemical and immunological studies on Aspergillus. III Chemical and immunological properties of glycopeptides obtained from Aspergillus fumigatus, Jap. J. Microbiol. 15-237 (1971).
7. W. N. Burnette, "Western Blotting": Electrophoretic transfer of proteins from sodium dodecyl sulfate-polyacrylamide gels to unmodified nitro-cellulose and radiographic detection with antibody and radioiodinated protein A, Anal. Biochem. 112-195 (1981).
8. E. V. Wilson and V. M. Hearn, Use of Aspergillus fumigatus mycelial antigens in enzyme-linked immunosorbent assay and counter-immuno-electrophoresis, J. med. Microbiol., 16:97 (1983).
9. V. M. Hearn and D. W. R. Mackenzie, The preparation and chemical composition of fractions from Aspergillus wall and protoplasts possessing antigenic activity, J. gen. Microbiol. 112:35 (1979).

10. P. W. Holloway, A simple procedure for removal of triton X-100 from protein samples. Anal. Biochem. 53:304 (1973).

11. C. E. Frasch, Removal of detergent by ultrafiltration, in: "Dialog", No.13, Amicon BV, Oosterhout (NB), Holland.

12. V. M. Hearn and D. W. R. Mackenzie, Antigenic activity of sub-cellular fractions of Aspergillus fumigatus, Medical Mycology, Zbl. Bakt. Suppl.8, Preusser, eds. Gustav Fischer Verlag, Stuttgart (1980).

13. V. M. Hearn and D. W. R. Mackenzie, The preparation and partial purification of fractions from mycelial fungi with antigenic activity, Mol. Immunol. 17:1097 (1980).

14. V. M. Hearn and D. W. R. Mackenzie, Mycelial antigens from two strains of Aspergillus fumigatus: an analysis by two-dimensional immuno-electrophoresis, Mykosen 23:549 (1980).

15. U. K. Laemmli,Cleavage of structural proteins during the assembly of the head of bacteriophage T4, Nature 227:680 (1970).

16. C. R. Merril, D. Goldman, S. A. Sedman, and M. H. Ebert, Ultra-sensitive stain for proteins in polyacrylamide gels shows regional variation in cerebrospinal fluid proteins, Science 211:1437 (1981).

17. J. K. Dzandu, M. E. Deh, D. L. Barratt, and G. E. Wise, Detection of erythrocyte membrane proteins, sialoglycoproteins and lipids in the same polyacrylamide gel using a double-staining technique, Proc. Natl. Acad. Sci. U.S.A. 81:1733 (1984).

18. R. A. Kapitany and E. J. Zebrowsky, A high resolution PAS stain for polyacrylamide gel electrophoresis, Anal. Biochem. 56-361 (1973).

19. H. Towbin, T. Staehelin, and J. Gordon, Electrophoretic transfer of proteins from polyacrylamide gels to nitrocellulose sheets: procedure and some applications, Proc. Natl. Acad. Sci. U.S.A. 76:4350 (1979).

20. K. Hancock and V. C. W. Tsang, India ink staining of proteins on nitro-cellulose paper, Anal. Biochem. 133:157 (1983).

21. J. R. Bartles and A. L. Hubbard, ^{125}I-Wheat germ agglutinin blotting: increased sensitivity with polyvinylpyrolidone quenching and periodate oxidation/reductive phenylamination,Anal. Biochem. 140-284 (1984).

22. M. Steinbuch and R. Audran, The isolation of IgG from mammalian sera with the aid of caprylic acid, Arch. Biochem. Biophys. 134-279 (1969).

23. E. V. Wilson and V. M. Hearn, A comparison of surface and cytoplasmic antigens of Aspergillus fumigatus in an enzyme-linked immunosorbent assay (ELISA), Mykosen 25:653 (1982).

24. H. Schønheyder and P. Andersen, Fractionation of Aspergillus fumigatus antigens by hydrophobic interaction chromatography and gel filtration, Int. Arch. Allergy Appl. Immunol. 73:231 (1984).

25. H. Schønheyder, C. Møller-Hansen, P. Andersen, and A. Stenderup, Serum antibodies to Aspergillus fumigatus in Danish farmers, Sabouraudia 23:93 (1985).

26. R. Matthews, J. P. Burnie, A. Fox, and S. Tabaqchali, Immunoblot analysis of serological responses in invasive aspergillosis, J. Clin. Pathol. 38:1300 (1985).

ANTIGEN DETECTION IN INVASIVE ASPERGILLOSIS

Bertrand Dupont

Pasteur Institute Hospital
Paris, France

Aspergillus in a ubiquitous mold. There are hundreds of species in the
genus Aspergillus, but only a few have been reported to cause diseases.
Aspergillus fumigatus accounts for the bulk of invasive disease. A.
flavus is the second most important species and appears to be more frequent in tropical areas. A. niger, A. nidulans and a dozen other species
are less frequently involved.

Aspergillosis is usually acquired by inhalation of airborne spores.
Although spores are continually inhaled.in the respiratory tract, the
disease is rare and requires local or general predisposing factors. The
pathogenic spectrum of Aspergillus is varied. Colonization of a preexisting
cavity may result in aspergilloma. Allergy or hypersensitivity predominate
in extrinsic asthma, allergic bronchopulmonary aspergillosis, and extrinsic
allergic aveolitis. Invasive pulmonary aspergillosis may lead to widespread
disseminated disease. Acute invasive pulmonary aspergillosis leads to
death in 2 to 3 weeks and dissemination to other organs occurs in one third
of cases (Table 1) (8,15,29). Chronic necrotizing aspergillosis also called
semi invasive aspergillosis, is a distinct entity (4,10), and aspergillosis
in chronic granulomatous disease has its own features (5). Aspergilloma,
acute bronchitis and empyema are other forms of pulmonary aspergillosis.
Distinct clinical forms of the disease are of importance in determining the
amount of antigen excreted, the capacity to induce antibody, and the rapidity
of death.

Invasive aspergillosis occurs most frequently in patients with
acute leukemia, bone marrow transplantation and in kidney, heart, and liver
recipients. Cytotoxic and corticosteroid therapy are major predisposing
factors (8,11,12,15,19.29) which have resulted in increasing morbidity and
mortality due to invasive aspergillosis (9).

Two major pitfalls are the difficulty in diagnosing aspergillosis and
the low efficacy of available antifungal treatments. Blood culture are almost always negative despite dissemination demonstrated at autopsy (Table 1).
The diagnosis is certain with histological demonstration of the fungus and
a positive culture. However, biopsy is often contraindicated by thrombo-
cytopenia and the poor general condition of the patient. Demonstration
of Aspergillus in the respiratory tract lacks sensitivity. Presence of
hyphae and/or positive culture in a compromised host are highly suggestive
of infection especially if the sample is obtained from bronchoalveolar

Table 1. Clinical features of invasive
aspergillosis in three large
series of patients.

	Young et al. 1970 (1)	Meyer et al. 1973 (2)	Fisher et al. 1981 (3)
Number of patients	98	93	91
Malignant underlying disease	90	92	90
Pulmonary localization	92	90	83
Only pulmonary	60	68	63
Blood culture positive	0	0	0
Proven cases by histology and/or culture	98	93	78

lavage or distal brushing, if the patient is a child or if the species
isolated is A. fumigatus or A. flavus (16,17,24). About 2% of normal patients
have Aspergillus in sputum and this can be as high as 10% in patients with
chronic bronchopulmonary diseases. Serology for antibody is often negative;
this can be due to the fact that samples are drawn too early before antibodies are produced, to immunosuppression, to circulating immune complexes,
or to immune complexes sequestered in infected tissues. Causes of positive
or negative test for antibody are shown in Table 2. Detection of circulating
antigens has been developed to improve diagnosis of aspergillosis (6).
Studies performed in several laboratories have examined the detection of
Aspergillus antigens in serum, urine and bronchoalveolar lavage fluid.
These studies were performed in experimentally infected animals and in
patients. Table 3 summarizes publications in which an animal model was
used. These studies are difficult to compare because the methodology of
antigen detection has varied; in addition, while in most studies antigen
was detected in serum, in other reports antigen was found in urine (13) and
or in bronchoalveolar lavage fluid (1). Various animal species have been
used, either intact or more often immunosupressed with corticosteroids and/or
cyclophosphamide. Sensitivity and specificity of the tests are often not
specified. The day of positivity of antigen detection compared to time of
death of animals is seldom indicated. Nevertheless, in the whole these
tests have a good sensitivity particularly RIA and ELISA, and a good
specificity close to 100%.

Absence of standardization of methods and reagents is also evident in
studies performed on patients (Table 4). For example, serum is tested
either with or without dissociation of immune complexes. Although sensitivity of the tests is slightly lower than in animal models, specificity
is still excellent. The majority of tested patients had acute leukemia.

Table 2. Aspergillosis - Causes of positive or
negative tests for antibody.

Presence of antibody	Absence of antibody
Aspergilloma	Antibodies not yet produced
Allergic bronchopulmonary aspergillosis	Intense immunosuppression
	Antibody-antigen complexes
Semi-invasive aspergillosis	Low antibody response
Aspergillosis in chronic granulomatous disease	Antigen used not cross reacting with infecting antigen
Invasive aspergillosis	Technical problems
Chronic bronchitis	
Bronchial ectasis	
Cystic fibrosis	

Table 3. Aspergillosis – Antigen detection in experimentally infected animals.

Authors (ref.)	Test used	Antigen used	Biological fluid	Animals number +/tested	% Sensitivity	% Specificity
White et al. 1977	CIE	NS	Serum	Mouse individually 3/14 pooled NS	21 100	NS
Lehmann et al. 1978	CIE	Carbohydrate	Serum urine	Rabbit 8/8	100	100
Rippon et al. 1978	HA ID	Crude	Serum	Rabbit NS	NS	NS
Richardson et al. 1979	CIE ELISA	Crude	Serum	Mouse 23-27/37 Rabbit 1/1	62-73 100	100 (8)
Lehmann et al. 1979	CIE	Carbohydrate	Serum	Mouse 11/27	41	100 (36)
Weiner et al. 1979	RIA	Carbohydrate	Serum	Rabbit 40/51	78	100 (101)
Shaffer et al. 1979	RIA	Carbohydrate	Serum	Rabbit 4/6	67	NS
Andrews et al. 1981	RIA	Carbohydrate	Broncho alveolar lavage	Rabbit 10/11 2-4/10	90 20/40	97 (36)
Sabetta et al. 1985	ELISA	Carbohydrate	Serum	Rabbit 5/6	83	NS

CIE : counterimmunoelectrophoresis – HA : hemagglutination – ELISA : enzyme -linked immunosorbent assay
ID : immunodiffusion – NS : not specified – RIA : radioimmunoassay
Specificity : number between brackets indicates number of animals tested

Table 4. Aspergillosis - Antigen detection in patients.

Authors (ref.)	Test used	Antigen used	Biological fluid	Number +/ tested	% Sensitivity	% Specificity
Lehmann et al. 1978	CIE	Carbohydrate	Serum	1/1	100	100
Shaffer et al. 1979	RIA	Carbohydrate	Serum	3/3	100	100 (21)
Lehmann et al. 1979	CIE	Carbohydrate	Serum	6/60	10	100 (216)
Weiner 1980	RIA	Carbohydrate	Serum	4/7	57	100 (70)
Andrews et al. 1981	RIA	Carbohydrate	Broncho alveolar lavage	4/4	100	94 (35)
Weiner et al. 1983	RIA	Carbohydrate	Serum	7/9	77	100 (41)
Sabetta et al. 1985	ELISA	Carbohydrate	Serum	11/19	58	100 (42)

See footnotes Table 3

In six studies serum was the biological fluid tested. In Andrews' work bronchoalveolar lavage fluid was used which failed to discriminate between, invasive aspergillosis and aspergilloma (2). The amount of detected antigen is in nanograms and ranges from a few up to hundreds of nanograms. The most sensitive tests are ELISA and RIA with a lower limit of sensitivity of about 5 to 10 nanograms/ml in concentrated samples.

Following this rapid overview of the problem data will be summarized from the study we performed with Dr. J.E. Bennett. This work consisted in detecting a galactomannan, extracted from A. fumigatus, both in serum and in urine of experimentally infected rabbits and humans with aspergillosis (6).

Preparation of galactomannan was carried out as follows: A. fumigatus was grown in a shaken chemically defined broth medium for 4 days. Mycelium was ground in a Braun homogeneizer. After centrifugation the supernatant was ten-fold concentrated. Proteins were removed with ammonium sulfate. Part of the remaining proteins were retained with DEAE cellulose chromatography. The final step of purification consisted of chromatography on a column with Concanavalin A. The final product was a glactomannan that contained 0.7% nitrogen (3). Glactomannan detection was performed by ELISA and RIA. Our ELISA was a competitive inhibition test in which the antigen was coated on the plate, and the test sample was preincubated with a reference serum from a hyperimmunized rabbit. RIA used the galactomannan conjugated to ^{125}I tyramine (3).

Sensitivity was 10 to 40 ng/ml for ELISA and RIA. Sera tested for antigen were always diluted 1:4 and boiled for dissociation of immune complexes. Urine samples were dialyzed against PBS prior to testing.

Rabbit urine was collected on a plastic sheet under the cage floor. A screen was interposed between the floor and the plastic sheet to prevent stool contamination.

In experimental infection rabbits were injected intravenously with viable conidia. All rabbits survived an inoculum of 3×10^6 conidia or less, but 20 of 22 rabbits died within 7 days after inocula of at least 1×10^7 conidia.

Detection of galactomannan in rabbit sera was positive in only 4 of 12 lethally infected animals. Antigen concentrations ranged between 108 and 356 ng/ml.

Antigen detection was much more encouraging using urine of infected rabbits. As shown in Table 5 galactomannan was found in urine of all 13 fatally infected rabbits, but only in 3 of 16 surviving animals.

There was virtually no delay in onset of antigenuria. Antigen concentrations tended to rise as infection progressed, but were not closely correlated with the size of the original inoculum. Three rabbits survided the infection with an inoculum of 5.8×10^6 conidia. For these three animals antigenuria converted from positive to negative. We looked for the possibility that urine antigen reflected growth of Aspergillus in urine. Ten urine samples from four rabbits given 55 million spores were tested. All contained antigen. Uncentrifuged as well as the centrifuged sediment of 20 ml urine were cultured. A single urine grew one colony of A. fumigatus in one tube culture, and no other urine contained this fungus.

The possibility that growth of Aspergillus in the kidney accounted for all the urine antigen was also examined. This hypothesis appeared unlikely in view of the rising urine antigen concentration despite stable kidney colony counts.

Table 5. Galactomannan in urine of rabbits injected
intravenously with live _Aspergillus_ conidia.

Inoculum conidia X 10^6	Number of rabbits	Antigen in urine Presence	Antigen in urine Maximum ng/ml	Death	Day of death post inoculation
65	4	+	660	YES	1 - 4
21.6	4	+	1990	YES	2 - 4
10	4	+	1900	YES	5 - 7
5.8	1	+	640	YES	6
5.8	3	+	270	NO	
2.8	5	−	0	NO	
1.2	4	−	0	NO	
1	4	−	0	NO	

Presence of antigen in urine led to a study of galactomannan molecular
weight. Dialyzed and undialyzed urine from one infected rabbit were eluted
through a Sephadex G 200 column. Collected fractions were analyzed by
ELISA. Urine antigen had a disperse molecular weight. When compared with
linear dextran the molecular weight is centered around 18000 to 20000
daltons. Such a molecular weight is compatible with glomerular filtration.
It allows urine concentration, for example, in a Minicon B 15.

Because galactomannan was present in urine of infected rabbits and
because urine is easy to obtain, antigen detection in urine was studied
in 12 patients with aspergillosis proven by culture and histological
examination (7).

Antigen was detected in serum in only 2 of 12 patients and galactomannan concentrations were near the limit of sensitivity of the assay.

Antigen detection in patients' urine was much more relevant to diagnosis
of aspergillosis. Fifty-one urines were collected from 13 patients with
biopsy or autopsy proven aspergillosis. Galactomannan was detected in urine
from 7 patients. Antigen concentrations were in the range of 1 to 83 ng/ml.
Urines were 10-fold concentrated prior to testing (Table 6). Patient 3
was especially interesting. This patient had urine antigen each of the 9
days immediately preceding surgical excision of a lung lesion. Serum
antigen was negative prior to surgery. In the first 24 hours after excision,
urine antigen assays gave only borderline values. Daily urines for the next
11 days were clearly negative. The patient promptly improved post-
operatively. This case may be relevant to cases 10 and 11 who had no antigenuria, when first tested 2 and 3 days after surgical excision of lung
lesion. In patient 8 negative antigenuria may have resulted from prolonged
intravenous amphotericin B that preceded urine assay, although hyphae were
visible in the prostate at autopsy. Patient 9 without antigenuria had
only a small palatal lesion that responded to amphotericin B. Patient
13 had the hyper IgE syndrome and had an aspergilloma due to A. terreus.
He was negative 3 weeks before excision, and a few hyphae were present in
pulmonary tissue around the fungus ball cavity. Patient 12 seems to be a
true false negative. We was a child with a chronic granulomatous disease
with lung, skin and brain involvement.

In this series, 6 of 8 fatal cases had antigenuria. In one other
series 8 patients were suspected of having invasive aspergillosis, but an
alternative diagnosis was proven in all but one who remained undiagnosed,
despite lung biopsy. None of the eight patients had urine galactomannan.

Table 6. Antigen detection in urine of patients with
aspergillosis (adapted from Dupont et al. (7).

Case	Underlying disease	Species isolated	Number +/ Tested
1	ALL	A. fumigatus	4/4
2	Hemolytic anemia corticosteroids	A. fumigatus	1/1
3	Hodgkin's	A. fumigatus	9/21
4	Breast cancer corticosteroids	A. fumigatus	6/6
5	Behçet corticosteroids	A. fumigatus	2/2
6	BMT	A. fumigatus and A. glaucus	1/1
7	All, corticosteroids	A. fumigatus	1/1
8	Hemolytic anemia corticosteroids	A. fumigatus	0/2
9	AML	A. flavus	0/3
10	SLE azathioprin	A. fumigatus	0/5
11	Aplastic anemia corticosteroids	A. flavus	0/2
12	CGD	A. fumigatus	0/2
13	Hyper IgE	A.terreus	0/1

ALL : acute lymphocytic leukemia
BMT : bone marrow transplant
AML : acute myelogenous leukemia
SLE : Systemic lupus erythematosus
CGD : chronic granulomatous disease

   This work indicates that it is worthwhile to look for Aspergillus
antigen in urine. This biological fluid may be more rewarding for antigen
detection than serum.

   In conclusion, two points should be emphasized: First, prospective
studies are needed in which several parameters are precisely defined:
diagnostic criteria, Aspergillus species involved, extent of lesions,
progression of disease, time of fever, time of pulmonary infiltrates, time
of antifungal therapy or surgical treatment, and time of antigen detection
compared to time of death. Such a study requires serial collection of samples
and a bank for sera and urine for all patients exposed to aspergillosis.
The second point is to insist on the necessity for standardization of
methods and reagents in order to compare results between different
laboratories.

REFERENCES

1. C.P. Andrews and M.H. Weiner, 1981, Immunodiagnosis of invasive pulmonary
   aspergillosis in rabbits, Am. Rev. Resp. Dis., 124:60-64.
2. C.P. Andrews and M.H. Weiner, 1982, Aspergillus antigen detection in
   bronchoalveolar lavage fluid from patients with invasive aspergillosis
   and aspergillomas, Am. J. Med., 73:372-380.
3. J.E. Bennett, A.K. Battacharjee, C.P.J. Glaudemans, 1985, Galactofuranosyl groups are immunodominant in Aspergillus fumigatus galactomannan, Molecular Immunol., 22:251-254.
4. R.E. Binder, L.J. Failing, R.D. Pugatch, G. Mahasaen, and G.L. Snider,
   1982, Chronic necrotizing pulmonary aspergillosis: a discrete clinical
   entity, Medicine, 61:109-129.

5. M.S. Cohen, P.E. Isturiz, H.L. Malech, R.K. Root, C.M. Wilfert, L. Gutman, and R.H. Buckley, 1981, Fungal infection in chronic granulomatous disease. The importance of phagocyte in defence against fungi, Am. J. Med., 71:59-66.

6. L. de Repentigny and E. Reiss, 1984, Current trends in immunodiagnosis of candidiasis and aspergillosis, Rev. Infect. Dis., 6:301-312.

7. B. Dupont, M. Huber, S.J. Kim, J.E. Bennett, 1987, Galactomannan antigenemia and antigenuria in aspergillosis: studies in patients and experimentally infected rabbits, J. Infect. Dis., 155:1-11.

8. B.D. Fischer, D. Armstrong, B. Yu, and J.W. Gold, 1981, Invasive aspergillosis. Progress in early diagnosis and treatment, Am. J. Med., 71:571-577.

9. D.W. Frazer, J.I. Ward, L. Ajello, and B.D. Plikaytis, 1974, Aspergillosis and other systemic mycoses. The growing problem, Jama, 242:1631-1635.

10. W.B. Gefter, T.R. Weingrad, D.M. Epstein, R.M. Ochs, and W.T. Miller, 1981, "Semi invasive" pulmonary aspergillosis, Radiology, 140:313-321.

11. S.L. Gerson, G.H. Talbot, S. Hurwitz, B.L. Strom. E.J. Lusk, and P.A. Cassileth, 1984, Prolonged granulocytopenia: the major risk factor for invasive pulmonary aspergillosis in patients with acute leukemia, Ann. Intern. Med., 100:345-351.

12. T.L. Gustafson, W. Schaffner, G.B. Lavely, C.W. Stratton, H.K. Johnson, and R.H. Hutcheson, 1983, Invasive aspergillosis in renal transplant recipients: correlation with corticosteroid therapy, J. Infect. Dis., 148:230-238.

13. P.F. Lehmann and E. Reiss, 1978, Invasive aspergillosis: antiserum for circulating antigen produced after immunization with serum from infected rabbits, Infect. Immun., 20:570-572.

14. P.F. Lehmann and E. Reiss, 1979, Aspergillus fumigatus antigenemia: detection of antigen in mice and in human patients. Bull. Soc. Fr. Myc. Med., 8:57-64.

15. R.D. Meyer, L.S. Young, D. Armstrong, and B. Yu, 1973, Aspergillosis complicating neoplastic disease, Am. J. Med., 54:6-15.

16. O. Morin, M. Miegeville, P. Germand, CH. Bouillard, and J.H. Bourhes, 1985, Aspergilloses pulmonaires invasives chez les malades a "hauts risques," Bull Soc. Fr. Myc. Med., 14:181-194.

17. M.A. Nalesnik, R.L. Myerowiz, R. Junkins, J. Lenkey, and D. Herbert, 1980, Significance of Aspergillus species isolated from respiratory secretions in the diagnosis of invasive pulmonary aspergillosis, J. Clin. Microbiol., 11:370-376.

18. M.D. Richardson, L.O. White, and R.C. Warren, 1979, Detection of circulating antigen of Aspergillus fumigatus in sera of mice and rabbits by enzyme-linked immunosorbent assay, Mycopathologia, 67:83-88.

19. M.G. Rinaldi, 1983, Invasive aspergillosis, Rev. Infect. Dist., 5:1061-1077.

20. J.W. Rippon and D.N. Anderson, 1978, Experimental mycosis in immunosuppressed rabbits, Mycopathologia, 64:97-100.

21. J.R. Sabetta, P. Minister, and V.T. Andriole, 1985, The diagnosis of invasive aspergillosis by an enzyme-linked immunosorbent assay for circulating antigen, J. Infect. Dis., 152:946-953.

22. P.J. Shaffer, G. Kobayashi, and G. Medoff, 1979, Demonstration of antigenemia in patients with invasive aspergillosis by solid phase (protein A-rich Staphylococcus aureus) radioimmunoassay, Am. J. Med., 67:627-630.

23. P.J. Shaffer, G. Medoff, and G. Kobayashi, 1979, Demonstration of antigenemia by radioimmunoassay in rabbits experimentally infected with Aspergillus, J. Infect. Dis., 139:313-329.

24. T.R. Treger, D.W. Visscher, M.S. Bartlett, and J.W. Smith, 1985, Diagnosis of pulmonary infection caused by Aspergillus: usefulness of respiratory cultures, J. Infect. Dis., 152:572-576.

25. M.H. Weiner, 1980, Antigenemia detected by radioimmunoassay in systemic aspergillosis, Ann. Intern. Med., 92:793-796.
26. M.H. Weiner and M. Coats-Stephen, 1979, Immunodiagnosis of systemic aspergillosis. Antigenemia detected by radioimmunoassay in experimental infection, J. Lab. Clin. Med., 93:111-119.
27. M.H. Weiner, G.H. Talbot, S.L. Gerson, G. Felice, and P.A. Cassileth, 1983, Antigen detection in the diagnosis of invasive aspergillosis. Utility in controlled, blinded trials, Ann. Intern. Med., 99:777-782.
28. L.O. White, M.D. Richardson, H.C. Newham, E. Gibb, and R.C. Warren, 1977, Circulating antigen of Aspergillus fumigatus in cortisone treated mice challenged with conidia: detection by counterimmunoelectrophoresis, FEMS Microb. Letters, 2:153-156.
29. R.C. Young, J.E. Bennett, C.L. Vogel, P.P. Carbone, and U.T Devita, 1970, Aspergillosis. The spectrum of disease in 98 patients, Medicine, 49:147-173.

ASPERGILLOSIS IN IMMUNOCOMPROMISED PATIENTS

PART I: THE PROBLEM OF DIAGNOSIS

R.A. Barnes and T.R. Rogers

Charing Cross and Westminster Medical School
London, U.K.

Fungal infection is an increasing problem in immunocompromised patients and is associated with a high mortality.
The use of more aggressive cytotoxic regimens and bone marrow
transplantation (BMT) for the treatment of leukemia and inborn
errors of metabolism has brought about an increase in the
numbers of patients with profound immunosuppression. Improved
supportive care and better antibacterial chemotherapeutic
agents have resulted in their prolonged survival. We have
observed a marked increase in the incidence of invasive
aspergillosis in BMT and remission-induction leukemics at our
hospital. Moreover, fungal infection is implicated as a cause
of fever in 34% of neutropenic patients. Establishing a
diagnosis in such patients is notoriosly difficult. Antibody
production is poor in the immunocompromised host and serological tests for Aspergillus antibody have not proved
clinically useful. Thus, diagnosis has depended on isolation
of Aspergillus from transbronchial specimens obtained at
bronchoscopy, often a hazardous procedure in the critically ill
patient.
    We retrospectively analyzed 11 patients who died of proven
aspergillosis. Possible risk factors such as duration of
neutropenia, steriod and antibiotic treatment were studied and
the method used to establish diagnosis was documented.
    Of 11 patients who died of aspergillosis, antemortem
diagnosis was established in only 5 cases, and in only 3
patients was it diagnosed more than 48 hours before death.
Prolonged neutropenia ($\geq$21 days) was observed in 73% of the
cases; all received long-term, broad spectrum antibiotics, and
64% were treated with extended courses of high-dose corticosteroids.
    The failure to confidently diagnose aspergillosis except
at post-mortem has prompted us to evaluate an assay for the
detection of Aspergillus antigen in serum and other body
fluids.

ASPERGILLOSIS IN IMMUNOCOMPROMISED PATIENTS. PART II:

EVALUATION OF AN ENZYME-LINKED IMMUNOSORBENT ASSAY (ELISA)

FOR THE DETECTION OF ASPERGILLUS ANTIGEN

K.A. Haynes[*], R.A. Barnes[*], E.V. Wilson[+],
V.M. Hearn[+], D.W.R. Mackenzie[+] and T.R. Rogers[*]

[*]Charing Cross and Westminster Medical School and
[+]Mycological Reference Laboratory, Public Health
Laboratory Service, London, U.K.

We evaluated an IgG inhibition ELISA, developed at the Mycological Reference Laboratory, in a retrospective study of 15 bone marrow transplant and leukemic patients. The group included 8 patients with proven invasive aspergillosis, 2 with clinically suspected fungal infection and 5 with other systemic mycoses.

Aspergillus fumigatus NCPF 2109 was grown and water soluble (WS) and bound fraction (BF) antigens were extracted as described by Wilson and Hearn (Wilson E.V. and Hearn V.M., 1983, J. Med. Microbiol., 16:97). The WS antigen was used to coat micro-titre plates and the BF was diluted (in antibody/antigen negative serum) to give antigen controls from 1250 ng/ml to 1.25 ng/ml. Samples (diluted 1:5 in phosphate buffered saline with 0.005% Tween 20) and controls were heated at 80ºC for 30 min, and then placed in uncoated micro-titre plates containing a strongly positive Aspergillus antiserum (inhibitor-serum). After incubation, the samples and controls were transferred to the washed WS-coated plates and a standard ELISA protocol to detect bound human IgG was followed. An antibody control (i.e. inhibitor serum in antibody/antigen negative serum) was included on every batch of plates and the % inhibition of this antibody control by the BF antigen series produced a standard curve from which test antigen levels were determined. All positive specimens were tested again to verify results.

All 8 proven cases of invasive aspergillosis had at least one specimen that was antigen positive by ELISA. No specimen from the control group was antigen positive by ELISA. These data indicate that the sensitivity and specificity of the ELISA is high; however, the testing of specimens from a larger number of proven cases of invasive aspergillosis and other systemic mycoses would have allowed a more confident appraisal. We are thus evaluating the ELISA in a prospective study of neutropenic patients.

Immunocompromised patients are particularly at risk of infection with fungi including <u>Aspergillus</u> species. Therefore, this ELISA could prove an effective tool in the management of infection in this patient group and others at similar risk. However, our data showed that antigenemia was transient; hence, if the presence of antigen is to be detected with confidence, specimens should be taken daily and regular ELISA runs performed. This would help to ensure the early diagnosis of invasive aspergillosis.

IMMUNOCHEMICAL STUDIES OF CANDIDA AND ASPERGILLUS ANTIGENS: TAXONOMIC AND DIAGNOSTIC SIGNIFICANCE

R.M.F. Guinet, S.M. Bruneau and H. de Montclos

Centre d'Immunochimie Microbienne, Institut
Pasteur de Lyon, 77 rue Pasteur, 69365 Lyon
Cedex 07, France

Various quantitative immunoelectrophoretic methods were applied to the taxonomic study of the genus Candida and to the serodiagnosis of aspergillosis.

Extracts were produced from blastoconidia of 16 Candida species, and three polyvalent antisera were raised in rabbits against extracts of Candida tropicalis, C. albicans and C. guilliermondii. Crossed-line immunoelectrophoresis (CLIE) was used to compare the extracts from the different species. The cross-reactive antigens were numbered and a cross-reactivity percentage calculated. In each antigen-antibody system, low, medium or high cross-reactivity groups of species were revealed.

Enzymatic activities were detected directly on precipitates formed in CLIE such as α and β esterases, phosphatase, leucine aminopeptidase, glucosidase, and isocitrate dehydrogenase. Cross-reactions were studied using CLIE and rocket-line immunoelectrophoresis (RLIE). This study showed that phosphatase and glucosidase allowed rapid detection of related species, while esterase and leucine aminopeptidase could be used as a species markers. Quantitative immunoelectro-phoresis appeared to be a useful taxonomic tool.

For the serodiagnosis of aspergillosis, mechanically produced extracts and culture filtrates were prepared from Aspergillus fumigatus ATCC 13073 and MRL 2140, and the corresponding polyvalent antisera were raised in rabbits. Crossed immunoelectrophoresis showed 65 and 56 precipitating antigens with mycelial extracts for ATCC and MRL strains, respectively, and 45 and 48 with culture filtrates. Crossed-immunoelectrophoresis was used to compare culture filtrates and extracts of 10 different strains of Aspergillus fumigatus and strains of other Aspergillus species and some important differences were observed. Line adsorption immunoelectrophoresis was used to study the precipitating antibodies contained in sera from 18 patients with aspergilloma, 20 patients with invasive aspergillosis and 50 control patients. Reference patterns using extracted antigens and culture filtrates showed 44 and 37 precipitation lines, respectively. Large amounts of antibodies of various specificities were found in aspergilloma in contrast with invasive aspergillosis. Interestingly, one specific antigen induced antibodies in 80% of patients with invasive aspergillosis with the culture filtrate antigen-antibody system. The purification of this antigen is under development in our laboratory and could allow sensitive and specific serodiagnosis of aspergillosis.

INTEREST IN SEROLOGICAL TECHNIQUES FOR THE DETECTION AND IDENTIFICATION OF PHOMA EXIGUA, THE AGENT OF POTATO GANGRENE

L. Hingand, M. Aguelon*, S. Le Coz, C. Kerlan,
B. Jouan and J. Dunez*

Institut National de la Recherche Agronomique,
Station de Pathologie Végétale. Centre de Recherches
de Rennes. F35650 Le Rheu and
*Station de Pathologie Végétale, Centre de
Bordeaux, BP 131, 33140 Pont de la Maye, France

Potato gangrene is an important storage disease of potatoes. The responsible fungus is Phoma exigua. The two varieties of this fungus, P. exigua and P. foveata can induce necrotic lesions on tubers, but P. exigua var. foveata is responsible for most damage. Early detection of infection and precise and rapid identification of the fungus variety are of major interest. So far, differentiation of the varieties have relied on biochemical tests and specific production of a brown pigment in the P. foveata cultures. Our attempts were to develop serological tests both for the detection of P. exigua and differentiation of the 2 varieties.

Two types of antisera were produced against mycelial soluble antigens and pycniospores (crude pycniospore suspensions obtained by sonication of a culture). The antisera could differentiate the spores belonging to each of the varieties of the fungus. In practice, the spores of the fungus could be detected by indirect immunofluorescence in the pellet resulting from low speed centrifugation of the washing water of infected tubers (Hingand et al., 1983, Agronomie, 3:51-56). The antiserum raised against the P. exigua var. foveata pycniospores did not cross-react with its mycelium.

Antisera were prepared against the mycelium of the two varieties. The mycelium produced in liquid medium was filtered and freeze-dried. The resulting powder was homogenized in Tris buffer, centrifuged at 10,000 rpm for 10 min, and the resulting supernatant was concentrated by dialysis against PEG 20,000. When tested against the 2 fungal varieties, cross reactions developed and the antisera did not permit differentiation between the 2 varieties. However, antibodies from these antisera were successfully used in ELISA to detect the fungus in infected tissues (M. Aguelon, J. Dunez, 1984, Ann. Appl. Biol., 105:463-469.

DETECTION OF ASPERGILLUS GLYCOPROTEIN IN SERA OF PATIENTS WITH INVASIVE ASPERGILLOSIS BY AN ENZYME-LINKED IMMUNOSORBENT ASSAY

P. Le Pape* and J. Deunff**

*Laboratoire de Parasitologie, UFR de Medecine
de Nantes; **Laboratoire de Parasitologie, UFR de
Medecine de Rennes 1, rue Gaston, Veil, 44045 Nantes

Invasive aspergillosis is encountered with increasing frequency in the immunocompromised host which results in a high rate of mortality. The antemortem diagnosis of this opportunistic mycosis is very difficult. Consequently, enzyme-linked immunosorbent assays and radioimmunoassays for circulating antigen have been developed (M.D. Richardson, L.O. White, R.C. Warren, 1979, Mycopathologia, 67:83; J.R. Sabetta, P. Miniter, V.T. Andriole, 1985, J. Infect. Dis., 152:946; M.H. Weiner, G.H. Talbot, S.L. Gerson, 1983, Ann. Intern. Med., 99:777).

The present study focused on the development of an inhibition ELISA for circulating glycoprotein antigen of Aspergillus. The glycoproteins could be detected at concentrations of 5 to 10 ng/ml in buffer and 10 to 20 ng/ml in serum. The ELISA was used to test 28 sera from 10 patients with invasive aspergillosis. This group consisted of 5 men and 5 women whose ages ranged from 14 to 75 years. Invasive aspergillosis was proven histologically at autopsy on 5 patients. Nine infections were caused by A. fumigatus and 1 by A. flavus. Ten normal controls were also analyzed. The ELISA used is similar to that described by Yolken; microplates were coated in this study with A. fumigatus antigen prepared by affinity chromato-graphy on Concanavalin A-Sepharose. The antibodies were isolated by affinity chromatography using the glycoproteins of the antigen. Patient sera were tested for antibody to Aspergillus by immunoelectrophoresis and indirect hemagglutination.

In sera from healthy controls, the inhibition ranged up to 8.7%. In contrast, when the sera of 10 patients were tested for invasive aspergillosis, the ELISA detected antigen in 9 patients. In the tenth patient the inhibition was low (11%), but the inhibition of serum just before the onset of the aspergillosis was 6.3%. Inhibition of positive sera ranged from 15 to 66%. Only 1 patient was positive for antibody.

AN ENZYME IMMUNOASSAY FOR THE DETECTION OF TENUAZONIC ACID,

A TOXIN FROM THE RICE FUNGAL PATHOGEN PYRICULARIA ORYZAE

M.H. Lebrun*, P. Poncet[o], F. Gaudemer[+], M. Boutar[+]
and A. Gaudemer[+]

Laboratoire de Cryptogamie* (CNRS LA86) Université
Paris-Sud, Orsay; Centre de transfusion[o], Institut
Pasteur, Paris; Laboratoire de Chimie de
Coordination Bioorganique[+] (CNRS LA255), Université
Paris-Sud, Orsay, France

Pyricularia oryzae is a fungal pathogen of rice. This
fungus produces a phytotoxic metabolite derived from isoleucine
which is called tenuazonic acid (N. Umetsu, J. Kaji, K. Tamari,
1973, Agr. Biol. Chem., 37:451) (TA, Fig. 1). This toxin is a
protein synthesis inhibitor (T. Muramatsu, N. Umetsu, K.
Matsuda, K. Tamari, 1974, Agr. Biol. Chem., 38:2049). In order
to detect TA in the infected plant we have developed an enzyme-
linked immunosorbent assay (ELISA). We have synthesized an
analog of TA with a terminal amino group (LYST, Fig. 2).

Fig.1 Tenuazonic acid (TA)

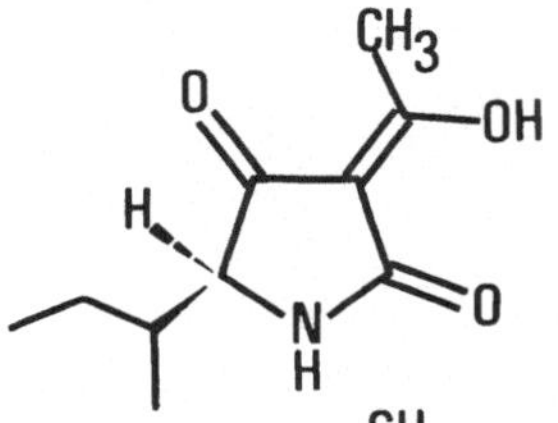

Fig.2 LYST

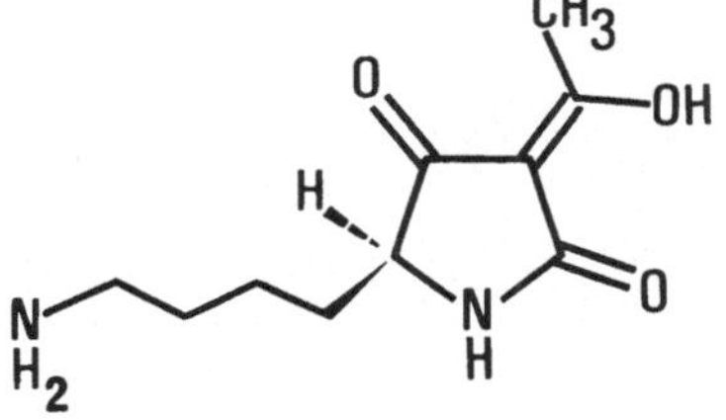

This analog has been coupled with glutaraldehyde to
different proteins, including bovine serum albumin (BSA),
chicken ovalbumin (OVA), thyroglobulin (Tr) and bovine
gammaglobulin (BGG). Since TA and LYST have a strong UV
absorption at 280 nm and are good iron chelators (M.H. Lebrun,
P. Duvert, F. Gaudemer, C. Deballon, A. Gaudemer, P. Boucly,

1985, J. Inorg. Biochem., 24:154), we have estimated the number of LYST linked to the proteins described above by UV absorption of the conjugates and by the visible absorption of conjugates treated with iron (max=440 nm). We thus find that BSA-LYST conjugates have an average of 15 LYST per protein. Antibodies were produced in rabbits against BSA-LYST antigens and purified by affinity chromatography on a LYST-Sepharose column. These antibodies recognize both the carrier protein (BSA) and TA, but they do not show significant binding to other proteins (OVA for example). LYST is poorly recognized by these antibodies.

The ability of different soluble TA analogs to inhibit the binding of antibodies to OVA-LYST conjugates on a solid phase have been used to study the antibody specificity. Fixed rabbit antibodies were revealed by goat anti-rabbit antibodies coupled to peroxidase. Analogs with different hydrophobic (n-propyl, iso-propyl, n-butyl, iso-butyl) lateral chains were as much recognized as TA (sec-butyl). Analogs with modifications of the TA heterocycle or single amino acids from which TA and LYST are derived (isoleucine and lysine) failed to inhibit the specific immunological reaction. The immunoassay was optimized to detect 0.2 ng TA (ID30). Fungal culture filtrates or in-fected leaf extracts did not affect the quantification of TA by ELISA since the same amount of toxin could be measured by HPLC.

VALUE AND LIMITS OF ELISA IN ASPERGILLOSIS: SEROLOGICAL DIAGNOSIS IN CLINICAL FORMS AND CONTROL OF A HIGH RISK POPULATION

C. Pinel, R. Grillot and B. Lebeau

Service de Parasitologie, CHU de Grenoble
BP 217X, 38043 Grenoble Cedex, France

Our purpose was to test the efficacy of IgG detection against _Aspergillus_ _fumigatus_ antigen by an enzyme-linked immunosorbent assay (ELISA) and to compare the results with those obtained by counterimmunoelectrophoresis (CIE) and indirect immunofluorescence (IIF). Our method was standardized using 157 sera. To sensitize the microtiter plates we used a somatic and metabolic antigen. The adsorbance values were compared with a reference positive control serum (aspergilloma).

The positive threshold value was determined with 44 sera of healthy women (examined for toxoplasmosis serology). The specific threshold value was fixed from ELISA measurements of 113 patients sera with proved candidiasis, different parasitosis and pulmonary tuberculosis.

To test the usefulness of this method, we studied 57 sera from 36 non-immunosuppressed patients with aspergilloma, and 10 sera from a child with chronic granulomatous disease. No false negative or limit values were obtained and the sensitivity was better than in CIE. As other authors have reported, we noted the weak performance and contradictory results of IIF.

Our ELISA method was studied in the serological diagnosis of invasive pulmonary aspergillosis (IPA) in granulocytopenic patients (13 sera from 6 patients with proved IPA). The results confirmed the utility of this method for the diagnosis of IPA (3 positive cases with ELISA and none with CIE). This method has been used for the serological control of a high risk population with the following results:

- In the 23 granulocytopenic subjects ELISA sensitivity seemed better and it allowed the detection of one case of a probable IPA;
- In 8 children with cystic fibrosis, ELISA positive values corroborated by high IIF titers (antigen IIF: sections of rabbit kidney infected with _A. fumigatus_ spores) are a good means of serodiagnosis for aspergillosis;
- In 27 non-immunosuppressed pneumopathic patients with suspicion of aspergillosis the specificity of this method is lower (risk of some false positive values).

Our prelinimary results allow us to conclude that ELISA with our antigen is a useful technique in aspergillosis diagnosis provided that it is employed with CIE in non-immunosuppressed subjects and with IIF in compromised patients.

A RAPID DOUBLE ANTIBODY BIOTIN-STREPTAVIDIN

ELISA FOR ASPERGILLUS FUMIGATUS GLYCOPROTEIN ANTIGEN

M.D. Richardson, L.A.McTaggart, and G.S.Shankland

Medical Mycology Unit

University of Glasgow, Scotland

Most enzyme immunoassays for the detection of fungal antigenaemia in immunocompromised patients rely upon immunoreactants which are covalently labelled with enzymes. However, assays can also be formatted which use immunoreactants linked to low molecular weight cofactors, such as biotin. The biotin labelled immunoreactants are subsequently bound to enzymes by means of proteins which have an extraordinary affinity for biotin. One such protein is streptavidin. The rapid and essentially irreversible binding of biotin to streptavidin, along with the ability to attach several biotin molecules to a single antibody molecule, makes the system particularly attractive for improving sensitivity in immunoassay diagnosis.

Despite the development of CIE, RIA and ELISA methods for antigen detection in invasive aspergillosis over the past ten years, none has gained widespread acceptance. In an attempt to increase the sensitivity, specificity and speed of existing enzyme immunoassays for detection of Aspergillus antigenaemia we have developed a rapid and sensitive double antibody biotin streptavidin ELISA for glycoprotein antigens where each component stage consisted of a one hour incubation period. The basic components of the assay performed on a polyvinyl solid phase were an antigen capture antibody step: a bound antigen detection antibody step: a biotinylated anti-IgG step which was then followed by the addition of a preformed streptavidin-biotin-horseradish peroxidase complex; finally followed by the addition of an appropriate substrate.

In assays where a rabbit anti-A.fumigatus capture antibody and a human-A.fumigatus detection antibody were used various antigen types of A.fumigatus could be detected at levels of 4ug/ml in buffer. Antigen was not detected when diluted in normal human serum. In assays where a human capture and a rabbit detection antibody were used both cytoplasmic and culture filtrate antigens could be detected at levels of 0.4ng/ml in buffer and 4ng/ml in serum. Detection in serum did not require any form of pretreatment. The human capture-rabbit detection assay format was specific for A.fumigatus. Cross-reactivity between the A.fumigatus immunoreactants and antigenic extracts of A.clavatus, A.niger, A.nidulans, A.terreus and A.flavus was not detected.

PEROXIDASE ANTI-PEROXIDASE STAINING FOR THE SPECIFIC IDENTIFICATION

OF FUNGAL ANTIGENS IN TISSUE SECTIONS

G.S. Shankland and M.D. Richardson

Medical Mycology Unit
University of Glasgow, Scotland

Histological recognition of opportunistic fungal infections provides definitive proof of invasion by a fungus which might have been mistaken for a laboratory contaminant or part of the commensal flora. Histology can be a rapid method for the diagnosis of a fungal infection and may be important in recognising the presence of an opportunistic infection in biopsy material from seriously ill patients.

The appearance of fungal lesions in tissue depends on variables such as the organ involved, the underlying disease and the immunological competence of the host. Immunological imaging techniques help to alleviate some of the problems encountered when attempting to identify fungi in tissue sections by their morphology and the tissue reactions they elicit. Similarities among tissue forms of opportunistic fungi have resulted in attempts to employ immunohistological techniques to enhance the recognition of fungal elements in clinical specimens.

The antigenic properties of the fungal cell walls and cytoplasmic contents in tissue sections are not altered by fixation or storage. The recognition of these antigens by specific immune sera has enabled fungal genera to be identified in a peroxidase anti-peroxidase staining system employing tissue sections from proven cases of aspergillosis, candidosis and mucormycosis. Since control sections from verified cases of fungal infections are not always available, to demonstrate any cross-reactivity amongst the antisera used for staining tissue sections, an agar block technique was developed. This method simulates fungal hyphae growing three dimensionally through tissue. The technique is easy to perform and unlike material from animal models of systemic infection, the amount of fungus present in the blocks can be controlled.

It is possible to differentiate the genera Aspergillus, Candida and Rhizopus from each other. However, cross-reactions among species within the same genus were evident. Interspecific cross-reactions were studied using the new agar block technique. These cross-reactions were also investigated by an ELISA method. The technique has potential as a rapid, sensitive and practical procedure for identification of specific pathogenic fungi in tissue sections.

AN ASSESSMENT OF THE ASSAY FOR SERUM ANTIBODIES TO ASPERGILLUS FUMIGATUS CATALASE IN THE DIAGNOSTIC LABORATORY

H.C. Schönheyder

Institute of Medical Microbiology, Bartholin
Building, University of Aarhus, DK-8000 Aarhus
C. Denmark

Aspergillus fumigatus catalase has been identified as an important antigen in pulmonary aspergillosis (PA) associated with a fungus ball or a lung infiltrate, and a rapid immuno-electrophoretic assay for catalase antibodies ("bubblography") has been developed (H. Schonheyder, P. Andersen, J.C. Munck Petersen, 1985, Eur. J. Clin. Microbiol., 4:299). The diagnostic sensitivity for PA, excluding allergic bronchopulmonary aspergillosis, was 88% and the diagnostic specificity was 94%. The catalase antibody assay is well suited to the diagnostic laboratory, and a 2-year experience with the assay is reviewed with special reference to the correlation between catalase-specific antibodies and precipitins.

During a 2-year period (1984-86) sera from patients were tested for Aspergillus fumigatus antibodies by bubblography and by double immunodiffusion (ID) using culture filtrate and somatic antigen preparations (H. Schonheyder, P. Andersen, A. Stenderup, 1982, Acta Pathol. Microbiol. Immunol. Scand. Sect. B, 90:273). The test group included 137 males and 73 females; the mean age was 47 years, range 1 to 90 years. Sera were tested in serial two-fold dilutions in order to determine the end-point titre. When more than one sample was available from a patient the primary one was used for comparison, with the exception of one patient with chronic necrotizing pulmonary aspergillosis (CNPA) in whom a seroconversion was observed (P. Andersen, H. Schonheyder, S. Oster, 1985, Mykosen, 28:595). For comparison of precipitin and catalase antibody titres, reactive (+ve) sera from the previous study were also included (H. Schonheyder, P. Andersen, J.C. Munck Petersen, 1985, Eur. J. Clin. Microbiol., 4:299).

Thirty-five sera were +ve in one or both tests and 21 of these (60%) were positive for catalase antibodies, whereas 40% (N=14) reacted in the precipitin assay only. A higher proportion of the +ve patients from the Department of Thoracic Medicine showed reactivity of catalase antibodies (15/19) than from other Departments (Oncology, Infectious Diseases, General Medicine and Pediatrics) (6/16); this difference reached statistical significance (χ^2 =0.02, Fisher's exact test). This underlines the paramount importance of defining the patient population in which serological surveys are carried out. The detailed clinical histories of the 14 patients wich precipitins alone are not available, but it must be observed that chronic granulomatous disease was a cuncurrent disorder in 4 of the cases.

Precipitin and catalase antibody titres in 71 individual patients were compared and a positive correlation between precipitin and catalase antibody titres was apparent (Spearman R=2p<0.001). Still, there was a notable discrepancy between the reactivity of individual samples in the 2 assays, especially of samples with ID titres in the intermediate range (4-16). A preliminary evaluation of the clinical histories of patients with disproportionately high catalase antibody activity points to a recent onset, a progressive course, and cavitary lesions being common denominators. This should, however, be evaluated in a prospective, preferably multicentre study.

In conclusion, the catalase antibody assay cannot stand on its own. Despite a distinctive correlation of precipitin and catalase antibody titres, marked differences were observed in individual cases. As hyphal tissue infiltration may be a prerequisite for the antibody response to the catalase antigen, it is conceivable that the catalase antibody test may help to delineate a group of patients with active, cavitating disease.

DETECTION OF ASPERGILLUS FUMIGATUS ANTIGENIC COMPONENTS

RECOGNIZING IgG, IgM, IgA OR IgE SPECIFIC ANTIBODIES BY ELIFA

H. Thoannes, J. Poirriez and J.M. Pinon

Laboratoire de Parasitologie-Mycologie, Hopital
Maison Blanche, 45, rue Cognacq Jay, 51092 Reims
Cedex, France

The ELIFA (Enzyme-Linked Immuno-Filtration Assay) method
(J.M. Pinon et al., 1985, J. Immunol. Methods, 77:15) may be
used for an analytical, immunologic study of the Aspergillus
fumigatus antigenic components. Hyperimmune sera (from
patients with allergic bronchopulmonary aspergillosis or
aspergilloma) permit identification of 6 antigenic fractions
which induce a triple immunological response. Production of
IgG, IgM and/or IgA against the same epitopes has been
determined. On the other hand, 4 allergenic components
(stimulating an IgE response) (J.L. Longbottom, 1983, Clin.
Exp. Immunol., 53:354) were detected by ELIFA and are completely different from the 6 previous fractions. The C
antigenic fraction (which supports a chymotryptic activity) (J.
Biguet et al., 1967, Rev. Immunol., 31:317; J.P. Dessaint et
al., 1976, Clin. Immunol. Immuno-pathol., 5:314) was revealed
only by IgG antibodies.
Within a mixed Aspergillus fumigatus antigen extract
(membrane and cytoplasmic), it appears that 3 types of
antigenic determinants exist: epitopes inducing IgG-IgM-IgA,
allergenic epitopes IgG-IgE starter, and epitopes stimulating
only IgG antibodies including the C fraction.
Analytical methods such as gel filtration, polyacrylamide
gel electrophoresis and isoelectrofocusing permit identification of molecular weights and isoelectric points of these
different functional, antigenic components.

ANTIGENS AND ALLERGENS OF THE GENUS CHRYSOSPORIUM (EMMONSIA) AGENT OF ADIASPIROMYCOSIS

A. Tomsikova

Institute of Microbiology, School of Medicine
Charles University Plzen, Czechoslovakia

Four species of dimorphic fungi can cause adiapiromycosis: Chrysosporium parvum var. crescens, C. parvum, Emmonsia ciferrina and E. brasiliensis.

We compared the antigenic and allergenic activities of cellular components, the production of exoantigens and the relationship of 26 strains belonging to the named species and to C. pannorum, C. pruinosum, and of 3 mutants of C. parvum var. crescens obtained by ultraviolet irradiation. The experiments were made on 95 rabbits that were immunized by complete somatic, metabolic antigens and mannoproteins prepared from the named species. The formation of specific IgM and IgG antibodies was studied by 4 serological reactions, the development of allergy by means of skin tests, and passive anaphylaxis.

Differences in the antigenic activity among the genera, species and different strains of the same species were determined according to the formation of specific antibodies. The antigenic activity was correlated with virulence, the capability and rate of conversion into the parasitic phase, and the macromorphology. The allergenic activity was different in the genera, species and single strains of species, in the mutants, in the parasitic and saprophytic phase. The mutants showed a strong reactivity. The whole genus of Chrysosporium has one common antigen. Specific antigens have been found in the different species or strains. Because of common antigens between these test fungi and dermatophytes, Candida albicans, and Cryptococcus neoformans, serological identification must use preadsorbed sera or specific antigens.

Coccidioides
and
Histoplasma

WALL-ASSOCIATED ANTIGENS OF COCCIDIOIDES IMMITIS

Garry T. Cole[1], S. H. Sun[2], J. Dominguez[2], L. Yuan[1],
M. Franco[4], and T. N. Kirkland[3]

Department of Botany, University of Texas, Austin,
Texas, 78713[1], Veterans Administration Hospital,
San Antonio, Texas 78284[2] and San Diego, California,
92161[3], and Estadual Paulista, Botucatu, Brazil [4]

INTRODUCTION

The cell wall of most pathogenic fungi has been shown to be a reservoir
of immunoreactive macromolecules (Hall et al., 1978, San Blas, 1982, 1985;
Huppert, 1983, Hearn, 1984; Brawner and Cutler, 1986a,b; Ponton and Jones,
1986). Coccidioides immitis, causative agent of 'valley fever'
(coccidioidomycosis), is an almost ideal model for investigating the
importance of wall-associated antigens to host response. Evidence has been
presented that clinically significant biological activities are elicited in
laboratory animals by their exposure to the fungal cell wall or solubilized
wall extracts of C. immitis. For example, animals exposed to wall products
have exhibited delayed-type hypersensitivity, (Ward et al., 1975), production
of macrophage migration inhibition factor and lymphocyte transformation, (Cox
and Vivas, 1977, Cox et al., 1981), immunoprotection against intranasal
challenge, (Lecara et al., 1983), and response of both tube precipitin and
complement-fixing antibodies (Collins et al., 1977).

Coccidioides is a diphasic microbe which forms septate hyphae and
arthroconidia in its saprobic phase, and morphologically distinct spherules
and endospores in its parasitic cycle (Huppert and Sun, 1980; Huppert et al.,
1982). Cell types of both phases have been grown in vitro and isolated from
cultures in sufficient quantities for comparative studies of wall composition
(Cole et al., 1985a,b). Early reports of the wall chemistry of each cell type
of C. immitis revealed that quantitative and certain qualitative differences
exist between arthroconidia, spherules, and endospores (Tarbet and Breslau,
1953; Wheat et al., 1977, 1978; Hector and Pappagianis, 1982; Cole et al.,
1983; Cole et al., 1985b). This was not an unexpected discovery, both because
of the morphogenetic disparity of the saprobic and parasitic cycles, and the
availability of evidence that differences occur in host response to the three
cell types (Drutz and Huppert, 1983; Frey and Drutz, 1986). A logical
explanation for the latter phenomenon is that during initial contact and
subsequent interaction between the fungus and host tissues, different spectra
of chemical components are presented to the host, which in turn, generate
different responses. Our approach to studies of the participation of wall-
associated macromolecules in such complex fungal-host relations has been to
(1) initially develop a reliable procedure for isolation of wall components
(Cole et al., 1982,1983), (2) make use of an established antigen-antibody
reference system for comparative studies of antigen composition of different
wall fractions (Huppert et al., 1978,1979), and (3) screen the antigenic
fractions using both humoral and cellular immunoassays for selection of

immunoreactive components and their further fractionation, purification and characterization (Cole et al., 1986,1987).

In this paper we illustrate an antigen/antibody reference system, which is a modification of the earlier system described by Huppert and coworkers (1978,1979), and demonstrate methods for identification and selection of immunoreactive components of crude fractions of C immitis. We have focused on a water-soluble fraction of the arthroconidial wall (Cole et al., 1987) as well as the culture supernatant of the saprobic (mycelial) phase. The antigen composition of the soluble conidial wall fraction (SCWF) was previously examined using the established antigen/antibody reference system (Cole et al., 1983). SCWF has already proved to be immunoreactive in a lymphocyte proliferation assay (Cole et al., 1986; 1987). Several specific antigens of SCWF are identified which appear to contribute significantly to the immunoreactivity of this wall fraction, and partial purification of one of these antigens is reported. We also describe the isolation of an antigen which is present in SCWF and culture supernatants of the saprobic and parasitic phases. The antigen (Ag), referred to as AgCS (Cole et al., 1985a), may be immunodiagnostic for coccidioidomycosis.

MATERIALS AND METHODS

Isolation and Fractionation of Conidial Wall and Culture Supernatant Components

The saprobic phase of C. immitis (strain C634) was grown on media consisting of 1% glucose, 0.5% yeast extract (GYE; Difco), and 1.5% Ionoagar No. 2 (Oxo Ltd., London). Conidia were vacuum-harvested (Cole et al., 1982, 1983) from cultures which had been incubated for 40–60 days at $30^{\circ}C$. The cells were suspended in distilled water (approx. 10^8 conidia/ml) either containing $1\times10^{-3}M$ phenylmethylsulfonyl fluoride (PMSF, Sigma) to inhibit proteinase activity (Turini et al., 1969) during the isolation procedure, or in the absence of PMSF. The steps used for isolation of SCWF have been described (Cole et al., 1983; Cole and Sun, 1985). The mycelial phase of C. immitis was grown in GYE broth at $30^{\circ}C$ for 3–16 days in shake culture as previously described (Cole and Sun, 1985). The culture supernatant was dialysed exhaustively against distilled water (6000–8000 MW cutoff). Both the water-soluble wall material and retentate of the dialysed culture supernatant were lyophilized and stored at $-20^{\circ}C$ in preparation for further fractionation.

Sephacryl fractionation. A Sephacryl S-300 (Pharmacia) chromatographic column (318 ml bed volume) was used for fractionation of SCWF and the culture supernatant as previously described (Cole et al., 1987). Fractions were monitored at 254 nm. Selected fractions were pooled, dialysed against dH_2O ($4^{\circ}C$, 48h, 4 changes of dialysate), and lyophilized in preparation for subsequent analysis. Molecular weight (MW) standards (Sigma) with a range of 200 kilodaltons (kd) to 6.5 kd were eluted in the same manner as the samples but monitored at 280 nm.

Acetone extraction. Fraction 2 (FR2,Fig. 5) from several Sephacryl S-300 fractionations were pooled, dialysed against dH_2O, lyophilized and then extracted with cold ($4^{\circ}C$) absolute acetone. Each fraction (approx. 4.0 mg) was first solubilized in 200 ul filtered dH_2O ($4^{\circ}C$) to which 100 ul of absolute acetone was added. The suspension was left on ice for 10 min and then centrifuged (13,000 rpm, Beckman Microfuge II). The supernatant was further extracted by addition of 100 ul acetone and treated as above. The supernatant obtained from this 50% acetone cut was again extracted by addition of 400 ul cold acetone. The pellet obtained from this extraction (75% acetone cut) was dried under N_2 gas and stored at $-20^{\circ}C$.

HPLC. The dried, acetone-extracted pellet (75% acetone cut) of Sephacryl FR2 was resolubilized in filtered dH$_2$O and further fractionated on a Superose 12 HR 10/30 gel fitration column (Pharmacia) by high pressure liquid chromatography using a Hewlett-Packard Model HP 1090 liquid chromatograph equipped with a diode array detector (Fig. 15). The sample was eluted with 0.05 M phosphate buffer (pH 7.2) containing 0.15 M NaCl at a flow rate of 0.250 ml/min and pressure of 215 psi. Fractions were monitored at 224 nm.

Fraction 3 of the 16-day mycelial culture supernatant obtained by Sephacryl S-300 gel filtration (elution profile not shown) was further fractionated by ion exchange chromatography (DEAE, 170 ml bed volume) using 20 mM citrate buffer (pH 6.5) followed by desorption with 0.2 M citrate buffer (pH 6.6). Fraction 2 obtained from this chromatographic separation (elution profile not shown) was subjected to HPLC fractionation. The sample was passed through a DEAE column (Waters; Protein Pak DEAE 5PW) followed by the Superose 12 HR 10/30 column in series. The sample was eluted with 0.05 M phosphate buffer (pH 6.6) containing 0.15 M NaCl at a flow rate of 0.250 ml/min and pressure of 215 PSI. Fractions were monitored at 214 nm (Fig. 16).

Immunoelectrophoresis and Immunodiffusion

The antigen content of various wall fractions of C. immitis has been previously analysed using two-dimensional immunoelectrophoresis (2D-IEP) and advancing line immunoelectrophoresis (AL-IEP) techniques (Cole et al., 1983; Cox et al., 1984; Reiss et al., 1985; Cole et al., 1985a). Immunodiffusion (ID) tests were performed according to the method of Huppert and Bailey (1963,1965a,b) for detection of antigens which reacted with complement fixation (CF) and tube precipitin (TP) antibodies.

IEP analysis of SCWF and Sephacryl/HPLC fractions. Identification of specific antigens (Ags) is based on a coccidioidin/anti-coccidioidin (CDN/anti-CDN) reference system which was developed by Huppert and coworkers (1978,1979). Coccidioidin is a skin test-active substance that has been used for many years to survey human populations for exposure to arthroconidia of C. immitis (Smith et al., 1948). The reference antigen was prepared as two separate products (Huppert, 1983): a pooled, broth culture filtrate of several strains of Coccidioides mycelia (F fraction), and a toluene lysate of the washed, cellular retentate (L fraction). Anti-CDN (F+L) was derived from a hyperimmunized burro and used in immunoelectrophoresis procedures to demonstrate the antigenic components of CDN, as outlined by Huppert et al. (1978,1979). Since the antiserum used in this original reference system was near depletion, we obtained antiserum from a newly immunized burro. The antigen used for immunization was coccidioidin (CDN, accession no. XV Q5C) and was prepared in the same manner as reported by Huppert et al. (1978,1979). The same strains were also used, which included C100, C157, C277, C309 and C338. The immunization protocol was essentially the same as that used by Huppert and coworkers (1978,1979) except that plasmapheresis was employed to obtain plasma from which immunoglobulins were isolated. The new CDN/anti-CDN reference system was used to monitor antigenic content of the fractions examined in this investigation.

Tandem 2D-IEP was used to identify antigens on the basis of their fusion with numbered precipitin peaks (Figs. 10-12) in the CDN/anti-CDN reference system as previously described (Cole et al., 1983). The AL-IEP method employed in this paper has also been reported (Reiss et al., 1985, Cole et al., 1986). The reference antigen incorporated into the intermediate gel (i.e., XV Q5C; Figs. 6-8) was substituted in certain AL-IEP plates with human serum (Figs. 13,22,23). In these cases, the intermediate gel was composed of 1 part serum to 1 part 2% agarose. The serum used was either normal human serum (NHS; Fig. 22), or serum from coccidioidomycosis patients with high CF antibody titer (CF+; Figs. 13,23) (Huppert et al., 1977; Huppert, 1983).

This procedure, together with immunodiffusion, served as the humoral-immunoassays for screening wall-associated antigens.

IEP analysis of heat sensitive antigens. To test for the presence of heat sensitive antigens, lyophilized fraction 3a (FR3a) from the Sephacryl S-300 fractionation of SCWF (Fig. 5) and lyophilized fraction 3 from HPLC separation of the 16-day mycelial culture supernatant (Fig. 16) were resolubilized in phosphate-buffered saline (PBS), pH 7.4, and incubated in a water bath at 60°C for 15-30 min. The heat-denatured fractions were then added to the anodal well of the 2D-IEP plate for tandem analysis (Figs. 12 and 21, respectively).

IEP examination of other antigen preparations. Spherulin, which was used for complement fixation tests (Scalarone et al., 1974), was obtained from Berkeley Biologicals (Berkeley, California). The skin test antigen (64D2.5) was originally obtained from Dr. C. E. Smith and is composed of combined filtrates of several strains grown on modified asparagine broth (Smith et al., 1948). The HS (heat stable) antigen (Kaufman et al., 1985) was kindly provided by Dr. L. Kaufman, Centers for Disease Control, Atlanta, Georgia. All antigens were examined in tandem 2D-IEP plates for identity based on the reference system (Figs. 24-26).

Immunodiffusion tests. The reference antisera were obtained from patients with coccicioidomycosis identified as IDCF-Ab25 and IDTP-Ab37, respectively (accession numbers from patient sera collection, Medical Mycology Research Laboratory, VA Hospital, San Antonio, Texas). The reference antigens for IDCF and IDTP were the F and L fractions of CDN, respectively. The reference systems were titrated to produce a precipitin line in the midregion between the antigen and antibody wells (Fig. 9).

Cell Immunoreactivity

SCWF and Sephacryl S-300 fractions were tested for biological activity (Cole et al., 1987). Draining lymph node cells were obtained from BALB/c mice which had been immunized three times with attenuated spherules. The lymph node cells were cultured with or without antigen, and uptake of [^{3}H] thymidine was measured at 72 to 90 h as an estimate of cell proliferation. The responses are expressed (Fig. 5) as Δ cpm (counts per minute with antigen - counts per minute without antigen). Standard errors were less than 10% of the mean. The concentration of antigen used to obtain the data in Fig. 5 was 10 ug/ml culture media which was the minimal concentration used in the immunoreactivity assay (Cole et al., 1987).

Gel Electrophoresis

Sodium dodecyl sulfate-polyacrylamide gel electrophoresis (SDS-PAGE) was performed using a 12.5% slab gel and the discontinuous buffer system described by Laemmli (1970). Lyophilized samples were dissolved in a buffer that contained 2.3 % SDS, 0.4 M 2-mercaptoethanol, 0.002% bromophenol blue, 50% glycerol, and 62.5 mM Tris-HCl (pH 8.6), and then heated at 100°C for 2 min. For each sample, approximately 50ug of protein was applied to the gel. Electrophoresis was performed at a constant voltage (80 V) at room temperature for 6h. The polypeptide banding pattern was revealed by Coomassie blue R-250 staining (Bio-Rad Laboratories, Richmond, California), and was compared to a molecular weight standard kit (Bio-Rad; 14,400-92,500 daltons).

RESULTS

CDN/anti-CDN Reference System

The maximum number of precipitin peaks in the original (Huppert et al.,

1978) 2D-IEP analysis of coccidioidin (CDN) against anti-CDN-precipitated immunoglobulin (Ig) was revealed using Ig dilutions of 1:2, 1:5 and 1:10 in electrophoresis buffer (barbital, pH 8.6). Dilutions of 1:5 and 1:10 are shown in Figs. 1 and 2, respectively. Neither the protein concentration of the Ig in the upper gel, nor the concentration of the reference antigen were reported by Huppert and coworkers (Huppert et al., 1978). A total of 21 precipitin peaks are labelled in Figs. 1 and 2. The distinct antigen-antibody reactions are designated as Ag1, Ag2, etc. according to the system described by Huppert et al. (1978). In Figs. 3 and 4, anti-CDN immunoglobulins recently

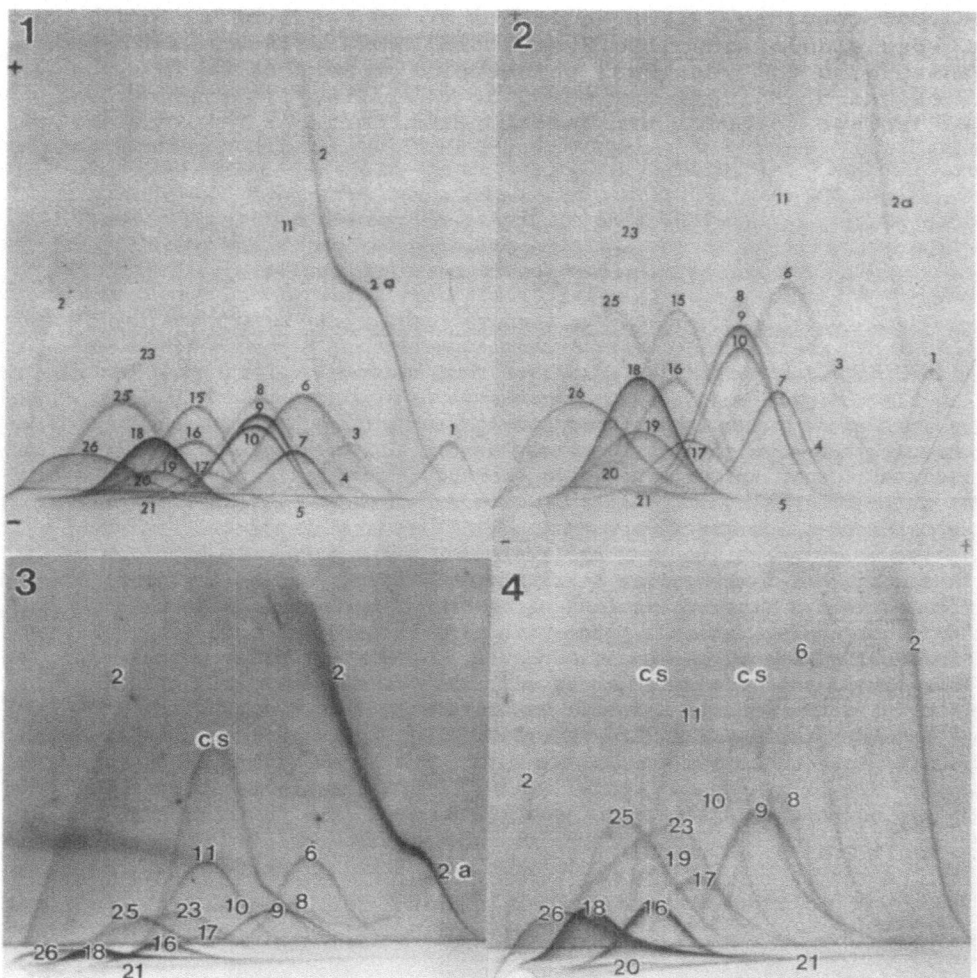

Figs. 1,2. 2D-IEP of coccidioidin (cathodal well) against anti-coccidioidin-precipitated Ig (in upper gel) diluted to 1:5 and 1:10, respectively (original CDN/anti-CDN reference system; after Huppert et al., 1978).
Figs. 3,4. 2D-IEP of coccidioidin (cathodal well) against newly-raised antiserum (as above) diluted 1:10 and 1:15, respectively.

derived from a hyperimmunized burro were incorporated into the upper gel at 1:10 and 1:15 dilutions, respectively, of precipitated Ig reconstituted to 5 mg protein/ml electrophoresis buffer (barbital, pH 8.0). The reference antigen (CDN, accession no. XV Q5C) was placed in the cathodal well at concentrations of 50ug/20ul PBS, and 100ug/20ul PBS, respectively. A total of 16 precipitin peaks are distinguished using these two Ig dilutions, and an additional 5 peaks were revealed at Ig dilutions of 1:5 and 1:20 (not shown). The majority of antigens originally demonstrated in CDN, therefore, are detected in the new system and comparable patterns of precipitin peaks are revealed. In both systems, Ag18 is a prominent precipitin and could be used as a reference peak to locate and identify other antigens (Huppert et al., 1978). In both the original and new reference systems, Ag2 is revealed with a prominent anodal shoulder (c.f. Figs. 1,3). Huppert and coworkers (1978) also demonstrated that Ag12 bore an anodal shoulder, but the precipitin peak was revealed in their electrophoresis plates only when an Ig dilution of 1:2 was used in the upper gel. However, its position and shape are comparable to the previously-reported AgCS (Figs. 3,4), which we had suggested was an antigen not represented in the original reference system (Cole et al., 1985,1986a). Differences in heights of corresponding precipitin peaks of the two reference systems could be a function of the Ig and/or CDN concentrations used. Certain antigens in the new system (e.g. Ags 10 and 17 in Fig. 4), do not show the same position (relative to Ag18) as in the original reference system. However, we have identified these peaks on the basis of their shape and position relative to other adjacent peaks. More important, however, is the high degree of reproducibility of the pattern of precipitin peaks at all dilutions of Ig tested in the new reference system, which was also a feature of the original CDN/anti-CDN reference system (Huppert et al., 1978,1979).

Composition and Immunoreactivity of Sephacryl Fractions of SCWF

A typical elution profile resulting from Sephacryl S-300 fractionation of the soluble conidial wall material is shown in Fig. 15. The elution profile of the standard mixture of blue dextran ($2x10^6$ daltons) and selected proteins demonstrates that the column adequately separated components of the starting material with a molecular weight range of 200-6.5 kd. The vertical bars represent immunoreactivity of immune LN cells to each column fraction as well as the original soluble wall material (SCWF) applied to the column. Each sample was examined in the cell proliferation assay at a concentration of 10 ug/ml cell culture media. FR1a was collected in the void volume eluted from the column after sample application. The blue dextran standard was also eluted in the void volume (Fig. 5). Fraction 1a had the highest protein/carbohydrate (P/C) ratio. The next highest P/C ratios were found in FR2, 3a, and 3b. These were also the most immunoreactive fractions obtained from the Sephacryl column, and each was more stimulatory in the immune LN assay than the starting material (SCWF). These studies demonstrate that the SCWF antigens which stimulated immune lymph node cells can be resolved to some degree by size exclusion chromatography. The most active antigens are in the 90-6.5 kd MW range, with a second peak of activity (i.e., FR5) in the low MW range (~5kd). Both the highest and lowest MW fractions (FRs 1a,6) are less antigenic. There was no obvious correlation between protein/carbohydrate ratios and antigenicity for immune LN cells.

AL-IEP of Sephacryl Fractions of SCWF

Comparative studies of the antigen composition of FRs 2,3a, and 3b by AL-IEP (Figs. 6-8) have revealed both quantitative and qualitative differences. Using burro Ig dilutions of 1:5 (Fig. 6) and 1:15 (Fig. 7) in the upper gel, and a titration of antigen added to the anodal wells (a-e), it is possible to detect at least 10 antigens in FR2. Both Ag2 and AgCS are recognized as components of this fraction. In fact, the complexity of the antigenic composition of FR2 is comparable to that of the original material (SCWF)

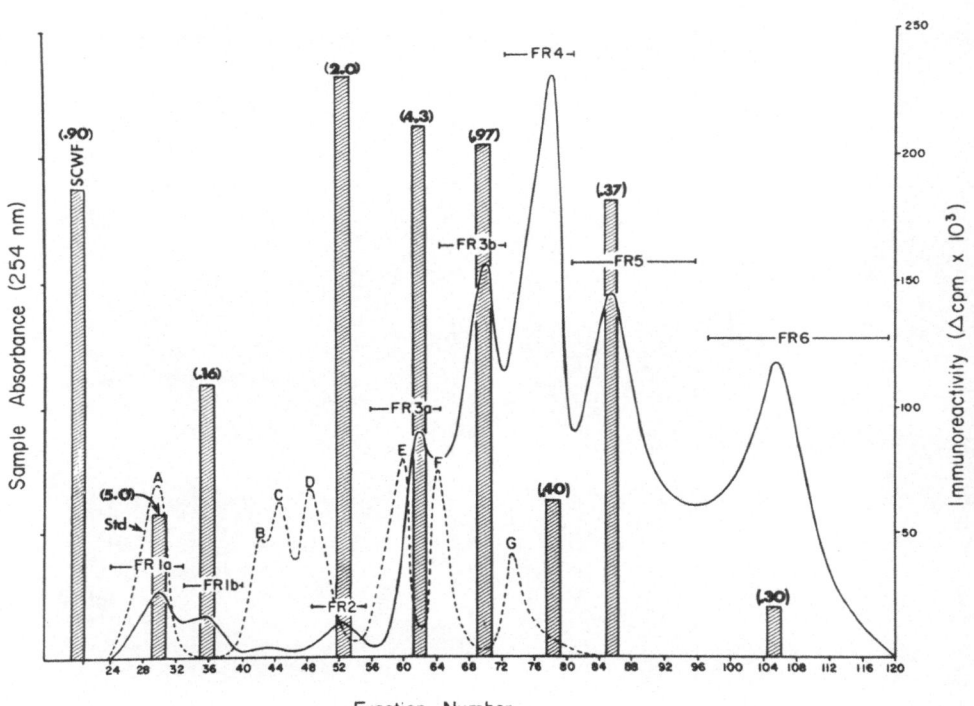

Fig. 5. Chromatographic elution profile of SCWF (solid line) from Sephacryl S-300 column. Sample fractions monitored at 254 nm. Elution profile of standard (Std) mixture (dotted line) monitored at 280 nm is superimposed. Bars represent immunoreactivity of SCWF and each pooled column fraction in immune lymph node proliferation assay at sample concentration of 10 ug/ml (Cole et al., 1987). Numbers in parentheses at top of each bar represent protein/carbohydrate ratios, calculated by the Lowry method (Lowry et al., 1951) for protein, and the phenol-sulfuric acid method (Dubois et al., 1956) for carbohydrate. Components of standard mixture (A-G) are blue dextran (2 x 10^3 kd), β-amylase from sweet potato (2 x 10^2 kd), alcohol dehydrogenase from yeast (1.5 x 10^2 kd), BSA (66 kd), carbonic anhydrase from bovine erythrocytes (29 kd), cytochrome C from horse heart (12.4 kd), and aprotinin from bovine lung (6.5 kd).

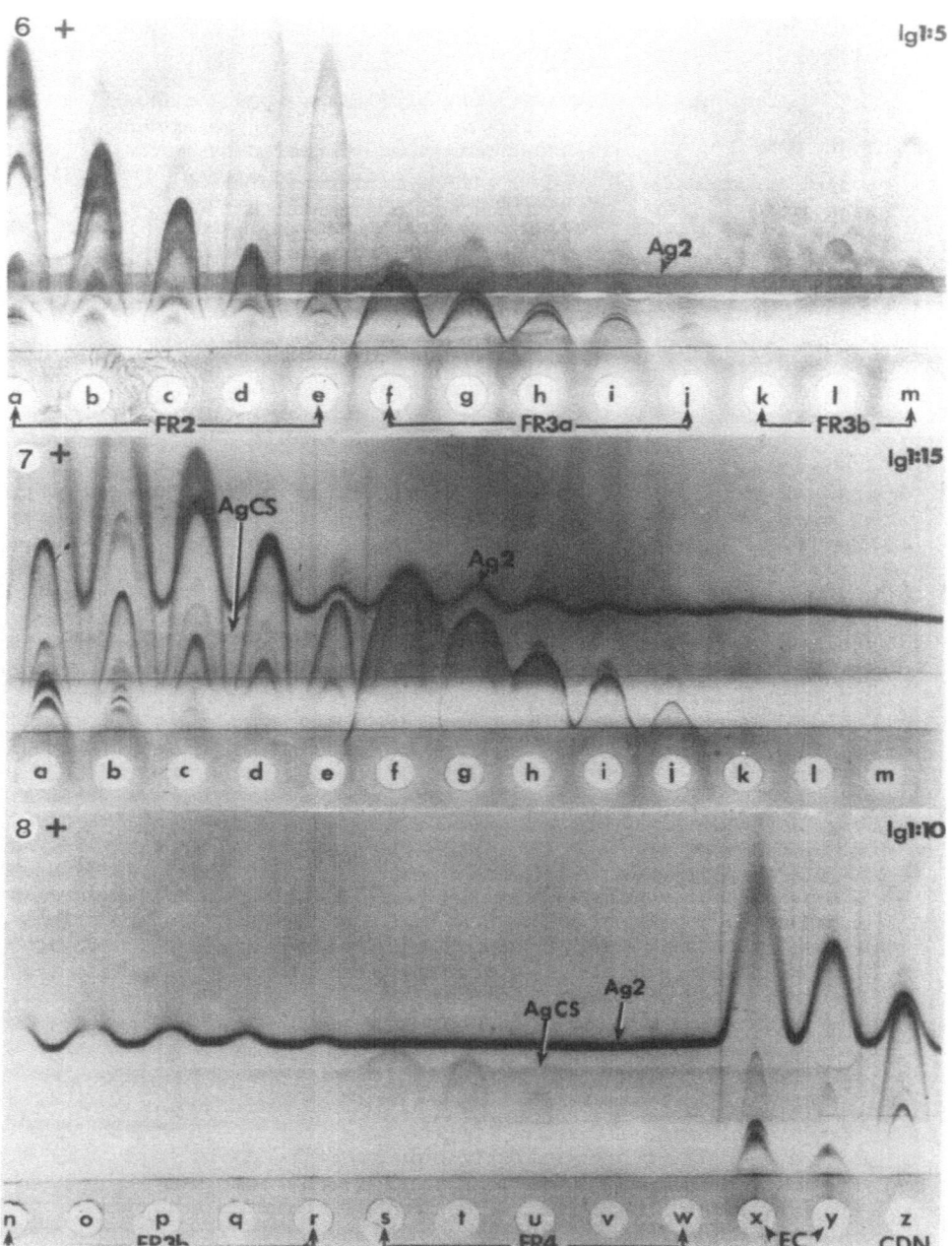

Fig. 6–8. Advancing line–IEP gels showing precipitin peaks above anodal wells which contain Sephacryl fractions (FR) 2, 3a, 3b, and 4. The lyophilized fractions were resolubilized in PBS. Anodal wells a–e (FR2) and f–i (FR3a) in Figs. 6 and 7 contain antigen concentrations of 20, 10, 5, 2.5, and 1.25 mg/ml buffer, respectively. Anodal wells k–m (FR3b) in Figs. 6, 7 and n–r (FR3b) in Fig. 8 contain antigen concentrations of 20, 10, and 5 mg/ml, respectively. Anodal wells s–w (FR4) contain antigen concentrations of 20–1.25 mg/ml as above. Anodal wells, x, y in Fig. 8, contain resolubilized endospore cytosol (EC) at concentrations of 20 and 10 mg/ml. Anodal well z in Fig. 8 contains the reference antigen (CDN) at 10 mg/ml. Reference antigens are labelled Ag2 and AgCS.

applied to the Sephacryl column. On the other hand, the AL-IEP gels of FRs 3a and 3b (Figs. 6,7) are less complex, and reveal at least 7 and 3 precipitin peaks, respectively. The immunoelectrophoresis gel of FR4 shows only two precipitin peaks using an Ig dilution of 1:10 (Fig. 8). One of these peaks is AgCS. As previously demonstrated, (Cole et al., 1985a; Cole and Sun, 1985), the advancing line of AgCS is not distorted by 'rocket' electrophoresis of the PBS-solubilized endospore cytosol fraction (Fig 8), but is sharply distorted by CDN (F+L). The method of preparation of the cytoplasmic fraction of C. immitis endospores has been described (Cole et al., 1985b). While Ag2 is a component of FR2, 3a, and 3b (Figs 6,7), albeit minor component of the last two fractions, it is not detected in FR4.

In summary, Sephacryl fractionation of SCWF consistently yields some peaks containing a large number of antigenic molecules (i.e., FR2) and others with a smaller number (i.e., FRs 3a,3b,4). Such reproducible fractionation was useful for selection of immunoreactive components and their subsequent purification.

Identification of Heat-sensitive Antigens in SCWF

When FR3a was examined in the immunodiffusion reference system together with CDN, a distinct, fused precipitin line formed between these two samples and the reference antibody (IDCF$_{Ab}$, Fig. 9). Fusion also occurred with the CF precipitin line between the reference antibody and antigen (IDCF$_{Ag}$). After separately heating FR3a and CDN to 60°C for 15 min prior to immunodiffusion, the CF precipitin line disappeared between each sample and IDCF$_{Ab}$.

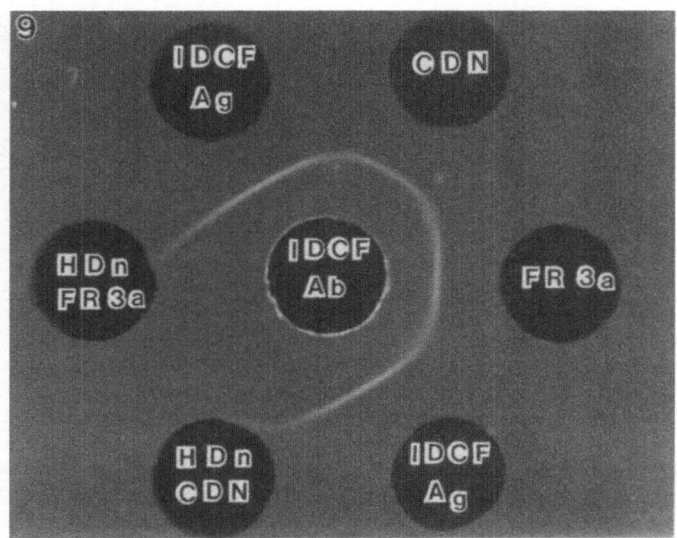

Fig. 9. Reactivity of Sephacryl fraction 3a (FR3a), heat-denatured FR3a (HDn FR3a), coccidioidin (CDN), and heat-denatured CDN (HDn CDN) in the immunodiffusion (ID) assay for complement-fixation (CF) precipitin reactions. The reference CF antibody and CF antigen are labelled IDCF$_{Ab}$ and IDCF$_{Ag}$, respectively.

Using tandem 2D-IEP, we identified two heat-sensitive antigens which may be associated with CF activity. A control gel is shown in Fig. 10 in which coccidioidin (R) in the cathodal well and PBS (B) added to the anodal well have been subjected to tandem 2D-IEP. Some distortion of the precipitin peaks of Ags 16, 23, and 26 is evident (c.f. Fig. 4), because the anodal well was not sealed with agarose prior to electrophoresis (Cox et al., 1984; Cole et al., 1986). However, the degree of peak distortion in the control gels was consistent and such distortion could easily be distinguished from tandem peaks. In Figs. 11 and 12, FR3a and heat-denatured (HDn) Fr3a have been added to the anodal wells, respectively. Electrophoresis was performed exactly as in Fig. 10. Three tandem peaks are visible in Fig. 11 resulting from the fusion of Ags 11, 18, and 23 (arrowheads) in CDN (R) with those of FR3a. However, only Ag11 showed a persistent tandem after FR3a was heated (Fig. 12).

In Fig. 13, the intermediate gel contains CF+ serum from a coccidioidomycosis patient. After the reference antigen (R) was subjected to 2D-IEP, the prominent precipitin peaks of Ags 18 and 23 were no longer visible in the upper gel. They precipitated in the intermediate gel as a result of

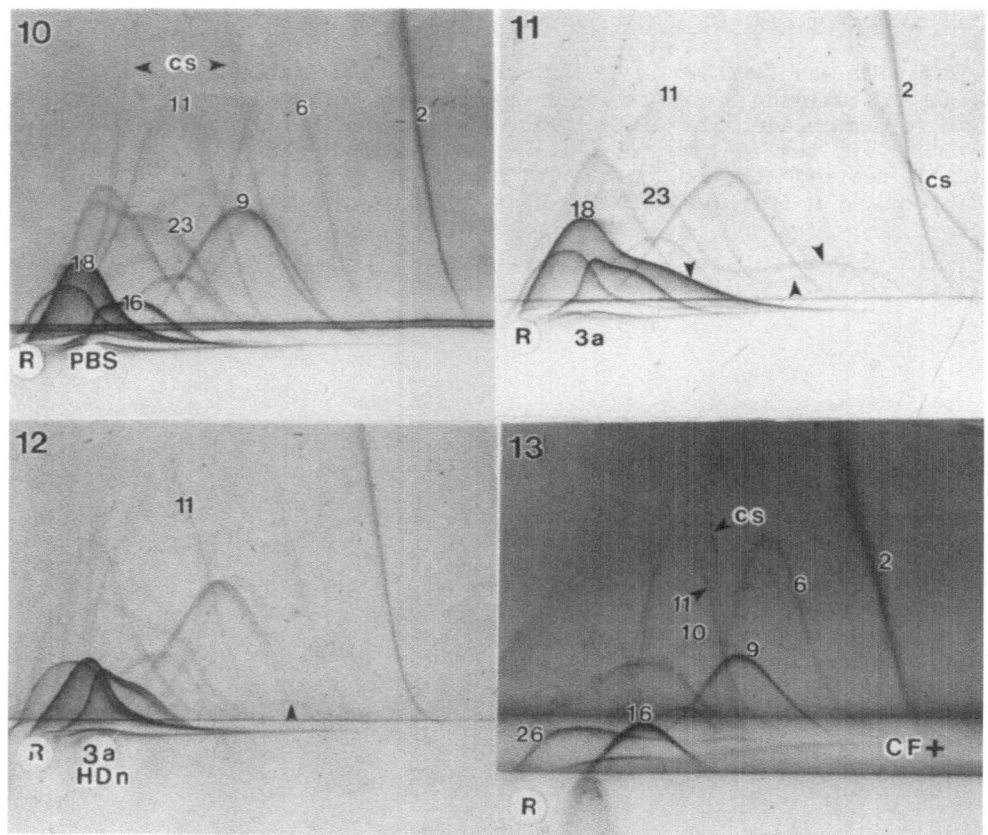

Figs. 10-12. Tandem 2D-IEP gels (Ig dilution 1:15). Fig. 10: Control gel with CDN (R) in tandem with PBS (B). Fig. 11: CDN (R) in tandem with Sephacryl fraction 3a (FR3a). Fig. 12: CDN in tandem with heat-denatured FR 3a (HDn3a). Arrowheads in Figs. 11, 12 locate tandem peaks. In each case, except for PBS, anodal cells contain 100 ug of sample.

Fig. 13. 2D-IEP plate of CDN in which intermediate gel contains complement fixation positive (CF+) serum of coccidioidomycosis patient. Ig dilution in upper gel is 1:15.

specific antibody/antigen reactions. On the other hand, Ag11 did not precipitate in the intermediate gel and is visible in the upper gel (Fig. 13). The combined application of tandem 2D-IEP and intermediate gel adsorption techniques is useful for antigen identification and estimation of humoral activity against patient sera.

Isolation and Detection of Ag11

Examination of the protein composition of SCWF has been performed by SDS-PAGE and the results have been reported (Cole et al., 1987). A major polypeptide band of SCWF is revealed in Fig. 14 with a MW of 36 kd. Sephacryl S-300 separation of SCWF has yielded fractions which show quantitative and qualitative differences in their protein composition based on SDS-PAGE (Cole et al., 1987). In the case of Sephacryl FR2 (Fig. 14), the 36kd protein has been concentrated. Separation of the 36 kd component of FR2 from most other proteins of this fraction has been achieved by cold acetone extraction (Fig. 14). When the acetone extracted fraction was then applied to a Superose 12 gel filtration column and subjected to HPLC, a partially purified protein was isolated (Fig. 15). The HPLC fraction was demonstrated to be the 36kd protein by SDS-PAGE. Subsequent fractionation, using a Sephadex G-50 minicolumn (Pharmacia), of the acetone extracted Sephacryl FR2 has demonstrated that the 36kd protein can be separated from the 28kd protein (Fig. 4) and then purified by HPLC.

The acetone extract of Sephacryl FR2 shown in Fig. 14 was examined in tandem with the reference antigen in 2D-IEP gels. In order to observe tandem peaks with all antigens of the reference system, tandem gels were examined using Ig dilutions of 1:5, 1:10, and 1:15 in electrophoresis buffer. The only reference antigen showing a tandem with the acetone extract (36kd concentrate) was Ag11 (Fig. 18).

Isolation and Detection of AgCS

The 16-day mycelial culture supernatant was subjected consecutively to Sephacryl S-300 gel filtration, DEAE ion exchange chromatography, and HPLC using DEAE and Superose 12 HR 10/30 gel filtration columns in series as described above. The elution profile resulting from HPLC separation is shown in Fig. 16A. The fraction whose retention time (RT) is 72.258 min was collected from several identical HPLC separations, pooled, and concentrated by ultrafiltration (Amicon) using a 5000 MW-cutoff filter. The spectra for the up-slope, peak, and down-slope of the HPLC fraction (i.e., RT=71.3, 72.0, 73.0, respectively) were identical for wavelengths between 200 nm and 400 nm (Fig. 16B). The chromatograms for the compound(s) eluted from the column between 69.2 and 77 min, and detected using absorbance wavelengths of 214, 224, 254, and 280 nm, essentially overlapped (Fig. 16C). The protein composition of the HPLC fraction (RT=71.3 to 73.0) was examined by SDS-PAGE which revealed a major band with a MW of approximately 19 kd(Fig. 17).

The antigen composition of the crude 16-day mycelial culture supernatant fraction obtained from Sephacryl S-300 gel filtration was first examined using tandem 2D-IEP gels (Fig. 19). The major antigen detected was AgCS. Minor antigen components included Ag2, 11, and 23. The antigen composition of the HPLC fraction (Fig. 16C) of this same original supernatant is shown in Fig. 20. The only two antigens detected were AgCS and Ag23. The major antigen was AgCS which corresponds with the major polypeptide band (MW=19 kd) in the SDS-PAGE gel (Fig. 17). In order to eliminate reactivity of Ag23 in the 2D-IEP gel, the HPLC fraction was heated to 60°C for 15 min prior to immunoelectrophoresis. Only the heat stable AgCS was detected after this treatment (Fig. 21) In an attempt to purify AgCS, the same HPLC fraction was passed through a Sephadex G-50 column and eluted with distilled water. A low molecular weight fraction was concentrated which was revealed as the 19kd MW polypeptide band by SDS-PAGE (Fig. 17), and AgCS by 2D-IEP.

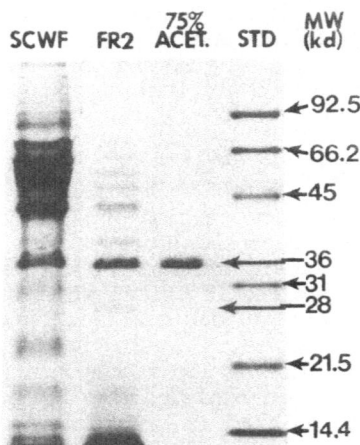

Fig. 14.   SDS—PAGE gel of SCWF, Sephacryl fraction 2 (FR2), and resolubilized
pellet of 75% acetone cut of FR2.   STD, protein standards.

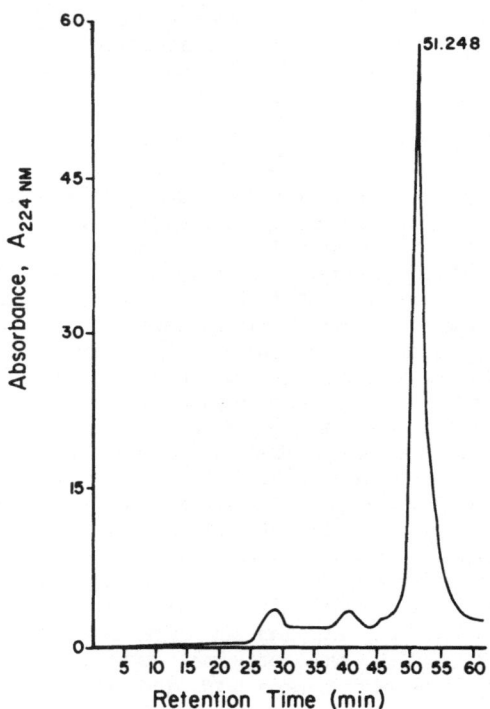

Fig. 15.   HPLC chromatogram of resolubilized pellet of 75% acetone cut
obtained from extraction of Sephacryl fraction 2 (FR2) of SCWF.   Fraction
monitored at 224 nm.

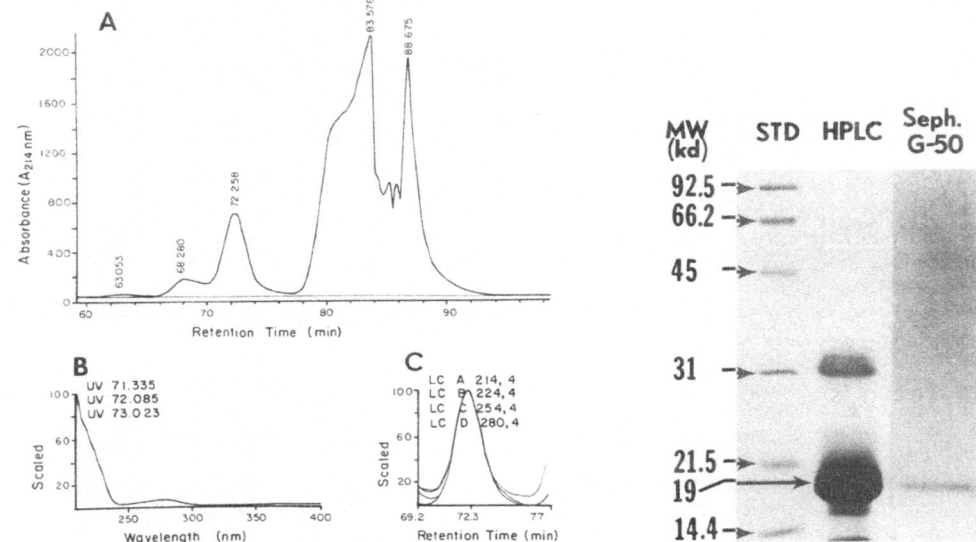

Fig. 16A–C. HPLC chromatogram (Fig. 16A) of 16-day mycelial culture supernatant fraction 2. The latter was obtained from low pressure liquid chromatographic fractionation on a DEAE (170 ml bed volume) column (elution profile not shown). Overlaid spectra of the HPLC fraction eluted between retention time (RT) of 71.3 and 73.0 min are shown in Fig. 16B. Overlaid chromatograms of the same HPLC fraction using absorbance wavelengths of 214, 224, 254, and 280 nm are revealed in Fig. 16C.

Fig. 17. SDS–PAGE of HPLC fraction of 16-day mycelial culture supernatant with RT=72.258 shown in Fig. 16, and Sephadex G-50 low molecular weight fraction obtained from the separation of this same HPLC fraction. STD, protein standards.

Immunoreactivity of Ag11 and AgCS

The prescence of Ag 11- and AgCS-specific antibodies in patient sera was tested in 2D-IEP plates provided with intermediate gels containing either normal human serum (NHS; Fig. 22) or patient sera with a high CF titer (CF+, Fig. 23). Both Ag11 and AgCS precipitin peaks are well defined in the upper gel of the control plate (Fig. 22). In Fig. 23, however, the precipitin peak of AgCS is not visible in the upper gel, while the Ag11 peak appears similar to the control. It seems that AgCS elicited a strong humoral response in this patient with high CF titer, while Ag11 elicited little detectable antibody response.

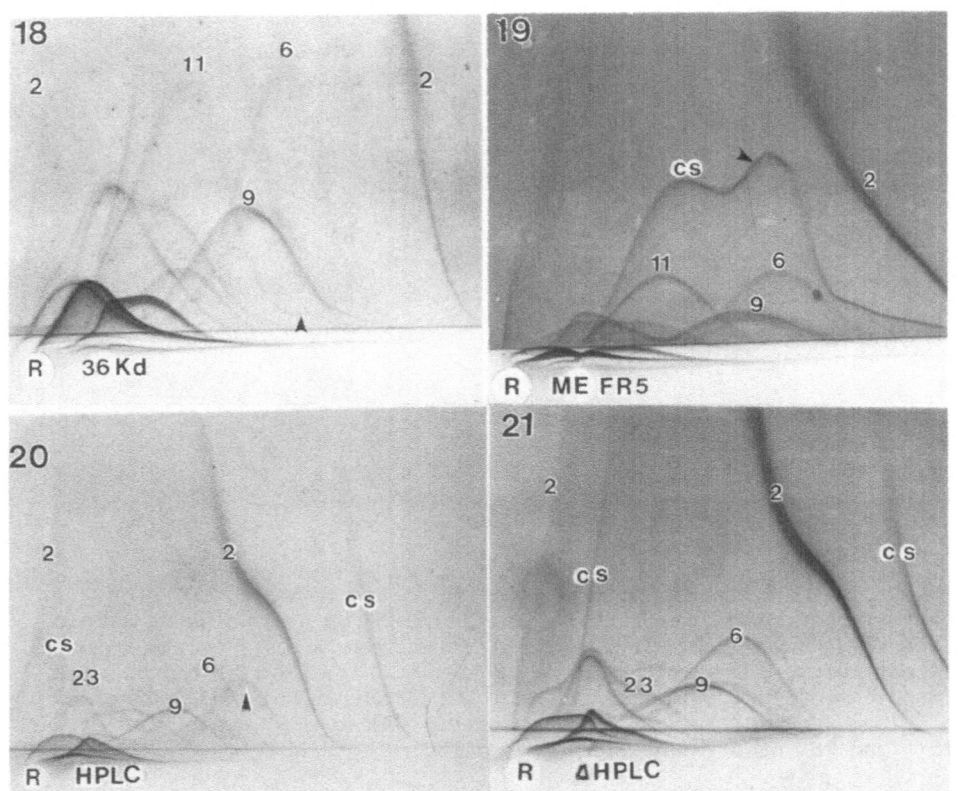

Figs. 18-21. Tandem 2D-IEP gels (Ig dilutions 1:15, 1:10, 1:10, and 1:10, respectively). Fig. 18:CDN (R) in tandem with resolubilized pellet of 75% acetone cut of Sephacryl FR2 (36 kd). Fig. 19: CDN (R) in tandem with Sephacryl S-300 fraction 3 obtained from chromatographic fractionation of the 16-day mycelial culture supernatant (elution profile not shown). Fig. 20: CDN (R) in tandem with HPLC fraction (RT=72.258 min; c.f. Fig. 16). Fig. 21: CDN (R) in tandem with heat-denatured (60°C, 15 min) HPLC fraction (Δ HPLC) shown in Fig. 20.

Detection of AgCS in Other Products of C. immitis

Spherulin, which has been used clinically as a test antigen for complement fixation activity, was placed in the anodal well of a tandem 2D-IEP plate (Fig. 24). A prominent tandem peak with AgCS is visible. A tandem peak with AgCS also formed with the skin test antigen (Fig. 25). Of particular interest is the formation of a prominent tandem peak between the reference AgCS and heat stable (HS) antigen (Fig. 26) reported by Kaufman et al. (1985). We have also examined the HS antigen in 2D-IEP gels containing the anti-CDN reference serum (unpublished) and demonstrated that AgCS is the major component of this preparation.

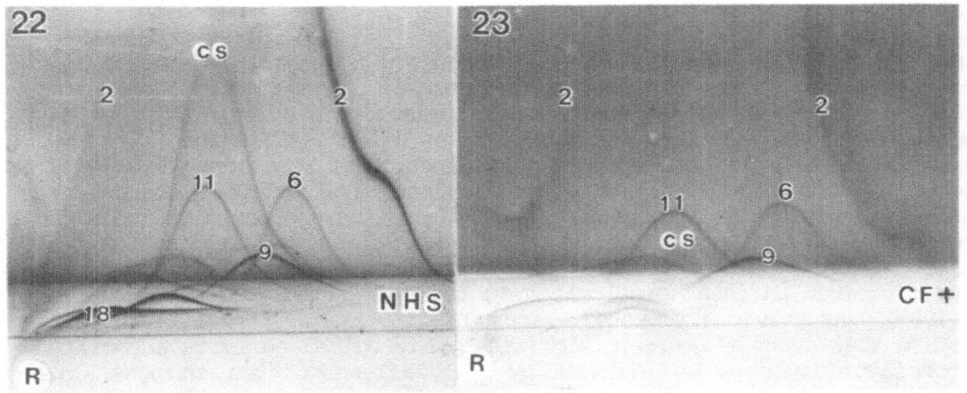

Figs. 22,23. 2D-IEP plates of CDN in which intermediate gels contain normal human serum (NHS, Fig. 22), and complement-fixation positive patient serum (CF+, Fig. 23). Ig dilution in upper gel is 1:15.

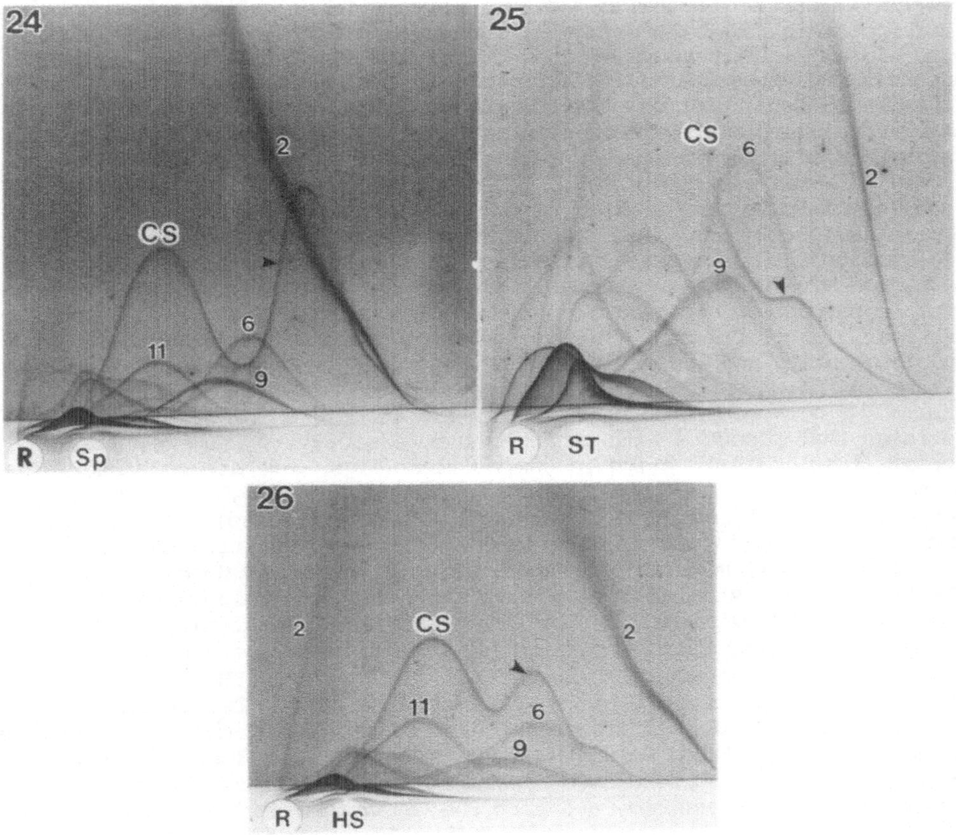

Figs. 24-26. Tandem 2D-IEP gels (Ig dilution in upper gels are 1:10 (Figs. 24, 26) and 1:15 (Fig. 25). Fig. 24. CDN(R) in tandem with spherulin (Sp). Fig. 25. CDN (R) in tandem with skin-test antigen (ST). Fig. 26. CDN (R) in tandem with heat-stable antigen (HS).

DISCUSSION

Huppert and coworkers (1978,1979) recognized the value of an antigen-antibody reference system for monitoring successive steps during physicochemical fractionation in attempts to isolate single antigens. The CDN/anti-CDN reference system has proved to be essential for our standardization of antigen preparations, selection of fractions which will be subjected to further purification procedures, and evaluation of the purity of isolated immunoreactive components. The logic of using coccidioidin as the reference antigen for identification of immunoreactive wall macromolecules is supported by our recent observation, based on immunoelectron microscopy, that most coccidioidin antigens are located in the cell envelope (Cole et al., 1986, 1987). As previously stated, accumulated evidence is available in the literature that every clinically significant biological activity associated with coccidioidomycosis is elicited by cell wall-associated antigens (Cole et al., 1983). Coccidioidin has also been shown to elicit most clinically significant immunological responses in coccidioidomycosis (Huppert, 1983). It follows, therefore, that CDN is a valid reference antigen complex for identification of wall antigens in our investigations. Using essentially the same procedures as Huppert and coworkers (1978,1979) for preparation of CDN and isolation of anti-CDN immunoglobulins from a hyperimmunized burro, we have reproduced the original CDN/anti-CDN reference system.

In this paper we have outlined procedures for identification of immunoreactive macromolecules in a crude, water-soluble wall fraction obtained from arthroconidia (SCWF), and from the mycelial culture supernatant of C. immitis. In an earlier study, we showed that SCWF contains potent antigens which elicit the highest levels of response in the immune LN proliferation assay of all soluble fractions tested. Since the goals of our investigations are to purify, locate and characterize antigens of Coccidioides which elicit humoral and/or cellular immune responses, the soluble conidial wall fraction has proved to be an excellent starting material. The method of isolation of SCWF does not involve cell rupture or death, so there is little exposure to intracytoplasmic antigens. In order to avoid possible enzymatic alteration of antigenic components as a result of activity of wall-associated proteinases, the extracion of SCWF was performed in the presence of PMSF. This enzyme activity inhibitor was not used during attempts to isolate the wall-associated and extracellular proteinase. The antigens in SCWF are available in reasonable quantity and the water-soluble material can be easily fractionated using size exclusion chromatography. Fractionation of SCWF on a Sephacryl S-300 column has been monitored using the AL-IEP technique and we have demonstrated that products with enhanced immunoreactivity (Fig. 5), and in most cases, less complex antigenic composition (Figs. 6-8) are obtained. The high degree of reproducibility of the chemical and immunological composition of SCWF and Sephacryl fractions obtained from successive isolates (Cole et al., 1983,1986) is particularly important and further justifies use of the soluble conidial wall material for analysis of C. immitis antigens.

Immunodiffusion studies of SCWF have shown that the soluble conidial wall fraction is reactive with both TP and CF antibodies (Cole et al., 1987), which are the primary immunodiagnostic tests for coccidioidomycosis (Huppert, 1983). After chromatographic fractionation of SCWF (Fig. 5), only FR3a retained reactivity with CF antibody (Fig. 9) (Cole et al., 1987). Two precipitin peaks in a 2D-IEP gel of FR3a disappeared after heating and were identified as Ags 18 and 23 (Figs. 11,12). The heat-denatured FR3a also lost reactivity to the CF antibody in the ID reference system. Coccidioidin has also been shown to be reactive with CF antibody and that the CF antigen in CDN is heat labile (Fig. 9) (Huppert, 1983; Cox and Britt, 1985). It is possible that one or both of the heat labile antigens of FR3a is (are) equivalent to the CF antigen.

Using a combination of chromatographic and cold acetone extraction procedures, we isolated a component of Sephacryl FR2 (Fig. 5) which has been

410

identified as a 36kd protein by SDS-PAGE (Fig. 14), and Ag11 by tandem 2D-IEP (Fig. 18). Our initial investigations have indicated that Ag11 shows little reactivity to patient sera in the humoral immunoassay (Fig. 13,23). On the other hand, preliminary studies of this partially purified protein have shown that is is a wall-associated proteinase capable of digesting collagen and elastin over a wide pH and temperature range. Results of these preliminary studies, which will be presented in a future communication, are summarized in Table 1.

Isolation and fractionation of SCWF yielding the proteinase-containing Sephacyl FR2 were performed in the absence of PMSF. Both elastolytic and collagenolytic activities have been previously reported for C. immitis, (Nziramasanga and Lupan, 1985; Lupan and Nziramasanga, 1986), but isolation characterization or localization of the enzyme(s) has(have) not been performed.

Table 1

Properties of C. immitis Proteinase
Isolated from SCWF[1]

Substrates:	Human collagen, elastin, hemoglobin, IgG and sIgA, and bovine casein
MW (SDS-PAGE):	36 kd
pH Optimum:	8.0-9.0
Temperature Optimum:	35-40°C at pH 8.0
Stability:	pH range 4.0 - 9.0; Temp. to 50°C
Isoelectric Point:	4.5
Inhibitors:	DFP[2], TPCK[3], α-1-antitrypsin, PMSF[4]

[1]Soluble conidial wall fraction

[2]Diisopropylphosphorofluoridate (Sigma)

[3]N-tosyl-L-phenylalanine chloromethyl ketone (Sigma)

[4]Phenylmethylsulfonyl fluroide (Sigma)

The mycelial culture supernatants of C. immitis (strains C634 and C735). have been shown to contain detectable amounts of AgCS, based on 2D-IEP analyses, as early as 3 days after inoculation of GYE broth with arthroconidia (SH Sun, J Dominguez, and GT Cole, Abst. F-30, Am. Soc. Microbiol. Meeting, Atlanta, 1987). After growth of the mycelial culture for 16 days, AgCS was shown to be a major antigenic component of the culture supernatant (Fig. 19). For this reason, the 16-day culture supernatant was used for chromatographic fractionation and isolation of this antigen. AgCS was originally described by us in earlier reports (Cole et al., 1985a,b; Cole and Sun, 1985; Cole et al., 1986). It was shown to be cell wall-associated and was not detected in cytosol preparations of either saprobic or parasitic cell types. In addition, studies of the antigen content of cell wall preparations of Blastomyces dermatitidis, Histoplasma capsulatum and Candida albicans, using the CDN/anti-CDN reference system, did not reveal AgCS (Cole et al., 1985a). The significance of these observations was substantiated by our findings in this study that AgCS is the major component of the heat stable (HS) antigen reported by Kaufman et al. (1985), as well as a component of the biologically-active products, spherulin and skin test antigen (Figs. 24-26). AgCS has also been shown to be heat stable at 60°C for 30 min. Kaufman et al. (1985) further reported that the HS antigen is the "most useful coccidioidin antigen for specifically immmunoidentifying C. immitis cultures", but "antibodies to HS are infrequently found in human sera from patients with coccidioidomycosis and are thus of little serodiagnostic value." The reason for the infrequent occurrence of HS antibody to this prominent antigen in human coccidioidomycosis sera is not known. It was suggested that the antibody may be transiently present or present at concentrations below the level of sensitivity of the authors' detection procedure (Kaufman et al., 1985). In preliminary studies, using chromatographically-purified AgCS in the enzyme-linked immunosorbent assay (ELISA), we have demonstrated a high level of sensitivity and specificity for antibody to AgCS in several coccidioidomycosis patient sera compared to heterologous sera from patients with other systemic mycoses. The results of this study will be presented in a future communication. It is still possible, therefore, that AgCS is a potentially valuable immunodiagnostic antigen. Further purification of this immunoreactive fraction using hybridoma techniques is in progress.

ACKNOWLEDGEMENTS

    This work was supported by a National Institutes of Health grant (AI 19149) to the senior author.

REFERENCES

Brawner, D. L., and Cutler, J. E., 1986, Ultrastructural and biochemical studies of two dynamically expressed cell surface determinants on Candida albicans, Infect. Immun., 51:327.

Brawner, D. L., and Cutler, J. E., 1986, Variability in expression of cell surface antigens of Candida albicans during morphogenesis, Infect. Immun., 51:337.

Cole, G. T., Chinn, J. W., Pope, L. M., and Starr, P., 1985a, Characterization and distribution of 3-0-methylmannose in Coccidioides immitis, in: "Proc. 4th Int. Conf. on Coccidioidomycosis," H. Einstein and T. Catanzaro, eds., Nat. Found. Infect. Dis. Publ., Washington, D.C., pp. 130-145.

Cole, G. T., Kirkland, T. N., and Sun, S. H., 1987, An immunoreactive, water-soluble conidial wall fraction of Coccidioides immitis, Infect. Immun., 55:657.

Cole, G. T., Pope, L. M., Huppert, M., Sun, S. H., and Starr, P., 1983, Ultrastructure and composition of conidial wall fractions of Coccidioides immitis, Exp. Mycol., 7:297.

Cole, G. T., Pope, L. M., Huppert, M., Sun, S. H., and Starr, P., 1985b, Wall composition of differenct cell types of Coccidioides immitis, in: "Proc. 4th Int. Conf. on Coccidioidomycosis," H. Einstein and T. Catanzaro, eds., Nat. Found. Infect. Dis. Publ., Washington, D. C., pp. 112-129.

Cole, G. T., Starr, M. E., Sun, S. H., and Kirkland, T. N., 1986, Antigen isolation in Coccidioides immitis, in: Microbiology-1986," L. Leive, ed., Am. Soc. Microbiol. Publ., Washington, D. C., pp. 159-164.

Cole, G. T., and Sun, S. H., 1985, Arthroconidium-spherule-endospore transformation in Coccidioides immitis, in: "Fungal Dimorphism," P. J. Szaniszlo, ed., Plenum Publ., New York, pp. 281-333.

Cole, G. T., Sun, S. H., and Huppert, M., 1982, Isolation and ultrastructural examination of conidial wall components of Coccidioides and Aspergillus, Scanning Electron Microsc.., IV:1677.

Collins, M. S., Pappagianis, D., and Yee, J., 1977, Enzymatic solubilization of precipitin and complement fixing antigen from endospores, spherules and spherule fraction of Coccidioides immitis, in: "Coccidioidomycosis: Current Clinical and Diagnostic Status," L. Ajello, ed., Symposia Specialists, Miami, pp. 429-444.

Cox, R. A., and Britt, L. A., 1985, Antigenic heterogeneity of an alkali soluble, water-soluble cell wall extract of Coccidioides immitis, Infect. Immun., 50:365.

Cox, R. A., Huppert, M., Starr, P., and Britt, L.A., 1984, Reactivity of alkali-soluble, water-soluble cell wall antigen of Coccidioides immitis with anti-Coccidioides immunoglobulin M precipitin antibody, Infect. Immun., 43:502.

Cox, R. A., Mead, C. G., and Pavey, E. F., 1981, Comparisons of mycelia- and sperule-derived antigens in cellular immune assays of Coccidioides immitis-infected guinea pigs, Infect. Immun., 31:687.

Cox, R. A., and Vivas, J. R., 1977, Spectrum of in vivo and in vitro cell-mediated responses in coccidioidomycosis. Cell Immunol., 31:130.

Drutz, D. J., and Huppert, M., 1983, Coccidioidomycosis: factors affecting the host-parasite interaction, J. Infect. Dis., 147:372.

Dubois, M., Goiller, K. A., Hamilton, J. K., Rebers, P. A., and Smith, F., 1956, Colorimetric method for determination of sugars and related substances, Anal. Chem., 28:530.'

Frey, C. L, and Drutz, D. J., 1986, Influence of fungal surface components on the interaction of Coccidioides immitis with polymorphonuclear neutrophils, J. Infect. Dis., 153:933.

Hall, N. K., Deighton, F., and Larsh, H. W., 1978, Use of an alkali-soluble, water-soluble extract of Blastomyces dermatitidis yeast phase cell walls and isoelectrically focused components in peripheral lymphocyte transformations, Infect. Immun., 19:411.

Hearn, V. M., 1984, Surface antigens of intact Aspergillus fumigatus mycelium: their localization using radiolabelled protein A as marker, J. Gen. Microbiol., 130:907.

Hector, R., and Pappagianis, D., 1982, Enzymatic degradation of the walls of spherules of Coccidioides immitis, Exp. Mycol., 6:136.

Huppert, M., 1983, Antigens used for measuring immunological reactivity, in: "Fungi Pathogenic for Humans and Animals, Part B, Pathogenicity and Detection: I, "D. Howard, ed., Marcel Dekker, New York, pp. 219-302.

Huppert, M., Adler, J. P., Rice, E. H., and Sun, S. H., 1979, Common antigens among systemic disease fungi analyzed by two-dimensional immunoelectrophoresis, Infect. Immun., 23:479.

Huppert, M., and Bailey, J. W., 1963, Immunodiffusion as a screening test for coccidioidomycosis. I. The accuracy and reproducibility of the immunodiffusion test which correlates with complement fixation, Am. Clin. Pathol., 44:364.

Huppert, M., and Bailey, J. W., 1965, The use of immunodiffusion in coccidioidomycosis. II. An immunodiffusion test as a substitute for the tube precipitin test, Am. J. Clin. Pathol., 44:369.

Huppert, M., Krasnow, I., Vikovich, K. R., Sun, S. H., Rice, E. H., and Kutner, L. J., 1977, Comparison of coccidioidin and spherulin in complement fixation tests for coccidioidomycosis, J. Clin. Microbiol., 6:33.

Huppert, M., Spratt, N. S., Vukovich, K. R., Sun, S. H., and Rice, E. H., 1978, Antigenic analysis of coccidioidin and spherulin determined by two-dimensional immunoelectrophoresis, Infect. Immun., 20:541.

Huppert, M., and Sun, S. H., 1980, Overview of mycology, and mycology of Coccidioides immitis, in: "Coccidioidomycosis: A Text," D. A. Stevens, ed., Plenum Publ., New York, pp. 21–46.

Huppert, M., Sun, S. H., and Harrison, J. L., 1982, Morphogenesis throughout saprobic and parasitic cycles of Coccidioides immitis, Mycopathol., 78:107.

Kaufman, L., Standard, P. G., Huppert, M., and Pappagianis, D., 1985, Comparison and diagnostic value of the coccidioidin heat-stable (HS and tube precipitin) antigens in immunodiffusion, J. Clin. Microbiol., 22:515.

Laemmli, U.K., 1970, Cleavage of structural proteins during the assembly of the head of bacteriophage T4., Nature (London), 227:680.

Lecara, G., Cox, R. A., and Simpson, R. B., 1983, Coccidioides immitis vaccine: potential of an alkali-soluble, water-soluble cell wall antigen, Infect. Immun., 39:437.

Lowry, O. H., Rosebrough, N. J., Farr, A. L., and Randall, R. J., 1951, Protein measurement with the Folin phenol reagent, J. Biol. Chem., 193:265.

Lupan, D. M., and Nziramasanga, P., 1986, Collagenolytic activity of Coccidioides immitis, Infect. Immun., 51:360.

Nziramasanga, P., and Lupan, D. M., 1985, Elastase activity of Coccidioides immitis, J. Med. Microbiol., 19:109.

Ponton, J., and Jones, J. M., 1986, Analysis of cell wall extracts of Candida albicans by sodium dodecyl sulfate-polyacrylamide gel electrophoresis and western blot techniques, Infect. Immun., 53:565.

Reiss, E., Huppert, M., and Cherniak, R., 1985, Characterization of protein and mannan polysaccharide antigens of yeasts, moulds, and Actinomycetes, Curr. Top. Med. Mycol., 1:172.

San-Blas, G., 1982, The cell wall of fungal human pathogens: its possible role in host-parasite relationships, Mycopathologia, 79:159.

San-Blas, G., 1985, Paracoccidioides brasiliensis: Cell wall glucans, pathogenicity, and dimorphism, Curr. Top. Med. Mycol., 1:235.

Scalarone, G. M., Levine, H. B., Pappagianis, D., and Chaparas, S. D., 1974, Spherulin as a complement-fixing antigen in human coccidioidomycosis, Am. Rev. Resp. Dis. 110:324.

Smith, C. E., Whiting, E. G., Baker, E. E., Rosenberger, H. G., Beard, R. R., and Saito, M. T., 1948, The use of coccidioidin, Am. Rev. Tuberc. 57:330.

Storti, R. V., Coen, D. M., and Rich, A., 1976, Tissue-specific forms of actin in the developing chick, Cell, 8:521.

Tarbet, J. E., and Breslau, A. M. 1953, Histochemical investigation of the spherule of Coccidioides immitis in relation to the host reaction, J. Infect. Dis., 92:183.

Turini, P., Kurooka, S., Steer, M., Corbascio, A. N., and Singer, T. R., 1969, The action of phenylmethylsulfonyl fluoride on human acetyl cholinesterase, chymostrypsin and trypsin J. Pharmacol. Exp. Theraeutics, 167:98.

Ward, E. R., Cox, R. A., Schmitt, J. A., Huppert, M., and Sun, S. H. 1975, Delayed-type hypersensitivity responses to a cell wall fraction of the mycelial phase of Coccidioides immitis, Infect. Immun. 12:1093.

Wheat, R. W., Su-Chung, K. S., Ornellas, E. P., and Scheer, E. R., 1978, Extraction of skin test activity from Coccidioides immitis mycelia by water, perchloric acid and aqueous phenol extraction, Infect. Immun., 19:152.

414

Wheat, R. W., Tritschler, C., Conant, N. F., and Lowe, E. P., 1977, Comparison of *Coccidioides immitis* arthrospore, mycelium and spherule cell walls, and influence of growth medium on mycelial cell wall composition, *Infect. Immun.*, 17:91.

IMMUNOCHEMICAL ANALYSIS OF HISTOPLASMIN PROTEINS AND POLYSACCHARIDE

Errol Reiss and Sandra L. Bragg

Division of Mycotic Diseases
Center for Infectious Diseases
Centers for Disease Control
Public Health Service
US Department of Health and Human Services
Atlanta, Georgia 30333

Introduction

Histoplasmin is the supernatant fluid obtained after static cultivation of the mycelial form of Histoplasma capsulatum for 3 to 6 months at ambient temperature (22 to 25°C) on asparagine-glucose-glycerol-salts medium (Smith et al, 1948). Three types of histoplasmin exist -- skin test reagents from the mycelial and yeast forms and serological histoplasmin. The method of preparing skin test histoplasmin was established long ago (Shaw et al, 1950) and single lots were used widely for many years. The skin test antigen was standardized by its ability to elicit a reproducible cutaneous hypersensitivity response in sensitized humans. The antigens in histoplasmin responsible for the skin test have been inferred but no direct structure-activity evidence exists, because monomolecular antigens have not been available. Recently, histoplasmin from yeast form cultures has been approved for human use as a skin test antigen (Scalarone et al, 1985). The bulk of accumulated information concerns mycelial histoplasmin and there are far fewer data about the characteristics of the yeast form product. For that reason, and because of space limitation, the following discussion applies primarily to mycelial histoplasmin.

More flexible conditions apply for producing serological histoplasmin because (i) rigid control for in vivo human use is not obligatory; (ii) there is more information about individual precipitinogens, and (iii) the responsible antigens have been partially purified and characterized. Advances in understanding histoplasmin antigens from the serological standpoint should provide a stimulus to define the content of skin test histoplasmin.

Uses of histoplasmin

The diagnostic usefulness of histoplasmin skin tests is limited. There are numerous healthy positive reactors in the endemic areas and, on the other hand, a distinct possibility exists for cutaneous anergy in disseminated histoplasmosis. That is, positive results occur without active disease and vice versa. Furthermore a single skin test in a previously exposed person may elevate circulating antibodies, thus obscuring the interpretation of serologic tests. Skin testing is justified in epidemiologic surveys and as a diagnostic aid in patients with a recent history of travel in an endemic area, in laboratory workers outside the endemic area and in children not old enough for ordinary environmental exposure. Within the endemic area a delayed cutaneous response to histoplasmin can serve as a general test of immunocompetence.

The stimulation of antibody concentrations by skin tests with mycelial histoplasmin in previously sensitized persons was first appreciated by Prior and Saslow (1952). Later, Heiner (1958), demonstrated the appearance of the immunoprecipitate containing the M-antigen in double agar-immunodiffusion after a recent skin test. Complement fixing antibody titers against histoplasmin were systematically evaluated in healthy skin test-positive persons (Campbell and Hill 1964). A single skin test stimulated antibody concentrations to increase in 7 of the 12 subjects (58.3%). Titers, initially negative, rose to between 1/8 and 1/256 (mean 1/67) during a 30-day period.

Cultures

There is no objective justification to use a pool of several H. capsulatum isolates to obtain more potent histoplasmin. The majority of strains produce the H-factor (Ehrhard and Pine, 1972; Standard et al, 1984). A single culture, i.e., 6623 (ATCC 26320) produces an M-rich histoplasmin, whereas 6624 (ATCC 12700) makes histoplasmin that is H-rich and M-poor. The resulting histoplasmins can be blended to obtain a balance of H and M for serological tests. The 6624 histoplasmin can be a starting material to produce a truly M-deficient skin test reagent.

We attempted to trace the original cultures used to prepare H-42, the standard skin test histoplasmin for many years (Schwarz, 1981, pp. 151-154). This product originated in 1950 when Dr. Arden Howell, Jr. and colleagues in the US Public Health Service pooled 3 lots of histoplasmin (Shaw et al, 1950). These lots were produced by the classic method with cultures grown for 3 months. Lot H-36 originated from a 1938 pediatric isolate; the source of lot H-38 was unspecified, and lot H-39 was from the first H. capsulatum canine isolate. Whether these isolates were deposited in a culture collection could not be determined.

Preparation

The standard method of static mycelial form growth on the Smith et al, (1948) medium for 3 or more months is sanctioned by the U S Food and Drug Administration (1977). Biosafety level 3 practices and facilities are recommended for mold form cultures of H. capsulatum (U S Department of Health and Human Services, 1984). Batch cultures should be verified as nonviable after the addition of Merthiolate (Standard and Kaufman, 1982). Cell free supernatant fluids (= unpurified histoplasmin) are concentrated 10-fold by Amicon ultrafiltration over a YM10 membrane. The retentate, histoplasmin, is filter sterilized and stored at 4°C. Unpurified histoplasmin is reputed to be very stable both for skin tests (Schwarz 1981, p. 151) and for serological purposes. Purified histoplasmin factors are not nearly as stable, especially the M antigen, Pine et al (1981). Lyophilization is not recommended.

Medium effects. The yield of M-antigen in histoplasmin is sensitive to the nitrogen source in asparagine-glucose-glycerol-salts medium. When L -asparagine is replaced by the D, L - enantiomer and supplemented by acid hydrolyzed casein (0.4%), the yield of M-antigen was increased 4-fold (Ehrhard and Pine, 1972). Glycerol in the synthetic medium is not assimilated and may persist during concentration.

418

Release of histoplasmin. Although the classic method calls for at least 3 months growth of a static culture, Ehrhard and Pine (1972) resuspended a 14-days-old mycelium of high H and M producing strain 6623 in PBS. When the culture was reincubated on a gyratory shaker, rapid disintegration of the hyphae ensued, accompanied by the release of M antigen. An active enzyme - catalyzed autolysis may explain why some strains produced larger amounts of M antigen. Broth filtrates of H. capsulatum were shown to contain β -D - glucosidase, β -1, 3-D glucanase, and α -D-mannosidase activity. No chitinase activity was detected (Davis et al, 1977). Thus, some requirements for an autolytic process are present. Resuspension in PBS of young mycelium can provide an alternative route to obtain M-enriched histoplasmin.

Precipitating Antigens

The major histoplasmin antigens that participate in immunoprecipitin reactions have been identified and partially purified (Fig. 1). Heiner (1958) first designated the most readily detected precipitinogens as C, H, M, X, and Y, based on agar double immunodiffusion reactions with sera from human histoplasmosis patients. The C antigen was named because it is held in common with blastomycin and coccidioidin. The M precipitin arc, Heiner observed, occurred in acute histoplasmosis but sometimes also appeared after a single skin test in previously sensitized persons. Antibodies against the H antigen were less common than anti-M and more often were associated with active, proven, histoplasmosis. The occurrence of the remaining x and y factors has not been confirmed but Heiner's major findings (based on 7 proven and 257 suspected cases) have been confirmed and extended.

Tompkins (1965) applied the term YP to an antigen which he extracted from yeast form H. capsulatum into the organic phase of a biphasic phenol-water solvent system. Later, the New York State Department of Health which continued to use this antigen referred to it as Y, creating some ambiguity with Heiner's Y. The latter antigen is a different and minor factor. An additional precipitinogen in histoplasmin characterized by Green et al (1976) was called "non-M". The immunochemical analysis of histoplasmin factors has proceeded intermittently with modest progress. Knowledge in this field has yet to be consolidated.

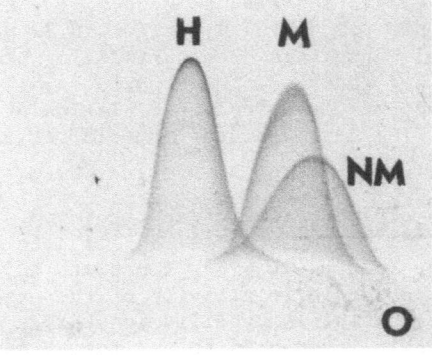

Figure 1. Two dimensional crossed rocket immunoelectrophoresis of histoplasmin (CDC) showing the M, non-M (NM), and H immunoprecipitin arcs. The anode was to the left (1st dimension) and top (2nd dimension). The conditions used were those of Weeke, 1973.

Anion and Cation Exchange Chromatography

Anion exchange chromatography on columns of diethylaminoethyl (DEAE) cellulose has been refined since the first application to purify histoplasmin was described (Greene et al, 1960; Dickerson and Busey, 1968; Pine et al, 1977). The conditions set in earlier reports were simultaneous elution with a pH gradient, and one of increasing ionic strength (ie, from pH 8.0 to pH 4.8 with the phosphate buffer molarity increasing from 0.01 to 0.3 M). Under these conditions the M antigen was first to appear in the eluate in the range of pH 6 to 8 followed by the H antigen eluting in the range of pH 5.1 to 5.5. Later, a modification was devised whereby preliminary gel permeation chromatography excluded turbid material and nucleic acid (Pine et al, 1977). Two cycles of anion exchange chromatography were necessary for maximum resolution of sample components. Firstly, holding the pH constant at pH 7.0 with Tris buffer, a linear NaCl gradient from 0.05 M to 0.3 M was applied. The M antigen eluted promptly when the molarity reached 0.05 M. Because it was more negatively charged, the H antigen eluted in a broader zone beginning at 0.15 M. Secondly, a pH step gradient was used consisting of phosphate buffer going from pH 8.0 to 4.0. The elution profile in this second cycle was quite similar to the results of Greene et al (1960) and Dickerson and Busey (1968).

Anion exchange columns achieved the intended result--good separation of the H and M factors, but a major drawback remained. The common fungal antigen had affinity for DEAE and desorbed in a broad zone dispersed in the chromatogram. Recall that this substance called C antigen by Heiner (1958) and later "Bd" by Green et al (1976) is a heat stable, probable polysaccharide or glycoprotein that shares antigenic determinants with similar antigens present in Blastomyces dermatitidis, Coccidioides immitis, and Paracoccidioides brasiliensis (Azuma et al, 1974; Green et al, 1976; Heiner 1958). Galactomannan-containing antigens from H. capsulatum mycelia tend to bind to DEAE-gels (Reiss et al, 1974). Later, more evidence will be introduced that the "C", "Bd" and galactomannan-containing polymers are the common fungal antigen.

The H, M, and "non-M" antigens of histoplasmin bind to carboxymethyl-agarose (CM-Sepharose CL-6B) at pH 3 in 0.05 M citrate buffer (Green and Pine, 1985, Figure 2). At this pH the antigens have a net positive charge and bind to the $-O-CH_2COO^-$ groups of the ion exchanger. The order of elution of the antigens, when a linear salt gradient was applied, was as follows: (1) C-antigen appeared in the breakthrough volume with the column buffer; (2) additional C antigen appeared when the salt concentration reached 0.08 M; (3) M-antigen, 0.1-0.13 M salt; (4) "non M", 0.2 M; (5) H antigen, 0.29 M. A clear advantage over previous anion exchange procedures using DEAE was that the bulk of the C antigen was not adsorbed. Rechromatography of the isolated M, non M, and H- containing peaks resulted in depletion of carbohydrate in the purified products. The carbohydrate: protein ratios were: M, 0.13; non-M, 0.15; H, 0.09.

Better than 90% of the applied protein was recovered. Retention of serological activity depended on the storage pH. If left at pH 3, almost all activity was lost after 30 days at 4°C. Immediate adjustment of the pH to 8.0 stabilized the serologic activity of the products. The total activity, measured by radial immunodiffusion (RID), was 75 to 100% of the unpurified histoplasmin. Compared to the former method of purification, using DEAE anion exchange, the specific activity (RID) was increased as a result of the CM-Sepharose procedure. In respect to the M antigen a 6-fold enhancement was achieved and the H-antigen was 2.5-fold more active. No cross contamination was observed when the CM-Sepharose column-purified antigens were analyzed by silver-stained sodium dodecyl sulfate-polyacrylamide gel electrophoresis (SDS-PAGE). On this basis, the H and M antigens appeared to be monomolecular with molecular weights of 108 kilodaltons (kDa) and 147 kDa, respectively.

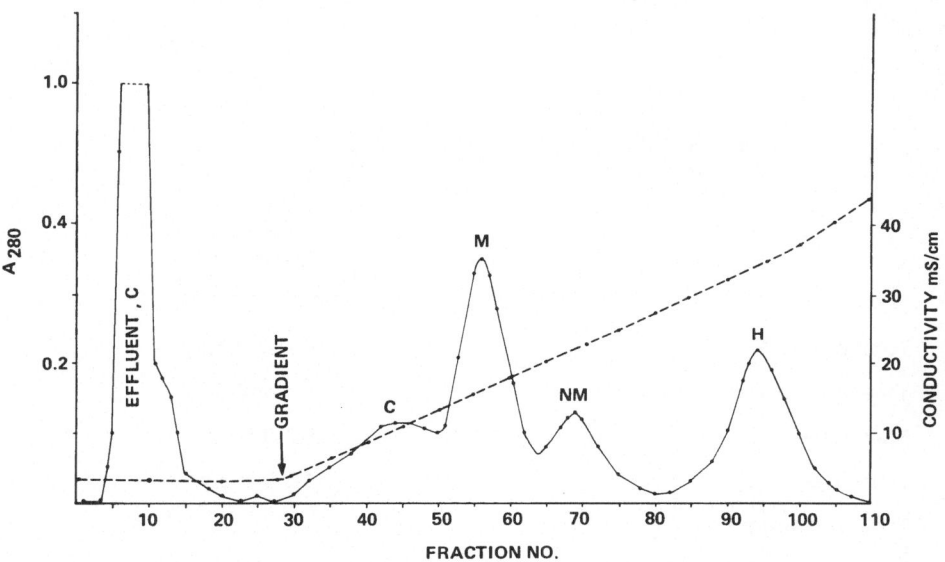

Figure 2. Elution profile of histoplasmin on CM-Sepharose CL-6B column (1 x 30 cm). Sample applied in 0.05 M citrate buffer, pH 3.0, and eluted with a 0-0.55 M linear salt gradient. Fraction vol, 3.8 ml. Source: reprinted with permission of Green and Pine (1985) and Springer-Verlag.

The common fungal antigen C contains carbohydrate and is either a polysaccharide or glycoprotein that is released through the cell wall during growth (Reiss et al, 1974, 1977). This antigen is stable to boiling and Pronase, and it is labile to periodate oxidation (Brock et al, 1984; Reiss et al, 1986). The C antigen can be purified from histoplasmin by affinity for the lectin concanavalin A (Con A, Ellsworth et al, 1977). Con A binds to glycosidic moieties that contain a 3,4,6-arabino α-D-glycopyranosyl structure including nonreducing end units and 2-O-substituted α-D-mannopyranose units (So and Goldstein, 1968). The material appearing in the effluent from a CM-Sepharose CL-6B column before application of the salt gradient (Figure 2) was chromatographed on a 2 x 11 cm column of Con A-Sepharose according to Ellsworth et al, 1977. Two 50 mg salt-free samples were the starting materials. Only 2.5 mg from each batch appeared in the buffer effluent, whereas 0.2 M methyl- α-D-mannopyranoside eluted an average of 37.5 mg of the C antigen, thus accounting for 75% of the applied sample. Table 1 shows that the C antigen is a galactomannan with a significant content of amino sugar. Evidence for 6.72% content of an amino sugar is based on the nitrous acid deamination of the intact polymer. The 2, 5-anhydrohexoses formed interact with 3-methyl-2-benzthiazolone hydrazone to form a blue product. (Smith and Gilkerson, 1979). At present we are seeking confirmation by an independent gas-liquid chromatographic method.

The C antigen is antigenic for mice, and monoclonal antibodies (MAb) such as CB4 (an IgM) are readily produced (Reiss et al, 1986). The MAb CB4 reacts with the C antigen in agar gel double immunodiffusion (ID), in indirect enzyme immunoassay on nitrocellulose (dot EIA, Figure 3), and in the enzyme-linked immunoelectrotransfer blot (EITB) assay with 5-20% gradient SDS-PAGE.

Table 1. Compositional analysis of C antigen, purified by affinity for Con A,[a] and of the C (Ag)-CB4 (MAb) immunoprecipitate

Constituent	C % wt/wt ± SD	C-CB4 Precipitate μg/ppte. ± SD[f]
Total protein[b]	1	291 ± 11.3
Total neutral sugar[c]	94.4	nd[g]
Mannose[d]	18.1 ± 1.2	52.9 ± 9.9
Galactose[d]	4.5 ± 0.2	12.9 ± 2.4
% neutral sugar recovered as Gal, Man	22.6	nd
Amino sugar[e]	6.72 ± 1.4	37.7 ± 12.3
Man : Gal : Amino sugar (molar ratio)	10 : 2.5 : 3.0	10 : 2.4 : 5.7

[a] Ellsworth et al, 1977.

[b] Coomassie blue dye binding, triplicates (C antigen), A_{280nm} absorbance for MAb.

[c] Phenol-sulfuric acid method.

[d] 5-6 replicates, trifluoroacetic acid hydrolysis, 2 M, 3 hr, 100°. Peracetylated aldononitriles chromatographed on 0.25 mm x 50 m fused silica column of RSL 150 (Alltech, Inc.) for 4 min at 175°, then programmed at 1°/min to 200° and held for 8 min. α-methyl xyloside internal standard (Guerrant and Moss, 1984).

[e] triplicates, Smith and Gilkerson, 1979.

[f] 1640 ug Mab + 1000 ug C Ag combined; then 2 hr, 37°C, overnight 4°C, wash 3 times PBS.

[g] nd, not determined.

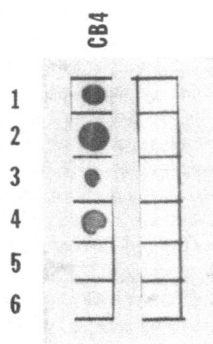

Figure 3. Indirect dot EIA shows the reaction of MAb CB4 (IgM) against Con A purified C antigen of histoplasmin and against the common fungal antigen present in exoantigens of the major systemic, dimorphic fungi: (1) C antigen, 2 mg/ml; (2) Paracoccidioides brasiliensis; (3) Coccidioides immitis; (4) Blastomyces dermatitidis; (5) A-protein purified from B. dermatitidis; (6) bovine serum albumin, 1 mg/ml. Conditions of the EIA were as follows: 1 ul of each antigen or exoantigen (neat) was applied to nitrocellulose; buffer used as diluent and for washing was PBS + 0.3% Tween 20. All incubations were for 1 hr at 23°C followed by 3 washes. The MAb CB4, as ascites fluid, was diluted 1/1000; the indicator antibody was a 1/1000 dilution of mu-chain specific anti-mouse IgM peroxidase (Cappel). Color was developed for 10 min with 3,3'-diaminobenzidine-H_2O_2. Blot strip on the right received PBS in place of MAb. See Standard et al, 1984, for exoantigen production.

The precipitin reaction in ID showed that although the M-factor was purified by gel permeation and anion exchange chromatography the C-antigen persists throughout. Dot EIA reveals that MAb CB4 also binds to an epitope common in the coccidioidin, blastomycin and paracoccidioidin exoantigens, but not to the purified A antigen of B. dermatitidis or to the mannan of Candida albicans. The C-antigen after SDS-PAGE is dispersed along the gel and is visible in the replica immunoblot. This "smeared " reaction is readily identified by EIA with sera from mice and rabbits immunized with histoplasmin and, to a lesser extent, with sera from infected humans (Reiss et al, 1986).

The C antigen obscures diagnostic specificity because it is common to the major genera of systemic dimorphic fungal pathogens and thus its removal is warranted. This lack of specificity does not mean that the C antigen and its respective antibodies are of no diagnostic significance. The advent of a means to prepare pure C antigen makes it possible to compare the antibody titer to disease activity. The structure analysis of C antigen continues to focus on the constituents responsible for crossreactivity, but especially to find out whether some component, such as the amino sugar-containing moiety is responsible for a degree of antigenic specificity.

Wheat et al (1986) found evidence for antigenuria in all of the 16 patients they studied who had extrapulmonary histoplasmosis. Antigenuria was determined by a double antibody sandwich radioimmunoassay. The urine antigen was heat stable, but was inactivated by periodate oxidation or by mixed glycosidases. Con-A agarose effectively adsorbed the antigen from urine. Although these negative selection methods do not unequivocally identify the antigen, the properties are consistent with those of the C factor and deserve further delineation. In contrast to antigenemia in either candidiasis or aspergillosis, which is a fleeting phenomenon (de Repentigny and Reiss, 1984), antigenuria in histoplasmosis was measurable in some patients for periods of 2 months (Wheat et al, 1986).

The H antigen is an acidic protein that is the first and most abundant antigen to be released from mycelial forms as judged by the exoantigen method (Standard et al, 1984). In contrast, anti-H-precipitins appear only gradually in acute histoplasmosis, if they are seen at all. Sweet et al (1979) studied sera from 40 cases of chronic histoplasmosis for precipitins using ID, counterimmunoelectrophoresis (CIE), and Laurell rocket immunoelectrophoresis (Weeke 1973). Approximately 32 to 39% of the sera produced the H-precipitate and no differences were observed when the 3 methods were compared. The assumption is that the H-precipitin is usually found in patients with active and progressive disease (Utz et al, 1976).

The behavior of the H-antigen during ion exchange chromatography is discussed below. SDS-PAGE gels and the corresponding enzyme-linked immunoelectro transfer blots (EITB) show that the H antigen is a single protein with a molecular mass of 108 to 110 kDa and a pI of 4.5 (Figures 4,5, also see Green and Pine 1985).

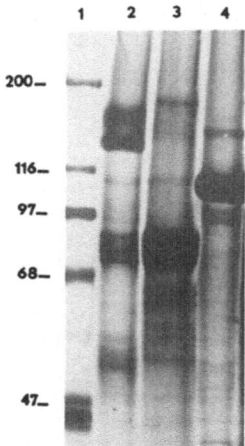

Figure 4. SDS-PAGE of purified H and M antigens of histoplasmin in 5 to 20% linear density gradient resolving gel. Antigens were treated with 1 to 2.5% SDS in 9 M urea and heated at 65° C for 15 min (Tsang et al, 1983). Lane 1, size marker proteins (Bio-Rad Inc.); lane 2 shows the M antigen with major proteins at 150 kDa and less abundant ones at 70-75 kDa; lane 3; after treatment with 0.1 M dithiothreitol at 100°C, 10 min, the 150 kDa proteins disappear and the 70-75 kDa ones are more concentrated; lane 4, H antigen. There was no change in the pattern after dithiothreitol treatment.

The M protein is released into histoplasmin during growth and autolysis of the mycelial form. Only some H. capsulatum cultures are a good source of M-rich histoplasmin. The exoantigen method, utilizing mild aqueous extraction of young mycelia (10-to 14-days-old), reveals the presence of M-antigen. Thus, M-antigen is released in small amounts during growth, or else localized autolysis plasticizes the hyphal apices causing leakage of the antigen. The basic requirements exist for an autolytic system in yeast form H. capsulatum (Davis et al, 1977). Measurement of autolytic enzymes should be extended to include mycelial form cultures.

Yeast form cultures, perhaps because they resist autolysis, are not good sources of M-antigen. Yet in vivo, anti-M precipitins appear early in acute human histoplasmosis (Utz et al, 1976; Kaufman and Reiss, 1986), and they persist into convalescence and beyond (Tompkins, 1965; Wheat et al, 1982). The presence of M-antibodies in healed histoplasmosis may be in response to leakage of minute amounts of antigens from granulomas containing dormant but probably viable yeast forms. Goodwin and Des-Prez (1978) commented that pulmonary granulomas are not always stable and over the years they may expand concentrically and leak antigens into the surrounding lung parenchyma.

For the above reasons, the M antigen will continue to be a major H. capsulatum antigen for immunochemical analysis. Ion exchange chromatography has shown that M-antigen is a protein that at pH 8.0 has a weak negative charge and desorbs briskly from DEAE-cellulose. In immunoelectrophoresis the M-antigen migrates only a short distance from the origin towards the anode. SDS-PAGE of M purified on gel permeation and anion exchange columns showed that it consisted of a doublet of proteins at 150 kDa and also, in a reduced amount, as a doublet in the 70 to 75 kDa range (Reiss et al, 1986; Figure 5). Isoelectric focusing of the M antigen followed by SDS-PAGE in the second dimension indicated the major protein had a pI of 4.6 and a mass of 150 kDa. Green et al (1985) purified M on a cation exchange column and estimated its pI at 4.7 with a mass of 147.5 kDa.

The results of the two laboratories were in good agreement. Yet when the western immunoblot EITB assay was carried out on M-antigen electrophoresed by SDS-PAGE, antibodies from immunized mice and rabbits or sera from infected humans reacted only with the 70-75 kDa doublet (Figure 5). Moreover, MAbs (ie, EC2 IgG) produced from mice immunized with histoplasmin only reacted with the 70-75 kDa doublet. The major component of M antigen, under the conditions of SDS-PAGE was a nonimmunoreactive dimer (Reiss et al, 1986).

Evidence supporting this view was provided when 0.1 M dithiothreitol reduction of the M antigen caused the 150 kDa doublet to relocate completely to the 70-75 kDa position. Further evidence was obtained when the characteristic M-immunoprecipitate was excised from agarose gels, washed, and electrophoresed by SDS-PAGE (either dithiothreitol-reduced or untreated). The reducing conditions dissociated the immune complex, revealing the 70-75 kDa doublet.

Further studies are necessary to determine the naturally occurring molecular weight of the M-antigen. Our hypothesis is that the 70-75 kDa doublet associates to form a nonimmunoreactive dimer in which the immunodominant epitopes are occluded. If that is so, reductive alkylation could dramatically increase the concentration of serologically active M antigen in histoplasmin.

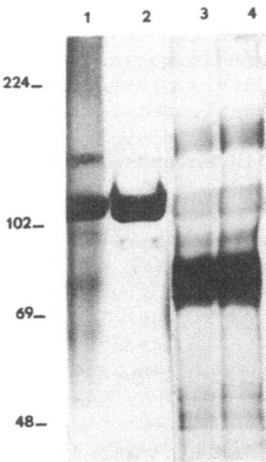

Figure 5. EITB assay showing the reactions of immune sera or MAbs with the H and M antigens of histoplasmin. Lane 1, Reaction of rabbit anti-H-serum with the H-antigen purified by DEAE. Note a diffuse zone of reactivity due to persistence of C antigen in this chromatographic step; lane 2, Purification of H antigen using CM-Sepharose excludes C antigen (same antibodies used); lanes 3 and 4, Reaction of DEAE-purified M-antigen with MAb EC2. On left, prestained size marker proteins (Bethesda Research Laboratories, Inc). See Reiss et al, (1986) for methods.

Purified M-antigen was treated with $NaIO_4$ to inactivate residual C-factor so that anti-M could be quantitated in a competitive binding microtiter inhibition-EIA (Brock et al, 1984). The ability of this EIA to detect anti-M was improved after periodate oxidation. Before treatment only 7 of 12 (58%) of M precipitin-positive sera had a significant EIA reaction, whereas that percentage increased to 88% after treatment. Inactivation of periodate sensitive C-antigen also contributed to lower background absorbance in the EIA. Although the microtiter EIA is feasible, emphasis in our laboratory has shifted to the EITB, or immunoblot assay, as a prototype diagnostic test. The latter test has the potential to simultaneously measure antibodies to 3 or more histoplasmin factors while maintaining the sensitivity of the microtiter EIA.

Sera from rabbits immunized with histoplasmin and from some human histoplasmosis patients react with a protein designated "non-M" that in IEP trails the anodic mobility of M. The classic method of immunization with immunoprecipitates excised from IEP gels cannot effectively separate M from non-M but this problem was circumvented by immunizing rabbits with non-M arcs produced by the reaction between an anti-H serum (that was also reactive with non-M) and the M-antigen (also containing non-M, Green et al, 1976). Our current research (Figure 1) shows that non-M is antigenically distinct from both H and M. The significance of non-M in the disease process has not been investigated since its description in 1976. The M_r (SDS-PAGE) of non-M was determined to be 92.9 kDa (Green et al, 1985).

For the sake of completeness, the Y factor is mentioned in order to discuss humoral responses in histoplasmosis. Recall that the Y antigen is extracted from intact yeast forms of H. capsulatum into the phenolic phase of a phenol-water extraction. Tompkins (1965) detected precipitins against the Y-factor in disseminated histoplasmosis. The antigen has been used by the New York State Department of Health for many years as part of a battery of antigens for double immunodiffusion. The chemical composition of Y is unknown.

Summary of humoral responses in histoplasmosis. Figure 6 proposes that precipitin responses to the major histoplasmin and Y factors in acute histoplasmosis are sequentially switched on according to the severity of disease. This idea is intriguing but it is based only on anecdotal evidence rather than close monitoring in the context of a longitudinal study of a large series of acute histoplasmosis patients.

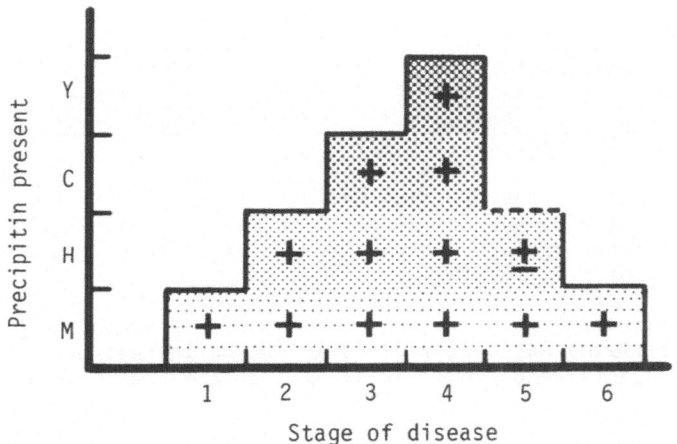

Figure 6. Hypothetical composite of precipitin responses in histoplasmosis. Anti-M is first to arise early in the acute self-limited phase (1) and persists through convalescence (5) and remission (6). H-precipitins are less frequent but appear in moderate pulmonary (2) and cavitary or extra-pulmonary (3) disease. Anti-H disappears during convalescence. C-precipitins are infrequent and indicate extrapulmonary involvement. In disseminated disease (4) anti-Y precipitins are encountered. No data exist about correlation of "non M" and disease activity. This information is anecdotal and is not supported by studies in which all histoplasmin factors were tested in a series of histoplasmosis patients.

Classic serologic methods were applied in the Indianapolis, Indiana USA urban histoplasmosis outbreak of 1979 (Wheat et al, 1982). Among 269 patients with suspected or proven histoplasmosis the complement fixation test was positive in 54.3% compared with only 30% positive in the ID test. M-precipitins remained detectable until 18 months after onset. H-precipitins which correlated positively with dissemination, were present in only 13% of patients. This reaction disappeared 9 months after the onset of symptoms. The failure in this study of ID tests to detect a higher rate of cases than 30% is unexplained inasmuch as in the same outbreak 50% of school children with subclinical or mild infection produced precipitins. Kaufman and Reiss (1986) asserted that precipitins occur in about 70% of patients with histoplasmosis proven by culture or histopathology.

Additional hospital-based studies are needed to compare classic serology to more sensitive yet specific formats such as EITB tests. Frequent monitoring of patients, coupled with the availability of good clinical histories to determine the category of histoplasmosis, is essential for success. From an immunochemical perspective, the goal of producing monomolecular antigens representing M, H, C, Y and "non-M" will make it possible to independently assess the contribution of each to the composite humoral and T-cell mediated host responses in histoplasmosis.

Acknowledgments

The authors would like to thank Randall J. Kuykendall for technical assistance, Nancy A. Lee for her secretarial assistance, and Dr. Libero Ajello for a critical review of the manuscript.

References

Azuma, I, Kanetsuna, F., Tanaka, Y., Yamamura, Y., and Carbonell, L. M., 1974, Chemical and immunological properties of galactomannans obtained from Histoplasma duboisii, Histoplasma capsulatum, Paracoccidioides brasiliensis, and Blastomyces dermatitidis, Mycopathol. Mycol. Appl. 54: 111-125.

Brock, E. G., Reiss, E., Pine, L., and Kaufman, L., 1984, Effect of periodate oxidation on the detection of antibodies against the M-antigen of histoplasmin by enzyme immunoassay (EIA) - inhibition, Current Microbiol. 10: 177-180.

Campbell, C. C. and Hill, G. B., 1964, Further studies on the development of complement-fixing antibodies and precipitins in healthy histoplasmin-sensitive persons following a single histoplasmin skin test. Amer. Rev. Resp. Dis. 90: 927-934.

Davis, T. E. Jr., Domer, J.E., and Li, Y. T., 1977, Cell wall studies of Histoplasma capsulatum and Blastomyces dermatitidis using autologous and heterologous enzymes, Infect. Immun. 15: 978-987.

Department of Health and Human Services, U. S., 1984, "Biosafety in Microbiological and Biomedical Laboratories," J. H. Richardson, W. E. Barkley, eds., U. S. Government Printing Office, Washington, D.C.

Dickerson, Q. H. Jr., and Busey, J. F., 1968, Chromatographic separation of h and m antigens from histoplasmin, Proc. Soc. Exper. Biol. Med. 128: 654-658.

Ehrhard, H.-B. and Pine, L., 1972, Factors influencing the production of H and M antigens by Histoplasma capsulatum: Development and evaluation of a shake culture procedure, Appl. Microbiol. 23: 236-249.

Ellsworth, J. H., Reiss, E., Bradley, R. L., Chmel, H., and Armstrong, D., 1977, Comparative serological and cutaneous reactivity of candidal cytoplasmic proteins and mannan separated by affinity for concanavalin A, J. Clin. Microbiol. 5: 91-99.

Food and Drug Administration, U. S., 1977, Skin test antigens, proposed implementation of efficacy review, Fed. Register, 42, 52674-52687.

Goodwin, R. A. Jr., and Des-Prez, R. M., 1978, State of the art - histoplasmosis, Amer. Rev. Resp. Dis. 117: 929-955.

Green, J. H., Harrell, W. K., Gray, S. B., Johnson, J. E., Bolin, R. C., Gross, H. and Malcolm, G. B., 1976, H and M antigens of Histoplasma capsulatum: Preparation of antisera and location of these antigens in yeast-phase cells, Infect. Immun. 14: 826-831.

Green, J. H. and Pine, L., 1985, Preparation of h and m antigens of Histoplasma capsulatum free of heterologous antigens, Current Microbiol. 12: 209-216.

Greene, C. H., De Lalla, L. S., and Tompkins, V. N., 1960, Separation of specific antigens of Histoplasma capsulatum by ion-exchange chromatography, Proc. Soc. Exper. Biol. Med. 105: 140-141.

Guerrant, G. O. and Moss, C. W. 1984, Determination of monosaccharides as aldononitrile, O-methyloxime, alditol, and cyclitol acetate derivatives by gas chromatography, Analyt. Chem. 56: 633-638.

Heiner, D. C., 1958, Diagnosis of histoplasmosis using precipitin reactions in agar gel, Pediatrics, 22: 616-629.

Kaufman, L. and Reiss, E., 1986, Serodiagnosis of fungal diseases, pp. 446-466, in: "Manual of Clinical Laboratory Immunology, 3rd ed.," N. R. Rose, H. Friedman, and J. L. Fahey, eds., Amer. Soc. Microbiol., Washington D. C.

Pine, L., Gross, H., Bradley-Malcolm, G., George, J. R., Gray, S. B., and Moss, C. N., 1977, Procedures for the production and separation of h and m antigens in histoplasmin: Chemical and serological properties of the isolated products, Mycopathol. 61: 131-141.

Pine, L., Smith, S. J., Gross, H., Barbaree, J. M., and Malcolm, G. B., 1981, Studies on the thermal degradation of the h and m antigens of lyophilized histoplasmin, Sabouraudia, 19: 55-70.

Prior, J. A. and Saslaw, S., 1952, Effect of repeated histoplasmin skin tests on skin reactivity and collodion agglutination, Amer. Rev. Tuberc. 66:588.

Reiss, E., 1986, Histoplasma capsulatum, pp 77-102, in: "Molecular Immunology of Mycotic and Actinomycotic Infections", Elsevier Science Publishing Co., Inc., New York.

Reiss, E., Knowles, J. B., Bragg, S. L., and Kaufman, L., 1986, Monoclonal antibodies against the M-protein and carbohydrate antigens of histoplasmin characterized by the enzyme-linked immunoelectrotransfer blot method. Infect. Immun. 53: 540-546.

Reiss, E., Miller, S. E., Kaplan, W. and Kaufman, L., 1977, Antigenic, chemical, and structural properties of cell walls of Histoplasma capsulatum yeast-form chemotypes 1 and 2 after serial enzymatic hydrolysis. Infect. Immun. 16: 690-700.

Reiss, E., Mitchell, W. O., Stone, S. H., and Hasenclever, H. F., 1974, Cellular immune activity of a galactomannan-protein complex from mycelia of Histoplasma capsulatum, Infect. Immun. 10: 802-809.

Scalarone, G. M., Restrepo, M. A., Shacht, L. R., Rupp, G. H., and Le Roy, M. D., 1985, Lack of antibody induction by Histolyn-CYL, a new skin-testing reagent for histoplasmosis. J. Med. Vet. Mycol. 23: 335-338.

Schwarz, J., 1981, "Histoplasmosis", Praeger Publishers, New York, pp. 151-154.

Shaw, L. W., Howell, A. Jr., and Weiss, E. S., 1950, Biological assay of lots of histoplasmin and the selection of a new working lot, Pub. Health. Rep. 65: 583-609.

Smith, R. L. and Gilkerson, E., 1979, Quantitation of glycosaminoglycan hexosamine using 3-methyl-2-benzthiazolone hydrazone hydrochloride, Analyt. Biochem. 98: 478-480.

Smith, C. E., Whiting, E. G., Baker, E. E., Rosenberger, H. G., Beard, R. R., and Saito, M. T., 1948, The use of coccidioidin, Amer. Rev. Tuberc. 57: 330-360.

So, L. L., and Goldstein, I. J., 1968, Protein-carbohydrate interaction XIII, The interaction of concanavalin A with α-mannans from a variety of microorganisms, J. Biol. Chem. 243: 2003-2007.

Standard, P. G., and Kaufman, L., 1982, Safety considerations in handling exoantigen extracts from pathogenic fungi, J. Clin. Microbiol. 15: 663-667.

Standard, P. G., Kaufman, L., and Whaley, S. D., 1984, Exoantigen test - Rapid identification of pathogenic mould isolates by immunodiffusion, in: "Immunology Series No. 11," Procedural Guide, U. S. Centers for Disease Control, Atlanta, GA.

Sweet, G. H., Cimprich, R. S., Cook, A. C. and Sweet, D. E., 1979, Antibodies in histoplasmosis detected by use of yeast and mycelial antigens in immunodiffusion and electroimmunodiffusion, Amer. Rev. Resp. Dis. 129: 441-449.

Tompkins, V. N., 1965, Soluble antigenic constituents of yeast phase Histoplasma capsulatum, Amer. Rev. Resp. Dis. Suppl. 92: 126-133.

Tsang, V. C. W., Peralta, J., and Simons, A. R., 1983, Enzyme-linked immunoelectrotransfer blot techniques (EITB) for studying the specificities of antigens and antibodies separated by gel electrophoresis, Meth. Enzymol. 92:377-391.

Utz, J. P., Becker, A., Buechner, H. A., Campbell, G. D., Einstein, H. E., and Seabury, J. H., 1976, The pulmonary mycoses: diagnostic and therapeutic guidelines, pp. 1-28, E. R. Squibb and Sons, Inc., Princeton, NJ.

Weeke, B., 1973, Rocket immunoelectrophoresis, Scand. J. Immunol. Suppl. 1: 37-46.

Wheat, J. French, M. L. V., Kohler, R. B., Zimmerman, S. E., Smith, W. R., Norton, J. A., Eitzen, H. E., Smith, C. D., and Slama, T. G., 1982, The diagnostic laboratory tests for histoplasmosis--analysis of experience in a large urban outbreak, Ann. Intern. Med. 97: 680-685.

Wheat, L. J., Kohler, R. B., and Tewari, R. P., 1986, Diagnosis of disseminated histoplasmosis by detection of Histoplasma capsulatum antigen in serum and urine specimens, N. Engl. J. Med. 314: 83-88.

IMMUNE RESPONSES TO SUBCELLULAR ANTIGENS OF HISTOPLASMA CAPSULATUM

Ram P. Tewary, R.B. Kohler and L.J. Wheat

S.I.U. School of Medicine, Springfield, IL
and Indiana University, School of Medicine
Indianapolis, IN USA

INTRODUCTION

Histoplasmosis is a respiratory fungal infection with significant morbidity and mortality, and also has been recently recognized, with increasing frequency, in patients with acquired immunodeficiency syndrome (AIDS). It is caused by Histoplasma capsulatum – a dimorphic fungus which exists in the mycelial form in nature and in the yeast form in susceptible hosts. At least, 40 million human infections with Histoplasma have been recognized in the USA and this number is increasing annually by approximately 500,000 new cases (1,4). Histoplasmosis has been reported from at least 60 countries and appears to have a circumglobal distribution (7).

During the last three decades, the pioneering work of several investigators has resulted in the development of many serological tests for diagnosis of histoplasmosis (5,6,10,11,26). The use of these techniques has led to the detection of numerous, otherwise unrecognized cases. However, the usefulness of these tests is hampered by the high frequency of false positive and false negative reactions.

Numerous attempts have also been made for the development of a suitable Histoplasma vaccine for immunoprophylaxis of certain high risk individuals (8,19,20,22,25). So far, these efforts have not been as rewarding.

During the last ten years, the work in our laboratory has been primarily directed toward the study of immune responses to different subcellular antigens from Histoplasma capsulatum. In this communication, we present some of our findings on subcellular antigens of Histoplasma. A brief description of our work on protective immunity and delayed hypersensitivity, will be followed by the presentation of our findings on two serologic tests for diagnosis of histoplasmosis.

Protective Immunity and Delayed Hypersensitivity

Exposure to H. capsulatum with or without apparent signs of the disease results in the development of delayed hypersensitivity (DH) to Histoplasma antigens. The development of DH is thought to be associated with acquired immunity but the nature of the relationship between DH and protective immunity is not clearly defined.

In an attempt to better understand the mechanism of acquired immunity to histoplasmosis in humans, we have used a murine model of self-limiting infection to study different host responses after subcutaneous (s.c.) immunization by sublethal infection with yeast cells of H. capsulatum (21). Protective immunity in immunized donors and recipients of immune splenocytes was evaluated after a lethal challenge with H. capsulatum. Delayed hypersensitivity reactions to histoplasmin and Histoplasma ribosomes were determined by footpad reactivity and oil-induced-peritoneal exudate cells (PEC) were used to study the production of macrophage migration inhibition factor (MIF). Mice were immunized s.c. by sublethal infection with 1×10^5 yeast cells of H. capsulatum. At different intervals from day 1 to 15 weeks post-immunization, splenocytes from a set of 10 immunized mice were transferred i.v. to 10 syngeneic recipients (10^8 cells/mouse). At specified time intervals, five recipients and five immune donors were tested for DH reactions to histoplasmin and Histoplasma ribosomes by the liquid displacement technique using a double-arm buret filled with safranin colored 70% alcohol (21). Peritoneal exudate cells from separate sets of five immune donors and five recipients were used to assay the production of MIF by the agarose droplet technique (21). The protective immunity at different stages of immunization and cell transfer was assessed by 30 day survival, and recovery of viable organisms from spleen following an intravenous challenge with 2×10^6 yeast cells of H. capsulatum. The immunized animals showed a significant footpad reactivity ($p < 0.001$) at 4 days to both histoplasmin and ribosomal antigen (50 µg protein) which peaked between 21 and 28 days, declined thereafter and reduced to the control level at 84 days (Fig. 1). Significant inhibition of macrophage migration ($p < 0.001$) was observed in the presence of both antigens, as early as four days post immunization, remained approximately at the same level up to 42 days and showed decline thereafter. All animals survived when the immunized donors were challenged with the lethal dose of H. capsulatum between 7 and 105 days post-immunization.

A similar relationship between delayed hypersensitivity and protective immunity was observed when syngeneic recipients of immune splenocytes were tested. Adoptive transfer of spleen cells from immunized donors up to 7 days did not confer protective immunity to syngeneic recipients, although at this stage significant DH reactions ($p < 0.05$) were observed in the recipients and their peritoneal cells produced a high level of MIF (Fig. 2). Both DH reaction and MIF production could not be detected when donor spleen cells were transferred at 105 days after immunization but protection to lethal challenge conferred by transferred splenocytes remained at 90-100%. The intensity of DH reaction did not correlate with the level of protective immunity during the observation period, since a high level of protection was demonstrated at a time (42-105 days) when a definite decay in the DH reaction was observed.

Blastogenic Responses of Lymphocytes from Immunized Mice

We studied the lymphoproliferative responses of spleen cells to histoplasmin and Histoplasma ribosomal antigens and to the mitogens, concanavalin A (ConA), phytohemagglutinin (PHA) and lipopolysaccharide (LPS) in normal and immunized mice (C3H/HeN) at different intervals after immunization (10^5 live yeast cells of Histoplasma, s.c.). Lymphoproliferative responses to histoplasmin in vitro was compared with that induced by Histoplasma ribosomes (24). Spleen cells from normal and immunized animals (5 mice per group) were prepared in RPMI 1640 with 5% heat-inactivated normal mouse serum at 10^7 cells per ml and the lymphocyte transformation assay was performed as previously described (24). The results were expressed as counts per minute (^{3}H-thymidine incorporation) in controls subtracted from counts per minute in experimental cultures.

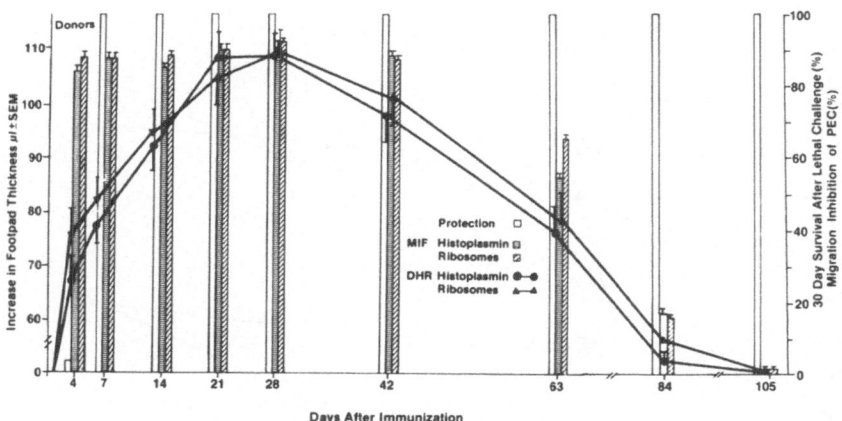

Fig. 1. Immunity, footpad reactivity and MIF production by PEC from mice immunized by sublethal infection with Histoplasma (10^5 yeast cells, s.c.). Mean ± SEM of 5 mice/time period for DH and MIF and 20 mice/group for immunity (10 mice/group/experiment). From Mykosen 29:116, 1986.

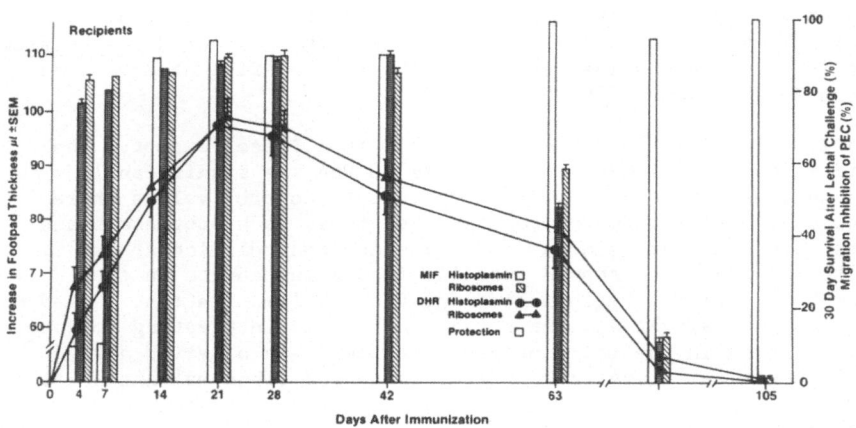

Fig. 2. Immunity, footpad reactivity and MIF production by PEC from mice receiving immune spleen cells. Mean ± SEM of 5 mice/time period for DH and MIF and 20 mice/group for protective immunity (10 mice/group/experiment). From Mykosen 29:116, 1986.

Four days after immunization, the mean response to ConA of spleno-cytes from immune mice was significantly depressed (p <0.001) as compared with the response of control mice (Fig. 3). The response remained depressed at day 7 and 14, at day 21; however, it was not significantly different from controls. The response of spleen cells from immune mice

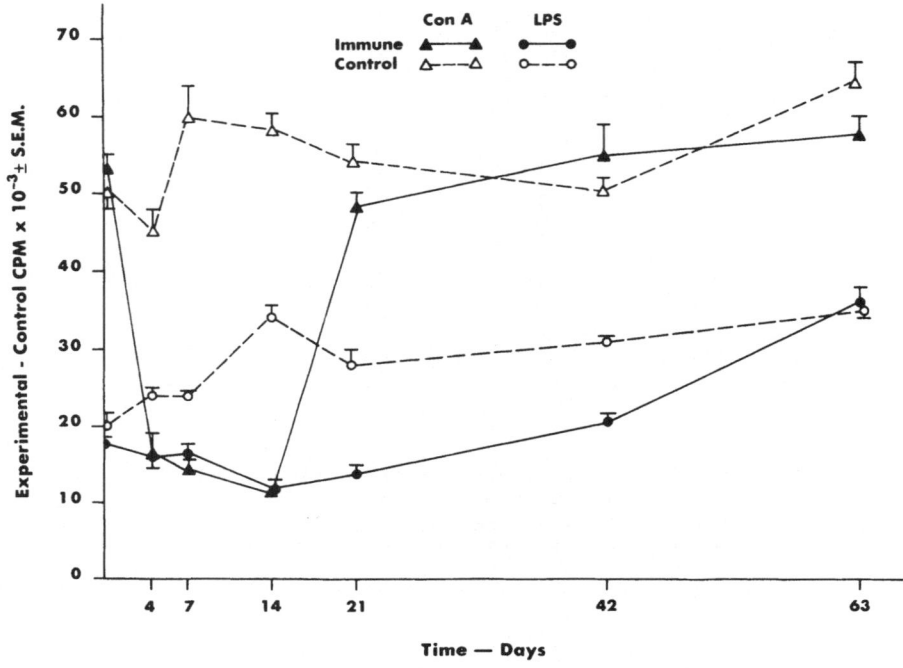

Fig. 3. Blastogenic responses of spleen cells to ConA and LPS from normal mice and mice immunized by sublethal infection with yeast of H. capsulatum (10^5 cells, s.c.). Each value represents the mean of five observations ± SEM. (From Infect. Immun. 36:1013, 1982).

to stimulation by LPS was also depressed, but to a lesser extent. The blastogenic response of immune spleen cells to PHA was significantly suppressed between 4 and 21 days, but returned to control values there-after (Fig. 4.). The lymphoproliferative responses to histoplasmin and Histoplasma ribosomes were similar in immune and control mice at day 0. The cells from immune mice showed steadily increasing counts in the presence of both antigens beginning at day 4 after immunization, peaked at day 42 and showed a slight decline on day 63. It is interesting to note that no suppression in specific antigenic response was observed after exposure of immune spleen cells to histoplasmin and ribosomal antigens.

Enzyme-Linked Immunosorbent Assay for Detection of Anti-Histoplasma
Antibodies

We have previously shown that immunization of mice with Histoplasma ribosomes and a membrane-rich fraction elicits a high degree of immunity against histoplasmosis (8,22,25). The immunity is primarily mediated by T lymphocytes and their active proliferation in the recipients is necessary for the expresesion of immunity to histoplasmosis (23). However, the role of antibodies in protection to Histoplasma still remains as a controversial issue.

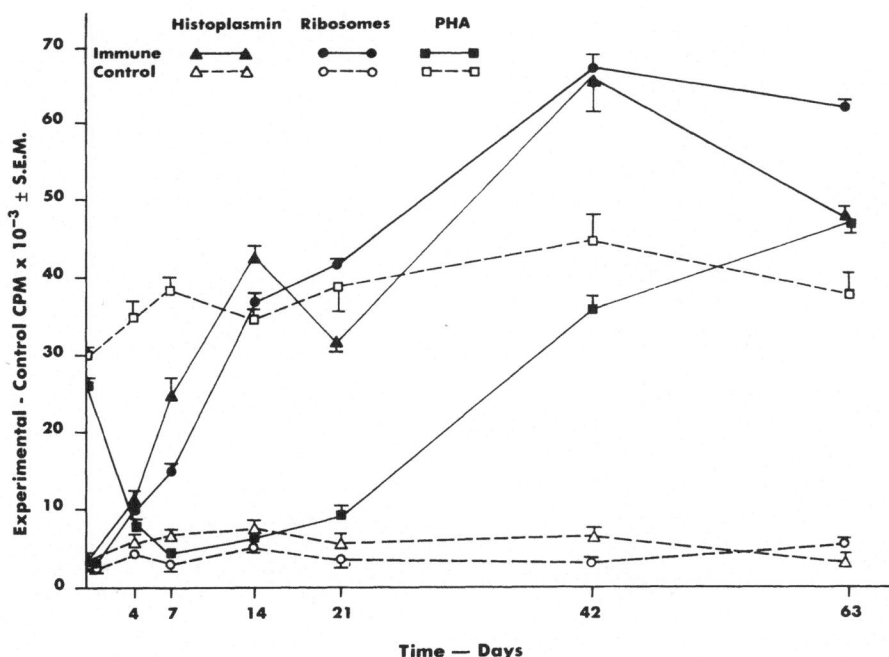

Fig. 4. Blastogenic responses of spleen cells to histoplasmin, Histoplasma ribosomes, and PHA from normal mice and mice immunized by sublethal infection with yeast cells of H. capsulatum (10^5 cells, s.c.). Each value represents the mean of five observations ± SEM. (From Infect. Immun. 36:1013, 1982).

In subsequent studies, we developed an indirect sandwich enzyme-linked immunosorbent assay (ELISA) for the detection of anti-Histoplasma antibodies in human sera. Furthermore, we have used this test to determine the relative distribution of antibodies to histoplasmin and Histoplasma ribosomes in sera from proven cases of histoplasmosis and other fungal infections and tuberculosis. The experimental details of the ELISA will be published elsewhere (Diagnostic Immunology). In brief, the wells of microtiter plates were precoated with rabbit anti-Histoplasma IgG, Fab'2 fraction followed by the addition of histoplasmin and ribosomal antigens. Serial dilutions of the test sera and a fixed dilution (1:100) of five control sera were added to separate wells. The goat anti-human IgG (γ chain specific) alkaline phosphatase conjugate was added to each well. After incubation at 37° C, paranitrophenyel phosphate (substrate) was added to each well and the absorbance was read at 405 nm in a Titertek Multiskan ELISA reader. The results were expressed as ELISA titer = Highest dilution with absorbance at 405 nm greater than 1.4 X mean of five controls for ribosomes and 1.8 X mean of five controls for histoplasmin.

Sera from 64 patients with histoplasmosis were tested for antibodies to histoplasmin and Histoplasma ribosomes. Anti-ribosomal antibodies were detected in 62/64 (97%) whereas only 48/64 (75%) were positive with histoplasmin (Fig. 5). Cross-reactions were observed when sera from patients with blastomycosis, coccidioidomycosis and paracoccidioidomycosis were tested with either antigens. In blastomycosis cross-reaction was seen in 3/6 sera with ribosomes and in 2/6 with histoplasmin. Cross-reactions

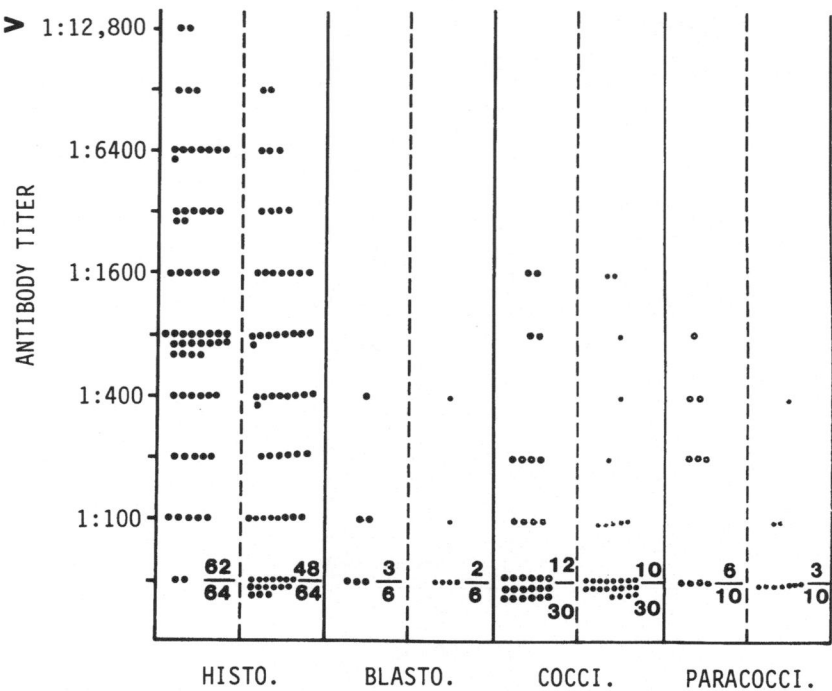

Fig. 5. Antibodies to Histoplasma ribosomes (open circles) and
histoplasmin (closed circles) by ELISA in sera from patients with
histoplasmosis, blastomycosis, coccidioidomycosis and
paracoccidioidomycosis. Titers ≥ 1:100 are considered positive.

were seen in 12/30 coccidioidomycosis sera with ribosomes and 10/30 with
histoplasmin. In paracoccidioidomycosis 6/10 sera cross-reacted with
ribosomes and 3/10 with histoplasmin. Cross-reaction in cryptococcosis
was seen in 1/4 cases with ribosomes and 2/4 with histoplasmin (Fig. 6).
In candidiasis, 8/13 sera cross-reacted with ribosomes and 4/14 with
histoplasmin. Cross-reaction was seen in 6/15 sera with ribosomes and
4/14 with histoplasmin in patients with aspergillosis. In tuberculosis,
7/14 sera cross-reacted with ribosomes and 4/14 with histoplasmin. The
ELISA results of intial testing were reproducible on repeat testing, with
a correlation coefficient of 0.87 with ribosomes and 0.92 with
histoplasmin.

Detection of Histoplasma Antigen in Serum and Urine Specimen

An alternative approach to immunodiagnosis, the detection of fungal
antigen, is very helpful for the diagnosis of cryptococcosis and also
appears to be promising for the diagnosis of candidiasis and aspergillosis. These tests do not rely on patients' immune responses, which is
especially important in immunocompromised patients whose humoral immune
responses may be attenuated or delayed. In view of these findings, we
sought to develop a serological test which is based on the detection of
Histoplasma antigen rather than antibodies. In this report, we describe
our findings on the development of a solid-phase radioimmunoassay for the
detection of Histoplasma antigen in serum and urine specimens (27).

In our radioimmunoassay, polysterene tubes were coated with rabbit
anti-Histoplasma IgG and 0.1 ml of undiluted urine or serum specimen was
added to the tubes. Finally, the antigen that adhered to the solid phase

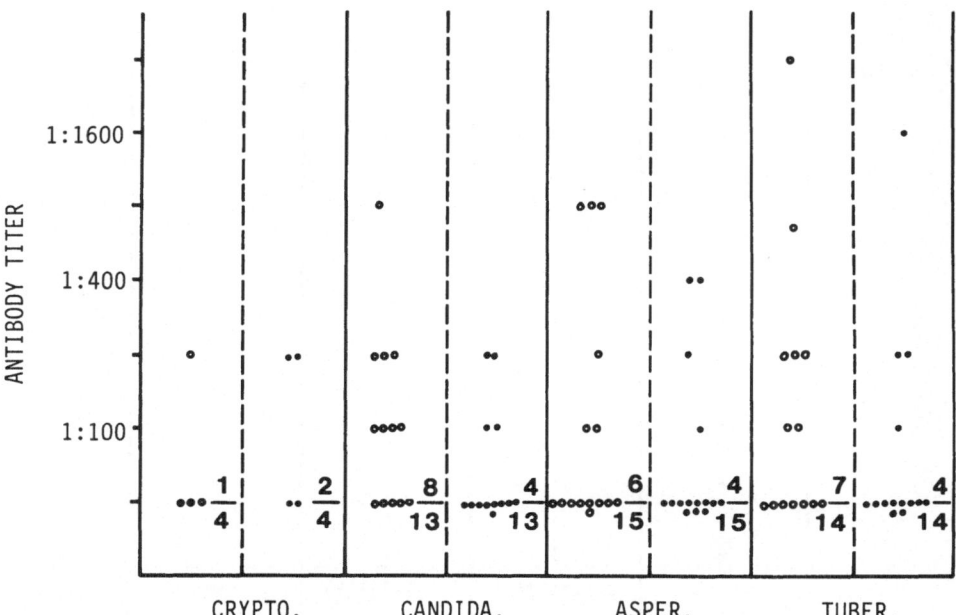

Fig. 6. Antibodies to Histoplasma ribosomes (open circles) and histoplasmin (closed circles) by ELISA in sera from patients with cryptococcosis, candidiasis, aspergillosis and tuberculosis. Titers $\geq$ 1:100 are considered positive.

antibody was measured by incubation with ^{125}I-labelled rabbit IgG antibody to Histoplasma. Fresh urine and serum samples from the same two laboratory employees were tested each day, the assay was performed (controls). All values were divided by 1.5 times the mean of the two normal specimens (serum/urine) and expressed as radioimmunoassay units (RU). The values $\geq$ 1.0 RU were considered positive for H. capsulatum antigen.

Urine samples from 88 patients (94 episodes) with histoplasmosis and 295 controls were tested by the radioimmunoassay for H. capsulatum antigen. Antigen was detected in urine of 20/22 (91%) epidoses of disseminated histoplasmosis that occurred in 16 patients, and in urine of 6/32 patients (19%) with self-limited disease, 2/32 patients (6%) with cavitary disease, and 4/8 patients (50%) with sarcoid-like manifestations of histoplasmosis (Fig. 7). Two patients with disseminated histoplasmosi did not have antigenuria, but their assay results (0.96 and 0.92 RU) were only slightly below the cutoff point for positive values. The first patient had mild illness but Histoplasma was isolated from the bone marrow. He had improved considerably before ketoconozole treatment was started. The second patient had acquired immunodeficiency syndrome (AIDS) who had been treated for three earlier episodes of disseminated histoplasmosis, that were confirmed by recovery of H. capsulatum from bone marrow and blood, as well as by detection of the antigen in the urine. None of the 295 control urine specimens from patients with other fungal and bacterial infections were positive for Histoplasma antigen (Fig. 8).

Antigen assays were also performed on heat inactivated sera that were available from 67 episodes of histoplasmosis in 61 patients. Antigen was

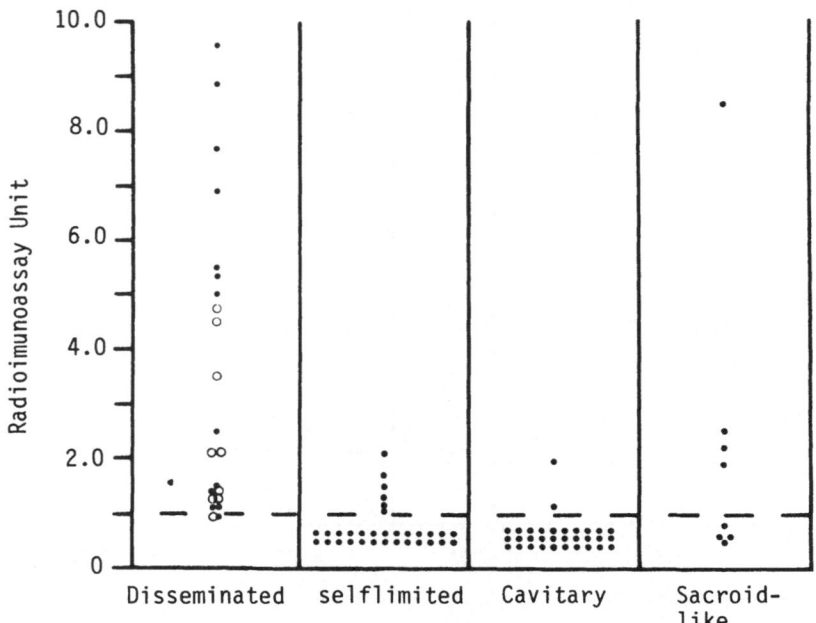

Fig. 7. H. capsulatum urinary antigen in patients with different clinical manifestations of histoplasmosis. Values above the broken horizontal line are considered positive. Open circles represent relapses of disseminated histoplasmosis. From Wheat et al., N. Engl. J. Med. 314:83, 1986.

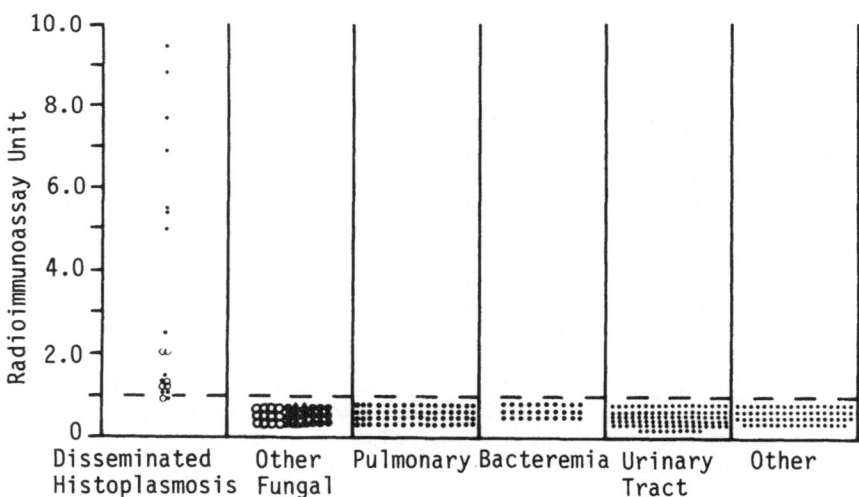

Fig. 8. H. capsulatum urinary antigen in controls. Control patients with disseminated fungal infections are indicated by open circles, invasive fungal pulmonary infections by closed triangles and localized fungal infections by closed circles. From Wheat et al., N. Engl. J. Med. 314:83, 1986.

detected in sera of 11/22 episodes in 16 patients with disseminated histoplasmosis, in none of the 12 antigenuric patients with other forms of histoplasmosis, in one of the specimens from the 33 patients with histo-plasmosis but no antigenuria. No Histoplasma antigen was detected in any of the 50 control sera. Mean antigen levels in the serum of patients with mild forms of disseminated histoplasmosis were lower than those in pa-tients with more severe disease (0.7 + 0.1 vs. 6.9 + 5.8 RU; respective-ly). Antigenemia was observed in 11/13 episodes of more severe disease but none of the milder episodes. There was a poor correlation in the antigen levels in urine and serum specimens.

Both antigenemia and antigenuria decreased during or after initiation of antifungal therapy and reoccurred in patients who had a relapse. Thus, it appears that determination of Histoplasma antigen levels in urine and serum specimens may also be useful in monitoring the efficacy of antifun-gal therapy for disseminated and possibly other forms of histoplasmosis.

To determine the reliability of the RIA technique, 21 urine specimens that contain antigen and 20 negative specimens from patients with histo-plasmosis were retested. The test was reproduced in 18 of the 21 positive specimens (86%); the negative findings were reproduced in all 20 control urine specimens (100%; correlation coefficient, R = 0.946). The positive values that were not reproduced in 3 specimens had small quantities of the antigen when first tested and almost reached the positive levels on retes-ting.

Histoplasma urinary antigen was not altered by heating at 100° C for 15 min, or by treatment with pronase, trypsin, pepsin, DNase and RNase for 4 h at 37° C. However, the antigen was degraded by digestion with mixed glycosidases and treatment with sodium periodate. The antigen was adsorb-ed by incubation of a positive urine with Sepharose-4B-concanavalin A. The molecular weight of the antigen varied from 10,000 to 60,000 daltons as determined by molecular seive chromatography on a Sephadex G100 column using elution profiles of several known molecular weight standards. The urinary antigen was purified by Seopharose-4B-ConA affinity chromatography and was found to contain carbohydrate and protein (1.4:1 ratio). The antigen bound to the surface of Sepharose-4B-ConA elicited production of antibodies to H. capsulatum in experimental animals. Adsorption with Histoplasma cell wall or whole yeast cells inhibited the ability of ^{125}I-labelled anti-H. capsulatum antibodies to detect the urinary antigen. We conclude that the urinary antigen is a glycoprotein (molecular weight = 10,000 to 60,000 daltons) which is present on the surface of intact cells.

DISCUSSION

Immunity and delayed hypersensitivity (DH) reactions in tuberculosis are considered as the prototype of immune responses in chronic bacterial and fungal infections. Tuberculin hypersensitivity and tuberculous immu-nity are thought to be similar cell-mediated immunological responses with different lymphokine mediators (13). For example, it has been demonstrat-ed that macrophages can at the same time be responsive to lymphokines me-diating DH reactions and unresponsive to lymphokines mediating inhibition of BCG multiplication (2). Hence, the immune response can be modulated in such a manner that one activity is blocked while another is preserved. The biological significance of the DH reactions is not clearly defined, even thought the DH reaction is considered as an index of cellular immu-nity. Mitsuyama et al. (16), while studying the correlation between delayed footpad reactivity and resistance to local bacterial infection, demonstrated that the biological significance of the delayed footpad reactions lies in its ability to eliminate locally injected bacteria by enhancing the local resistance through macrophage accumulation. The fact that enhanced resistance was expressed against an unrelated organism, (S. typhimurium) at the reaction site coincides with the the non-specificity of acquired cellular response in the expression phase of Listeria-immune mice (14). Both delayed footpad reaction and resistance could be transferred locally with non-adherent spleen cells. Although Patel (18) found that resistance against L. monocytogenes was increased at the injection site of a specific antigen in the immunized animals, he found no correlation between the magnitude of the DH reactions and the level of resistance thus induced. Furthermore, there was increased resistance to infection, even when no DH could be detected. Thus, there was a dissociation between DH reactivity and the level of protective immunity against L. monocytogenes after injection of specific antigen.

In our studies, DH reactivity developed at 4 days after immunization, peaked between 21 and 28 days which at this stage correlated with the protective immunity assessed by intravenous challenge. It is of interest that DH declined thereafter but resistance to intravenous challenge remained approximately at the same level up to 105 days, the maximum period observed in this study. These observations are remarkably similar to that of Jerrels and Osterman (9) in C3H/HeN mice infected with Rickettsia tsutsugamushi. In these mice, DH developed slowly and peaked at 21 days post-immunization and correlated with expression of maximum resistance to intraperitoneal challenge. Delayed hypersensitivity reactions declined thereafter, but resistance to intraperitoneal challenge remained up to 28 days, the maximum period observed.

Mitsuyama et al. (15) observed no MIF activity in Listeria-immune mice before day 10 post-immunization, whereas immunization with killed Listeria in complete Freund's adjuvant showed inhibition of macrophage migration without delayed footpad reactivity. Delayed footpad reactivity and local resistance were conferred by cells from immune mice 6 days after immunization; at this stage no macrophage migration inhibitory activity was observed. In our study, the earlier and sustained levels of MIF activity as compared to DH reactions could be attributed to differential reactivity of the lymphokines responsible for the two phenomena.

The DH reactions and MIF activities in our studies could be transferred to syngeneic recipients by immune spleen cells. In addition to manifestation of these correlates of the cell-mediated immune response, the recipient mice showed a 90-100% protection to a lethal challenge when cells were transferred from 14 to 105 day immune mice. We have shown previously that the immunity to histoplasmosis conferred by adoptive transfer is mediated by T lymphocytes (12,23). It is evident from these studies that the development of DH preceeds the expression of protective immunity to histoplasosis in mice and the immunity persists longer than DH reactions after a single immunizing dose. The mechanisms involved in this apparent dichotomy of the two processes warrant further investigation.

Suppression in cellular immune responses of mice in association with the generation of potent immunosuppressor activity by spleen cells, has been demonstrated, after i.v. infection with $5-10 \times 10^5$ yeast cells of H. capsulatum (3,17). Maximum suppressor activity of DH responses to SRBC and histoplasmin and impairment of ConA- and histoplasmin-induced blasto-genic responses were observed during the first 3 weeks. The suppressive activity disappeared after 8 weeks. In our studies, we observed sup-pressed blastogenic responses to ConA and PHA at 4 days after s.c. immun-ization of mice by sublethal infection (1×10^5 yeast cells) and the suppression disappeared after 21 days (24). In contrast, no suppression in blastogenic response was observed in splenocytes after exposure to histoplasmin and Histoplasma ribosomes. It is of interest, that a similar transitory suppression of nonspecific immune response was observed in C57BL/6 mice one week after intratracheal inoculation with 5×10^5 live yeast cells of H. capsulatum, but there was no suppression in the blastogenic response of splenocytes to Histoplasma antigen (D.A. Nickerson, unpublished data). In contrast, the suppression in both specific and nonspecific immune responses were seen after intravenous inoculations with the same number of viable Histoplasma. Thus, it appears that the generation of suppressor cell activity primarily depends on the route of inoculation of mice with H. capsulatum.

Serological tests are routinely used for diagnosis and in epidemiological studies of histoplasmosis. However, serological cross-reactions occasionally create serious problems in differential diagnosis of histoplasmosis and other systemic fungal infections (5,6,10,11,26). We have recently developed a reliable indirect sandwich ELISA for the detection of antibodies to histoplasmin and Histoplasma ribosomes in human sera. To our knowledge, this is the first study where ribosomes, a defined subcellular fraction of Histoplasma yeast cell, has been used as an antigen to detect antibodies to the organism. Anti-ribosomal antibodies were detected in 97% (62/64) of human sera from proven cases of histoplasmosis, whereas only 75% (48/64) of the sera were positive with histoplasmin (culture filtrate), indicating the superiority of ribosomal fraction, as the test antigen. Significant cross-reactivity, however, was observed in sera from patients with blastomycosis, coccidioidomycosis, paracoccidioidomycosis, cryptococcosis, candidiasis and tuberculosis with both histoplasmin and ribosomal antigens.

Our continuing search for a better serological test has resulted in the development of a radioimmunoassay (RIA) for detection of H. capsulatum antigen, rather than antibodies, in serum and urine specimens (27). This RIA is a reliable new method for diagnosis of disseminated histoplasmosis, that does not rely on patients' immune status, which is especially important in immunocomprised hosts, whose humoral immune responses may be attenuated or delayed.

H. capsulatum antigen was present in urine specimens from 20 of 22 episodes of disseminated histoplasmosis. In the 22 episodes, 9 were relatively mild and occurred in patients with subacute or chronic illnesses. The antigen was present in urine specimens during 7 of the 9 episodes of mild dissemination and in all 13 episodes of severe disease. In the two patients with negative findings, the antigen values were only slightly below the cutoff point for positive results (0.92 and 0.96 RU). One of these two patients had AIDS and had experienced three earlier episodes of disseminated histoplasmosis. During the fourth episode, H. capsulatum was isolated only from the cerebrospinal fluid (CSF), the patient had clinical manifestations of chronic meningitis and the antigen was detected in the CSF. The second patient of disseminated histoplasmosis with minimal antigenuria (0.96 RU), presented with fever and a weight loss of 4.5 kg, but no other findings indicative of dissemination. At the time, when Histoplasma was isolated from the bone marrow, the patient had become afebrile and started to gain weight. These findings indicate that antigen may not appear in the urine of some patients with focal sites of dissemination.

H. capsulatum antigen was also present in the serum of patients with severe form of disseminated histoplasmosis. Antigenemia occurred infrequently (<5%) in patients with other forms of histoplasmosis. However, sera were not treated to dissociate antigen from immune complexes - a procedure that may increase the sensitivity of the technique for antigen detection.

441

Detection of the antigen appears to be useful for monitoring the efficacy of treatment and for excluding the possibility that relapsing infection has occurred in patients with disseminated disease. In 8 of 9 episodes of disseminated histoplasmosis, antigenuria cleared after effective treatment. Antigenemia also cleared after successful therapy in 3 patients and remained positive at lower levels (1.05 RU) in a patient with AIDS. Persistent antigenemia after amphotericin B therapy preceded culture positive relapses in 3 other patients. Antigenuria was observed during nine relapses of disseminated histoplasmosis that occurred in six patients. The demonstration of antigenuria was the first laboratory evidence for relapse and the basis for beginning treatment with amphotericin B in our last four patients with relapsing infection. These findings indicate that the levels of antigenuria and antigenemia may be helpful in monitoring the efficacy of treatment, and for identification of relapses in patients with disseminated histoplasmosis.

SUMMARY

We have investigated the immune responses to Histoplasma subcellular antigens in patients with histoplasmosis and in mice with experimental infection. Maximum delayed hypersensitivity (DH) to Histoplasma ribosomes and histoplasmin as well as protection against a lethal challenge (i.v.) were observed at 3-4 weeks after immunization of mice by a sublethal infection with yeast cells of Histoplasma. DH reactions gradually declined thereafter, reaching the control level at 84 days. In contrast, no significant alteration in protective immunity was observed up to 105 days. Splenocytes from immunized mice also exhibited blastogenic responses to ribosomes and histoplasmin. Sera from histoplasmosis patients had antibodies to histoplasmin and ribosomes but signfiicant cross-reactions were observed in patients with other respiratory infections. Furthermore, we have developed a reliable radioimmunoassay (RIA) for detection of Histoplasma antigen in urine and serum specimens from patients with disseminated histoplasmosis. Both antigenemia and antigenuria decreased after initiation of antifungal therapy and recurred in patients who had a relapse. We conclude that the RIA for detection of H. capsulatum antigen in urine, blood and spinal fluid offers a new, rapid method for diagnosing disseminated histoplasmosis, monitoring therapy and identifying relapses.

ACKNOWLEDGEMENTS

We thank Drs. Demosthenes Pappagianis, University of California, Rebecca Cox, San Antonio State Chest Hospital and Angela Restrepo, Corporation of Biological Investigations, Medellin, Colombia for providing human sera from patients with coccidioidomycosis and paracoccidioidomy-cosis. We also thank Dr. Jerry A Colliver for statistical analysis of the data and Mrs. Barbara Reichert for excellent secretarial assistance.

This investigation was supported by grants from the Veterans Administration, National Institute of Allergy and Infectious Diseases (AI-16158, AI-22283) and from the research funds of the Southern Illinois University School of Medicine (CRC #24-86).

REFERENCES

1. Ajello, L., 1977, Systemic mycoses in modern medicine, Contr. Microbiol. Immunol., 3:2.

2. Anderson, D.W. and Crowle, A.J., 1981, Evans blue dye adjuvant enhances delayed hypersensitivity while blocking immunity to Mycobacterium tuberculosis in mice, Infect. Immun., 31:413.

3. Artz, R.P. and Bullock, W.E., 1979, Immunoregulatory responses in experimental disseminated histoplasmosis: lymphoid organ histopathology and serological studies, Infect. Immun., 23:884.

4. Bullock, W.E. and Deepe, G.S., 1983, Editorial: Medical Mycology in crisis, J. Lab. Clin. Med., 102:685.

5. Campbell, C.C., 1967, Serology in the respiratory mycoses, Sabouraudia, 5:240.

6. Campbell, C.C., 1968, Use and interpretation of serologic and skin tests in the respiratory mycoses: current considerations. Dis. Chest, 54:49.

7. Emmons, C.W., Binford, C.H., Utz, J.P. and Kwon-Chung, K.J., 1977, Histoplasmosis, in: "Medical Mycology", 3rd ed., p. 305-341, Lea and Febiger, Philadelphia, PA.

8. Feit, C. and Tewari, R.P., 1974, Immunogenicity of ribosomal preparations from yeast cells of Histoplasma capsulatum, Infect. Immun., 10:1091.

9. Jerrels, T.R. and Osterman, J.V., 1982, Host defenses in experimental scrub typhus: Delayed-type hypersensitivity response of inbred mice, Infect. Immun., 35:117.

10. Kaufman, L., 1971, Serological tests for histoplasmosis: their use and intrpretation, in: "Histoplasmosis. Proceedings of the Second National Conference, L. Ajello, E.W. Chick and M.L. Furcolow, eds., p. 321-326, Charles C. Thomas, Publisher, Springfield, IL.

11. Kaufman, L. and Clark, M.J., 1974, Value of concomitant use of complement fixation and immunodiffusion tests in the diagnosis of coccidioidomycosis, Appl. Microbiol., 28:641.

12. Khardori, N., Von Behren, L.A., Chaudhary, S., McConnachie, P. and Tewari, R.P., 1983, Cellular mediators of anti-Histoplasma immunity. I. Protective immunity and cellular changes in spleens of mice immunized by sublethal infection with yeast cells of Histoplasma capsulatum, Mykosen, 29:103.

13. Klun, C.L., Neiburger, R.G. and Youmans G.P., 1973, Relationship between mouse mycobacterial growth inhibitory factor and mouse migration-inhibitory factor in supernatant fluids from mouse lymphocyte cultures. Res. J. Reticuloendothel. Soc., 13:310.

14. Mackaness, G.B., 1964, The immunological basis of acquired cellular resistance, J. Exp. Med., 120:105.

15. Mitsuyama, M., Nomoto, K., Akeda, H., and Takeya, K, 1980, Enhanced elimination of Listeria monocytogenes at the site of delayed footpad reaction, Infect. Immun., 30:1.

16. Mitsuyama, M., Nomoto, K. and Takeya, K., 1982, Direct correlation between delayed footpad reaction and resistance to local bacterial infection, Infect. Immun., 36:72.

17. Nickerson, D.A., Havens, R.A. and Bullock, W.E., 1981, Immunoregulation in disseminated histoplasmosis: characterization of splenic suppressor cell populations, Cell. Immunol., 60:287.

18. Patel, P.J., 1980, Expression of antibactertial resistance at the site of a delayed hypersensitivity reaction, Infect. Immun., 29:59.

19. Salvin, S.G. and Smith, R.B., 1959, Antigens from the yeast-phase of Histoplasma capsulatum. III. Isolation, properties, and activity of a protein-carbohydrate complex, J. Infect. Dis., 105:45.

20. Salvin, S.B., 1965, Constituents of the cell wall of the yeast-phase of Histoplasma capsulatum, Amer. Rev. Resp. Dis. Suppl., 92:119.

21. Sharma, D., Khardori, N., Chaudhary, S., Von Behren, L.A,, McConnachie, P. and Tewari, R.P., 1986, Cellular mediators of anti-Histoplasma immunity. II Protective immunity and delayed hypersensitivity in mice immunized by sublethal infection with yeast cells of Histoplasma capsulatum, Mykosen, 29:116.

22. Tewari, R.P., 1975, Immunization against histoplasmosis, in: "The Immune System in Infectious Diseases", E. Neter and F. Milgrom, eds., p. 441-452, 4th Intl. Convoc. Immunol, S. Karger, Basel.

23. Tewari, R.P., Sharma, D.K., and Mathur, A., 1978, Significance of T lymphocytes in immunity elicited by immunization of ribosomes or live yeast cells of Histoplasma capsulatum, J. Infect. Dis., 138:605.

24. Tewari, R.P., Khardori, N., McConnachie, P., Von Behren, L.A. and Yamada, T., 1982, Blastogenic responses of lymphocytes from mice immunized by sublethal infection with yeast cells of Histoplasma capsulatum, Infect. Immun., 36:1013.

25. Tewari, R.P. and LaFemina, R.L., 1983, Immunogenicity of subcellular fractions from yeast cells of Histoplasma capsulatum, Japanese J. Med. Mycol., 24:179.

26. Wheat, J., French, M.L.V., Kamel, S., and Tewari, R.P., 1986, Evaluation of cross-reactions in Histoplasma capsulatum serologic tests, J. Clin. Microbiol., 23:493.

27. Wheat, L.J., Kohler, R.B. and Tewari, R.P., 1986, Diagnosis of disseminated histoplasmosis by detection of Histoplasma capsulatum antigen in serum and urine specimens, N. Engl. J. Med., 314:83.

RECOGNITION OF <u>COCCIDIOIDES</u> <u>IMMITIS</u> ANTIGENS WITH MONOCLONAL ANTIBODIES

S.J. Kraeger[1], D.J.P. Gennevois[2], R.A. Cox[3], and A.E. Karu[2]

[1]University of California, School of Public Health, Berkeley, CA; [2]Naval Biosciences Laboratory, Oakland, CA; [3]San Antonio Chest Hospital, San Antonio, TX

This paper summarizes recent observations on the antigenic specificity and suitability for diagnostic use of seven IgM monoclonal antibodies (MAbs) prepared in 1984 with <u>C. immitis</u> Silveira spherules and endospore/spherule culture filtrate (ESCF) as immunogens.

<u>Nature of the Antigens</u>. All seven MAbs reacted with a heat-stable methanol-precipitable extract of <u>C. immitis</u> endospores, suggesting that the antigenic determinants include carbo-hydrate. Denaturing electrophoresis and immunoblotting of ESCF resolved protein-containing bands with apparent molecular weights of 66, 65, 56, 45, and 32 kDa that were recognized by MAbs E35 and S82, suggesting shared epitopes. Coccidioidomycosis patient sera reacted with these same bands. The same bands bound fluorescein-labeled Concanavalin A (ConA), but no other lectins. Competition between ConA and E35 suggested that glucose or mannose is close to, or part of, the epitope of E35. Non-denaturing gel filtration resolved the antigens in ESCF into a broad distribution of sizes which showed three separate patterns of reactivity with the different MAbs (Figure 1). Deproteinized extracts of the largest carbohydrate-containing antigens (fractions 21-38 and 42-43 in Figure 1) were too complex for magnetic resonance analysis, but fractions 55 and 56, which bound only MAbs E35 and S82, contained a single oligosaccharide of -D-glucose.

<u>Specificity of the Antibodies</u>. Only tow MAbs showed more than 20% cross-reactivity with the other major systemic fungi in radio- and enzyme immunoassays (RIA, EIA). All of the MAbs reacted with spherules of six different <u>C. immitis</u> strains and ESCF from the Silveira strain, but four MAbs did not react with the ESCF from strain 46, which is of low virulence. Fluorescence microscopy showed specific differential staining of spherules, hyphae, and arthroconidia by the MAbs. For example, Figure 2 shows staining of the ends of arthroconidia by MAb E36. The MAbs also bound differently to these morphologic forms in RIAs.

<u>Diagnostic Applications</u>. All seven MAbs reacted with the <u>C. immitis</u> skin-testing antigens and the mycelial lysates that are currently used to detect precipitin-forming and complement-fixing antibodies. Experiments are under way to determine whether any of the antigens are diagnostic for coccidioidomycosis. We have developed a latex agglutination (LA) test to detect <u>C. immitis</u> antigens; to date the detection limit is 1 ng of <u>C. immitis</u> ESCF at 20 ng/ml in buffer. Detection thresholds for other pathogenic fungi are summarized in Table 1. We are attempting to optimize sensitivity and specificity when sera are tested by this method.

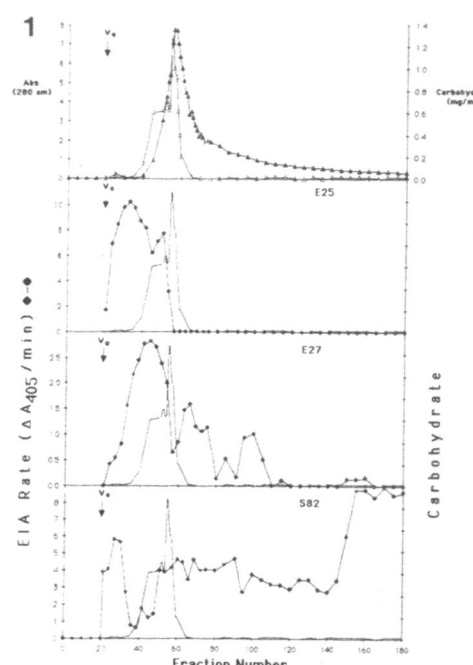

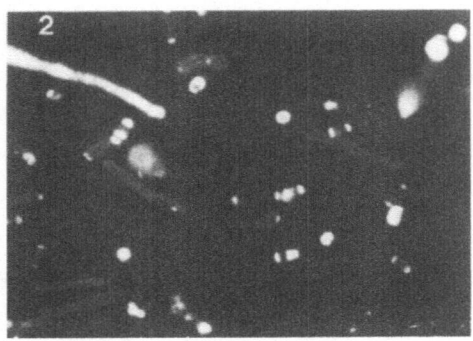

Table 1.

SENSITIVITY OF MICROPLATE LATEX AGGLUTINATION TEST

ANTIBODY	C. IMMITIS SILVEIRA	BLASTOMYCES DERMATITIDIS	HISTOPLASMA CAPSULATUM
	MINIMUM ANTIGEN DETECTABLE (NG)		
E3	16	>250	N.D.
E25	4	16	62
E27	8	31	125
E35	1	>500	>500
S56	4	125	N.D.

N.D. = NOT DETECTABLE: NO CROSS-REACTION

Fig. 1. Sephacryl S-400 gel filtration profile of C. immitis ESCF. The top panel shows the protein and carbohydrate content of each fraction. The succeeding panels show the reactivities of three MAbs with equivalent amounts of carbohydrate from each column fraction in an EIA. The carbohydrate profile is reproduced in each panel for reference. V_O is the void volume.

Fig. 2. Fluorescence micrograph of C. immitis arthroconidia reacted with MAb E36 and fluorescein-labeled goat anti-mouse globulin, showing differential staining of the arthroconidia.

Table 1. Minimum detectable dilutions of culture filtrates (ng of car-bohydrate) in a microplate LA test. Polystyrene beads (0.3 m) coated with the indicated MAbs were mixed with culture filtrates diluted in 0.05 ml of 0.85% NaCl-0.05 M glycine-NaOH (pH 8.2) with 0.1% bovine serum albumin, in microtiter "U"-shaped wells. Agglutination was scored after incubating the plates overnight at 22 C.

In summary, the determinants recognized by all seven MAbs appear to be primarily carbohydrates, found on several biochemically separable species. Each MAb reacted differently with various antigens or fungal particles in EIA, RIA, immunofluorescence, and LA tests. MAbs in this panel detect taxono-mic and developmental differences in isolates of C. immitis and other pathogenic fungi, and some can be used to localize antigens on the surface of fungal cells. Some of these MAbs are uniquely specific for C. immitis antigens, and their potential usefulness in diagnostic assays for coccidioidomycosis deserves further study.

Acknowledgements. We thank Dr. O. Hindsgaul, University of Alberta, for performing proton and carbon magnetic resonance analyses, and Dr. D. Baker, Bioresponse Inc., for valuable discussions. [supported by ONR Contract N00014-81-C-0570 (SJK, DG, AEK), and NIAID Grant A121431 (RAC)]

Actinomycetes

ANTIGENS OF THERMOPHILIC ACTINOMYCETES

Patrick Boiron

Unite de Mycologie
Institut Pasteur
Paris

INTRODUCTION

In recent years, an increasingly large number of cases of hyper-sensitivity pneumonitis (HP) have been reported. Several antigens from bacteria, fungi, protozoa, and birds droppings have been implicated, from a variety of sources including vegetables, wood, water reservoirs, dairy and grain products. The allergens are usually disseminated as aerosol dusts that can be inhaled. Predominant among the causative microorganisms are the thermophilic actinomycetes, especially Faenia rectivirgula, Thermoactino-myces sacchari, T. vulgaris, T. candidus, and Saccharomonospora viridis (Table 1). These organisms can survive in environments that have high temperatures (60-80°C) and humidity. Other species of the thermophilic actinomycetes: T. dichotomica, T. intermedius, Saccharopolyspora hirsuta, have so far, not been implicated as etiologic agents of HP, but they can also be isolated from the same environments (Table 1).

The need for physicians to become increasingly aware of this disease expression is the result of 3 factors:
        - recognition that HP can take place with increasing frequency, not only during occupational exposure, but also at home and in office. A variety of new microbial antigens implicated as the cause of HP are beeing described with regularity;
        - in its chronic form, the disease can result in cumulative and irreversible damage to lung tissue. Therefore a thorough differential diagnostic procedure is vital to rule out other lung diseases and to monitor specific immune parameters that may predict flares of the disease;
        - a better definition of the mechanism of immunopathology of HP may suggest better forms of anti-inflammatory therapy.

Two of the more common clinical examples of HP due to thermophilic actinomycetes are farmer's lung (Reynolds, 1982) and bagassosis (Boiron, 1985). Whereas the majority of HP follows occupational exposure (Do Pico, 1984), a condition resembling farmer's lung also occurs in occupants of homes and offices. This latter type of disease is caused by exposure to forced air heating, cooling or humidification systems when they become contaminated by thermophilic actinomycetes (Stankus and Salvaggio, 1983) (Table 2).

Table 1. Thermophilic Actinomycetes

Implicated in Hypersensitivity Pneumonitis

<u>Faenia</u> <u>rectivirgula</u>	(Krassilnikov and Agre 1964) Kurup and Agre 1983
<u>Thermoactinomyces</u> <u>sacchari</u>	Lacey 1971
<u>T.</u> <u>vulgaris</u>	Tsiklinsky 1899
<u>T.</u> <u>candidus</u>	Kurup, Barboriak, Fink and Lechevalier 1975
<u>Saccharomonospora</u> <u>viridis</u>	(Schuurmans, Olson and San Clemente 1956) Nonomura and Ohara 1971

Not Implicated in Hypersensitivity Pneumonitis

<u>T.</u> <u>dichotomica</u>	(Krassilnikov and Agre 1964) Cross and Goodfellow 1973
<u>T.</u> <u>intermedius</u>	Kurup, Hollick and Pagan 1981
<u>Saccharopolyspora</u> <u>hirsuta</u>	Lacey and Goodfellow 1975

Table 2. Diseases Associated with Thermophilic
Actinomycetes

Diseases	Antigen Source	Implicated Actinomycetes
Farmer's lung	Moldy hay	<u>F.</u> <u>rectivirgula</u> <u>T.</u> <u>vulgaris</u> <u>T.</u> <u>candidus</u>
Bagassosis	Bagasse	<u>T.</u> <u>sacchari</u> <u>T.</u> <u>vulgaris</u>
Humidifier and air-conditioning lungs	Humidifier water and ventilation ducts	<u>T.</u> <u>vulgaris</u> <u>T.</u> <u>candidus</u> <u>S.</u> <u>viridis</u>
Mushroom worker's lung	Mushroom moldy compost	<u>F.</u> <u>rectivirgula</u> <u>T.</u> <u>vulgaris</u> <u>T.</u> <u>candidus</u>

The clinical presentation of HP depends on the degree of exposure to the organic dusts (Fink, 1986; Salvaggio and De Shazo, 1986). The acute form is easily recognized because of the direct relationship between the patient's respiratory symptoms and antigen exposure. But the outset may, however, be more insidious. Respiratory and systemic symptoms include low-grade fever, chills, chest pain, cough, and dyspnea beginning 4 to 6 hours after exposure to the sensitin (Table 3). The symptoms may persist for 12 hours or so. Remission of symptoms after avoidance of the conta-minate area is an important diagnostic feature. If exposure is continuous and protracted, the syndrome may progress to a chronic form of disease. Then, symptoms are characterized by a progressively increasing cough, dyspnea on exertion, fever, weight loss, and fatigue which may be easily mistaken for chronic bronchitis or emphysema. In later stages, the spectrum of disease shifts to lung fibrosis, accompanied by a restrictive defect on pulmonary function testing, and finally to terminal respiratory failure (Slavin, 1985).

Table 3. Clinical Features of Hypersensitivity
Pneumonitis

	Acute Form	Chronic Form
Chills and fever	+	0
Dyspnea	+	+
Productive cough	+	+
Weight loss and fatigue	+	+
Alveolar-capillary block	+	+
Airway obstruction	0	late
Pulmonary fibrosis	0	late

PATHOGENICITY OF ANTIGENIC FACTORS

Several antigenic factors appears to be important in disease mani-festation. Of necessity, antigenic particles must be less than 5 μm to be deposited in the alveolar-air exchange portion of the lung. To induce an immunologic response, they need to have a sufficient molecular weight, an appropriate number and configuration of epitopes, and to be recognized as foreign by the host's immune mechanism. Spores of most thermophilic actino-mycetes satisfy all of these criteria. Reed (1974) reviewed the necessary antigenic conditions for granuloma formation involved in HP. The non-digestability of antigen by lysosomal enzymes might be important, leading to the persistence of antigen in alveolar macrophages. Most of the parti-culate antigens involved in HP appear to be non-digestable and may there-fore stimulate granuloma formation.

HP disease is characterized by the development of high concentrations of precipitins against the offending antigens. Demonstration of antibodies in the IgG, IgM, and IgA classes against specific antigens is helpful in the diagnosis. However, the significance of antibodies must be interpreted cautiously because specific antibodies can be detected in almost all patients, with and without clinical symptoms. Their presence is only an indication that prior exposure and sensitization have occurred. When antibodies are not detected in exposed symptomatic people, one must exclude the possibility that insensitive methods or inappropriate antigens are at fault.

Several serological methods are available for demonstrating antibodies (Doll and Bozelka, 1983), including agar gel double diffusion, complement fixation, hemagglutination, immunofluorescence, counter-immunoelectrophoresis, immuno-electrodiffusion and ELISA. Intensive efforts have been made to find a laboratory test that would discriminate patients with HP disease from asymptomatic exposed persons. The ELISA appears to be a sensitive, specific and quantitative test that correlate better with the clinical diagnosis than the precipitin tests. However, although patients with HP, like farmer's lung or bagassosis, have significant ELISA titers, asymptomatic patients, and even normal controls, also have detectable levels.

All the tests developed thus far, give results that are poorly comparable. The most important reason for lack of correlation is antigenic variability. Antigenic extracts differ from strain to strain. Even different batches from the same strain can vary considerably. Standardization of antigens will be possible only when methods to prepare reliable and reproducible antigens will be developed for use in serological assays. Standardization of such antigens will be achieved by growing selected strains under controlled conditions and use of suitable methods to purify the antigens.

STRAINS SELECTION FOR ANTIGENS PRODUCTION

Considerable variation exists among quality and quantity of antigens produced from strains of thermophilic actinomycetes. The extracellular antigens from 5 strains of F. rectivirgula were studied by Kurup and Fink* (1979) by crossed- immunoelectrophoresis using rabbit anti-F. rectivirgula serum. The study demonstrated common and specific antigenic determinants among the different strains. The number of precipitin arcs varied between 7 and 26, but only 1 was detected in all the strains, whereas 2 more were frequently detected. Quantitative differences in the antigenic constituents were detected among most strains. The difference in antigenic components of F. rectivirgula strains was also demonstrated because antibodies could not be completely absorbed by heterologous antigens. Because the 3 major antigens reacted with the sera of patients with farmer's lung, pratically no special strains of F. rectivirgula need to be selected for antigen production.

On the contrary, strains of T. candidus need to be carefully selected for producing culture filtrate antigens. Indeed, the study by Kurup (1985) of the antigens from 10 strains of T. candidus by fused rocket electrophoresis using rabbit anti-T. candidus serum, showed considerable differences in their reactivity. Tandem crossed immunoelectrophoresis demonstrated only a few antigens common to the different strains. Sera of 35 patients with HP were analyzed with these 10 antigenic extracts. No single antigenic preparation was able to detect antibodies in all sera, but together 3 antigenic extracts detected precipitins in all 35 patients' sera.

Table 4. Enzymatic Activities of Double Dialysis Antigens of 11 Strains of T. sacchari Detected by the API-Zym System.

Enzymes	Strains										
	1476	978	1001	980	981	986	998	1000	1002	1466	1005
Alkaline phosphatase	0	++	0	+	+	+	++	+	+	+	+
Esterase (C-4)	++	++	++	++	++	++	++	++	++	++	+
Esterase-lipase (C-8)	++	++	++	++	++	+/++	++	++	++	++	+
Leucine-aryla-midase	0	+	0	+	+	0	0	0	0	+	+/++
α-chymotrypsine	0	++	0	0	0	0	+/++	0	0	0	0
Acid phosphatase	++	++	++	++	++	+	+	++	++	++	+
Naphthol AS-BI phospho-hydrolase	+	++	+	0	0	0	++	+/++	+	+	0
α-glucosidase	0	++	0	+	+	0	+	+	0	+	+

0, negative reaction; +, positive reaction; ++, strongly positive reaction.

The enzyme activities of extracellular antigens from 11 strains of T. sacchari were analyzed (Boiron, 1986). The study showed high variations among the number and the concentration of activities detected among the strains, and also between batches (Table 4). The enzymes were not analyzed for their antigenic characteristics, but this study revealed that major difficulties remain in the preparation of good and reproducible antigens. Possibly, some of these enzymes, like F. rectivirgula chymotrypsin-like proteinase, against which precipitins are directed in farmer's lung (Gari et al., 1984; Nicolet et al., 1977), may play a role in the bagassosis.

Information on the antigenicity of the other species of thermophilic actinomycetes are almost non-existent. Some indications about the diffe-rences in their antigens were observed by Kurup (1985) who showed that sera from some patients that were unreactive against antigens prepared from laboratory strains, were positive with antigens extracted from organisms isolated from the patient's environment.

GROWTH CONDITIONS OF THERMOPHILIC ACTINOMYCETES

The majority of thermophilic actinomycetes have fastidious growth requirements, and the use of a complex natural medium for growth is necessary for most of them. Thus, when extracellular antigens are prepared, they contain macromolecular components derived from the medium itself. Standardization of such antigens is difficult because they may give false-positive and false-negative results when tested with patients' sera. Treuhaft (1981) reported that some individuals reacted serologically with yeast extract present in the growth medium. This reactivity most likely obscures specific reactions of patients' serum with thermophilic actino-mycetes antigens in 0.1 % of the cases. A potentially more serious drawback is that these unpurified antigens are unsuitable for use as in vivo reagents for skin testing or for inhalation challenge.

To provide antigenic material uncontaminated by substrate upon which the organisms have grown, Edwards (1971; 1972) developed the double dialysis method. By this means, high molecular-weight components are retained in the dialysis tubing, while the organisms are grown in the outer phase (dialysate). A second dialysis is used to eliminate traces of dialy-sable medium constituents from the extracellular antigens. But, invariably, some minute amounts of medium components are present in the isolated anti-gens. This may be due to the change in pore size of the dialysis tubing during autoclaving or by action of enzymes released by the growing orga-nisms. Another approach to obtain medium free antigenic extracts is the use of chemically defined medium. The antigens produced by F. rectivirgula and T. candidus in such media (Kurup and Fink, 1977; Treuhaft et al., 1981), were more reactive and specific than all previously known antigens. However, adequate synthetic media for antigen production are not yet available for growing other species of thermophilic actinomycetes.

To obtain consistent and reproductible antigenic extracts, the incu-bation temperature, aeration, inoculum size, and duration of incubation need to be carefully determined and standardized. The temperatures of incubation are 55°C for F. rectivirgula and T. sacchari, and 45°C for T. vulgaris, T. candidus and S. viridis. In stirred fermentors, optimum time for production of antigens is 5 days of growth for F. rectivirgula and S. viridis, 3 days for the other species. Incubation at higher tempe-ratures and prolonged incubation frequently breakdown the antigens.

Culture filtrate and cellular extracts prepared by a variety of methods·
from F. rectivirgula (Fletcher and Rondle, 1973) were compared for detecting
antibodies in the serum of patients with farmer's lung. All sera tested
reacted with concentrated culture fluids, but none with cellular extracts.

Several investigators have attempted to characterize (Vesterberg and
Holmberg, 1982) and to purify relevant antigens by fractionation methods.
Edwards (1972) used gel-filtration and ion-exchange chromatography, and
fractionated culture filtrate antigens of F. rectivirgula. Four major antigenic fractions were identified and labeled according to their electrophoretic mobility by immuno-electrophoresis. Two antigenic components of
F. rectivirgula were separated from culture filtrate antigens by preparative
isoelectric focusing and used in immunodiffusion (Kurup et al., 1981).
Recently, using gel-filtration and affinity chromatography, Kurup et al.
(1984) isolated a purified proteinic fraction of F. rectivirgula with a
molecular weight of approximately 16,000. This protein was used in an ELISA
and detected significant titers of antibodies in 13 of 15 patients with
farmer's lung and negligible titers in unexposed controls. This antigen
did not cross-react with antigens of other thermophilic actinomycetes.

For T. candidus, 2 proteinases were isolated by Roberts et al. (1983)
from culture filtrate antigens. They were separated from each other by ion-
exchange chromatography, and purified by gel-filtration. Proteinases can be
detected at low levels in some dust samples collected from moldy silage,
suggesting that direct inhalation exposure to these proteinases can occur.
The presence of anti-proteinase immunoglobulins in farmer's lung sera
indicates that these proteinases are in a position to have a direct role
in inducing lung sensitization. One of these proteinases, "TcP1", is not
inhibited by plasma proteinase inhibitors, ∝ 1 proteinase inhibitor and

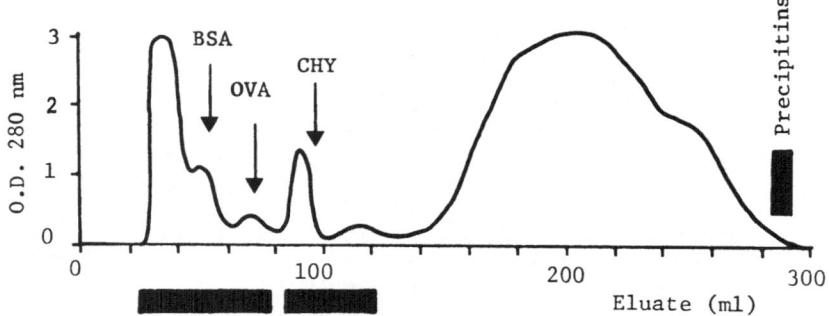

Figure 1. Gel-filtration of Double Dialysis Antigens of
T. sacchari. Molecular sieving was performed on
Sephadex G-100 (Pharmacia) column (1.5 by 85 cm) equi-
librated with PBS. Column fractions were analyzed by
UV absorption at 280 nm and by immuno-electrodiffusion
on cellulose acetate with rabbit anti-serum. BSA,
bovine serum albumine (molecular weight, 67,000);
OVA, ovalbumine (M.W., 43,000); CHY, chymotrypsin
(M.W., 25,000).

α 1-antichymotrypsin. The other proteinase, "TcP2", is inhibited by large excesses of the α 1 proteinase inhibitor and not at all by α 1-antichymotrypsin. These facts suggest that the lung does not have adequate defenses to inactivate these proteinases when they are inhaled. The use of the isolated proteinases in an ELISA demonstrated specific antibody in sera from patients with farmer's lung disease, whereas the unexposed control sera showed zero or negligible antibody concentration. These results will need confirmation with sera of exposed asymptomatic patients.

A protease ("thermitase") obtained from a T. vulgaris culture filtrate was purified chromatographically on Sephadex G-75, DEAE-cellulose, and Sephadex G-50 (Behnke et al., 1978). The purified enzyme demonstrated a molecular weight of 37,400 as determined by gel-filtration and by ultracentrifugation. The fine purification was achieved in a single step from the Na_2SO_4-precipitated crude product by means of isoelectrical focusing using granulated gel (Frömmel et al., 1978). This thermitase was not assayed for its antigenic properties.

T. sacchari antigenic extracts displayed a relatively complex array of antigens (Boiron, 1986). Twenty-five precipitin arcs can be detected by two-dimensional crossed rocket immuno-electrophoresis. Gel-filtration analysis resulted in 7 fractions of molecular weight varying between 105,000 and about 4,000 (Figure 1). The antigenic fractions were of high molecular weight, and most of the UV-absorbing material appeared to be associated with non-antigenic components of small molecular weight (Lehrer and Salvaggio, 1978).

If pure and standardized antigen are necessary for reliable and dependable serological results, the antigenic fractions purified until now cannot avoid false-negative reactions if used alone. All patients may not develop antibodies against these purified fractions. Under these circumstances and until disease-specific antigens are available, several purified antigens need to be used in parallel with crude antigens in serological diagnosis.

ANTIGENIC AND BIOCHEMICAL RELATIONSHIP AMONG THERMOPHILIC ACTINOMYCETES ANTIGENS

Because the thermophilic actinomycetes cause similar clinical diseases in susceptible individuals, it is conceivable that some of their antigens are identical or closely related. Antigens prepared by the double dialysis method from strains of thermophilic actinomycetes were tested against rabbit anti-sera by immunodiffusion and by crossed immuno-electrophoresis (Kurup et al., 1976; Kurup and Fink, 1979). The antibody produced against one strain of a species strongly cross-reacted with antigens from other strains of the same species, though the cross-reactions were more or less pronounced due to the variation of antigens production. Some degree of cross-reactivity was observed between F. rectivirgula and S. viridis, but both species did not show any antigenic relationship with Thermoactinomyces species. Antigens from Thermoactinomyces species cross-reacted with members within the genus. Interestingly, F. rectivirgula and S. viridis have a type IV cell wall, whereas Thermoactinomyces species have a type III cell wall (Lechevalier and Lechevalier, 1976).

These antigenic relationships are not rediscovered in any homology of physico-chemical properties of pure fractions isolated at this date from species of thermophilic actinomycetes (Table 5).

Table 5. Physico-chemical and Biochemical Properties of Pure Fractions Isolated from Thermophilic Actinomycetes Antigens.

Properties	Species and Pure Fractions					
	F. rectivirgula			T. candidus		T. vulgaris
				TcP1	TcP2	Thermitase
Molecular weight	51,000	29,000	16,000	29,000	52,000	37,400
Isoelectric point	3.8–4.0	3.5	3.8	6.0	4.2	9.0
Enzymatic activity	ND	ND	ND	+	+	+
Antibodies to fractions	+	+	+.	+	+	ND

ND: not done.

Table 6. Enzymatic Activities of Double Dialysis Antigens of Strains of Thermophilic Actinomycetes Detected by the API-Zym System.

Enzymes	Strains				
	F. rectivirgula (2)*	T. sacchari (11)	T. candidus (2)	T. vulgaris (2)	S. viridis (1)
Alkaline phosphatase	+	0/+/++	+	+	+
Esterase (C-4)	++	++	++	++	+
Esterase-lipase (C-8)	+/++	++	+	+	+
Leucine-arylamidase	++	0/+	0	0	+
∝ -chymotrypsine	++	0/++	0	0	0
Acid phosphatase	+/++	+/++	+	0	0
Naphthol AS-BI phosphohydrolase	+/++	0/+/++	0	0/+	0
∝ -glucosidase	0/++	0/+	0	0/+	+

*, number of strains;
0, negative reaction; +, positive reaction; ++, strongly positive reaction.

Using the API-Zym strip, the presence or absence of different enzymes in double dialysed antigens of thermophilic actinomycetes was assessed (Boiron, 1986) (Table 6). Our results showed slightly differences from those previously obtained by Hollick (1982): this could be a reflection of the different process to prepare the antigens (Hollick omitted the second dialysis step) or isolate variation. In spite of the variability of the enzymatic activities in the strains, it is interesting to observe that some enzymes are present in all species, e.g., esterases. It is possible that the esterase activity observed was due to protease since it is known that certain proteases can cleave ester bonds (Stryer, 1975). Development of enzymes prints or "zymograms" for proteolytic enzyme activity from the electrophoretograms of thermophilic actinomycetes double dialysis antigens showed characteristic patterns for F. rectivirgula, T. candidus and T. vulgaris (Roberts et al., 1977). Antigens preparations from T. sacchari and S. viridis did not show any proteolytic activity. However, further studies are needed, i.e., testing the effect of proteolytic activators, as trypsin or sulfhydryl reagents, to determine whether residual proteinase activity can be detected, and the effect of various pH on the electrophoresis run as well as on the reaction of the gel on the substrate.

RELEVANCE OF ENZYMATIC ACTIVITIES OF ANTIGENS TO IMMUNOPATHOGENESIS OF HYPERSENSITIVITY PNEUMONITIS

Some hypothesis proposed concerning the immunopathogenesis of HP ascribe an importance to "local cellular immunologic involvement" (Stankus et al.,1982). The following mechanism of immune pathogenesis is proposed (Salvaggio and De Shazo, 1986; Schuyler, 1985) (Figure 2).

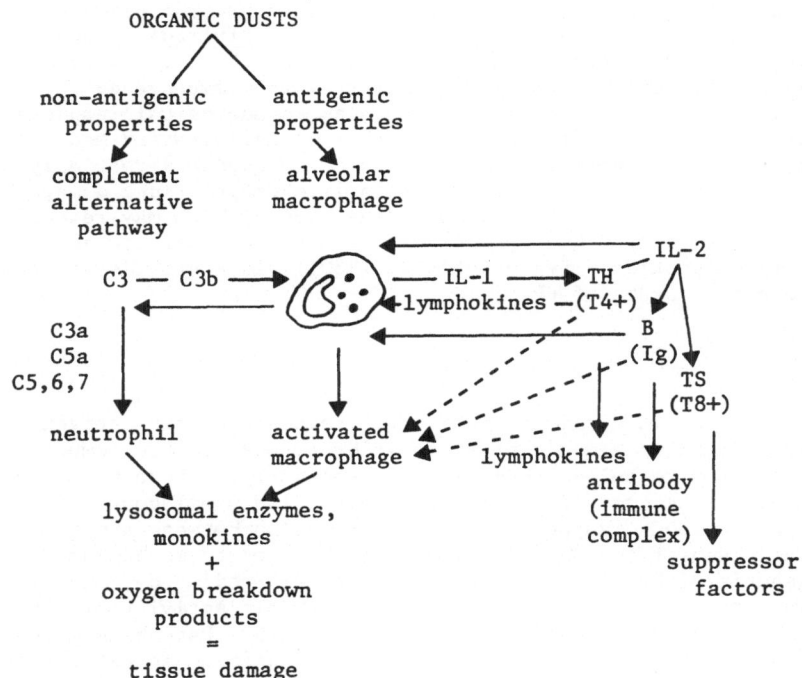

Figure 2. Immune Mechanism of HP. IL-1 (2), interleukin 1 (2); TH (S), helper (suppressor) T lymphocyte; B, B lymphocyte. Explanations are in the text.

457

Acute and/or repeated exposure to the offending antigen results in antigen deposition on the alveolar surface. The necessary stimuli for initial inflammatory reactions can be provided by non-specific direct activation of the alternative complement pathway by thermophilic actino-mycetes (Edwards, 1976) or by non-specific adjuvant property of actino-mycetes for the induction of specific T-cell mediated immunity (Bice et al., 1977). Antibody responses, which closely follow exposure, provide an addi-tional stimulus in the form of immune complexes that can bind complement, generate neutrophil chemotactic factors, and activate lung macrophages. These events and the arrival of polymorphonuclear leucocytes in the lung favor an acute inflammatory reaction characteristic of an Arthus-type. The phagocytosis of immune complexes may limit the progression of disease, especially if antigen exposure ceases. If exposure to antigen is prolonge- or recurrent, the Arthus response does not seem to persist, and the tissue damage could then be explained as a result of delayed hypersensitivity. Sensitized T lymphocytes secrete lymphokins such as MIF that activate the alveolar macrophages. Enzymes released from these macrophages would cleave the third and the fifth complement components leading to amplification of chemotactic activity for macrophages. Lysosomal hydrolases and oxidases from macrophages and neutrophils could contribute to cause tissue damages. The continuous accumulation of mononuclear phagocytes would contribute to formation of pulmonary granuloma and ultimately to fibrosis.

The frequent involvement of thermophilic actinomycetes in HP suggests that they may produce biologically active constituents which potentiate the immune response to their antigenic components. Several candidates for such activity have been proposed, including esterase and protease activities. The presence of anti-enzymes antibodies in serum of individuals sensitized to thermophilic actinomycetes leads to determine the actual contribution of enzymatic activities to the immunopathogenesis of HP. The demonstration that farmer's lung patients have antibodies to F. rectivirgula proteinase, lipase and phosphatase (Biguet et al., 1974; Walbaum et al., 1969) or to the T. candidus proteinases is evidence that they are immunogene. The enzymes may simply act as antigens and induce an immunological response. They may also cause inflammation by the release of inflammatory mediators from the complement components, C_3 or C_5, and/or lead to a direct enzymatic action on the respiratory tract, resulting in irreversible lung damage by lung tissue digestion (e.g., lung elastin digestion by TcP1 may release inflammatory mediators from macrophages attracted by chemotactic activity of elastin-derived peptides (Senior et al., 1980)). This provides another aspect in the pathogenesis of HP.

CONCLUSION

The difficulties in obtaining purified, disease-specific antigens for the reliable immunodiagnosis of HP could be related to the immunolo-gical criterion used to select them. Most of the antigens have been isola-ted and purified in accord with their reactivity with homologous anti-bodies. But the antibody responses do not distinguish between symptomatic and asymptomatic sensitized patients. The actual proposed mechanism of immunopathogenesis of HP give prominence to the cell-mediated immunity, especially in chronic form. Furthermore, the enzymatic activity of anti-gens, which may also be involved in the etiology of HP, have been given insufficient attention as a criterion for purified antigen selection. The isolation and purification of antigens based on their cell-mediated reac-tivity or enzymatic activity deserves further study.

SUMMARY

The progressive interest in HP is the result of the increased frequency with which this disease is recognized. Not only is HP related to occupational exposure, but it is also encountered at home and in work place. The cumulative effect of chronic exposure may include irreversible tissue damage and even, in the extreme, terminal respiratory failure.

Thermophilic actinomycetes constitute the predominent group of HP-inducing microorganisms. The serological response consists of precipitins directed against the offending antigens. But the presence of precipitins in the patients' serum only indicates prior exposure and not necessarily active disease.

Variability in the antigen preparations, and the large variety of immunological methods in use complicates the comparisons in immunodiagnosis tests. As a step forward, standardization of antigens should be attempted. Selection of strains and determination of the best growth conditions for antigen production are important because each thermophilic actinomycete has its own characteristics. Some purified antigens have been obtained from F. rectivirgula, T. candidus and T. vulgaris, but they detect antibodies both in symptomatic and asymptomatic exposed people, and, when used alone, they cannot avoid some false-negative reactions.

The basic data about the immunologic cross-reactions among thermophilic actinomycetes are analyzed from the acquired knowledge on the physico-chemical and enzymatic characteristics of the purified antigens derived from the different species. The importance of the enzymatic activities of the antigens in the immunodiagnosis and the immunopathogenesis of HP is discussed.

REFERENCES

Behnke, U., Schalinatus, E., Ruttloff, H., Höhne, W. E., and Frömmel, C., 1978, Charakterisierung einer Protease aus Thermoactinomyces vulgaris (Thermitase). 1. Untersuchungen zur Reinigung der Thermitase, Acta. Biol. Med. Ger., 37:1185.

Bice, D. E., Mc Carron, K., Hoffman, E. O., and Salvaggio, J. E., 1977, Adjuvant properties of Micropolyspora faeni, Int. Arch. Allergy Appl. Immunol., 55:267.

Biguet, J., Walbaum, S., and Vernes, A., 1974, Physico-chemical aspects and antigenic structure of Micropolyspora faeni, in:"Aspergillosis and Farmer's Lung in Man and Animal," R. de Haller and F. Suter, eds., Hans Huber Publishers, Berne.

Boiron, P., 1985, Aspects microbiologiques et serologiques de la bagassose, Doct. Pharm. Thesis, Paris.

Boiron, P., 1986, in preparation.

Doll, N. J., and Bozelka, B. E., 1983, Immunologic techniques utilized in the diagnosis of occupational lung disease, Clinics in Chest Medicine, 4:85.

Do Pico, G. A., 1984, Occupational lung disease in the rural environment, in: "Occupational Lung Disease," J. B. L. Gee, ed., Churchill Livingstone, New York.

Edwards, J. H., 1971, The production of farmer's lung antigens, Med. Lab. Technol., 28:172.

Edwards, J. H., 1972, The double dialysis method of producing farmer's lung antigens, J. Lab. Clin. Med., 79:683.

Edwards, J. H., 1972, The isolation of antigens associated with farmer's lung, Clin. Exp. Immunol., 11:341.

Edwards, J. H., 1976, A quantitative study on the activation of the alternative pathway of complement by mouldy hay dust and thermophilic actinomycetes, Clinical Allergy, 6:19.

Fink, J. N., 1986, Clinical features of hypersensitivity pneumonitis, Chest, 3:193S.

Fletcher, S. M., and Rondle, C. J. M., 1973, Micropolyspora faeni and farmer's lung disease, J. Hyg., Camb., 71:185.

Frömmel, C., Hausdorf, G., Höhne, W. E., Behnke, U., and Ruttloff, H., 1978, Charakterisierung einer Protease aus Thermoactinomyces vulgaris (Thermitase). 2. Einschritt-Feinreinigung und Proteinchemische. Charakterisierung, Acta. Biol. Med. Ger, 37:1193.

Gari, M., Recco, P., Pinon, J. M., and Seguela, J. P., 1984, Interet de la co-immunoelectrodiffusion sur acetate de cellulose dans le diagnostic du poumon de fermier pour la mise en evidence de systemes precipitants remarquables, Sem. Hop. Paris, 60:776.

Hollick, G. E., 1982, Enzymatic profiles of selected thermophilic actinomycetes, Microbios, 35:187.

Kurup, V. P., 1985, Serological diagnosis of hypersensitivity pneumonitis, in: "Microbiology," L. Leive, ed., ASM, Washington DC.

Kurup, V. P., Barboriak, J. J., Fink, J. N., and Scribner, G., 1976, Immunologic cross-reactions among thermophilic actinomycetes associated with hypersensitivity pneumonitis, J. Allergy Clin. Immunol., 57:417.

Kurup, V. P., and Fink, J. N., 1977, Extracellular antigens of Micropolyspora faeni grown in synthetic medium, Infection and Immunity, 15:608.

Kurup, V. P., and Fink, J. N., 1979, Antigenic relationships among thermophilic actinomycetes, Sabouraudia, 17:163.

Kurup, V. P., and Fink*, J. N., 1979, Antigens of Micropolyspora faeni strains, Int. Arch. Allergy Appl. Immun., 60:140.

Kurup, V. P., John, K. V., Ting, E. Y., Somasundaram, K., Resnick, A., and Marx, J. J., 1984, Immunochemical studies of a purified antigen from Micropolyspora faeni, Molecular Immunology, 21:215.

Kurup, V. P., Ting, E. Y., Fink, J. N., and Calvanico, N. J., 1981, Characterization of Micropolyspora faeni antigens, Infection and Immunity, 34:508.

Lechevalier, M. P., and Lechevalier, H. A., 1976, Chemical methods as criteria for the separation of nocardiae from other actinomycetes, The Biology of the Actinomycetes, 11:78.

Lehrer, S. B., and Salvaggio, J. E., 1978, Characterization of Thermoactinomyces sacchari antigens, Infection and Immunity, 20:519.

Nicolet, J., Bannerman, E. N., De Haller, R., and Wanner, M., 1977, Farmer's lung: immunological response to a group of extracellular enzymes of Micropolyspora faeni. An experimental and field study, Clin. Exp. Immunol., 27:401.

Reed, C. E., 1974, Allergic mechanisms in extrinsic allergic alveolitis, in: "Progress in Immunology II, vol. 4," L. Brent and J. Holborrow, eds., Amsterdam.

Reynolds, H. Y., 1982, Hypersensitivity pneumonitis, Clinics in Chest Medicine, 3:503.

Roberts, R. C., Nelles, L. P., Treuhaft, M. W., and Marx, J. J., 1983, Isolation and possible relevance of Thermoactinomyces candidus proteinases in farmer's lung disease, Infection and Immunity, 40:553.

Roberts, R. C., Zais, D. P., Marx, J. J., and Treuhaft, M. W., 1977, Comparative electrophoresis of the proteins and proteases in thermophilic actinomycetes, J. Lab. Clin. Med., 90:1076.

Salvaggio, J. E., and De Shazo, R. D., 1986, Pathogenesis of hyper-
    sensitivity pneumonitis, <u>Chest</u>, 89:190S.
Schuyler, M., 1985, Immune mechanisms in hypersensitivity pneumonitis,
    <u>in</u>: "Microbiology," L. Leive, ed., ASM, Washington DC.
Senior, R. M., Griffin, G. L., and Mecham, R. F., 1980, Chemotactic activity
    of elastin-derived peptides, <u>J. Clin. Invest.</u>, 66:859.
Slavin, R. G., 1985, Clinical aspects of hypersensitivity pneumonitis, <u>in</u>:
    "Microbiology," L. Leive, ed., ASM, Washington DC.
Stankus, R. P., Morgan, J. E., and Salvaggio, J. E., 1982, Immunology of
    hypersensitivity pneumonitis, <u>CRC Critical Reviews in Toxicology</u>,
    11:15.
Stankus, R. P., and Salvaggio, J. E., 1983, Hypersensitivity pneumonitis,
    <u>Clinics in Chest Medicine</u>, 4:55.
Stryer, L., 1975, "Biochemistry," W. H. Freeman and Co., San Francisco.
Treuhaft, M. W., Roberts, R. C., Hackbarth, C., and Marx, J. J., 1981,
    Characterization of synthetic medium antigens of <u>Micropolyspora</u> faeni
    and <u>Thermoactinomyces</u> candidus, <u>J. Allergy Clin. Immunol.</u>, 67:375.
Vesterberg, O., and Holmberg, K., 1982, Characterization of allergen
    extracts by two-dimensional electrophoretic techniques. <u>Micropoly-
    spora faeni</u> antigens, <u>Clin. Chem</u>, 28:993.
Walbaum, S., Biguet, J., and Tran-Van Ky, P., 1969, Structure antigenique
    de <u>Thermopolyspora polyspora</u>. Repercussions pratiques sur le
    diagnostic du poumon de fermier, <u>Ann. Inst. Pasteur</u>, 117:673.

INDEX

Acquired immune deficiency syndrome, see AIDS

Actinomycetes
   agar gel diffusion test 453
   antigens 448
   enzyme and pathogenicity 450
   thermophilic Actinomycetes 448

Adhesion
   Candida albicans 157

Agglutination
   latex, candidosis 205, 211, 219

AIDS
   and Cryptococcus neoformans 3, 269
   cryptococcal antigen 286, 307

Allescheria boydii 6

Alternaria
   allergens 237

Alternaria alternata 239

Allergens
   detection 223, 240
   identification 223, 240
   of Alternaria 237
   of Aspergillus 223
   of Botrytis 262
   of Candida 259
   of Chrysosporium 391
   preparation 223, 241

Allergic bronchopulmonary aspergillosis 223, 307, 367

Antibody
   monoclonal, hybridoma techniques 19, 99
   polyclonal 11, 171

Antibody detection
   CIE 14, 141, 149, 377
   ELISA 21, 81, 203, 339, 381, 382, 383, 385, 386, 390
   immunodiffusion 12
   immunoelectrophoresis 4, 397
   immunoblots (see Western blots) 358

Antigen
  circulating 21
  preparation 8, 53, 112, 169, 206, 241, 275, 286, 336, 356,
    398, 418
  suppression 12, 24, 189

Antigenemia
  and monoclonal antibodies 100
  detection in invasive aspergillosis 367
  galactomannan in aspergillosis 371
  in candidosis 212
  in coccidioidomycosis 445
  in histoplasmosis 431
  mannan in candidosis 22, 170, 202, 203, 209, 212
  trichophytin in dermatophytosis 340

Arabinitol 178, 202

Arabinomannan 178

Aspergilloma 377

Aspergillosis 3, 225
  in immunocompromised patients 376, 377
  invasive 367, 382

Aspergillus
    A. flavus 367
    A. fumigatus 367
        antigens of immunocompromised patients 376, 377
        catalase and immunoassays 388
        ELISA and antigen detection 368, 381, 383, 385
        peroxidase-antiperoxidase 387
    A. nidulans 386
    A. oryzae 386
    A. terreus 386

Aspergillus
  allergens 223
  antigen preparation 346, 356
  antigenic differences 347
  antigenemia 367
  cell wall antigen 355
  exoantigen 345, 357
  galactomannan 213, 371
  glycoprotein antigen 355, 386
  IgE and IgG 261, 390
  immunocompromised host and diagnosis 381
  immunoelectrophoresis 145, 152, 224, 358, 377
  metabolic antigens 3, 355, 377
  somatic antigens 3, 355
  taxonomy 377

Autoimmune
  bagassosis
  Blastomyces dermatitidis 4, 420

Botrytis cinerea 262

Candida
    adherence 157
    allergens 259, 261
    antigen preparation 4, 141, 149
    antigenemia 170
    arabinitol 178
    cellular immune response 185
    chronic mucocutaneous candidosis 22, 185
    cytoplasmic antigens 25, 143, 149
    fibrinogen 161
    immunoassays 16, 141
    immunochemistry 72, 141
    immunodeficiency 185
    latex agglutination 205, 219, 211
    lectins and 57
    mannan 8, 53, 62, 72, 100, 141, 149, 170, 185
    monoclonal antibodies 19, 99, 145, 149, 200, 213
    serotypes 17, 141, 170, 213
    toxin 206
    wall antigens 20, 53, 58, 71, 99, 141, 150, 161, 185, 197,
        200, 213
    wall and enzymes 197
    yeast killer toxins 216

Candida albicans 3, 100, 141, 157, 169, 185, 197, 198, 200, 202,
    203, 205, 206, 209, 211, 212, 213, 217, 219, 377
    C. guilliermondii 206, 220
    C. parapsilosis 143, 198
    C. spp. 206, 377
    C. tropicalis 142, 206

Candidosis
    chronic mucocutaneous (CMC) 22, 185
    diagnosis 22, 205, 206, 211
    latex test 205, 211, 219
    systemic 5, 169, 202, 203, 206, 211, 212, 217, 219
    vaginal 205, 209

Catalase
    antibody to A. fumigatus 388

Cellular immune response
    allergens and 237
    Candida 185
    Cryptococcus 303

Cell wall 185, 259
    chemistry of 141
    dermatophytes 205
    electron microscopy 60, 71
    immunochemistry 71, 141
    lectin 57, 123

Coccidioides immitis
    antigen preparation 398, 445
    dimorphism and 395
    enzyme and pathogenicity 411, 445
    monoclonal antibodies 445
    2D-IEP

Cryptococcus neoformans
    AIDS and 21, 286, 307
    antibodies 273, 283, 301
    antigen identification 273
    capsular polysaccharide(s) 21, 265, 301
    cellular immunity 303
    CIE 283
    cytoplasmic antigens 283
    delayed hypersensitivity 284, 431
    detection of antibodies 17, 273
    detection of antigens 307
    humoral responses to 303
    immunocytochemistry and 75
    latex agglutination 265, 295
    migration inhibition 303
    monoclonal antibodies 273, 301
    polysaccharide 301
    serotypes 265, 273, 301, 306
    serotypes and epidemiology 265, 306
    soluble antigen 307

Dermatophytes
    antigens 311
    antigen production 311
    keratinase 334
    Microsporum 312
    monoclonal antibodies 311, 338
    serotypes and 311
    standardization of antigens 340
    trichophytin 340
    Trichophyton mentagrophytes 334, 336
    Trichophyton rubrum 312, 336, 338

ELIFA
    for detection of A. fumitatus antigens 390

ELISA
    aspergillosis 357, 385, 387
    Candida 169, 209
    for detection of Pityrosporum antibodies 339, 342
    for detection of Cryptococcus neoformans antibodies 301

Entomophthorales
    electron microscopy 122
    immunocytochemistry of 122

Exoantigens 111
    dimorphic fungi 111
    Faenia rectivirgula 448
    farmer's lung 448
    fibrinogen binding factor
    in aspergillosis 111

Fonsecaea pedrosoi
    antigen identification 319
    immunochemistry 319
    pathology 319
    thermal stress 320

Galactomannans 22, 99
    Aspergillus and 213, 371

Glucans
    invertebrate immunity and 121

Glucurunoxylomannan
    of Cryptococcus neoformans 273, 283

Glycoprotein antigen
    detection in aspergillosis 382

Histoplasma capsulatum 5, 111, 417, 431
    antigen purification 423
    antigenemia 436
    H antigen 421
    M antigen 421
    cell immune response 432
    cutaneous hypersensitivity 417
    delayed hypersensitivity 431
    detection 434
    histoplasmin 417
    immunoassay 422
    polysaccharide antigens 431
    protein purification 420

Histoplasmin
    immunochemical analysis 419

Hybridoma
    techniques for generating monoclonal antibodies 100

Hypersensitivity pneumonitis 448

IgE
    Alternaria 239
    Aspergillus and 224
    Botrytis 262
    Candida 260
    Cladosporium 239

IgG
    Aspergillus and 224, 261
    Candida 261
    Coccidioides 445
    Cryptococcus 288

Immunochemistry 57, 71

Immunocytochemistry
    electron microscopy 71, 151, 213, 237, 288

Immunoelectroosmophoresis 12

Immunoelectrophoresis 12, 145, 358, 397, 419

Immunoenzymology 82, 141

Immunofluorescence 18, 143, 213
    Candida antigen 217

Immunoperoxidase 19, 387

Immunoprecipitation 17, 143, 326, 404, 419

Immunosuppressors 191

Invertebrate immunity 121

Latex agglutination test
    for Cryptococcus neoformans antigen 295, 307
    for diagnostic of systemic candidosis 211, 219
    for diagnostic of vaginal candidosis 205

Lectin 57, 86, 123

Mannan 20, 100, 141, 149
    antigen of C. albicans 169
    cellular immune response in vitro and in vivo 185
    chemistry 41
    monoclonal antibodies 141, 200, 213
Monoclonal antibodies
    Candida and 20, 141, 149, 200
    Coccidioides and 445
    Cryptococcus and 273
    Dermatophytes and 312, 338
    techniques 99
    yeast killer toxin 216

Nomuraea rileyi 122

Plant pathogens 379

Polysaccharides
    chemical techniques 41
    cytochemical techniques 57
    immunoelectron microscopy 71
    galactomannan 371
    glucuronoxylomannan 273, 301
    glucan 126, 445
    mannan 141, 209

Proteases
    Candida and 198
    Coccidioides immitis and 411
    Dermatophytes and 334
    Entomopathogens and

Protoplast 286

Pyricularia oryzae 383

Pityrosporum antigens 342

Pityrosporum orbiculare
    antibody detection 339

Radioimmunodiagnosis 81, 225, 240

RAST 225, 240

Soluble antigens
    cryptococcal antigen in AIDS 307

Skin antigens 340

<u>Trichophyton</u> <u>mentagrophytes</u>
    antigens 312
    keratinase 334

<u>Trychophyton</u> <u>rubrum</u>
    antigens 312, 336

Vaginal candidosis
    latex agglutination test 205, 209
Yeast killer toxin
    monoclonal antibodies 216

Western blots
    techniques 17
    antigens of <u>Aspergillus</u> <u>fumigatus</u> 223, 355
    antigens of <u>Alternaria</u> and <u>Cladosporium</u> 237
    antigens of <u>Candida</u> <u>albicans</u> 141, 153
    antigens of <u>Fonsecaea</u> <u>pedrosoi</u> 319
    antigens of <u>Histoplasma</u> <u>capsulatum</u> 417